Ecological Studies

Analysis and Synthesis

Volume 28

Pond Littoral Ecosystems

Structure and Functioning

Methods and Results of Quantitative Ecosystem Research
in the Czechoslovakian IBP Wetland Project

Edited by
D. Dykyjová and J. Květ

With 183 Figures

Springer-Verlag Berlin Heidelberg New York 1978

Dr. Dagmar Dykyjová
Dr. Jan Květ
Czechoslovak Academy of Sciences
Institute of Botany
Department of Hydrobotany
Třeboň, Dukelská 145/ČSSR

ISBN 3-540-08569-6 Springer-Verlag Berlin Heidelberg New York
ISBN 0-387-08569-6 Springer-Verlag New York Heidelberg Berlin

Library of Congress Cataloging in Publication Data. Main entry under title: Pond littoral ecosystems. (Ecological studies; 28). Includes bibliographical references. 1. Pond ecology. 2. Pond ecology—Czechoslovakia. 3. International Biological Programme. I. Dykyjová, Dagmar. II. Květ, Jan, 1933—. III. Series. QH 541.5.P63P66. 574.5′2632.77-19008.

Typesetting, printing, and binding: Brühlsche Universitätsdruckerei, Lahn-Gießen.
2131/3130—543210

Preface

The Czechoslovak IBP Wetlands project was focused on intense quantitative ecological investigations of shallow littoral ecosystems of typical Central European fishponds, i.e., small man-made water bodies managed century-long for fish production. It was initiated in 1965 as a part of the national contribution to the International Biological Programme (IBP) by a small group of Czech botanists inspired by the idea of voluntary international scientific cooperation in studies of "biological basis of productivity and human welfare." During the subsequent ten years of research work, new colleagues or groups of different specialists joined our team. The final result was a bulk of complex ecological studies from two geographically and ecologically differing regions. The research program followed the basic original scheme of quantitative assessment of biological productivity through an analysis of the structure and functioning of ecosystems, as it had been adopted by the ecologically (PT and PF) as well as the physiologically (PP) orientated sections of the IBP. Simultaneously, broad international cooperation has been evolved with colleagues and laboratories in numerous countries, especially in Poland, Romania, USSR, Austria, the Netherlands, Denmark, Sweden, the United Kingdom, Belgium, Norway, India, the USA and Venezuela.

This international collaboration, first initiated by the Czechoslovak and Polish scientists, eventually led to the establishment of the international IBP working group for wetlands at the 1972 symposium at Mikołajki, Poland. This group is now preparing an international IBP synthesis volume on wetlands.

The results of our national investigations summarized in this volume also reflect several general aspects of wetland ecology. The volume does not present a final synthesis in the form of overall ecosystem budgets and models, but the editors have attempted to give as much as possible of condensed quantitative data as is needed for syntheses and ecosystem modeling. With respect to the present state of knowledge on the ecology of freshwater wetlands, such a point of view is probably more profitable for the reader than any sophisticated and refined models supported by an insufficient amount of sound data.

The editors of this volume, originally coordinators of the two research teams, working, respectively, in the South Bohemian and South Moravian wetlands, are greatly indebted to a number of persons who have facilitated the long-lasting interdisciplinary ecological investigations, from their modest start to the more elaborate final research. First, our thanks belong to Ivan Málek, first president of the Czechoslovak national committee for the IBP, and convener of the IBP's section on production processes, for his stimulating ideas and enthusiastic interest. On his initiative, the Czechoslovak Academy of Sciences secured both financial support and personel for the Czechoslovak participation in the IBP. This has also significantly enhanced the evolution of interdisciplinary ecological research in our country and initiated the transformation of ecology into a modern synthetic science, prepared to answer difficult questions of contemporary landscape management and conservation. We are equally indebted to Slavomil Hejný, corresponding member of the Czechoslovak Academy of Sciences and director of the Institute of Botany. He has been the initiator and leader of the basic research on plant ecology and primary productivity within our wetlands project, and has provided a wealth of background knowledge on the structure and management of biotic communities in fishponds. His deep interest and personal involvement in the field investigations, as well as his maximum support of all research activities have promoted the limited initial investigations to the level of intense research. All our colleagues who have prepared the individual contributions to this volume as a result of their participation in the research project deserve many thanks for their voluntary discipline and respect for the unifying concept of the book.

The coherence in the whole volume has required a great deal of technical skill and effort on behalf of the staff of the hydrobotany department of the Institute of Botany. The editors wish to express their gratitude to Mrs. Naďa Chmelařová, Mrs. Jaroslava Lukavská, and Miss Drahoslava Machová for their maximum patience in retyping the English text, drawing the figures and preparing the tables respectively.

The publication of this volume in the *Ecological Studies* series would not have been possible without the personal interest and kind assistance of Professor O. L. Lange, member of the editorial board of the series. His valuable critical comments on the draft manuscript are greatly appreciated. We are also most grateful to the publisher Dr. K. Springer and his highly qualified staff for the excellent graphic setup of the book, as well as for their patience with various technical shortcomings of the manuscript, and with us as editors.

Třeboň, January 1978

DAGMAR DYKYJOVÁ
JAN KVĚT

Contents

Section 2
Environmental Factors in Fishpond Littorals

Section 3
Primary Production and Production Processes in Littoral Plant Communities

Contributors

DVOŘÁK, J.	Institute of Botany, Czechoslovak Academy of Sciences, Praha, Průhonice/ČSSR
DYKYJOVÁ, D.	Institute of Botany, Department of Hydrobotany, Czechoslovak Academy of Sciences, Třeboň, Dukelská 145/ČSSR
ETTL, H.	Institute of Botany, Department of Hydrobotany, Czechoslovak Academy of Sciences, Brno, Mendlovo nám. 1/ČSSR
FIALA, K.	Institute of Botany, Department of Ecology, Czechoslovak Academy of Sciences, Brno, Stará 18/ČSSR
GLOSER, J.	Institute of Botany, Department of Ecology, Czechoslovak Academy of Sciences, Brno, Stará 18/ČSSR
HEJNÝ, S.	Institute of Botany, Czechoslovak Academy of Sciences, Praha, Pruhonice/ČSSR
HUDEC, K.	Institute of Vertebrate Zoology, Czechoslovak Academy of Sciences, Brno, Květná 8/ČSSR
HUSÁK, Š.	Institute of Botany, Czechoslovak Academy of Sciences, Praha, Průhonice/ČSSR
JANKOVSKÁ, V.	Institute of Botany, Department of Ecology, Czechoslovak Academy of Sciences, Brno, Stará 18/ČSSR
KOMÁREK, J.	Institute of Botany, Department of Hydrobotany, Czechoslovak Academy of Sciences, Třeboň, Dukelská 145/ČSSR
KOMÁRKOVÁ, J.	Institute of Botany, Department of Hydrobotany, Czechoslovak Academy of Sciences, Třeboň, Dukelská 145/ČSSR
KVĚT, J.	Institute of Botany, Department of Hydrobotany, Czechoslovak Academy of Sciences, Třeboň, Dukelská 145/ČSSR
MARVAN, P.	Institute of Botany, Department of Hydrobotany, Czechoslovak Academy of Sciences, Brno, Mendlovo nám. 1/ČSSR

OBRTEL, R.	Institute of Vertebrate Zoology, Czechoslovak Academy of Sciences, Brno, Květná 8/ČSSR
ONDOK, J.P.	Institute of Botany, Department of Hydrobotany, Czechoslovak Academy of Sciences, Třeboň, Dukelská 145/ČSSR
PELIKÁN, J.	Institute of Vertebrate Zoology, Czechoslovak Academy of Sciences, Brno, Květná 8/ČSSR
PŘIBÁŇ, K.	Institute of Botany, Department of Hydrobotany, Czechoslovak Academy of Sciences, Třeboň, Dukelská 145/ČSSR
PŘIBIL, S.	Institute of Botany, Department of Hydrobotany, Czechoslovak Academy of Sciences, Třeboň, Dukelská 145/ČSSR
REJMÁNEK, M.	Institute of Entomology, Czechoslovak Academy of Sciences, Praha 2, Viničná 7/ČSSR
REJMÁNKOVÁ, E.	Institute of Botany, Department of Hydrobotany, Czechoslovak Academy of Sciences, Třeboň, Dukelská 145/ČSSR
RYCHNOVSKÁ, M.	Institute of Botany, Department of Ecology, Czechoslovak Academy of Sciences, Brno, Stará 18/ČSSR
SKUHRAVÝ, V.	Institute of Entomology, Czechoslovak Academy of Sciences, Praha-4, u Háje/ČSSR
ŠMÍD, P.	Institute of Botany, Department of Ecology, Czechoslovak Academy of Sciences, Brno, Stará 18/ČSSR
ŠŤASTNÝ, K.	Institute of Landscape Ecology, Czechoslovak Academy of Sciences, Řičany, Bezručova 127/ČSSR
ÚLEHLOVÁ, B.	Institute of Botany, Department of Ecology, Czechoslovak Academy of Sciences, Brno, Stará 18/ČSSR
VÉBER, K.	Laboratory of Biotechnology, Institute of Microbiology, Czechoslovak Academy of Sciences, Třeboň, Opatovický mlýn/ČSSR
VELÁSQUEZ, J.	Universidad Central de Venezuela, Apartado 10098, Escuela de Biología, Caracas/Venezuela

Introduction to the Ecology of Fishpond Littorals

S. Hejný and J. Květ

During the history of mankind, people of all periods and races have attempted to draw clear-cut boundaries between land and water in the landscapes they inhabited. A balanced mosaic of terrestrial and aquatic ecosystems has always been regarded as an ideal pattern providing both safe and productive habitats for populations of the species *Homo sapiens* L. People have always feared undefined transitions between land and water, various kinds of wetlands and marshes, whose great ecological importance is only now becoming acknowledged.

Our ancestors, who created whole systems of fishponds in the wet or humid and often marshy regions of Central Europe eight to five centuries ago, must have been guided by this thinking, too. Their effort has been perpetuated by further generations engaged in the maintenance and management of the ponds. This book, which deals with various aspects of the ecology of Central European fishpond littorals, must start with a tribute to all ancient fishpond builders and managers, mostly unknown to us by their names but well known by the fruits of their hard work.

By contrast to all natural standing waters, a fishpond, though it may look like a lake, is a man-made complex ecosystem, with all consequences of this fact. Man not only designs and creates fishponds by damming a shallow valley or depression which he then fills up with water, but he also perpetuates the existence of each fishpond by proper management. Most ponds are managed for fish production, first of all of carp (*Cyprinus carpio* L.)—hence the name "fishpond"—whose ecological requirements are best fulfilled in shallow bodies of mostly open water. Such water bodies, if left to themselves, naturally fill up with mud, silt and detritus and relatively rapidly turn into marshes. Man must spend some energy and effort if he wants to prevent this development. Fishponds thus comprise unstable types of ecosystems requiring a constant energy subsidy for their prolonged existence.

Fishponds have hitherto been mostly studied as fish-producing units and/or as water reservoirs differing little from small and shallow lakes. This approach has yielded some outstanding results in production of plankton, benthos and fishes. Fishpond reed belts, i.e., parts of their littorals colonized by helophyte vegetation, are mostly narrow and relatively small in area, hence only little attention has been paid to them in these studies. Their share in the total production budget of a fispond may be relatively small (Straškraba, 1963, 1968), yet their own production and other aspects of their ecology may be of the utmost importance.

The permanent struggle for the existence of fishponds mainly takes place along the shores in the shallow littorals and particularly in the reed belts, where large amounts of organic matter are produced and mineral nutrients accumulated, and terrestrialization proceeds potentially most rapidly. It is therefore this critical zone of fishpond littorals that we have chosen for intense study in the Czechoslovak IBP Wetlands projects No. PT/5 and PP-P/3 (see IBP News, London, Nos. 13 and 14). Thoroughly to understand the ecological processes taking place in the fishpond littoral means to understand, to a large degree, how to manage a fishpond with the least effort, whether it be for fish production, water conservation, recreation or any other useful purpose.

The surface areas of Central European fishponds vary from a few hundreds of m^2 to several km^2, and their average and maximum depths only rarely exceed 1.5 and 4 m respectively. In the heart of Europe, Bohemia and Moravia—now constituting the Czech Socialist Republic which is the western part of Czechoslovakia—may be regarded as a country of classical fishpond management and fish-farming with long tradition. This is represented by such names as Vilém of Pernštejn (1435–1521), J. Štěpánek of Netolice (+ 1538), J. Skála Dubravius (1486–1553), M. Ruthard of Malešov (second half of the 15th century), J. Krčín of Jelčany (1535–1604), V. Horák (1819–1900), A. Frič (1832–1913) and J. Šusta (1835–1914), whose knowledge and experience are still appreciated by contemporary specialists. In the past, the regions of the most flourishing fish-farming and fishpond management were the wider surroundings of Bohdaneč and Chlumec nad Cidlinou in East and Northeast Bohemia where, however, many a fishpond gave way to sugar-beet cultivation in the 19th century; further the surroundings of Lednice and Pohořelice in South Moravia where, on the other hand, quite a few fishponds were renewed after the catastrophic drought in 1947, and—above all—the relatively flat basins of South Bohemia. Here, the Třeboň basin is particularly important with its elaborate waterworks connecting most of its large fishpond into one system. South Bohemia and South Moravia still represent the most important fish-producing districts of Czechoslovakia, while the still remaining eastern and northeastern Bohemian fishponds are also renowned for their reed (see Sect. 7.4).

As fishponds have mainly been constructed in wetland regions, their biotic communities do not differ substantially from those of the original natural wetlands; but some differences have always existed, and their extent has changed, with the ways of managing the fishponds.

According to the main purpose of its management, a fishpond belongs to one of the following four main categories:

1. Fry and fingerling ponds, mostly small (up to a few ha) and shallow (less than 1 m deep)

2. Hibernation ponds, mostly medium-sized (5–40 ha) and deep or with a certain flow of water during winter-time

3. "Main" production ponds, medium-sized to large (25–500 ha), for keeping one to three-year-old fish, nowadays mostly cropped every other year

4. Special ponds serving purposes additional to fish-farming: recreation, sports-fishing, game-keeping, water storage, sewage disposal, village ponds, etc. In some instances, these secondary purposes may predominate.

The management practices, aimed both at increasing the fish production and at maintaining the fishponds in existence, have passed through several stages during history (see Sect. 7.3). In addition to the natural environmental factors, the structure and functioning of biotic communities in fishpond littorals are therefore determined by the following factors:

1. Changes in the position of water level in the fishpond
2. Effects of mineral nutrients contained in the fertilizers and manure supplied to the fishpond either directly or (at present increasingly) indirectly with water flowing in from adjacent cultivated land, farms and human settlements
3. Control of fishpond vegetation, especially of helophytes, mostly by various mechanical means.

An analysis of these aspects of human influence and of its interactions with natural environmental factors is given in Section 7. Here, only a few essential remarks are given on:

Changes in Water Level. A specific feature of fishponds is the possibility to drain or fill them up at any time if enough water is available. While in shallow lakes situated in dry regions such processes occur spontaneously, in fishponds the changes are almost entirely in the hands of man: any site within the cadastral area of a fishpond may either be flooded or made dry at nearly any time. The actual position of the water level in a fishpond determines the distribution of ecophases along the bottom slope. The ecophase—an instantaneous characteristic of any point in a fishpond (Hejný, 1957, 1960, 1971)—is determined by the relative position of water level to bottom level. Four principal ecophases may be distinguished: hydrophase, littoral, limosal and terrestrial. Individual life forms of aquatic and littoral vascular plants as described by Hejný (1960), possess varying degrees of adaptation or tolerance to individual ecophases, as do various types of algae and microorganisms, or life forms of animals—as will be shown throughout this book. The frequency of incidence of individual ecophases naturally differs with bottom depth (measured from a fixed contour); this brings about a characteristic zonation of life forms, populations or communities of nearly all biota living in a fishpond. A more thorough analysis of ecophases and of their combination into ecocycles is given in Section 1.2.1.

The ecophases are to a large extent compatible with the frequently used limnological terms denoting zones of the littoral, as derived from the zonation observed on lake shores and applied—wrongly or rightly—also to fishpond littorals. Sites which are mostly in hydrophase, being flooded with more than about 1 m of water for most of the year, may be regarded as fishpond pelagial and sublittoral; those mostly in littoral to limosal ecophases constitute its eulittoral, and those where the terrestrial to limosal ecophases prevail for most of the time may be denoted supralittoral to epilittoral (cf. Hrbáček, 1966). These limnological terms, which express merely a certain analogy between a lake and a pond when filled, must be used with caution and mostly only for the littorals of relatively large and deep "main" fishponds. For most purposes, however, it is sufficient to distinguish three or four zones even in large ponds:

(1) the central open-water area, sometimes called pelagial or sublittoral; (2) the inner littoral which communicates more or less freely with the central part, and (3)

the outer littoral which is relatively isolated from the rest of the fishpond and communicates more or less intensely with the surrounding land. An additional zone, which may be called "central littoral" and is relatively independent of either open water or dry land, develops only along gently sloping fishpond shores, within sufficiently broad strips of dense helophyte vegetation or in reedswamps in fishpond bays. In the pelagial, planktonic communities usually predominate, while filamentous algae or submerged and floating macrophytes frequently take over in the sublittoral. Emergent macrophytes (helophytes) are dominant in the inner, central and largely also in the outer littoral where, however, marsh plants (uliginosophytes) tend to curtail their ecological dominance.

In practice, control of the water level in fishponds by man combines with the effects of alternating periods of dry and wet weather. The weather controls particularly the water level in "sky-fed" ponds depending entirely on local rainfall, while fishponds receiving most of their water from relatively large rivers or streams are the least weather-dependent.

Nutrient Regime in Fishponds. The character of the geological substrates and soils in the catchment area of a fishpond and the soils of the fishpond itself determine its natural trophy, i.e., its own mineral nutrient economy. The classical but vague categories of oligo-, eu-, hyper- and dystrophy used by Naumann (1931) and others apply to whole fishponds as well as to their various biotopes. So the South Bohemian fishponds in regions of poor soils may mostly be regarded as naturally oligotrophic or dystrophic, the latter receiving water largely from acid bogs. South Moravian fishponds, situated in a region of high soil fertility, may be classified as eutrophic by nature. Within a fishpond of a certain trophy, the spatial distribution of relatively greater or smaller nutrient release from the bottom into the water depends on the pattern of detritus and silt sedimentation and of accumulation of organic materials on the one hand, and on the decomposition and mineralization of organic matter and erosion of bottom sediments on the other. In the long term, accumulation tends to prevail in fishponds: hence the importance of low water level, or even of complete summer or winter drainage, during which the mineralization processes are enhanced, especially in the littoral. The scheme in Figure 1 illustrates, for an ideal fishpond, the distribution of biotopes (note the importance of the prevailing wind direction). In real situations, this pattern is modified by the morphometry of a fishpond. In littorals of fishponds with a divided shoreline, accumulation and erosion types of biotopes alternate according to the concavity or convexity of the shoreline. Both the structure and productivity of littoral biotic communities closely reflect this pattern. Sedimentation of silt and fine detritus prevails in the central part of each fishpond (i.e., pelagial, when flooded), but even here mud layers of different thickness arise, depending on depth, bottom configuration, and water movements. This pattern is particularly clear when a pond is drained during the growing season: the plant communities colonizing emerged shores and bottoms react sensitively not only to water supply, but also to the mineral nutrient supply and texture of the bottom sediments. A suitable classification of fishpond bottom soils has been proposed by Neuhäusl (1965).

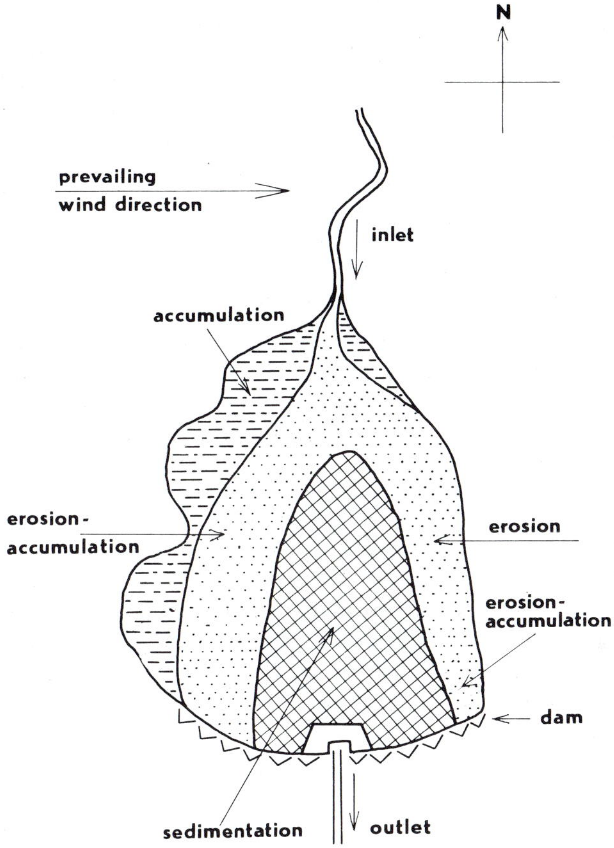

Fig. 1. Distribution of biotopes with prevailing accumulation, sedimentation and erosion in an ideal pond, oriented laterally to the prevailing westerly winds

With the advancing intensity of fishpond management, particularly in recent years, the differences in mineral nutrient regime between naturally oligotrophic to dystrophic and eutrophic fishponds have largely been abolished by the extensive application of large doses of fertilizers and lime to both the fishponds and adjoining land. According to the nutrient concentration in water practically all ponds may now be regarded, broadly speaking, as eutrophic to hypertrophic. Excessive eutrophication occurs namely in fishpond littorals on sites influenced by waste waters from farms, villages or small towns situated in the neighborhood of fishponds. According to the degree of saprobity, the water of most fishponds (using the scale of Šrámek-Hušek, 1946, 185–190) may be classified as β- or α-mesosaprobe, with β-polysaprobity occurring in heavily polluted whole ponds or fishpond bays for longer or shorter periods.

The original trophic conditions of fishponds still persist, to a greater or smaller degree, in their relatively isolated outer and central littorals with little communication with the main water body. The species composition of biotic communities therefore varies more between fishponds in these zones than in the pelagial or inner littoral. For the existence of these differences, however, the zonation of the littoral must not be disturbed by a too drastic control of the littoral vegetation.

Control of Littoral Vegetation. For a fishpond ecosystem, representing a set of successional stages requiring constant energy subsidy for its maintenance, a certain check on the advancing land-forming succession is essential. The control of littoral vegetation, particularly of the reed-belt helophyte communities, thus represents an indispensable tool in fishpond management. During its development, the intensity of this control has been increasing roughly in proportion with the increasing fertility and production in littoral habitats, brought about in turn by increasing eutrophication of the fishponds, as shown in Section 7.5.

The Central European fishponds studied here thus evidently represent specific sets of highly complex and artificially stabilized wetland ecosystems, with the proportions of flooded and drained land varying widely both in space and time. An ecosystem-type study of the entire set, conducted according to the postulates of the PT and PP sections of the IBP (see IBP News, London, No. 2, and Ellenberg and Ovington, 1964) would have been unmanageable. In view of its insufficiently understood importance for the existence of any fishpond, the fishpond littoral has been selected for detailed ecological study and has been characterized as a separate ecosystem some of whose components are closely connected with the rest of the fishpond, but others are only loosely or not at all, having possibly closer connection with the surrounding dry land. The application of the concept of ecophases and ecoperiods has made possible a further subdivision of the littoral into two or three distinct zones (see pp. 3–4). This approach is different from, but does not exclude, the customary limnological approach to the study of fishponds as water reservoirs. From this point of view, fishponds have been examined by the PF section of the IBP, in Czechoslovakia within the project PF/2 (see, e.g., Fott, 1972; Kořínek, 1972; Kořínková, 1971; Lellák, 1969).

Figure 2 provides a qualitative compartmental scheme of the structure and shows the most important relationships in the most typical ecosystem of the central fishpond littoral dominated by *Phragmites* and/or *Typha*. The whole system is divided pragmatically into four divisions: one pertaining to the distribution of incoming radiant energy and three spatial, i.e., the emergent (also including the rhizome and root systems of the emergent plants), aquatic, and bottom divisions. Each division may be treated as a separate subsystem, if required. The compartments have been defined so as to make their identification relatively easy. Energy flow through the same species population may occur in more than one compartment, e.g., gross and net production in plants, aquatic animal compartments (larvae), and aerial animal compartments (adult stages of the same insect species). In this scheme, the size of each compartment has been chosen arbitrarily, corresponding neither to the energy flow through that compartment, nor to the amount of energy stored in it. The arrows indicate the main pathways of energy flow within the system as well as the principal energy imports and exports to and from the system. Potential imports or exports, which may or may not take place, are indicated by the symbols in brackets (i) and (e). The energy sinks, presented as

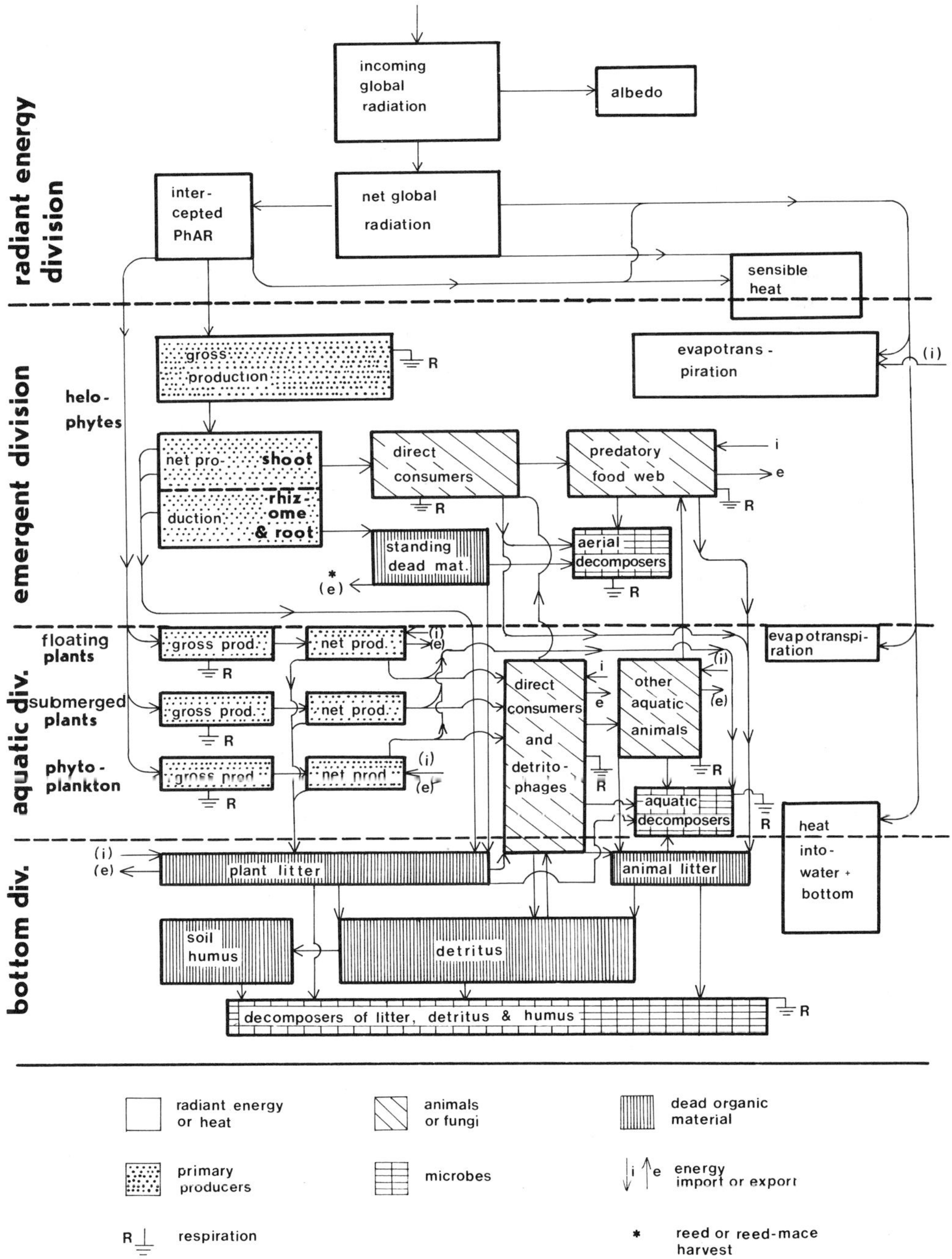

Fig. 2. Compartmental scheme of energy flow in a reed-belt ecosystem in the central fishpond littoral in littoral ecophase. (After J. P. Ondok and J. Květ)

respiratory energy losses, R, in this scheme, comprise all losses of energy from the system that are due to metabolic activities of the organisms involved. The scheme may be further subdivided, e.g., by splitting the emergent-plant net production according to species populations and/or plant organs; or simplified, e.g., by lumping together all aquatic plant synusia into one compartment or, for the ecosystem in terrestrial or shallow limosal ecophase, by eliminating the whole aquatic division and replacing it by an "understorey" terrestrial division. Incomplete schemes of energy flow may also be derived from this scheme, e.g., by considering only the plant net production in relation to intercepted PhAR, not taking into account the plant respiration and gross production, or by eliminating certain compartments whose participation in the overall energy flow may be regarded as negligible. Similar ecosystem schemes characterizing other zones and biotic communities of either the inner or outer littoral can be derived from this scheme, and various effects of fishpond management may also be incorporated. The scheme given in Figure 2 and the description of a fishpond reed-belt ecosystem by Pelikán et al. (1973) show, however, that an ecological study even of a mere fishpond reed belt will hardly ever be quite complete. Hence, the various facets of the ecosystem structure and functioning have received varying attention in our project, depending at least as much on the specialists available and interested as on the importance of the topics.

Our research was conducted in two regions typical of Central Europe, Hercynian South Bohemia and Pannonian South Moravia, taking the littoral of one typical "main" large production fishpond as a site of intense study in both regions. The two regions differ considerably in their climate as well as natural soil and water chemistry (see Sect. 2), but the two selected fishponds, Opatovický near Třeboň in South Bohemia, and Nesyt near Lednice in South Moravia, are managed in much the same manner. The similarities and dissimilarities between the two fishponds in the structure and functioning of their biotic communities may be regarded as typical and approximately characterizing the range of habitat variation encountered in Central European fishponds.

The structure of our IBP project, as well as of this book, clearly reflects that the work was initiated, coordinated and largely also carried out by botanists. For this reason, the ecological requirements, structure and functioning of the littoral vascular vegetation are treated in most detail (Sects. 1 and 3). Rather new for such studies is the attention paid to the algal vegetation of fishponds (Sect. 4). The microbial decomposition processes in wetlands are still incompletely understood and our studies contribute to the important efforts in this field (Sect. 5). Only some aspects of the varied animal life in fishpond littorals both above and below water have been studied (Sect. 6). For data on planktonic and benthic animal communities in fishponds, the reader is again referred to the results of the Czechoslovak IBP project No. PF/2. Of the management practices, those affecting the reed belt most deeply—but not destroying it—receive attention (Sect. 7). The study on reed propagation and cultivation also applies both to the proceeding destruction of the reed belts, and the difficulties in harvesting reed in fishponds. Conservation both of plant and animal communities in fishpond littorals is also briefly mentioned (Sect. 8).

References

Ellenberg, H., Ovington, J. D.: Produktions-Ökologie von Land-Lebensgemeinschaften im Rahmen des Internationalen Biologischen Programms. Ber. Geobot. Inst. Eidg. Tech. Hochsch. Stiftung Rübel Zürich **35**, 14–40 (1964)

Fott, J.: Observations on primary production of phytoplankton in two fishponds. In: Proc. IBP-UNESCO Symp. Productivity Problems in Freshwaters. Kajak, Z., Hillbricht-Ilkowska, A. (eds.). Warsaw and Cracow: PWN, 1972, pp. 673–683

Hejný, S.: Ein Beitrag zur ökologischen Gliederung der Makrophyten in den Niederungsgewässern der Tschechoslowakei. Preslia **29**, 349–368 (1957)

Hejný, S.: Ökologische Charakteristik der Wasser- und Sumpfpflanzen in den Slowakischen Tiefebenen (Donau- und Theissgebiet). Bratislava: Vyd. SAV, 1960

Hejný, S.: The dynamic characteristics of littoral vegetation with respect to changes of water level. Hidrobiologia **12**, 71–85 (1971)

Hrbáček, J.: Hydrobiologie. (In Czech). Prague: SPN, 1966

Kořínek, V.: Results of the study of some links of the food chain in a carp pond in Czechoslovakia. In: Proc. IBP-UNESCO Symp. Productivity Problems in Freshwaters. Kajak, Z., Hillbricht-Ilkowska, A. (eds.). Warsaw and Cracow: PWN, 1972, pp. 541–553

Kořínková, J.: Quantitative relations between submerged macrophytes and populations of invertebrates in a carp pond. Hidrobiologia **12**, 377–382 (1971)

Lellák, J.: The generation-rate of bottom fauna populations of the fishponds after wintering and summering. — Verh. Int. Ver. Theor. Angew. Limnol. **17**, 560–569 (1969)

Naumann, E.: Limnologische Terminologie. In: Handbuch der biologischen Arbeitsmethoden. Vol. IX, 8, 1931, pp. 1–476

Neuhäusl, R.: Vegetation der Röhrichte und der sublittoralen Magnocariceten im Wittingauer Becken. Vegetace ČSSR s. A. Vol. 1, Prague: Academia, 1965

Pelikán, J., Květ, J., Úlehlová, B.: Principal constituents and relationships in the reed-belt ecosystem at the Nesyt fishpond. In: Littoral of the Nesyt Fishpond. Květ, J. (ed.). Studie ČSAV 1973/15; 17–23, Prague: Academia, 1973

Šrámek-Hušek, R.: Introduction to Limnobiology (In Czech). Prague: Kropáč and Kucharský, 1946

Straškraba, M.: Share of the littoral region in the productivity of two fishponds in Southern Bohemia. Rozpr. Českosl. Akad. Věd., Řada Mat. Přír. Věd. **73** (13), 1–64 (1963)

Straškraba, M.: Der Anteil der höheren Pflanzen an der Produktion der stehenden Gewässer. Mittl. Int. Ver. Theor. Angew. Limnol. **14**, 212–230 (1968)

Fig. I. Early spring aspect of a typical small South Bohemian fishpond colonized by plant communities with prevailing *Glycerietum maximae* growing in organogenous bottom soils. (Photo: J. Ševčík)

Section 1
General Ecology and Inventarization of Biotic Communities

This chapter gives first a general geological, hydrological and pedological survey of the two areas of Czechoslovakia where our studies of the fishpond littoral ecosystems were conducted: the South Bohemian Třeboň Basin and the South Moravian region of Lednice (cf. Fig. I). The history of the construction and further human impact on the fishponds are briefly mentioned. Higher plant communities reflect this impact very well and have therefore been investigated quite thoroughly. A complete survey is given of the communities in the South Bohemian and South Moravian fishponds, with a short ecological characterization for each community type. An attempt is also made to define the association of certain algal groupings with some of the higher plant communities colonizing the fispond littorals. Vegetation maps based on aerial photographs and vegetational inventories are presented of the two specimen fishponds selected for intense study in the IBP: Opatovický in South Bohemia, and Nesyt in South Moravia. The principal animal populations inhabiting the fishpond littoral reed belts are also enumerated, and the various kinds and activities of decomposers living in fishpond littoral habitats are also characterized. Pollen analyses and identification of other plant remnants in peat bogs have made it possible to follow the postglacial history of wetland vegetation in the Třeboň basin; the main results of these investigations are briefly summarized.

1.1 General Characteristics of the Třeboň Basin and Lednice Region

Š. HUSÁK and S. HEJNÝ

The two regions in which our investigations were carried out, the Třeboň basin in South Bohemia and the Lednice region in South Moravia, differ markedly in physiographic and ecological conditions. The geographical location of the two regions is evident from Figure 1; for the principal data on the macroclimate see Table 1 in Section 2.1. In addition, the maps in Figures 2 and 3 illustrate the structure of the fishpond systems around Třeboň and Lednice.

1.1.1 Topography and Geomorphology

The Třeboň basin is situated in South Bohemia and its suboceanic climate is determined by the prevailing air passage from the Atlantic across Central Europe. The region is somewhat sheltered by the Šumava highlands (highest peak in its

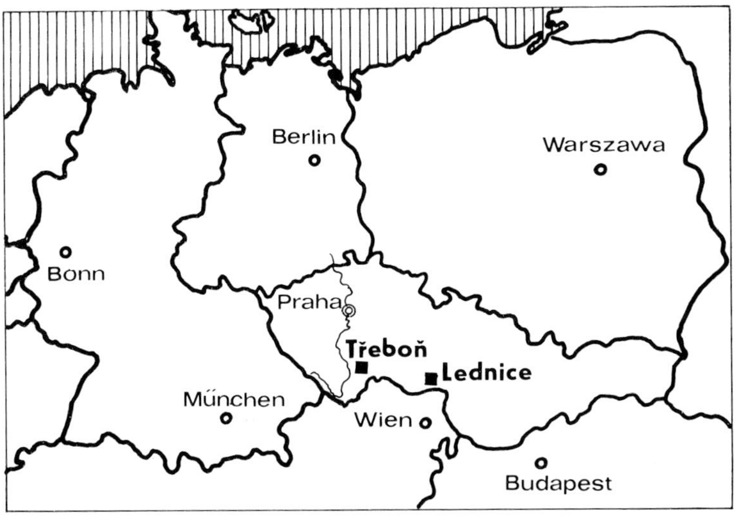

Fig. 1. Schematic map indicating the position of the Třeboň basin and of the Lednice region in Central Europe

Table 1. Largest fishponds in the Třeboň basin and in the Lednice region

Fishponds	Time of construction	Flooded area ha
Třeboň basin:		
Rožmberk	1584–1590	489
Horusický	1502–1531	415
Dvořiště	ca. 1363, enlarged ca. 1580	337
Velký Tisý	1502	317
Záblatský	15th century, enlarged ca. 1580	305
Staňkovský	1350–1400	241
Velká Holná	1350–1400	220
Svět	1573	201
Koclířov	1491–1495	192
Bošilecký	1350–1400	190
Opatovický	1350, enlarged 1495–1518 and ca. 1580	160
Kaňov	1502–1531	156
Ponědražský	15th century	139
Spolský	1571–1574	124
Lednice region:		
Nesyt	1350–1400	315
Mlýnský	1584–1600	107
Hlohovecký	1584–1600	104
Prostřední	1584–1600	49
Nový u Mikulova	renewed in 1950	37
Zámecký	ca. 1840	30

southern part: 1378 m), which separate Bohemia from Bavaria (Czudek, 1972). The Třeboň basin forms the axis of a larger Třeboň region. The altitude of the basin varies between some 400 and 500 m (Třeboň 430 m) and the relative elevations in the flat landscape mostly do not exceed 20–50 m (Demek, 1965). Fishponds constitute a characteristic landscape element in South Bohemia in general, and particularly in its two principal basins, those of Třeboň and České Budějovice. About 50% of the total area and 23% of the total number of fishponds in Czechoslovakia (about 415 km^2 and over 22,000 fishponds, respectively) are concentrated in South Bohemia. The average surface area of a Czechoslovak fishpond is 2.4 ha, but it is about 4 ha in the South Bohemian basins (Novotný, 1972). Twelve of Czechoslovakia's 20 largest fishponds are situated in the Třeboň basin; see Table 1 (Chábera and Šabatová, 1965; Novotný, 1972).

Apart from hills and loess deposits, river alluvia constitute a landscape element characteristic of South Moravia. The region of Lednice is situated at the northern edge of the Vienna Basin, which may be regarded as the northwestern outpost of the extensive Pannonian lowland plains. The alluvial plains around Lednice have been formed by the Dyje river near to its confluence with the Morava river. The relative elevations do not exceed 75 m. The Lednice fishponds (an established name, see also Project Aqua, Luther and Rzóska, 1971) are situated in a former bed of the Dyje river, now the valley of the Mikulovský potok brook. This is fringed by the Pavlovské vrchy hills (highest peak 550 m) on the northwest and by the hills of Valtice (highest peak 301 m) on the southeast (Demek, 1965; Czudek, 1972), both chains of hills belonging to the West Carpathian system. The prevailing climate may be regarded as subcontinental, with periods of weather determined by penetration of southeast European air masses alternating with periods determined by oceanic air masses from the Atlantic. The Lednice fishponds comprise 27 fishponds of varying size, of which the most important are listed in Table 1 (see also Heteša et al., 1973). The local climates of both the Třeboň basin and Lednice region are characterized in Section 2.1.1.

1.1.2 Basic Data on the Geological Substrate, Soils and Hydrology

Cretaceous sediments with locally superimposed tertiary sediments constitute the principal geological substrate in the Třeboň basin. River and stream alluvia and depressions are filled with quarternary deposits: alluvial silt and gravel, aeolic sands and peat. The character of these substrates determines that of the soils. In these, the content of clay particles generally increases with depth and the soil aeration is reduced accordingly (Novák, 1922; Spirhanzl and Káš, 1936). The soil nutrient content is generally poor: calcium deficiency is common, potassium is relatively sufficient only in the deep soil horizons, nitrogen content is low, that of phosphate is mediocre. Most soils are leached and show a tendency towards podzolization; their biological activity is poor. The soil reaction is mostly acid to highly acid (pH up to 3.3). Various types of podzols and sandy or peaty gleys prevail on sandy deposits, and pseudogleys, semigleys and peaty gleys on predominantly clay soils. Organomineral soils have developed on fishpond bottoms; they may be classified as *Phragmites* fen, limnic fen, sapropel, fishpond gyttja or dy (Neuhäusl, 1965; see also Sect. 2.4).

Running waters are represented by the rivers and streams of Lužnice (Fig. 2), Nežárka, Stropnice, Dračice and Koštěnický potok and by the artificial canal of Zlatá Stoka (Golden Canal) and artificial Nová Řeka (New River). The former was constructed by J. Š. Netolický in 1506 to 1520, the latter by J. Krčín in 1584–1589 in order to divert the surplus flooding water of the Lužnice river from the large Rožmberk fishpond. In addition to the main rivers, streams and canals, numerous subsidiary canals and ditches connect the fishpond systems in the Třeboň basin (see map in Fig. 2). This elaborate network of both natural and artificial running waters has made possible an efficient manipulation with large volumes of water in the Třeboň fishponds. The Třeboň basin, originally an inaccessible wetland area, has thus developed into a famous region of fish culture and forest plantations as well as into a landscape which has benefited from wise management.

The fishponds of Třeboň are filled with a naturally dystrophic water originating from large peatbogs occurring all over the Třeboň basin. This water becomes highly eutrophicated through fertilization and manuring both of the fishpond water and of the surrounding cultivated land. The chemical composition of the fishpond water is therefore highly variable depending, especially during the grow-

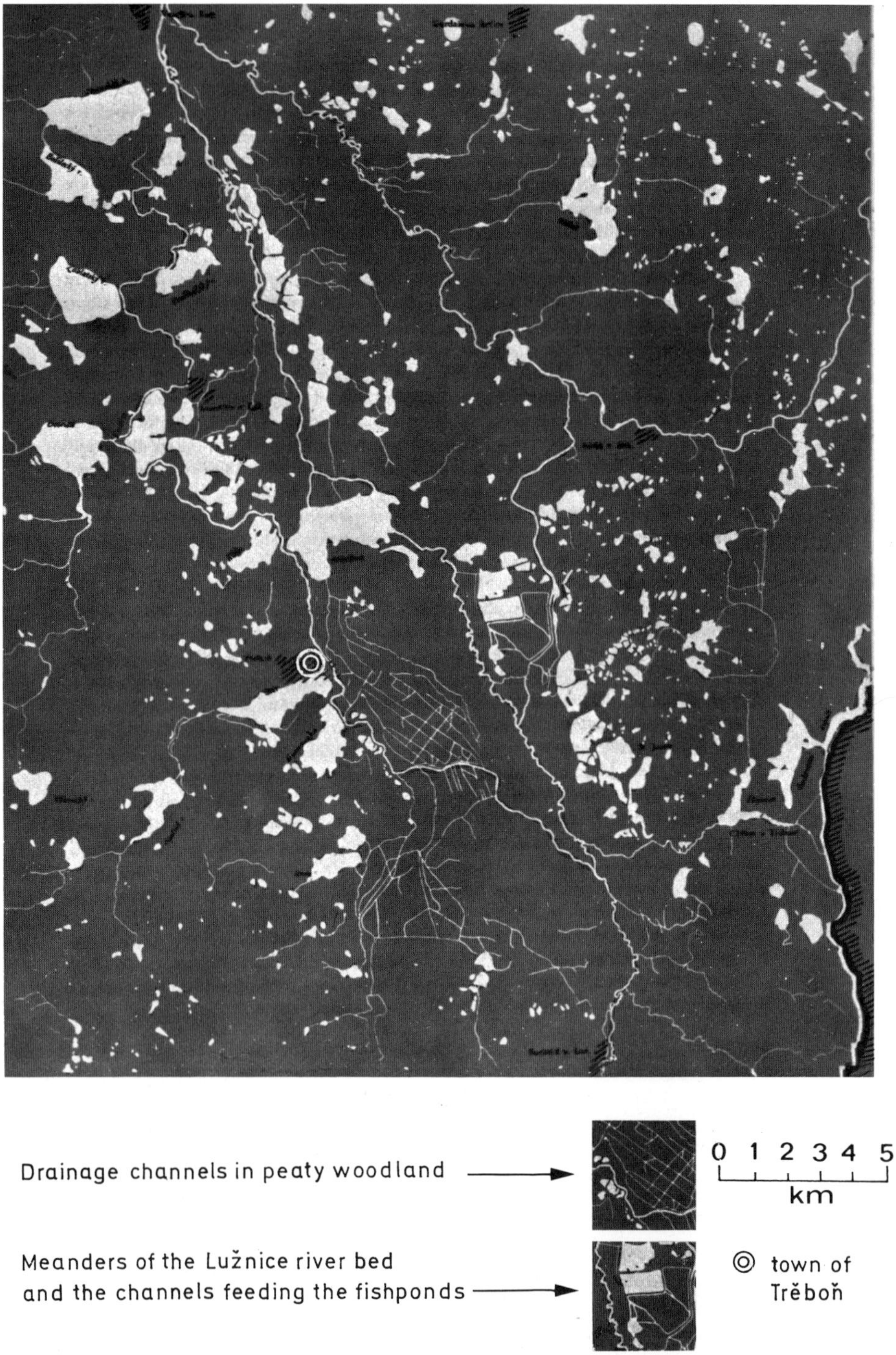

Fig. 2. Hydrological map of the network of both running and standing waters in the Třeboň basin

Table 2. Fish stock and production, fertilizer application and manuring, annual fodder supply to fish at Opatovický fishpond

	1964	1965[a]	1966	1967[a]	1968	1969[a]	1970	1971[a]	1972
Fish stock and production[b]:									
Carp	206.56	210.88	27.50	158.19	30.13	435,63	128.75	452.94	193.44
Other fish	22.19	13.00	2.38	20.06	3.94	6.25	4.31	127.40	6.00
Annual fish production[b]	185.31	288.69	134.75	64.88	228.06	179.00	230.44	291.13	207.06
Fertilizer application and manuring:									
Lime	187.00	8.26	175.00	261.88	208.13	34.38	186.88	—	150.00
Limestone	443.75	405.63	406.88	125.00	240.00	284.38	141.88	253.13	109.38
Superphosphate	226.25	258.13	190.63	158.00	100.00	168.13	103.13	43,75	18.75
Potassium salts + saltpeter	36.88	43.75	56.88	73.75	25.00	55.00	92.50	70.00	130.00
Manure	—	—	13.13	—	—	—	—	—	—
Dung-water	356.25	—	—	—	—	—	—	—	—
Soil-compost	95.63	15.63	1078.13	48.75	—	31.25	7.5	1375.00	—
Fodder supplied to fish[b]:									
Total[c]	17.5	526.88	29.38	436.25	23.75	1405.00	326.25	551.86	611.25

[a] Years of cropping fish stock.
[b] All data in kg per ha maximum flooded area (160.5 ha).
[c] Maize, rye, wheat, special mixtures.

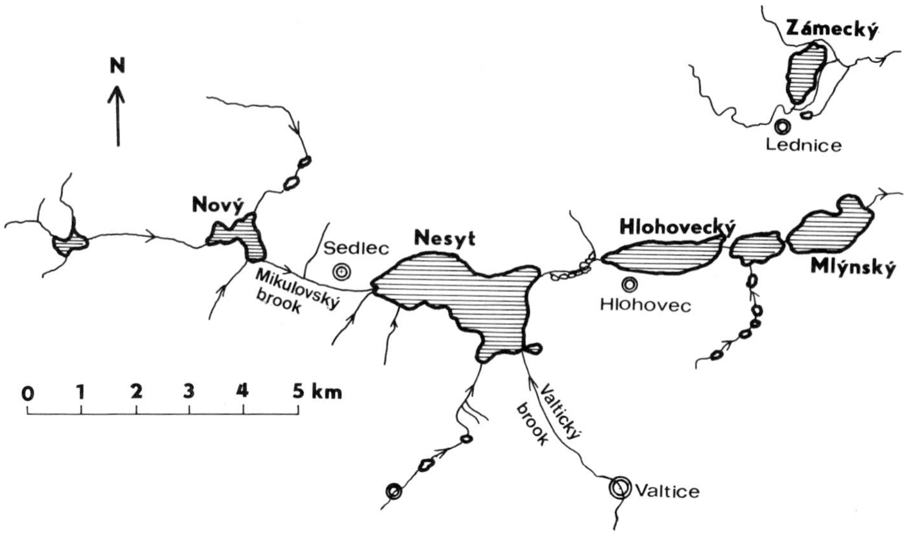

Fig. 3. Schematic map of the system of Lednice fishponds

ing season, on the intensity and timing of fertilizer application; for the data see Table 2. Detailed water analyses are presented in Section 2.3.

In the Lednice region, jurassic dolomites and limestones are superimposed on granite (Zapletal, 1950). The dolomite and limestone outcrops are visible in the surroundings of the Lednice fishponds in fields and vineyards. Characteristic of the region are marine sediments (see below) covered with holocene deposits which are up to 10 m deep. The processes of soil formation are affected by both the parent rock and the relatively dry climate. The soil nutrient pool is rich, the sorption complex is mainly saturated with calcium and magnesium. The soils are biologically highly active in spring; in summer drought usually limits their activity. Rendzinas and pararendzinas are the prevailing soil types on hillocks and on their slopes, chernozems prevail in flat areas and gley soils in depressions and along running waters (Pelíšek, 1964). Fishpond gyttja and sapropel predominate on fishpond bottoms. Saline soils occur in various parts of South Moravia, and their salinity is determined by a combination of various concentrations of $CaSO_4$, $MgSO_4$, and Na_2SO_4 with traces of chlorides. These soils develop mainly in contact with the tertiary marine sediments. From them, the salts are leached by ground water which is lifted up to the soil surface by capillarity. The water evaporates here, and gives rise to salt efflorescences. Results of chemical analyses of the soil and water in the Nesyt fishpond are presented in Sections 2.3 and 2.4. In relatively dry South Moravia, precipitation is mostly deficient during the growing season. All terrestrial ecosystems that are not supplied with additional water are therefore exposed to more or less severe drought for some time nearly every year. Wetland ecosystems develop only where the additional water supply is abundant and constant. Water is supplied to the main Lednice fishponds, including Nesyt, by the Mikulovský potok brook, the Valtický potok brook and by several smaller streams which usually dry out in summer (see Fig. 3). There is no larger stream in the immediate surroundings that would permanently ensure

Table 3. Morphometric data on the Opatovický (I) and Nesyt (II) fishponds and areas occupied by various types of littoral plant communities in the fishponds (situation in 1971)

I	I Opatovický	II Nesyt
1. Cadastral area (ha)	176	322
2. Maximum flooded area (ha)	160.5	315
3. Mean flooded area (ha)	~157	281
4. Diameter of a circle equal in area to (3) (km)	1.42	1.89
5. Length of shoreline (km)	9.40	9.55
6. Maximum depth at highest water level (m)	3.3	4.5
7. Maximum length of fishpond (km)	2.6	3.3
8. Maximum width of fishpond (km)	1.1	1.65
9. Length of dam (km)	1.2	0.1
10. Areas occupied by plant community types as mapped out in Figures 4 and 5 (ha):		
a) *Phragmites communis*	5.7	49.6
b) *Typha angustifolia*	1.6	3.3
c) *T. latifolia*	0.21	0.6
d) *Glyceria maxima*	9.5	0.5
e) *Schoenoplectus lacustris*	0.4	+
f) *Sparganium erectum*	0.01	–
g) *Phragmites comm.* + *Carex riparia*	—	7.2
h) *Carex gracilis* and diverse *Cariceta*	14.3	–
i) Halophyte and other marginal communities	–	16.3
j) *Phalaris arundinacea*	+	1.7
k) *Bolboschoenus maritimus*	3.9	3.0
11. Total area of littoral communities (ha)	35.7	82.2
12. The same as percentage of (1) (%)	20.3	25.5
13. Area occupied by the reed-belt plant communities (10a–g; ha)	17.2	61.2
14. The same as percentage of (2) (%)	10.7	19.4

sufficient water supply to these fishponds. They therefore often lack water. This problem is now being solved by the construction of a canal which will bring the water from the Dyje river into the Mikulovský potok brook. The supply of a chemically different water is likely to threaten the peculiar character of the slightly saline water in Nesyt as well as in the other main Lednice fishponds.

1.1.3 Description of the Opatovický Fishpond

This is one of the large "main" production fishponds in the Třeboň basin, in which mainly carp is cultivated (see map in Fig. 4). It was constructed in the 15th century by Š. Netolický on the site of former marshes and of the Opatovice hamlet; for a morphometric characterization of the Opatovický fishpond see Table 3. Its shoreline is divided, which favors the development of littoral helophyte communities in the bays; for their areas see also Table 3. The paved long fishpond dam, 7 m high, is connected with that of the adjacent Svět fishpond (with which Opatovický was once connected for a short time). The oaks reinforcing the dam were planted some 400 years ago. Two small brooks feed the fishpond, and additional water comes from the Zlatá Stoka (Golden Canal). The western shore is in direct contact with cultivated fields, while the southern to eastern shores are surrounded by woods planted, predominantly, with oak *(Quercus robur)*, spruce

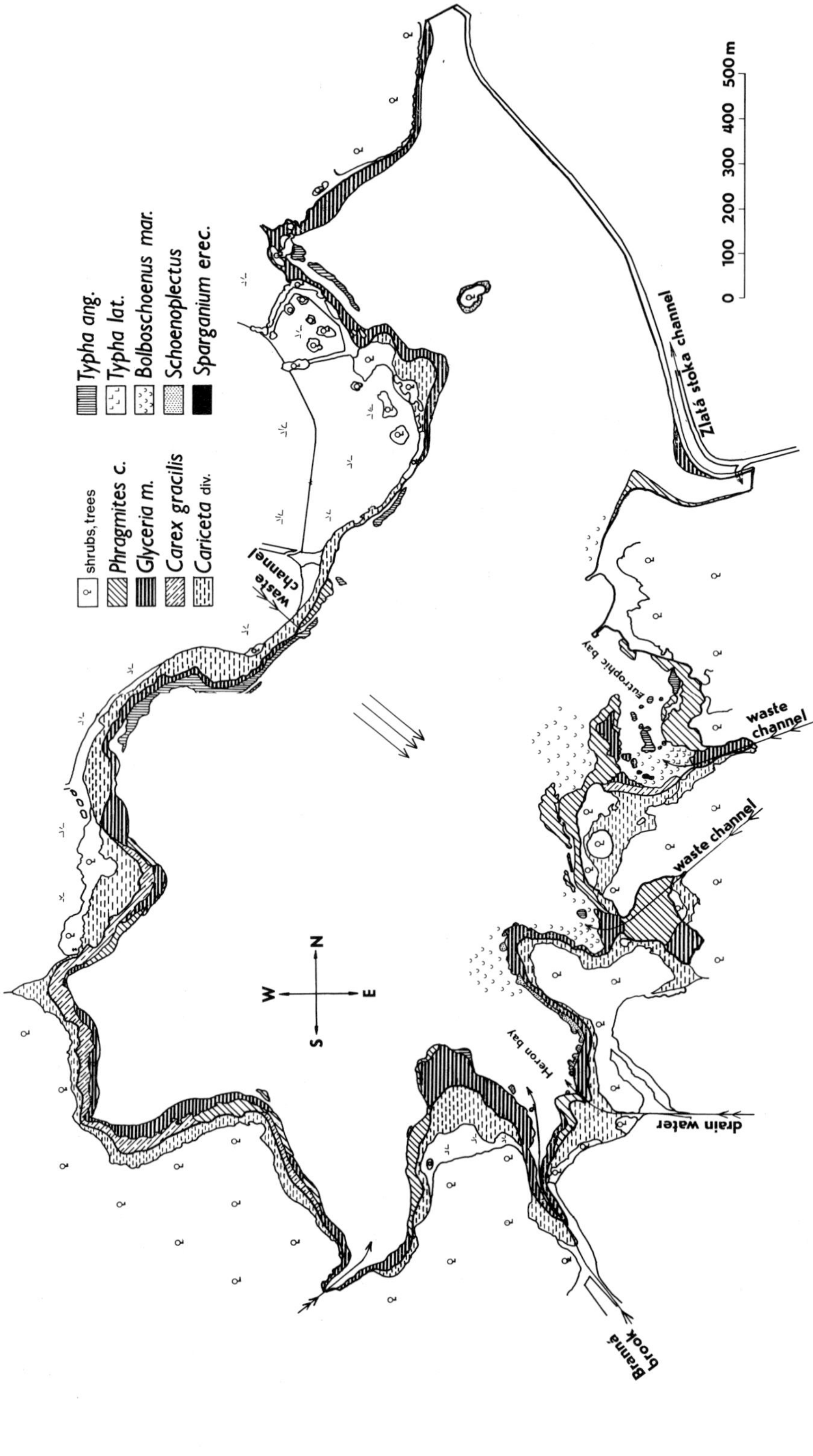

Fig. 4. Map of Opatovický fishpond showing the inflows (→), prevailing wind direction (→→), and main vegetation types of littoral communities, characterized by dominant species. Situation in 1971. For morphometric details see Table 3

(Picea excelsa), and pine *(Pinus silvestris)*, and with admixed aspen *(Populus tremula)* and willows (*Salix* spp.), this last mainly along the wood edges.

The woodlands of the Třeboň basin have been studied in detail by Březina (1975). In brief, acidophilus pine *(Pinus silvestris)* forests with oak *(Quercus robur)* or birch *(Betula alba)* and spruce *(Picea excelsa)* prevailed on dry sandy soil. A mixed oak-fir forest colonized depressions on clayey soil. Alder *(Alnus glutinosa)* swamp forest and willow carr *(Saliceta)* still occupy wet sites. Bog-pine *(Pinus rotundata)* forest with *Sphagnum* represents a permanent successional stage on deep peat, and spruce forest on shallow peat. Drained peat is nowadays mostly occupied by pine forest with blueberry *(Vaccinium myrtillus)* undergrowth (see also Jeník, 1974).

Willows (mainly *Salix cinerea, S.caprea, S.fragilis, S.aurita, S.pentandra* and others as well as their hybrids) have now invaded the wet meadows, no longer mown, which cover the flood-plain surrounding the fishpond. A rapid succession towards willow carr is taking place there but sedge communities still exist in many places (Jeník and Větvička, 1973). The shallow parts of the fishpond littoral are colonized by helophyte-dominated reedswamp communities which form a more or less continuous reed-belt along the shore except for the dam. Limited recreation takes place at the Opatovický fishpond; organic manuring of its water therefore ceased in 1968. Altogether five inlets still bring in wastes from a pig farm and from two sports and recreation centers. The incoming water from the Zlatá Stoka canal is also slightly eutrophicated. For analyses of the water in Opatovický fishpond see Section 2.3.

1.1.4 Description of the Nesyt Fishpond

Nesyt is the largest fishpond in Moravia (see map in Fig. 5). It forms the largest link in the system of the main Lednice fishponds (see map in Fig.3). It accumulates mainly the spring water from snow and rain coming in from a relatively large catchment area, situated in both Czechoslovak and Austrian territory, mostly of the Mikulovský potok brook which never dries out as its catchment area covers a part of the limestone Pavlovské vrchy hills. The morphometrical characteristic of Nesyt is given in Table 3. Its dam is only 100 m long. Several ditches feed drainage water to the fishpond, coming from slightly elevated (230–250 m) cultivated land along the northern shore. The steeper sandy shore to the east is locally eroded. The southern shore is flat, as also the western bay, part of which is in contact with the village of Sedlec. The village wastes pollute and eutrophicate the water in the eastern bay. The southeastern bay is also polluted with sewage water brought in by the Valtický Potok brook. Several sites on the western and southern shores of Nesyt host halophilous vegetation, rare in Czechoslovakia, and some 25 ha of reed in the western bay serve as a strict ornithological reserve. The whole Nesyt fispond is included in the Lednice fishponds nature reserve which was set up in 1953. Except for a few, mostly planted, trees and a narrow and short strip of willows along parts of the shoreline, there is no woodland immediately adjacent to the Nesyt shores. The fishpond is thus in contact with agricultural land along nearly the entire shoreline. Crop-farming is intense in the surroundings: wheat,

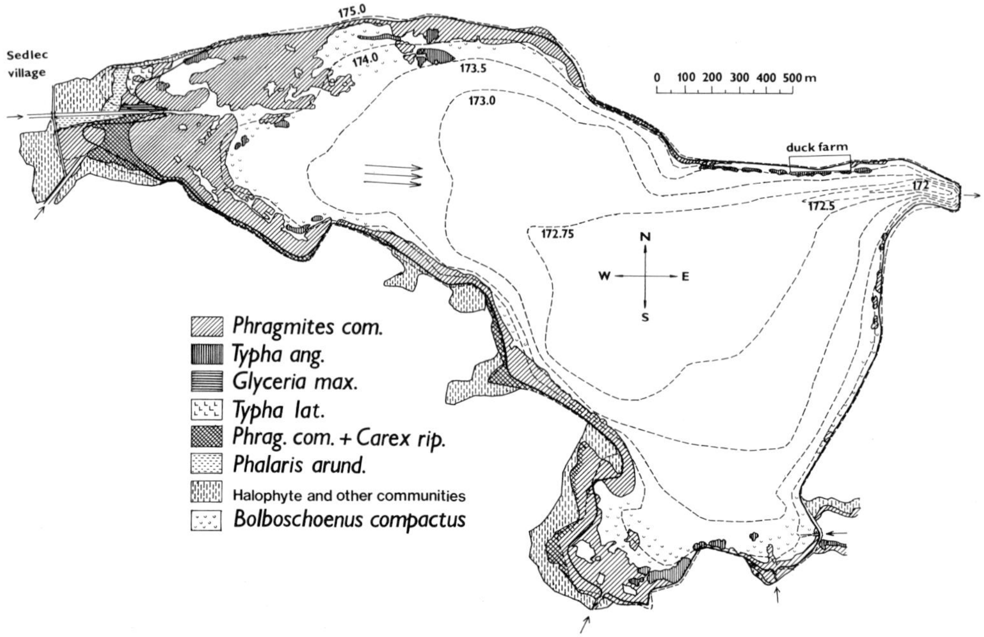

Fig. 5. Map of Nesyt fishpond showing contours of the bottom, principal inflows and outlet at the dam *(arrows)* and the vegetation types in the littoral and epilittoral, mostly characterized by the dominant species. Situation in 1971. For morphometric details see Table 3

maize, sugar beet, wine and vegetables are cultivated. The shoreline is fringed by a reed belt along the northern, western and southern shores. Extensive reed beds have developed in the two large bays of Nesyt, particularly in the western bay. Table 3 gives the areas which the various littoral communities occupied in 1971. It has to be noted that closed *Bolboschoenus*-dominated communities develop on relatively large areas only in years with a low water level. In addition, various communities of emerged bottoms develop, but these are not listed in Table 3. The changes in water level in Nesyt and Opatovický are given in Table 3.

The second half of this century has witnessed a hitherto unknown intensification of fishpond management aimed at increasing fish production. The net result is a decreased species diversity of the biotic communities (see Sect. 7.3). The survey of fish- and duck-farming as well as of fertilizer application and manuring at the Opatovický fishpond (Table 2), illustrates the intensification of fishpond management during the last 25 years. The system of fish-cropping during the investigations is evident from Table 2. Nowadays, a two-year rotation of the fish stock has become established in practically all the main production ponds. Until 1950, however, a three-year rotation was quite common. Up to the second half of the 19th century, fishponds used to be cropped even less frequently, every four to seven years. This development of fish farming also pertains to the Opatovický and Nesty fishponds. For details of the management of the fishponds and their littorals see Section 7.3.

References see pp. 93—95.

1.2 Higher Plant Communities

S. Hejný and Š. Husák

1.2.1 Variation of Plant Community Structures in Space and Time

Within a given set of climatic, edaphic and hydrochemical conditions, the water level and its changes in time determine the character and structural variation in both space and time of the plant as well as other biotic communities in a fishpond. Fishponds, with a water level controlled by man, are particularly suited for the study of the ecological effects both of a fluctuating and a stabilized water level. The concept of ecophases, ecoperiods and ecocycles was first elaborated with respect to vegetation dynamics in fishponds (Hejný, 1957, 1960) and later extended to other types of standing waters (Hejný, 1971). names of community types used in the further text correspond with those given in Section 1.2.2.

1. An *ecophase* may be described as an instantaneous "actual environment" with regard to water level and to specific ecological factors determining basic niches actually present in a biotope. In standing waters, the following ecophases may be recognized: hydrophase (= limnic phase), littoral, limosal and terrestrial ecophases. They are schematically presented in Figure 1, together with schematic drawings of higher plant life forms adapted best to each ecophase.

2. An *ecoperiod* is a sequence of ecophases during one growing season in temperate regions. The following types of ecoperiods may be recognized:

(a) Ecoperiods with stabilized water level reflected in a stable hydro-littoral gradient of ecophases in the littoral. Such ecoperiods provide the basic conditions for long-persisting perennial macrophyte communities in both open water and in the littoral. The reedswamp communities (foed. *Phragmition*) and those of tall sedges *(Caricion elatae* and *C. gracilis)* take particular advantage of the stable hydrological regime.

(b) Ecoperiods with a rapidly and distinctly sinking water level, giving rise to a littoral-terrestrial gradient of ecophases in the littoral. They give an impulse for replacement of aquatic plant communities by amphibious to terrestrial communities.

(c) Ecoperiods with a distinctly rising water level, which results in a terrestrial to littoral sequence of ecophases in the littoral. They stimulate the regeneration of amphibious and aquatic plant communities.

3. *Ecocycles* are determined by a sequence of ecoperiods over several years. Neither ecophases nor ecoperiods themselves limit or determine the selective

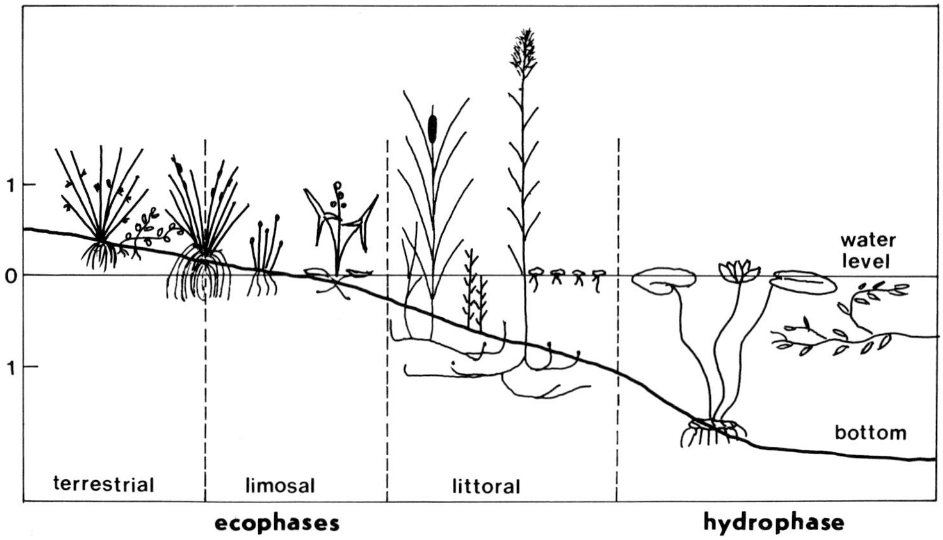

Fig. 1. Schematic illustration of ecophases in a fishpond littoral. *Vertical axis*: distance from actual water level in m

effects of the environment on plant species populations and communities. It is the duration and combination of ecophases, as links in a chain, forming together a definite type of long-term ecocycle, that determine the kind and degree of adaptation of the littoral and aquatic biota and of their communities to their habitats. In this respect, the "ecologically critical amplitude of water level" acquires great importance during a "critical ecoperiod" when subhydric soils emerge above the water surface and vice versa. A critical period, both in timing and duration, of emergence of the subhydric bottom soils leads to destruction of the hydric communities and to a build-up of terrestrial communities colonizing the bare emerged bottoms. During a critical period of submergence, on the other hand, a previously emerged bottom supporting terrestrial vegetation is flooded to such an extent that the terrestrial communities are destroyed and hydric communities develop anew after anabiosis under dry and aerobic conditions. A critical ecoperiod terminates one ecocycle and starts another. Ecophases and their duration are reflected in the types of macrophyte growth forms building individual synusia, which are the basic structural subdivisions of cenoses. Ecoperiods, with their gradients of ecophases, determine the actually dominant life form(s) of macrophytes and the complete structure of their communities. Ecocycles of various kinds are characterized by their typical succession series of communities and, to a certain degree, codetermine the type of a standing waterbody.

All biotic components in fishponds are adapted to a periodicity which also includes summer and winter drainage. Hejný (1960) has worked out a system of life forms of aquatic and marsh plants, reflecting their adaptation to water level and its fluctuations. This system is used in this book in addition to broader terms denoting plant-life forms such as "helophytes" or "hydrophytes".

Summer drainage has great influence on the whole complex of fishpond biocenoses because the dry conditions during an entire growing season (from April to September or October) affect the majority of links in the network of biocenoses (= ecosystem compartments).

Negative Effects. Negatively affected are all species, populations, life forms and communities adapted to a purely aquatic life. The negative influence depends on the duration and periodicity of summer drainage. This suppresses all species and communities of aquatic plants; among them, species capable of producing terrestrial forms survive for a short time, while the species whose seeds and rhizomes are buried in the bottom can survive the whole period of drainage in anabiosis. Summer drainage, however, also damages all plant species and communities living at the water-land boundary. This effect is either directly connected with the retreat of the shoreline, or it is indirect, due to the fall of water level to deeper soil horizons. Epilittoral communities situated beyond the normal shoreline are also affected, as are the species and communities associated with the fishpond dam. Hygrophilous trees growing in the epilittoral such as *Populus* and *Salix* species can be heavily damaged during summer drainage. Only the communities situated in waterlogged ecotopes around pond-bottom springs and along streams of running water in bays are in a more favorable position.

Positive Effects. Positively influenced are all species, populations and communities that can complete their life cycles in the relatively short time available. The plants are rooted in emerged wet or waterlogged subhydric soils (see Sect. 3.4). A regular incorporation of the following two basic ecological gradients into the biocenological network is necessary to understand the specific character of the biotic communities in fishponds: (1) both regular and irregular changes of water level; (2) repeated emergence of the subhydric soils and repeated retreat of the shoreline, as well as fall of the underground water table in ecotopes near the shoreline.

The possibility of drainage at any time and for periods of any length means that fishponds, as opposed to other types of standing waters with a long-lasting ecocycle (Hejný, 1971), have no fixed vegetational zonation as it is common in lakes (see Introduction). In fishponds, stabilization of biotic communities is only relative and temporary during hydro-littoral ecoperiods.

The changing water level and the possibility of its control at any time result in periodical alternations of different types of biotic communities in a fishpond during an ecocycle. These communities range from those of emerged bottoms, through those of littorals, to those of open water. The shifts of the shoreline bring about a differentiation in the development of unstable macrophyte cenoses, and an either inward or outward shift of the littoral beyond its usual "stabilized" boundaries. The changing actual position of the shoreline and the different duration of various ecophases in each part of a fishpond fulfill the fundamental requirements of the development, optimal structure and duration of various unstable fishpond phytocenoses. Two different complexes of cenoses develop during an ecocycle: (a) unstable stenoecious cenoses develop during critical ecoperiods of a temporary emergence of parts of the bottom, of summer drainage, and after refilling a fishpond; (b) stable euryecious cenoses establish themselves during long

Table 1. Changes in pond plant communities in years following summer drainage as the water level in a pond gradually returns to its normal position

Years after summer drainage[a]	Central part of fishpond		Vegetation dynamics in the littoral (general trends)
	Ecoperiod	Range of community types	
0 (summer drainage)	Limoso-terrestrial	*Bidention—Nanocyperion—Litorellion—Agropyro-Rumicion*	Onset of steneocious ephemerous communities of bare bottoms greatly differing in production, followed by their full development and subsequent decay. Degradation of reedswamp stands.
1	Limoso-littoral	*Oenanthion—Litorellion* (+nuclei of *Potamogetonion pusilli*)—*Nanocyperion—Agropyro-Rumicion*	Onset of stenoecious ephemerous littoral communities and of stenoecious forms of therophytes. Life of reedswamp stands returning to normal, appearance of their regeneration phase in the sublittoral.
2	Littoral-hydrophase	*Lemnion—Potamogetonion pusilli—Oenanthion*—regenerating *Phragmition—Litorellion*	Full development to disappearance of stenoecious littoral communities and of stenoecious cenoses of ephemers. Beginning of stabilization of cenoses in the pond.
3	Hydrophase	*Lemnion—Potamogetonion pusilli—Litorellion or Callitricho-Batrachion*	Disappearance of stenoecious forms. Stabilization of all cenoses in the pond.

[a] Each year corresponds with a certain ecoperiod and a range of plant community types (foederationes, see Sect. 1.2.2) in the central part of the pond and with a certain vegetation development in the littoral.

periods when a fishpond is filled with water maintained at a more or less constant level. These are the littoral-hydrophase ecoperiods.

The two cenotic complexes overlap in ecoperiods immediately following the refilling of a fishpond (Table 1), and the relatively stabilized cenoses exclude the unstable ones in ecoperiods during which the fishpond remains filled with water for several years.

1.2.2 Outline of Littoral Macrophyte Communities

1.2.2.1 General Remarks

Communities of aquatic vascular plants (macrophytes) represent the basic structural constituents of any fishpond ecosystem and of its subsystems. The fishpond macrophyte communities have been studied in far more detail than the communities of other organisms whose ecological niches are frequently defined in relation to the type of vegetation. A survey is therefore given here of the vascular plant communities occurring in fishponds and in their littorals in the two regions studied within the Czechoslovak IBP Wetlands project, namely in South Bohemia and South Moravia.

This survey does not provide the phytocenological characteristics of the macrophyte communities, neither does it discuss their syntaxonomical classification. In this book, devoted to the ecology of fishpond littoral ecosystems, stress is laid on the general ecological characteristics of these communities classified into associations which are, in turn, grouped into foederationes (alliances) as the nearest higher-ranking units according to the principles of the Zürich-Montpellier school of vegetation classification. The ecological characteristic pertains to the importance and role of each community type in fishponds, to its relationship (if it is known) to the cultivation of carp and other fish in the ponds and, particularly, to its relationship—either positive or negative—to fishpond management. This management may be regarded as a system of man-induced factors affecting all constituents of a fishpond ecosystem. Our survey comprises not only the littoral communities of perennial macrophytes, but all communities occurring within a wide range of conditions from open water to the outer littoral; a brief account is also given of communities colonizing emerged fishpond shores and bottoms. The role of these communities in fishponds will stand out quite clearly in this context.

The survey presented here has the single purpose indicated above. The complementarity of phytocenology with production and applied ecology should result in more exact definitions of suitably chosen basic units needed for either vegetation classification (associations) or for studying nature's functioning (ecosystems). This joint effort should lead both phytocenology and ecology towards a common goal, which is a clear assessment of the ecological and structural range, and an understanding of the functioning of each plant-community type in fishpond ecosystems.

Our survey does not comprise those types of communities which are not immediately linked with the land-water ecotone. The vegetational units playing an inferior role in fishponds are mentioned only briefly.

The principal types of natant and littoral plant communities are characterized as they appear in nature in their various topic and dynamic forms, which reflect, respectively, their variation in space and time. This variation affects, to a varying degree, the production and energy budgets of individual compartments in the ecosystems comprising these communities.

An ecologist may not be familiar either with the structural range and variation, or with all developmental links between the communities belonging to the ecosystem he is studying, and his compartmentalization of that ecosystem may thus become arbitrary. In this case, however, he will be neglecting numerous important theoretical aspects of the study of the ecosystem; he will also be deprived of a great many possibilities of manipulating the ecosystem. Equally deprived will be a phytocenologist if his classification of vegetational structures and assessment of their variation is not supported by data on the controlling environmental factors and on the role of a given community type in energy flow and material cycling in the biosphere.

In the following survey, the individual vegetation units (foederationes and their associations) are listed according to the order which is usual in phytocenological writings, employing, in principle, the approach of the Zürich-Montpellier phytosociological school. The following terms and abbreviations are used in this chapter, consistent with their use by Holub et al. (1967), who give a survey of vegetation units represented on the territory of Czechoslovakia, from which the present survey of the Czechoslovak fishpond vegetation is derived.

Indicating (Group of) Species: this comprises the species with the highest constancy values (IV, V) and with the greatest density in stands of a certain community type. At the same type, the type is identifiable by certain character and differential species, though they may possess only low constancy values.

Character Species (ch): these are species with optimum occurrence in a given community type.

Differential Species (dif): these species usually indicate certain edaphic or microclimatic conditions of habitats, or chorological circumstances. In this way, the species differentiate between two or more related community types, which are most frequently classified as different subassociations.

Dominant Species (dom): these species exhibit the greatest relative cover degree in a community or in its individual synusia.

Edifying Species (edif), also called ecological dominants: these are dominant species which have a decisive effect on the structure and development of a community and on its environment.

Locally Prevailing Species (loc): these are species which become dominant only under specific local conditions.

The other abbreviations used pertain to the known distributions of the communities in Czechoslovakia and in Europe: CZ: Czechoslovakia; Eu: Europe; A: Asia; N: North; E: East; S: South; W: West; C: Central; B: Bohemia; M: Moravia; S: Slovakia (the last three abbreviations only in combination with the designations of the four cardinal points. The geographical division of Czechoslovakia used here follows the administrative division as shown in the map in Figure 2; however, the whole of the Bohemian-Moravian upland, administratively divided between South Bohemia (SB) and South Moravia (SM), is referred to as South Bohemia (SB, unless otherwise stated) because of the similarity of ecological conditions with SB rather than SM.

The literature consulted for the geographical distribution of the communities both in Czechoslovakia and Europe, as well as in Asia, is not being referred to.

1.2.2.2 Survey of Individual Vegetation Types

— Foederatio *Lemnion minoris* Koch et R. Tüxen 1954 ap. Oberdorfer 1957

Communities of duckweeds and ecologically related plants occurring on the water level and confined to the surface pellicle; they often appear periodically. They do not root in the bottom, except for terrestrial forms, and therefore may migrate over the whole fishpond, moved by wind and water currents. As independent units, they occur in open water; as synusia components, they are found in associations ranging from the eulittoral to central parts of ponds.

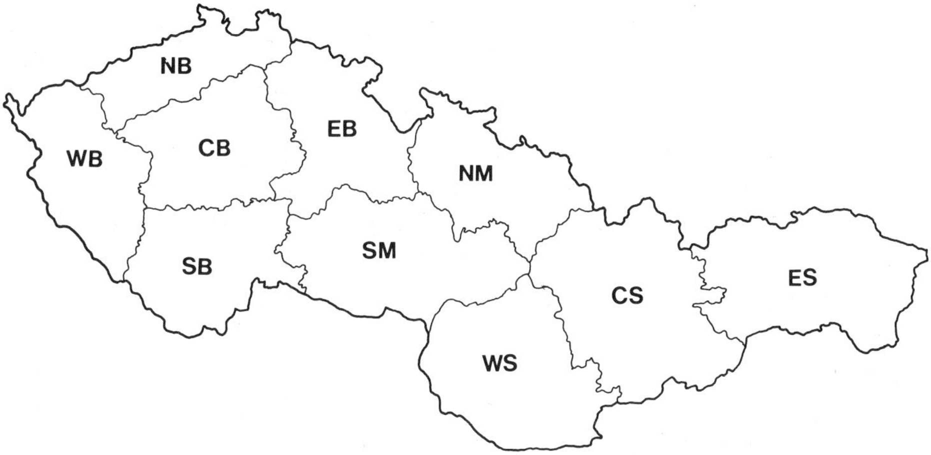

Fig. 2. Division of Czechoslovakia into districts. *Left to right: B*, Bohemia; *M*, Moravia; *S*, Slovakia; *C*, central; *N, E, S, W*, northern, eastern, southern, western

Stands of duckweed communities of the foederatio *Lemnion minoris* are structurally the simplest cenoses of macrophytes in fishponds because of their restricted species outfit and because of their capability to "move".

Despite the small number of constituent species and their structural simplicity these communities belong to the most variable units of aquatic vegetation both in space and time. This variation is affected particularly by the number of warm and windless days in the period between May and early summer. The communities are frequently directly affected by the chemical properties of the water; hence, they may be of value as indicators in fishpond management.

The rapid changes in species composition, rapid onset of populations, their early decay, and both spatial and temporal alternations of ecotopes in one pond allow the existence of different simpler stand structures, for which the following classification is put forward:

1. Topic forms: communities confined to certain ecotopes in ponds.

(a) Eulittoral form: lives as a synusium in communities of the foed. *Phragmition communis*, or, periodically, in inundated communities of tall sedges, *Magnocaricion*.

(b) Sublittoral form: usually forming a continuous zone as a more or less narrow strip along the waterward edges of reed-belts. This is the primary form of existence of independent *Lemnion* communities.

(c) Open-water "pelagial" form: usually existing as synusium in communities of rooted aquatic plants, but often also occurring as an independent community.

(d) Terrestrial form: usually occurring after emergence of subhydric soils in limosal ecophase as an independent community rooted in soil, gradually degenerating with decreasing water content in the soil. It can also exist temporarily as a synusium in communities of foed. *Bidention tripartiti*.

(e) Bay form: usually existing in fishpond bays as an independent community, structurally more complicated than the previous forms and usually mixed with other topic forms (a–c).

2. Dynamic forms: communities which can change their topological position and structure in time in the ecotopes mentioned above.

(a) Diffuse form: plants distributed diffusely over the water surface or in various ecotopes so that they do not touch one another, or forming numerous scattered patches of small colonies. This form usually builds the initial "pioneer" stands of duckweeds.

(b) Stabilized form: plants building stands, touching one another and forming more or less dense synusia or communities persisting for appreciably long periods.

(c) Explosive form: stands appearing suddenly in the growing season, usually of very short-term or ephemerous duration. They may arise only as a succession stage following the decay of stands of foed. *Potamogetonion pusilli*, but they usually develop suddenly under optimal temperature conditions. Hence they may arise through the propagation of various topic forms.

(d) Pulsating form: stands growing vigorously under very good conditions and retreating under unfavorable conditions. In a series of successive vegetation periods, these stands either spread or retreat from the sublittoral to the open water surface and vice versa. This pulsation should not be mistaken for an ephemerous occurrence (form 2c).

Forms of the pleustonic associations of foed. *Lemnion minoris* occur in fishponds of all types. The explosive and pulsating forms spread as a result of fertilization and particularly of organic manuring of fishponds; they are thus hemerophilous in character. These two types may be regarded as prospective in view both of their high production rates (see Sect. 3.1.7) and the possibility of maintaining them permanently. Duckweeds can be used as supplementary fodder, especially for poultry. On the other hand, extremely dense duckweed communities covering the entire water surface in eutrophicated fishponds for long periods represent a danger for the whole complex of biotic communities in a fishpond. The duckweed cover reduces their structural diversity and may even destroy them. Particularly unwanted is the shading effect of duckweeds on phytoplankton.

The foed. *Lemnion minoris* comprises the following associations:

Lemnetum trisulcae Soó 1927

Indicating species: *Lemna trisulca* (ch, dom).

Habitats: Mesotrophic fishponds with clean water. A community frequently in contact with *Batrachio circinati-Alismatetum graminei* which usually forms submerged stands of the eulittoral type (form). In eutrophicated fishponds, the duckweed communities occur as pulsating forms covering a larger or smaller part of the open water surface, often very densely. Distr.: Cz: SB, EB, SM, SS; Eu: all regions (?); A:C.

Riccietum fluitantis Slavnić 1956

Indicating group of species: *Riccia fluitans* (ch), *Ricciocarpus natans* (dif)

Habitats: Rather small oligotrophic fishponds, often in contact with *Potamogetoneto natantis-Nymphaeetum candidae*. With stronger eutrophication, *Riccietum fluitantis* retreats into the zone of tall sedges and reed beds in the outer littoral (depth 5–26 cm). Temporarily it also occurs in eutrophic ponds as an explosive type on the whole water surface, but this is only a transient phenomenon in the season following summer drainage. In this case, *Ricciocarpus natans* is regularly the only dominant species.

Distr.: Cz: SB, WB, NB; Eu: C, W, N.

Riccietum rhenanae Knapp et Stoff. 1965

Indicating species: *Riccia rhenana* (ch)

Habitats: In mesotrophic and eutrophic ponds only in districts rich in limestone and alkaline soils. In fishponds, this association occurs only temporarily during ecoperiods of shallow water.

Distr.: Cz: SB, SM, SS; Eu: C.

Wolffietum arrhizae Miyawaki et J. Tüxen 1960 occurs only in alluvial waters. Distr.: Cz: SM, SS; Eu: C, W, N.

Lemnetum minoris Th. Müller et Görs 1960

Indicating species: *Lemna minor* (ch, dom)

Habitats: Relatively nutrient-poor, cool mire waters, slightly acid to slightly alkaline. In fishponds, the occurrence of this community represents the first sign of eutrophication.

Distr.: Cz: all regions from colline to submontane and montane regions (in SM occurs regularly in alluvial waters but has not been observed in fishponds); Eu: all regions; A: C, N, W.

Lemno-Spirodeletum Koch 1954

Indicating group of species: *Spirodela polyrhiza* (dom), *Lemna minor* (ch)

Habitats: Relatively more eutrophicated waters, esp. fishponds with a tendency to hypertrophy, small village ponds (very dense stands occur here).

Distr.: Cz: SB, CB, EB, SM, SS, CS, ES; Eu: all regions.

Lemnetum gibbae Miyawaki et J. Tüxen 1960

Indicating group of species: *Lemna gibba* (dom), *L. minor* (ch), *Spirodela polyrhiza* (ch)

Habitats: Predominantly in the lowlands and less frequent up to the submontane regions, in eutrophic to polluted fishponds. Difficult to identify because *L. gibba* has two different growth forms: a "thin" flat form and a "thick" gibbous form. The flat form occurs under conditions of stress, e.g., in relatively nutrient-poor waters, the aerenchymatous gibbous form indicates optimum conditions of temperature and nutrient supply. Most of the analyzed communities of *Lemnetum minoris* therefore appear to belong to *Lemnetum gibbae* formed by the flat form of *L. gibba* (see also de Lange, 1975).

Distr.: Cz: SB, CB, EB, SM, WS, CS, ES; Eu: C, W.

Salvinio-Spirodeletum Slavnić 1956 has been observed in Cz only in alluvial waters in ES. Distr.: Cz: ES; Eu: W, C, E; A: C.

— Foederatio *Utricularion vulgaris* Passarge 1964

From the structural point of view this foederatio represents a link between the foederationes *Lemnion* and *Hydrocharition*. The communities are less motile than the *Lemnion* communities. Hence they are more or less fixed to certain sites, but the periodicity of their occurrence remains. In fishponds, the existence of these communities as synusia is confined to relatively well-irradiated sites, e.g., to diffuse reed-belt stands, especially to communities of *Glycerietum maximae* and of both *Typhetum* association or to diffuse stands of the ass. *Caricetum gracilis*. As independent communities, the *Lemno-Utricularietum* stands occur in the sublittoral and especially in quiet bays.

The foed. *Utricularion vulgaris* comprises the following associations:

Lemno-Utricularietum Soó 1928

Indicating group of species: *Utricularia vulgaris* (ch, dif), *U. neglecta* (ch, dif), *Lemna trisulca* (ch), *L. minor* (ch), *Spirodela polyrhiza* (ch)

Variability: The association occurs in two forms (subassociations): (i) *Lemno-Utricularietum utricularietosum vulgaris* Soó 1971; (ii) *L.-U. utricularietosum neglectae* Soó 1971, which may also be distinguished as independent associations in view of their geographical separation.

Habitats: Calm fishpond bays and loose margins of reed-bed communities on sapropel or reed-bed förna mixed with sapropel. Optimum living condition of either subassociation arise during warm growing seasons with a slightly lowered water level. These occur namely in sky-fed fishponds out of touch with any river or brook alluvium.

Distr.: Subass. (i): Cz: CB, EB. Subass. (ii): Cz: all regions. The whole ass.: Eu: C, W, N, E.

— Foederatio *Hydrocharition* (Vierhapper) Rübel 1933

with its associations:

Hydrocharietum morsus-ranae van Langendonck 1935

Indicating group of species: *Hydrocharis morsus-ranae* (ch, dom), *Utricularia vulgaris* (ch, loc.), *Spirodela polyrhiza* (ch)

Habitats: Calm fishpond bays, but only in fishponds with a short retention time, or in fishpond systems along rivers and creeks. Here usually communities small in area, but with a characteristic pattern, contrasting with large and dense communities of this association in calm alluvial waters. A highly sensitive vegetation type, especially to fishpond drainage both in summer and winter. Hence it is absent in sky-fed ponds.

Distr.: Cz: SB, EB, SM, WS, CS, ES; Eu: C, W, N, E.

Hydrochari-Stratiotetum (van Langendonck 1935) Westhoff 1942

Indicating group of species: *Hydrocharis morsus-ranae* (ch), *Stratiotes aloides* (ch)

Habitats: Heavily silted backwaters, oxbows and pools in alluvia of both SB (very rare) and SM (more frequent), hemerophobic in fishponds, being resistant neither to winter drainage nor to repeated summer drainage.

Distr.: Cz: SB, EB, SM, ES; Eu: C, W, N, E; A: W.

Ceratophylletum demersi (Soó 1927) Eggler 1933

Indicating group of species: *Ceratophyllum demersum* (ch, dom), *Lemna trisulca* (ch), *L. gibba* (ch), *Spirodela polyrhiza* (ch)

Habitats: In eutrophic fishponds with rich mineral (and slightly alkaline) soils covered with a thick layer of sapropel. Structurally, this community is very simple as a dense carpet of the dominant *Ceratophyllum* which fills homogeneously the entire water column. Lemnids usually build a synusium in the surface pellicle of water. The association has a great ecological amplitude from the sublittoral (depth 0.4–0.6 m) to the central pond area (depth 1–1.5 m), but only in small ponds. In large fishponds, this community type can only occur in accumulation biotopes, especially in calm bays. Under poorer soil conditions (e.g., in SB), it occurs periodically after a rapid change of water level, especially following a long littoral ecoperiod. This situation has developed only during the last decade as a result of a systematical eutrophication of ponds, and the community occurs here only in initial stands.

Distr.: Cz: all regions; Eu: C, N, W, E; A: C, W.

Communities of the foed. *Utricularion vulgaris* and *Hydrocharition morsus-ranae* occur in fishponds only in accumulation biotopes in bays and in periodically emerging reed-belt margins. To a certain extent, they are of "relic" nature capable only to maintain themselves rather then to spread. They seem rather hemerophobic, being intolerant of any management measures, but the ass. *Ceratophylletum demersi* represents an exception, having recently moved from river alluvia into the more eutrophicated fishponds.

— Foederatio *Nymphaeion albae* Oberdorfer 1957

Communities of distinct physiognomy, differing from others, belonging to the foed. *Potamogetonion lucentis* and *Potamogetonion pusilli*, by their floating leaves and perennial rhizomes. A specialized life form of aerohydatophytes determines the physiognomy of the communities. This basic life form is perennial (with the exception of *Trapa natans*). This is a good indication of the adaptation of aerohydatophytes to deeper water with a long-hydrophase ecocycle. Aerohydatophytes are therefore very well suited for life in backwaters, oxbows and lakes but in fishponds they occur only in habitats with a very long ecocycle without long periods of drainage either in winter or in summer. They are rather sensitive to certain intense management methods involving long or frequent drainage (*Myriophyllo-Nupharetum, Trapetum natantis*) or combining drainage and fertilizer application (*Potamogetoneto natantis-Nymphaeetum candidae, Nupharetum pumili*). The modern management of fishponds thus suppresses these stabilized cenoses which are replaced by annual types of euhydatophytes.

Myriophyllo-Nupharetum Koch 1926

Indication group of species: *Nuphar luteum* (ch, dom), *Myriophyllum verticillatum* (ch), *M. spicatum* (ch)

Habitats: Only in fishpond systems with water flowing through, especially in those along creeks or, rarely, along rivers. This association indicates ponds rich in deep sapropel soils with silt which are also rich in nutrients; the community is hardly resistant to repeated summer drainage and intolerant of winter drainage, occupying deeper water from 0.8–1.3 m in the central parts of ponds; in large ponds, it occurs more frequently in bays. In fishponds, this community type is rare and often lacks the *Myriophyllum* species. These occur regularly in lowland and colline alluvial waters, where also the community type is widespread.

Distr.: Cz: SB, EB, CB, NB, SM, WS, ES; Eu: C, W, N, E; A: C.

Potamogetoneto natantis-Nymphaeetum candidae (Hejný 1948)

Indicating group of species: *Potamogeton natans* (ch), *Nymphaea candida* (ch)

Habitats: A water-surface community, mainly in originally oligo- to mesotrophic fishponds. Favored substrates are peat or gravel covered with a thick layer of sapropel and dy at a depth of 0.8–1.5 m, with the water slightly stained with humic substances. In small fishponds with deep bottom sediments, stands of this community form a nearly continuous cover on the entire water surface. In combination and in contact with extensive stands of *Equisetum fluviatile*, communities of this type have contributed to a rapid filling of these small fishponds.

A pronouncedly hemerophobic community type which is sensitive to intense fishpond management, it retreats rapidly or disappears after heavy liming combined with winter drainage. First, *Nymphaea candida* loses its dominance, surviving only in a few irregularly spaced patches. It is replaced by *Potamogeton natans*, which is able to survive even in eutrophicated fishponds. Repeated winter drainage, however, suppresses the dominance of *Potamogeton* as well, and the community disintegrates. An association deserving full protection, nowadays surviving predominantly in fishponds of the Bohemian-Moravian upland.

A specimen vegetation relevé documents the association. Date and location: September 5, 1947; Kutinovksý fishpond near Vodňany (SB). Area: 20 m^2. Cover degree: 90% to 100%. Depth: 0.75 m.

Nymphaea candida	4	*Equisetum fluviatile*	2
Potamogeton natans	5	*Sagittaria sagittifolia*	1
P. acutifolius	1	*Utricularia neglecta*	+

Distr.: Cz: SB, NB, EB; Eu: C, N.

Nuphareto lutei-Nymphaeetum albae mainly occurs in Central Bohemian, Moravian and Slovakian alluvial waters in regions of calcareous and carbonate-rich soils and parent rocks.

Distr.: Cz: SM, WS, CS, ES; Eu: C, E; A: C.

Nupharetum pumili Oberdorfer 1967

Indicating group of species: *Nuphar pumilum* (ch), *Potamogeton obtusifolius* (loc)

Habitats: Deeper sapropel soils on sand in originally oligo- to mesotrophic fishponds. Rapidly retreating after fertilizer application or winter drainage.

A relic association deserving full protection, which occurs more rarely in fishponds than in systems of submontane alluvial waters.

Distr.: Cz: SB (rare); Eu: C.

Trapetum natantis Th. Müller et Görs 1960

Indicating species: *Trapa natans* s.l. (ch)

Habitats: On clayey soils with a thick layer of sapropel in fishponds, in deep water, from 0.6 to 1 m. The community retreats after summer and winter drainage or, rapidly, after early cutting in June. Neither fertilizing nor liming affect this community of eutrophic fishponds so that the association is hemerophilous under these conditions. The stands tend to be luxuriant and hence disliked by the fishpond managers.

Distr.: Cz: SB, EB, SM, WS, ES (scattered occurrence in all these regions); Eu: C, W, S, N; A: C.

Nymphoidetum peltatae (Allorge 1922) Th. Müller et Görs 1960

Indicating species: *Nymphoides peltata* (ch)

Habitats: On clayey soils with gravel covered with only a thin layer of sapropel in ponds; occurs rarely in deeper water and penetrates even into shallow littoral areas. The stands survive both summer and winter drainage as well as an early cut.

Distr.: Cz: SB, SM, WS; Eu: C, W, N, S; A: C.

The two associations just mentioned and combined by Oberdorfer (1957) into one ass. *Trapo-Nymphoidetum* require protection especially in fishpond reserves. They also have a very similar distribution both in SB (only in fishponds) and SM (only in alluvia).

Potamogetonetum graminei Koch 1926

Indicating group species: *Potamogeton gramineus* (ch, dom), *P. acutifolius* (ch), *P. angustifolius* (loc)

Habitats: Shallow sublittorals of mesotrophic fishponds with highly transparent water on sandy clay, rarely eulittorals along edges of *Caricetum elatae* communities. Frequently also in contact with the ass. *Batrachio circinati-Alismatetum graminei*, which is better developed in deeper water towards the center of a pond. *Potamogetonetum graminei* is more sensitive to intense liming, but it spreads rapidly after summer drainage and after dredging of the reed-belt, especially on bare soil surface.

Distr.: Cz: SB (rare), EB; Eu: C, W, N.

— Foederatio *Potamogetonion lucentis* Vollmar 1947 em. nom. Rivas-Martínez 1973

with its associations:

Potamogetonetum lucentis Hueck 1931

Indicating group of species: *Potamogeton lucens* (ch), *Batrachium circinatum* (ch, loc), *Alisma gramineum* (Fig. 3)

Habitats: Primarily mesotrophic ponds with clay or clay with gravel and highly transparent water, less than 1 m deep. On muddier bottom, *P. lucens* retreats and is replaced by advancing *P. natans*. This is an association with a good adaptability to contemporary fishpond management. The community can bear heavy fertilizing and re-establishes itself easily after summer drainage. The shoots of *P. lucens* die off in mid-summer. They never form dense stands in open water and their cutting is unnecessary, unless *Batrachium circinatum* forms a richly developed facies (Fig. 4) which may be regarded as a special association of *Batrachio circinati-Alismatetum graminei* Hejný h.l., illustrated by the following relevé:

Date and location: August 10, 1962; Bukový fishpond (SB).

Area: 100 m^2. Cover degree: 80% to 90%. Water depth: 0.8 m.

Batrachium circinatum	4	*Potamogeton pusillus*	1
B. trichophyllum	1	*P. lucens*	1
Alisma gramineum	2		

Distr.: Cz: SB, EB, SM, WS, ES; Eu: C, E, S.

Fig. 3. Floating community of *Potamogetonetum lucentis*. (Photo: K. Gregor)

Fig. 4. Typical South Bohemian fishpond with flowering populations of *Batrachium circinatum* and reed-belts of *Schoenoplectus lacustris*. (Photo: K. Gregor)

Elodeetum canadensis Eggler 1933

Indicating species: *Elodea canadensis* (ch, dom)

Habitats: *E. canadensis*, which can occur in ponds dispersed among other types of aquatic vegetation, lacks any pronounced phytocenological valence. In ecoperiods with littoral ecophase, it receives a stronger or weaker stimulus for sudden vegetative growth. This results in the formation, after one to two years, of dense stands of *E. canadensis* in deep water, filling the entire water column from littoral to deep water ecotopes (1.5 m). Repression of all the other components of higher aquatic vegetation favors this dense and structurally very simple alien association. It occurs in meso- and eutrophic fishponds for transient periods. The association is highly hemerophilous, resistant to fertilizing and cutting and very dangerous to the fish stock in summer and especially during the time of fish harvest because fishes are trapped and suffocated to death in the dense carpet of *Elodea* arising during water discharge. In winter, dense stands of *Elodea* are also dangerous to a hibernating fish stock because the decaying plants have a high consumption of oxygen. The control of this community is quite simple, by a combination of winter and summer drainage. Under such conditions, *Elodeetum canadensis* will retreat rapidly. It seems that in fishpond systems along running waters the incidence is more frequent than in sky-fed systems.

Distr.: Cz: SB, EB, SM, WS, ES (only alluvial waters in the last three regions); Eu, A: eurosiberian distr.

— Foederatio *Potamogetonion pusilli* Vollmar 1947 em. nom. Hejný h.l.

These communities are built up of water plants possessing only submerged leaves and very simple root systems (life-form of euhydatophytes). They can exist only in water and disappear after drainage of the hydric soil surface. Most of the constituent species are short-lived annuals, propagating rapidly both generatively by seeds, and vegetatively by hibernacula. Their communities are thus particularly well adapted to survival in fishponds. They react flexibly to various regimes of fluctuating water level and their occurrence is rather unstable. The communities belong to the following associations:

Potamogetoneto-Zannichellietum palustris Koch 1926

Indicating group of species: *Potamogeton pectinatus* (ch, dom), *P. pusillus* (ch), *Zannichellia palustris* ssp. *palustris* (dif)

Habitats: Eutrophic fishponds in regions of non-alkaline soils and with sandy-clay bottom, medium to strongly muddy (greater muddiness often indicated by *Zannichellia*), sky-fed ponds with fluctuating water level. Periodic occurrence after summer or winter drainage or, especially, in ponds with water level falling during spring-time. Long-lasting occurrence only in shallow ponds with deeper clayey soils, indicating fertilized and manured fishponds and bearing less transparent water. The community provides very good feed for carp and for wild ducks (especially the fruits of the plants). Its sudden appearance frequently indicates a transition from mesotrophy to eutrophy. Cutting is quite redundant as the stands die off in mid-summer. After heavy manuring this community replaces communities of the *Batrachio circinati-Alismatetum graminei*.

Distr.: Cz: SB, CB, EB, SM; Eu: C, W, N.

Potamogetoneto-Zannichellietum pedicellatae Soó 1947

Indicating group of species: *Potamogeton pectinatus* (ch, dom), *P. pusillus* (ch), *Zannichellia palustris* ssp. *pedicellata* (dif)

Habitats: Strongly eutrophic to hypertrophic ponds with intense organogenous accumulation of hydatophyte remnants; confined to alkaline waters and usually species-poorer than the previous association. It often indicates an occasional pollution with dung and occurs permanently in dung-polluted waters. Periodic occurrence in deep muddy soils after fall of water level. In view of their early spring development and disappearance towards the end of June, these communities need not be controlled. The decomposition of the constituent plants is rapid, yielding compost material of good quality. The communities provide very good feed for carp and for wild ducks.

Distr.: Cz: EB, SM, WS, ES; Eu: C, E; A: C.

Potamogetonetum trichoidis Freitag, Markus et Schwippl 1958

Indicating group of species: *Potamogeton trichoides* (ch, dom), *P. panormitanus* (ch), *P. pusillus* (ch)

Habitats: Mesotrophic to slightly eutrophic ponds, ecotopes with deep sandy or combined sandy-skeleton bottoms. It occurs in deeper water than *Potamogetoneto-Zannichelietum palustris* and is very poor in species, often consisting only of *P. trichoides*.

Distr.: Cz: SB, NB, EB, SM (rare), WS, CS; Eu: W, N, C.

Najadetum marinae (Oberdorfer 1957) Fukarek 1961 and *Najadetum minoris* Ubrizsy (1948) 1961
occur only in alluvial waters in the lowlands. Cz: SB, NB, EB, SM, WS, CS; Eu: W, N, C; A: C.

Potamogetonetum crispi Soó 1927

Indicating group of species: *Potamogeton crispus* (ch, dom), *P. berchtoldii* (ch), *Batrachium peltatum* (ch)

Habitats: In eutrophic to hypertrophic fishponds with a thick layer of organic mud arising from remnants of hydatophytes and, especially, from decomposing leaves, e.g., of *Alnus glutinosa*. In fishpond systems along running waters, the communities frequently indicate ecotopes silted with deep mud. This association has a typical autumn to spring development (from September to June). Hence the communities require no cutting but they are suitable for composting in mid-June, as *Potamogeton crispus* decomposes rapidly. Optimum of occurrence in deep water from 0.8 to 1.5 m.

Distr.: Cz: all regions; Eu: W, N, S, C, E; A: C.

Potamogetonetum obtusifolii (Sauer 1937) Carstensen 1955

Indicating group of species: *Potamogeton obtusifolius* (ch, dom), *P. acutifolius* (ch, loc)

Habitats: Calm and shallow bays with very deep sapropel in meso- to eutrophic ponds, optimum of occurrence in shallow water, 0.2–0.4 m deep. The communities indicate organogenous accumulation of material originating from the reed-belt (decomposing förna). The prevailing topic forms therefore form belts along the edges of reedswamp communities, especially along both *Typheta* and *Glycerietum maximae*.

Distr.: Cz: SB, WB, NB, CB, EB, WS; Eu: N, W, C.

Ceratophylletum submersi Soó 1928
occurs in districts of alkaline soils and waters in SM and ES, and is only very rare and fragmentary in ponds.

— Foederatio *Batrachion aquatilis* Passarge 1964

These communities consist of water plants with both submerged and floating leaves (small plants are hydatophytes) growing in shallow water with fluctuating water level. Ecoperiods with a limoso-littoral sequence provide the most important conditions for a full but short-term existence of these communities in deeper water. Periodicity of occurrence and a typical development from autumn to spring are the main features of the communities belonging to this foederatio.

Batrachietum aquatilis-peltatae Sauer 1947

Indicating group of species: *Batrachium aquatile* (ch, dom), *B. peltatum* (ch, dom)

Habitats: In meso- and eutrophic ponds with a loamy-clayey bottom and with a shallow mud layer. The community structure depends on the occurrence of a limoso-terrestrial ecoperiod during the summer to autumn of the preceding year, because a mass germination of *Batrachium aquatile* can only take place in limosal ecophase.

The communities therefore occur: (a) in the fishpond sublittoral and in bays, following a gradual fall of water level during the preceding autumn; (b) all over the water surface, after a deep fall of water level or summer drainage during the preceding summer. These communities are thus of very short-term duration in the central parts of fishponds, but they can survive longer along the shores and in bays if their bottoms emerge rather regularly. Juvenile phenological phases of the constituent species appear in autumn to winter, mature plants in the spring months. Full flowering-time of the community is from early May to June (earliest of all aquatic plant communities). The stands die off gradually during summer. Even then, however, some plants may survive amidst plant materials drifted along the shores. Cutting is redundant as a control measure, but the plant material is suitable for composting.

Distr.: Cz: all regions, but in SM, this association is frequently replaced by the vicariant ass. *Batrachietum rionii;* Eu: all regions (?).

Batrachio trichophylli-Callitrichetum cophocarpae Soó (1927) 1960

Indicating group of species: *Batrachium trichophyllum* (ch, dom), *Callitriche cophocarpa* (ch, dom)

Habitats: In oligo- to mesotrophic fishponds with a sandy gravel bottom. The period of occurrence is much the same as for the previous association, but fragments of communities of this association may persist for a longer time in the sublittoral of cool fishponds with highly transparent and nutrient-poor water.

Distr.: Cz: SB; Eu: C, E.

Batrachietum rionii Hejný et Husák ass. nova

Indicating group of species: *Batrachium rionii* (ch, dom), *B. baudotii* (ch, dom), *Potamogeton pectinatus* (ch)

Habitats: In fishponds with alkaline water in shallow sublittoral (depth 0.2 to 0.4 or even 0.6 m), usually in contact with ass. *Potamogetoneto-Zannichellietum pedicellatae*, which penetrates into deeper water. A fall of water level and emergence of the bottom during the previous summer to autumn again represents a condition necessary for the occurrence of this community type. The following relevé documents this new association.

Date and location: June 14, 1973; Nesyt fishpond (SM).
Area: 20 m^2. Cover degree: 90%. Water depth: 0.3 m.

Batrachium rionii 4	*Zannichellia palustris* 1
B. baudotii 3	*ssp. pedicellata*
Potamogeton pectinatus 1	*Lemna gibba* 1
P. pusillus 1	*L. minor* 1

Distr.: Cz: SM; Eu: C, S, E.

In the river alluvia of South and Southeast Slovakia, the association *Batrachietum aquatilis-polyphylli* Soó (1933) 1961 occurs rarely in alkaline waters.

Hottonietum palustris R. Tx. 1937

Indicating species: *Hottonia palustris* (ch)

Habitats: Hemerophobic in fishponds. Winter drainage is perhaps the controlling factor as *Hottonia palustris* has an autumn to spring development, with full development and flowering very early in May.

Distr.: Cz: SB, EB, SM, WS, CS, ES; Eu: W, C, N, E.

— Foederatio *Phragmition communis* Koch 1926

Reedswamp communities of distinct physiognomy occurring in littorals along shores of standing waters, formed by perennial ochthohydrophytes. They occur from lowlands to highlands in various kinds of mineral soil from sands to heavy clays and their litter gives rise to förna. The stands also colonize organogenous soils formed in the course of detritus accumulation in a pond, and are themselves the main participants in land formation.

The foederatio comprises a set of long-term successional stages with a capability to reinforce the shores by protecting them from erosion by wind, waves and split ice. The communities also filter the wash-out from the surroundings, brought into the pond at times of heavy rainfall. They provide important breeding biotopes for waterfowl, and raw material for industry; furthermore, they enclose and conserve organogenous material.

The following terminology has been worked out with respect to the forms and dynamics of reed-bed communities in fishponds:

The fishpond littoral itself represents a rather complex system in which various processes occur: accumulation, erosion, and both combined. This complexity is reflected in the structure of reed-bed communities. They comprise, in a gradation, all structural types of less complex communities ranging from open water to the littoral. In this complex of communities, the following basic forms (types) may be distinguished:

1. Topic forms: practically in any larger pond unaffected by amelioration and, particularly, in large or already partly silted ponds, the following topic as well as successional forms (seres) of littoral communities may be distinguished:

(a) Accumulation form: determined by an uninterrupted sedimentation of autochthonous organogenous material enriched with drifted allochthonous material. An accumulation sere of reedswamp communities develops in a gradual sequence from open water to the "normal" shoreline; a development of islets of loose soils (organogenous material) is typical. These are the basic sources of structural diversity in the build-up of mosaic complexes.

(b) Erosion form: develops along wind-exposed shores (mainly the SE or E shores) and is determined by a limited sedimentation both of autochthonous and organogenous drifted material which is constantly washed away. An erosion sere of communities forms a typical three-stage zonation, each phase having a simplified structure.

2. Dynamic forms: A. Dynamics in relation to sequence of ecophases:

(a) Initial stage: corresponds to the stabilized water-level régime of hydrolittoral ecoperiods. Only a loose stand of the ecological dominant is formed, mainly by offshoots which are frequently only sterile. Such stands are in contact with various synusia of aquatic plant communities.

(b) Invasion stage: corresponds to the stabilized régime of littoral ecoperiods. A homogeneous and stable set of populations of the ecological dominant exhibits an optimum development under optimum conditions and colonizes continuously relatively large areas. Increased occurrence of other species and of synusia of various plant communities is rarer here than in the previous stage.

(c) Terrestrial (terminal) stage: corresponds to the variable régime of littoral-terrestrial ecoperiods. Structurally more complicated stands of one or more ecological dominants are formed as a result of frequent drying of the ground and the dominant synusium becomes looser. Increased occurrence of synusia of the foed. *Caricion gracilis* or *Caricion rostratae* adds variety to the communities and enables their more detailed classification. The synusia of pleustonts are usually absent. The hitherto neglected synusia of *Bryophyta* acquire importance. The vertical stratification of terrestrial reed-beds is similar to that of other terrestrial communities.

The sequence of stages presented here can be followed in most fishpond littoral communities. In standing waters of lowland lakes and in river backwaters and oxbows, the situation is more complicated, but it will not be analyzed here.

B. Dynamics in relation to vigor and development

(a) Regeneration phase: a discontinuity, i.e., formation of separate reed-belt zones, takes place if the water level falls to a new shoreline and promotes the formation of a new regeneration stage of reedswamp, or if scraping off the marginal parts of a fishpond (see Sect. 7.3.3) also destroys the reed-belt so that the development of the *Phragmition* stands has to start anew on a new substrate (mainly sand or clay). In other instances, the ecological dominants can germinate en masse in the sedimentary zone, as long as a site remains in limosal ecophase for a certain time. In the second or third year after summer drainage, regeneration phases of reedswamp communities develop as synusia within communities of the foederationes *Bolboschoenion* and *Oenanthion;* these synusia may gradually change into initial or invasion stages of reedswamp. The regeneration phase may thus retain a continuity with the original reedswamp communities, unless this is interrupted by mowing (see Sect. 7.2).

(b) Degradation phase: when the ecoperiods deviate from the standard situation towards the prevalence of the limoso-terrestrial ecophases, the conditions become relatively unfavorable for the development of reedswamp communities, and their degradation sets in.

In these degradation phases (usually comprising one ecoperiod), the reedswamp communities lose their vigor, and the structure of the dominant synusium

Fig. 5. Reed-belt community of *Typhetum angustifoliae* in the Nesyt fishpond. (Photo: Š. Husák)

as well as of the whole community changes. The stand becomes looser, shorter and the dominant may become sterile, eventually not propagating even vegetatively. With the vigor both of ochthohydrophytes and uliginosophytes reduced, transitional synusia of communities belonging to the orders *Cyperetalia fusci* and *Bidentetalia tripartiti* join in, partly making use of the organogenous surface förna. These degradation phases are frequent during summer drainage, and are typical of emerged margins of fishponds.

The foederatio *Phragmition communis* comprises the following associations:

Phragmitetum communis (Gams 1927) Schmale 1939

Indicating species: *Phragmites communis* (edif)

Habitats: Littoral of meso- and eutrophic ponds on clay, loam and sand with a medium-thick layer of sapropel. The rather deep root system of *Phragmites* makes it possible for more species to participate in diffuse *Phragmites* stands, but in dense communities *Phragmites* is a very strong and vigorous ecological dominant, and such communities are very poor in species. These communities are, in fact, highly stable both in shallow and deep water because they tolerate both winter and summer drainage. They require relatively nutrient-rich soils, especially in calcium. In acidotrophic waters, the communities are more diffuse and may be destroyed by unfavorable conditions.

Distr.: Cz: all regions; Eu: all regions; A: all regions.

Schoenoplectetum lacustris (Allorge 1922) Schmale 1939

Indicating species: *Schoenoplectus lacustris* (edif)

Habitats: Above all littorals of mesotrophic fishponds on sandy soils with gravel, or on pure gravel or on deep clay without any or with only a thin layer of sapropel.

The compact and deep network of the *Schoenoplectus* rhizomes permits the co-existence only of those species which are rooted above the compact rhizomes (especially the synusia of the foederationes *Potamogetonion pusilli* and *Litorellion*).

In most fishponds in SB, the long and narrow barriers of *Schoenoplectus lacustris* act as foreposts of the reed-belt, forming dense stands only exceptionally. They damp the waves, being confined to the heavily eroded parts of the littoral.

Distr.: Cz: all regions; Eu: all regions; A: C, N, E.

Typhetum angustifoliae (Allorge 1922) Soó 1927

Indicating species: *Typha angustifolia* (edif) (Fig. 5)

Habitats: Littorals of meso- and eutrophic ponds, in clay, loam and sand with gravel. A thin or medium-thick layer of sapropel is present.

The moderately deep root system and diffuse network of rhizomes permit the presence of more species in the stands. This community is relatively stable, like the previous two, in deeper water (0.8–1 m) in fishpond littorals, where it is not damaged by summer or winter drainage and by great fluctuations of the water level.

Distr.: Cz: all regions; Eu: all regions; A: C, N, E.

Glycerietum maximae Hueck 1931

Indicating species: *Glyceria maxima* (edif) (Fig. 1 in Sect. 1)

Habitats: Littorals of eutrophic to hypertrophic ponds, in sandy loam to loose clay with a thick layer of dark sapropel. The shallow root system and dense network of rhizomes allow the existence of a limited number of other species rooted in this community, but the pleustonic synusia formed by the *Lemnion* and *Utricularion* communities are frequently richly developed.

This community indicates highly fertilized ponds, and may be expansive in ponds receiving organic material in excess (dung, brewery sludge, etc.). The expansion is also sudden after summer drainage or after a fall of water level and after scraping off the fishpond reed-belt. A rapid decomposition of the *Glyceria* leaves favors the development of *Chironomid* populations; from this point of view, the community is useful, especially for young carp populations. *Glycerietum maximae* is sensitive to waves and thus avoids exposed shores. It is also sensitive to overfilling of a pond, and breaks away from the bottom in water deeper than 0.7 m.

Distr.: Cz: all regions; Eu: all regions; A: C, N, E.

Typhetum latifoliae Soó 1927

Indicating group of species: *Typha latifolia* (edif)

Habitats: Littorals of eutrophic fishponds, in clay and loam with a medium-thick to thick layer of sapropel.

The moderately deep root system and the rather diffuse network of rhizomes permit the existence of a number of other species in the community. *Typha latifolia* retreats after a long period of high water level, after winter drainage and long drying-out of the bottom. The community thus retreats after too short periods of shallow water along the shores: hence, it is less stable in fishpond littorals than the previous associations of the foed. *Phragmition*.

Distr.: Cz: all regions; Eu: all regions.

Typhetum laxmanni (Ubrizsy 1961) Nedelcu 1968
occurs only in standing waters in the lowlands of SS and ES, and presently is spreading into pools on bare ground in the lowlands of SM.

Sparganietum erecti Roll 1938

Indicating group of species: *Sparganium erectum* (edif)

Habitats: Littorals of eutrophic ponds, in loose and deep clayey soils with a medium-thick layer of sapropel.

The deep root system and diffuse network of rhizomes of *Sparganium* would permit, theoretically, the co-existence of a number of other species in the community, but the loose and frequently deep soils largely eliminate their occurrence.

This community may persist for a long time in littorals with a permanent accumulation of loose mineral silt; in this case, it is a very good indicator of a rapid accumulation process, especially in small fishponds.

Distr.: Cz: all regions; Eu: W, N, C, E; A: C.

Acoretum calami Eggler 1933

Indicating group of species: *Acorus calamus* (edif)

Habitats: Littorals of eutrophic to hypertrophic ponds, but another ecotype of *Acorus calamus* forms stands in sand in dystrophic ponds. Prevailing occurrence in coarse sand to heavy clay, usually covered with a thick layer of sapropel. The shallow but compact network of rhizomes prevents the existence of most other species, except for the pleustonic synusia of the foed. *Lemnion*.

An alien species to Europe, *A. calamus* propagates only by rhizome offshoots. As a result, its regeneration capacity is small and becomes negligible after scraping off a shallow fishpond littoral. The community is highly sensitive to wave erosion and retreats rapidly after a long drying-out of the bottom. Recommendable as ameliorating type of community in ponds with liquid or solid organic pollution.

Distr.: Cz: all regions; Eu: W, N, C, E; A: C, E.

Equisetetum fluviatilis (Nowiński 1928) Steffen 1931

Indicating group of species: *Equisetum fluviatile* (edif), *Eleocharis palustris* (ch), *Ranunculus lingua* (ch)

Habitats: Littorals of dystrophic to mesotrophic ponds, in clay to fine sand with a deep layer of dark förna. A very deep root system with a dense network of fine roots and a compact network of rhizomes permit the co-existence only of shallow-rooted species. Synusia of the foed. *Lemnion*, *Hydrocharition* and *Cicution virosae* are frequently present.

The community retreats after more frequent drying of the bottom either in winter or summer as well as after intense liming and manuring. It can regenerate, though slowly, after scraping off the shallow fishpond littorals. It mainly indicates ponds with deeper deposits of peat, and does not occur in ponds in SM.

Distr.: Cz: all regions; Eu: all regions (?).

In earlier papers (e.g. Hejný, 1973a, b), the reedswamp communities belonging to the foed. *Phragmition communis* were classified in a "classical" way into floristically easily definable and broadly interpreted associations. But it was pointed out long ago (Hejný, 1960) that a detailed analysis of the ecophysiology of individual ecological dominants of reedswamps, and of their stands, would help in specifying the ranks and roles of their respective vegetation units. Yet recent phytocenological classifications of reedswamp communities have frequently been based on narrowly conceived vegetation units without specifying the reasons for introducing them.

In this chapter, the vegetational types (units) have been defined relatively narrowly for the following reasons: (a) the attenuation of photosynthetically active radiation differs both in closed and loose stands of the individual ecological dominants. Differences also exist in (b) stand density; (c) average shoot production; and (d) distribution of assimilate between the aboveground and underground stand parts as correlated with their synmorphological structure. This proves the relative independence of the associations as defined in the above text. (See Sect. 3.1 for a more detailed presentation of the differences.)

— Foederatio *Bolboschoenion maritimi* Dahl et Hadač 1941

Syn.: *Scirpion maritimi* Dahl et Hadač 1941

Communities of alkaline to brackish waters, in clay loam to clay with a higher content of soluble salts. Confined to the sublittorals to eulittorals of standing and, frequently, of periodic waters. The habitats are characterized by a regular fall of water level in the summer when they may become rather dry. The relatively short-term littoral ecophase and long-term limosal to terrestrial ecophase are reflected in the structure of the communities with a small participation of reedswamp species, and in an increased role of halophytic species populations. The following association of this foederatio occurs in fishponds:

Bolboschoenetum maritimi continentale Soó 1928

Syn.: *Astereto pannonici-Bolboschoenetum compacti* (Soó 1927) Hejný et Vicherek MSC

Indicating group of species: *Bolboschoenus maritimus* ssp. *compactus* (ch, dom), *Aster tripolium* ssp. *pannonicus* (ch), *Plantago maritima* ssp. *salsa* (ch), *Scorzonera parviflora* (ch), *Schoenoplectus tabernaemontani* (ch)

Habitats: Sublittorals to epilittorals of fishponds and other, especially periodic waters; in heavy thick clay in districts of saline soils. In a typical form, this community occurs down to the low sublittoral with the indicating group of species given above. In years of very low water level or in the second year following summer drainage, the stands penetrate deep into the pond in broad initial stages occurring in the whole asedimentary zone, being accompanied by communities of the ass. *Potamogetoneto-Zannichellietum pedicellatae*.

In eulittoral conditions the rare subass. *B.m.c.schoenoplectetosum tabernaemontani* also occurs, e.g., in the eulittoral of the Nesyt fishpond.

In contrast to stands in saline soils and alkaline waters, the communities of *Bolboschoenus maritimus* ssp. *maritimus* growing in aciditrophic soils have been classified as *Glycerio fluitantis-Oenanthetum aquaticae* subass. *bolboschoenetosum* (see under foed. *Oenanthion aquaticae*).

Distr.: Only in alkaline and/or saline districts Cz: SM, WS, CS, ES; Eu: C, N, E; A: C.

— Foederatio *Oenanthion aquaticae* Hejný 1948 apud Vicherek 1962

Communities of hydroochthophytes of the lower sublittoral, especially in ponds, periodic waters and young stages of oxbows. In fishponds occurring mostly only periodically after summer drainage but forming typical stands on large areas, distributed all over a fishpond, or in its central part.

These communities represent the basic type of unstable vegetation among fishpond communities of macrophytes. They are variable both in space and time. They come into being for only two to three years with a long (often 10 to 15 years) anabiosis of either vegetative or generative organs. Only some species live temporarily in loose drifts in the reed-belt. The following are the associations of this foederatio:

Glycerio fluitantis-Oenanthetum aquaticae (Eggler 1933) Hejný 1948

Indicating group of species: *Oenanthe aquatica* (ch, dom), *Glyceria fluitans* (ch, dom), *Bolboschoenus maritimus* ssp. *maritimus* (ch, dom)

Habitats: (a) in a typical form-subass. *typicum* Hejný in clay and clay loam of the erosion biotopes in the sublittorals of regularly partially emerging shores of sky-fed shallow ponds. In such habitats, this community lives almost permanently with occasional interruptions lasting only a few years; (b) subass. *agrostietosum stoloniferae* Hejný, where *Glyceria fluitans*, *Agrostis stolonifera* and *Alisma plantago-aquatica*, less so *Oenanthe aquatica*, are the prevailing species. The community characterizes especially ponds situated at the contact of tertiary lake sediments with old crystalline rocks and is therefore widespread from the colline to submontane regions. It frequently forms a link between poor communities of open water and epilittoral communities of the foed. *Agropyro-Rumicion crispi*. Under such conditions, it only exists for a short time but with regular recurrence as a transitional community dominated by *Glyceria fluitans* and *Agrostis stolonifera*, which has also been described as *Glycerio fluitantis-Agrostietum stoloniferae* R. Tüxen 1974. It has an important function, especially in littoral ecophase, because it is a basic source of carp feed occurring as periphyton on the richly developed foliage. The stands should therefore be protected and not cut. The two subass., *typicum* and *agrostietosum* frequently occur in ponds; (c) subass. *oenanthetosum aquaticae* (Eggler 1933) Hejný h.l. During summer drainage, initials of this subass., with dominant *Oenanthe aquatica*, arise in the central parts of fishponds in the sedimentary zone in limosal ecophase amidst prevailing communities of *Ranunculo scelerati-Rumicetum maritimi oenanthetosum aquaticae*.

In the second year after summer drainage, in the littoral ecophase a vigorous invasion stand of *Glycerio fluitantis-Oenanthetum aquaticae* develops in the same ecotope, diffusely over the whole water surface. *Oenanthe aquatica* is luxuriant here with tall individuals (1.5–2 m) and with inflated stems and petioles. Cutting or cleaning of the water surface may be needed in order to destroy the stands before the seeds are ripe.

This subass frequently occurs in SB ponds; (d) subass. *bolboschoenetosum maritimi* Hejný h.l. In large ponds with prevailing sandy soil over a clay subsoil, nearly pure stands of *Bolboschoenus maritimus* ssp. *maritimus* are found in the sublittoral after summer drainage or after a spring period with partially emerged shores. They only persist for two to three years and indicate the change of ecocycle in large fishponds. In a year when large parts of the sublittoral emerge *B. maritimus* flowers only sporadically, but forms explosively numerous offshoots with tubers. In the next year, the plants flower and fruit abundantly and the stand attains its optimum development. In the third year, *Bolboschoenus* starts to lose vigor. In the sublittoral, these communities are frequently in contact with the reed-belt, forming its peripheral sublittoral zone. Their mowing is not necessary as they may be regarded as harmless to fish farming.

This subassociation resembles *Bolboschoenetum maritimi* (Warming 1906) R.Tüxen 1937 em. 1969, both physiognomically and floristically, but it does not comprise the group of indicating species of the foed. *Bolboschoenion maritimi;* these are substituted by species of the foed. *Oenanthion aquaticae*, which prefer acidotrophic soils. The community is poor in vascular plant species and its structure is simple; in this respect it appears to represent a typical basal association.

Distr. of the whole association: Cz: all regions; Eu: all regions (?).

Rorippo amphibiae-Oenanthetum aquaticae (Soó 1928) Lohmeyer 1950 is hemerophobic in fishponds, achieving its full development in alluvial standing waters in lowlands.

Sagittario-Sparganietum emersi R.Tüxen 1953

Indicating group of species: *Sagittaria sagittifolia* (ch), *Sparganium emersum* (ch), *Alisma gramineum* (ch), *A. plantago-aquatica* (ch), *Butomus umbellatus* (loc, ch)

Habitats: Clayey to clay-and-gravel soils in shallow sky-fed ponds or in ponds along running waters. The community occurs in water 0.4–0.6 m deep. Its occurrence is less transient than that of the previous association, as it is more concentrated in fry ponds where it may also be found in their central parts. This community, in a typical "thermophilous" form, is frequently in contact with *Trapetum natantis* and usually also with *Nymphoidetum peltatae*. In a poorer form (without *Alisma gramineum*), in sandy sublittorals of the erosion biotopes, it is regularly in contact with *Potamogetoneto natantis-Nymphaeetum candidae*.

Distr.: Cz: SB, CB, EB, SM, NS, CS, ES; Eu: W, N, C, E; A: C, W.

Hippuridetum vulgaris Passarge 1955

Indicating group of species: *Hippuris vulgaris* (ch, dom), *Polygonum amphibium* (loc), *Rumex paluster* (loc)

Habitats: In shallow sublittorals and eulittorals of ponds and in shallow ditches in deep expanding clay with a moderately thick layer of sapropel, especially in the districts of mineral-rich and alkaline soils (frequent in the Nesyt fishpond). Best developed following water level fluctuations at a depth of 0.1–0.3 m. *Hippuris vulgaris* is highly adaptable to water-level fluctuations; it occurs from deep water along the reed-belt edges (about 1 m) to limosal ecophase in depressions, with *Veronica anagalloides* and *V. catenata*.

Distr.: Cz: EB, SM, WS, CS, ES; Eu: C, E, S.

Scirpetum radicantis Hejný h.l.

Indicating group of species: *Scirpus radicans* (ch, dom), *Eleocharis palustris* (ch), *Alisma plantago-aquatica* (ch)

Habitats: In clayey and sandy soils in mesotrophic fishponds, usually rich in peat sediments in the littoral. Occurs suddenly after summer drainage, usually in the sublittoral in ecotopes with spring water. During summer drainage, in its first year, the community develops during a prolonged limosal ecophase; during the second year, it passes through a littoral ecophase. Juvenile *Scirpus radicans* does not flower, like *Bolboschoenus maritimus*, and the two species may be mistaken for each other in the 1st year of their life. *Scirpus radicans* flowers in the second year but its populations usually form viviparous rosettes instead of inflorescences. This community is distinguishable physiognomically from others, especially from those with *B. maritimus*, by the curved stems of the dominant species and by its diffuse and conspicuous stands.

The following relevé documents the association.

Date and location: September 5, 1947; Flusárna pond (SB).

Area: 20 m^2. Cover degree: 100%.

Scirpus radicans 5	*Oenanthe aquatica* 2
Alisma plantago-aquatica 2	*Alopecurus aequalis* 1

Distr.: Cz: SB, CB; Eu: C.

Eleocharitetum palustris Ubrizsy 1948

Indicating species: *Eleocharis palustris* (ch, dom)

Habitats: Along shores of standing waters 0.2–0.6 m deep, colonizing dystrophic to eutrophic fishponds in various soils covered with a thin layer of mud.

Occurs in ponds only in the structurally very simple and rudimentary form of small colonies or association fragments. More typical and larger communities are observed in other types of standing waters, e.g., in periodic waters and pools.

Distr.: Cz: all regions (?); Eu: all regions (?); A: C.

Alismato lanceolati-Butometum umbellati (Timár 1957) Hejný 1969; Westhoff et Segal in Westhoff et den Held 1969

Indicating group of species: *Butomus umbellatus* (ch), *Alisma lanceolatum* (ch)

Habitats: Along shallow shores of standing and slowly running waters 0.3–0.4 m deep, in mineral soils rich in nutrients. This association prefers alluvial standing waters, but is also of importance in fry fishponds, especially in SM.

Distr.: Cz: SM, WS, CS, ES; Eu: C, E, S; A: C.

The next foederatio of *Glycerio-Sparganion* comprises communities of the ripal and subripal along running waters, which may merely be in contact with fishponds.

— Foederatio *Cicution virosae* (Hejný 1960) Segal apud Westhoff et den Held 1969

Coenoses of loose soils greatly contributing to sedimentation and land formation. They include an accumulation sere of communities characterized by stenoecious species whose root systems and whole ontogenesis are adapted to a "floating" organogenous substrate with a high gas content, and require littoral and limosal ecoperiods. The communities become fully developed only in conditions characteristic of a long accumulation process.

Under conditions of extensive fishpond management, the accumulation process progresses rapidly and a succession sere of sublittoral to eulittoral communities is formed. The associations of the foed. *Cicution virosae* are fully developed in two types of communities:

(a) Drift type. The drifts arise by accumulation of macrophyte and algal litter containing sclerenchymatous skeleton-like fragments; they form a relatively homogeneous disperse system along leeward shores of water reservoirs, mostly in direct contact with reedswamp communities. This type of *Cicution* communities still occurs commonly in lakes and fishponds. It displays the greatest variation in species composition, especially with respect to the presence of hydatophytes and tenagophytes.

(b) Islet type. Floating islets arise by detachment from the bottom of whole reed-bed areas, but they are mostly only small. These may migrate for a certain time before they eventually settle at leeward shores of a reservoir in dense hydatophyte stands or at reed-bed margins. *Typha latifolia* and *Glyceria maxima* are usually involved as components of these communities in fishponds.

Through the combination of both types, the processes of organic sedimentation and land-formation become quite rapid. But in the past, frequent summer drainage of ponds used to bring about a fast decomposition of the organogenous soils, down to the mineral bottom. When combined with burning, this practice cleared new stretches of open water. Modern intensive fishpond management (scraping of the shallow littorals, etc.) has gradually nearly destroyed the *Cicution* communities, whose relics deserve protection. They are now concentrated in the drift-belts fringing the inner edges of reed-belts in large fishponds.

The foederatio comprises the following associations occurring in fishponds:

Cicuto-Caricetum pseudocyperi Boer et Sissingh 1942

Indicating group of species: *Cicuta virosa* (ch, dom), *Carex pseudocyperus* (ch, dom), *Solanum dulcamara* (ch)

Habitats: Communities of loose soils occurring in drifts fringing the reed-belts in mesotrophic and eutrophic standing waters including fishponds. They also occur around springs in fishpond bottoms, forming here the subass. *solanetosum dulcamarae*.

On islets, colonized by the subass. *typicum*, comprising both dominant species during hydrolittoral ecoperiods, *Carex pseudocyperus* becomes dominant during limoso-terrestrial ecoperiods. This species is fully vigorous here, and produces mature seeds, while *Cicuta virosa* is less vigorous and sterile. The subass. *caricetosum pseudocyperi*, in which *Cicuta virosa* is absent, frequently occurs in eulittorals of small mesotrophic ponds surrounded by *Alnus* and *Salix* spp. groves.

Distr.: In lowlands to colline regions of Cz: SB, WB, CB, NB, NM, WS, ES; Eu: W, N, C.

Callaetum palustris (Van den Berghen 1952) Segal et Westhoff in Westhoff et den Held 1969

Indicating group of species: *Calla palustris* (ch), *Comarum palustre* (ch), *Carex rostrata* (ch), *Menyanthes trifoliata* (loc, ch)

Habitats: Communities of loose soils in dystrophic fishponds and in pools below fishpond dams in colline to submontane regions. Mostly destroyed in ponds during the last decade.

Distr.: Cz: SB, WB, NB, NM, WS; Eu: W, N, C.

The ass. *Carici-Menyanthetum* (Nowiński 1928) Soó (1938) 1955 occurs only in standing waters in lowland flood plains.

— Foederatio *Caricion elatae* Koch 1926

Physiognomically conspicuous communities comprising large complexes of tussocks formed mostly by perennial (0.3–0.4 m high) euochthophytes. They occur from lowlands to submontane regions, in clay and gravelly clayey soils with a marked gley horizon, in sublittorals of fishponds, especially on shores of large ponds. Their prolonged existence in the sublittoral is probably conditioned by a short-term hydrophase in spring followed by a long littoral ecophase for most of the growing season. The following associations are listed here:

Caricetum elatae (Kerner 1858) Koch 1926

Indicating group of species: *Carex elata* (edif), *Lysimachia vulgaris* (ch), *Scutellaria galericulata* (loc, ch)

Habitats: In mineral or organogenous soils in sublittorals (0.2–0.4 m deep) of oligotrophic to mesotrophic ponds. The community consists of three to four synusia. *Carex elata* alone, as edifying species, builds the mean bodies of tussocks 0.1–2.5 m in diameter and 0.5–2.4 m high (see Nekvasilová, 1973) (Fig. 6). Uliginosophytes live on the surface of the tussocks at their periphery as well as in their inner parts. Pelochthophytes (e.g., *Rumex maritimus*, *Chenopodium rubrum*, *Carex bohemica*) occur rather on destroyed flat tussocks. The open water between the tussocks is rich in synusia of pleustophytes (foed. *Lemnion* or *Sphagno-Utricularion*). The communities are usually colonized by nesting birds (*Larus ridibundus*, *Anas platyrhynchos*), which can destroy even high tussocks to small, flat and degenerated forms within a short time. *Caricetum elatae* represents a highly organized biocenological unit with a relatively high concentration both of plant and animal species. In this respect, this community resembles coastal salt-marsh communities. The amelioration by scraping off the shallow littorals of fishponds, undertaken in the last decade, has destroyed most of these communities. Their regeneration is slow, taking some 20–25 years.

The communities of the ass. *Caricetum elatae* are highly dangerous to small ponds, changing them into inaccessible swamps, very often with loose hydrosoils.

Distr.: From lowlands to colline regions. Cz: SB, WB, NB; Eu: W, N, C.

Cladietum marisci (Allorge 1922) Zobrist 1935
is one of the rarest communities in littorals of standing waters; it is fully protected and has not been observed in fishponds.

— Foederatio *Caricion rostratae* Bal.-Tul. 1963

Types of communities physiognomically similar to those of the previous foederatio, but their position in littoral ecotopes frequently depends on the presence of bottom springs, or on a long littoral ecophase. They usually prefer a thick humolithic (peat) layer. The communities occur from lowlands to highlands and belong to the following associations:

Caricetum rostratae Rübel 1912

Indicating group of species: *Carex rostrata* (edif), *Naumburgia thyrsiflora* (ch), *Equisetum fluviatile* (ch)

Habitats: In sandy and, especially, peaty-and-sandy soils, frequently with a thick layer of humolithic dy. In sublittorals and eulittorals of dystrophic and

Fig. 6. Tussocks of *Carex elata* occupying organogenous bottom substrate in shallow fishpond sublittoral. (Photo: K. Gregor)

mesotrophic standing waters. In fishponds, common in peaty regions, forming part of the following zonation sere (starting from open water): *Potamogetoneto natantis-Nymphaeetum candidae* → *Equisetetum fluviatilis* → *Caricetum rostratae*. The last two associations are rich in dominant *Equisetum fluviatile* and *Carex rostrata*. The two communities are physiognomically alike, being structurally more open than other communities of fishpond littorals. *Caricetum rostratae* communities also occur in fispond bays with peat sediments; they also frequently colonize wet sites below fishpond dams, from colline regions to highlands; usually a dominant littoral community in our mountain lakes.

Drainage in summer and winter, liming and amelioration have caused a rapid retreat of these communities in fisponds during the last decade.

Distr.: Cz: SB, WB, NB, CB, NM, WS; Eu: W, C, N.

Caricetum paniculatae Wangerin 1916 and *Caricetum appropinquatae* R. Tüxen 1947 do not occur in the SB and SM fishpond regions, being more characteristic of other types of standing waters.

— Foederatio *Caricion gracilis* Neuhäusl 1959

In comparison with the previous two foederationes, these communities mostly occur in the eulittoral with a very short hydrophase and littoral ecophase but with long limosal to terrestrial ecophases. They are found from lowlands to submontane regions, especially in soils on mineral sediments. They are structurally relatively homogeneous, being formed of a "carpet" of shoots, only rarely grouped into little tussocks. The associations of this foederatio are as follows:

Caricetum gracilis Almquist 1929

Indicating group of species: *Carex gracilis* (edif), *C. vesicaria* (ch), *Naumburgia thyrsiflora* (ch), *Stellaria glauca* (ch)

Habitats: In clayey and loamy soils with a thin layer of förna. In eulittorals of originally mesotrophic but usually eutrophicated ponds. Large communities occur here, consisting of dense populations of *Carex gracilis*. Communities consisting of more diffuse tussocks of *C. gracilis* develop only under extreme conditions in the sublittoral or in biotopes affected by wave action.

The development of the subass. *typicum* is conditioned by long limosal and terrestrial ecophases in erosion biotopes; the subass. *naumburgietosum thyrsiflorae* Blažková 1973 is conditioned by long littoral and limosal ecophases in accumulation biotopes.

The communities of *Caricetum elatae* and *C. rostratae* retreat in lavishly fertilized ponds, being replaced by the *Caricetum gracilis*, better utilizing the rich mineral nutrient supply. Good regeneration of *Caricetum gracilis* has also been observed in the eulittoral after scraping off its shallow parts.

Distr.: Cz: all regions (in SM more frequent in flood-plain standing waters than in fishponds); Eu: all regions (?).

Caricetum vesicariae Br.-Bl. et Denis 1926

Indicating group of species: *Carex vesicaria* (edif), *C. gracilis* (ch), *Eleocharis palustris* (ch), *Sparganium erectum* (ch)

Habitats: In deep clayey and loamy soils with a thin layer of mud, in sublittorals and eulittorals of mesotrophic and eutrophic ponds. This community is usually more open than *Caricetum gracilis*. *Carex vesicaria* is better adapted to a fluctuating water level. After emergence of subhydric soils, its juvenile populations regularly colonize their surface. Hence this community is most frequent in fry ponds. The *Caricetum vesicariae* is even more plastic than *Caricetum gracilis*, regenerating well after amelioration by scraping, not only in the eulittoral but also in the sublittoral after limoso-terrestrial ecoperiods.

Distr.: Cz: SB, CB, EB, SM, WS, CS, ES (in SM relatively rare in fishpond littorals, more frequent in flood-plain standing waters); Eu: W, N, C, E; A: C.

Caricetum ripariae Soó 1928

Indicating group of species: *Carex riparia* (edif), *Iris pseudacorus* (ch), *Teucrium scordium* (ch)

Habitats: In clayey soils and, especially, in fens rich in soluble salts. Large complexes occur in the eulittorals of alkaline standing waters, especially in the outer littorals of fishponds. The stands are adapted to flooding for a longer time in early spring as well as during the growing season. Adaptation to long-lasting hydro- and littoral ecophases as well as to a long terrestrial ecophase represent a specific feature of the *Caricetum ripariae* communities, which are characterized by a wide distribution in the Pannonian lowlands. They occur most frequently in littorals of fishponds with wide and regular fluctuations of water level.

In humid and cooler districts, e.g., in SB, the populations of *C. riparia*, forming only small colonies or community fragments, are adapted to sunny sites at the shoreline, in small and calm bays open to the south. They are frequently in contact with the *Glycerietum maximae* communities. In littoral ecotopes of fishponds in districts of alkaline soils, *Caricetum ripariae* is frequently in contact with the ass. *Meliloto-Caricetum otrubae.*

Distr.: Cz: CB, EB, SM, WS, ES; Eu: C, E; A: C.

Caricetum acutiformis (Soó 1928) Sauer 1937

Indicating group of species: *Carex acutiformis* (edif), *Lysimachia vulgaris* (ch)

Habitats: In loamy and clayey soils covered with a thick layer of mud. In eulittoral of mesotrophic standing waters, along the margins of fishponds, especially in bays bordered by the *Alnetum glutinosae.* Frequently occurs in narrow valleys harboring a number of small fishponds. Larger stands are rare in the eulittoral zone.

Distr.: Cz: SB, WB, CB, EB, SM, WS; Eu: W, C, E.

Caricetum melanostachyae Balázs 1943 was observed only in the alluvial lowlands of ES.

Resembling the occurrence of *Carex acutiformis* and *C. riparia* in mixed communities in the Pannonian lowlands (SM, WS, ES), *C. gracilis* and *C. riparia* occur in combined communities in the fishpond basins of SB.

The following general rule may be accepted in the fishpond littorals for these twin combinations of communities (= *Caricetum gracilis* + *Caricetum vesicariae; Caricetum ripariae* + *Caricetum acutiformis*). In South Bohemia *Carex vesicaria* is better adapted to a fluctuating water level than *C. gracilis.* In South Moravia, *Carex riparia* is better adapted than *C. acutiformis.*

The large formation of sublittoral and eulittoral communities of tall perennial macrophytes (foederationes *Phragmition communis, Caricion elatae, C. rostratae* and *C. gracilis*) occurring on the shores and in marginal parts of fishponds, represent relatively highly stabilized systems and, also, decisive stages of a succession which can alter a shallow pond into a complex of swamp communities during half a century. The succession would lead ultimately, in Central European conditions, to a flood-plain forest *(Alnetum glutinosae* or *Salici-Populetum).*

— Foederatio *Sphagno-Utricularion* Th. Müller et Görs 1960

Species-poor communities occurring in shallow and dystrophic standing waters. In fishponds, they are found only in dystrophic water, especially in peaty bays and along the borders of the *Caricetum elatae* or (*Sphagno-Caricetum filiformis.*) Small areas of open water with peaty detritus are the main ecotope in ponds of peaty districts. In the last decade, these communities have been reduced to small relics and urgently need protection.) Their regeneration after scraping off the peaty fishpond margins has not been observed. (Associations: *Sphagno-Utricularietum intermediae* Fijalkowski 1960 and *Sparganietum minimi* Schaaf 1925).

Distr.: Cz: SB, EB, NB; Eu: W, N, C.

— Foederatio *Litorellion uniflorae* Koch 1926

Species-poor communities of tenagophytes in the lower sublittoral of fishponds and pools. They occur in sandy soils and peaty sands and, secondarily, in soils silted with mud.

They require a long littoral ecophase during littoral-terrestrial ecoperiods. Primarily, they can be regarded as communities of oligotrophic standing waters. Stands found in fishponds are very poor in species, as their stenoecious species of indicative value gradually disappear. The following associations belong to this foederatio:

Eleocharitetum acicularis Koch 1926

Indicating group of species: *Eleocharis acicularis* (ch), *Litorella uniflora* (ch), *Limosella aquatica* (dif)

Habitats: Sandy and clayey soils with a very thin film of mud in the lower part of sublittoral; in fishponds usually in shallow water about 0.2–0.4 m deep. The following subassociations have developed in ponds:

subass. *typicum*, with equal dominance of *E. acicularis* and *L. uniflora*, occurring in ponds originating from shallow lakes, i.e., in large ponds in the central and southeastern parts of the Třeboň basin. After intense liming, manuring and winter drainage in the last decade, *L. uniflora* has suddenly retreated and the community changes into the subass. *eleocharitetosum acicularis*, representing a typical fishpond-culture derivative of the association. Pure stands of *E. acicularis* belong to its first regeneration stage after amelioration by scraping off poor sandy or clayey bottoms, exposing the bare subsoil. In fry ponds, this community with *E. acicularis* is highly important for the breeding fishes as a very fine and useful bed. In the lower sublittoral of large ponds the *Eleocharitetum acicularis* provides very good substrate for the development of rich periphyton.

Distr.: Cz: all regions, mostly as the subass. *eleocharitetosum acicularis;* Eu: all regions; A: C.

Pilularietum globuliferae (Ambrož 1939) R. Tüxen 1955 n.n. ex Th. Müller et Görs 1960

Indicating species: *Pilularia globulifera* (ch), *Juncus bulbosus* (ch)

Habitats: In very fine peaty sands in the low sublittoral of fishponds. This association was observed by Ambrož (1935, 1939) in fishponds near Třeboň during terrestrial ecoperiods in the thirties of this century. Since 1945, *Pilularia globulifera* has not been found here any more.

P. globulifera belongs to a set of highly stenoecious species, having retreated from Central Europe in the first half of this century, i.e., long before the start of the present intense management of fishponds.

Distr.: Cz: SB (probably extinct); Eu: W, N, C.

— Foederatio *Veronico-Juncion bulbosi* Segal MSC

Species-poor communities colonizing the depressions in eulittorals of standing waters, including fishponds. These depressions arise through mechanical or biotic damage to the humus-containing surface soil layers. The plants are thus living in the coarse sandy subsoil. They require a long littoral and a prolonged limosal ecophase. In contrast to the communities of the foed. *Litorellion*, the period of flowering lasts longer in the *Veronico-Juncion* communities. The pertinent association is:

Ranunculo-Juncetum bulbosi (Nordhagen 1921) Oberdorfer 1957

Indicating groups of species: *Ranunculus flammula* (ch), *Juncus bulbosus* (ch), *Veronica scutellata* (ch)

Habitats: Coarse sands on clayey subsoil covered with an evenly spread film of peaty mud, in the eulittorals of dystrophic standing waters and in fishponds in peaty districts. Occurs regularly in bare subsoil, exposed after scraping off the shallow fishponds littorals. This community is thus hemerophilous from its onset. It can also exist in very shallow depressions and on a damaged surface of the bottom soil.

Ranunculo-Juncetum bulbosi represents a pioneer association in peaty sands and in any other disturbed ecotopes in peaty districts, which are suitable for rapid colonization. The community occurs in very shallow water (from 0.1 to 0.25 m) and in limosal ecophase. Here, it forms stands only 0.1–0.15 m high. The optimum development of the community is confined to a littoral-limosal ecoperiod. But the dominant *Ranunculus flammula* can survive in deeper water (to 0.4 m) for a longer time, where it forms floating leaves. *Juncus bulbosus* can live in another growth form *(f. confervoides)* in highly acid water 0.5–1.5 m deep. This illustrates a situation whereby dominant species relatively very plastic in a whole scale of niches, build up only a very stenoecious community with a restricted set of niches.

Distr.: Cz: SB, NB; Eu: C, W, N.

— Foederatio *Bidention tripartiti* Nordhagen 1940

Communities relatively poor in species, with the ecological group of Pelochthotherophyta prevailing. Occurrence in emerged soils of fishpond bottoms, in village ditches, along margins of wetlands in woodlands and along banks of running waters. The communities colonize the sedimentary zones of pond bottoms with deeper organogenous soils rich in soluble salts and nitrogen compounds. In contrast to other types of standing waters in summer-dry fishponds, the communities occur in large and greatly variable stands. A limosal ecophase of medium duration and a long terrestrial ecophase from mid- to late summer are important for their development. These communities occur from the lowlands to submontane regions and may be classified into the following associations:

Ranunculo scelerati-Rumicetum maritimi Sissingh (1946) 1966

Indicating group of species: *Ranunculus sceleratus* (ch, dom), *Rumex maritimus* (ch), *R. paluster* (loc), *Epilobium hirsutum* (ch), *Chenopodium rubrum* (ch)

Habitats: Deep sapropel soils in the sedimentary zones of summer-drained fishponds; emerged fishpond shores.

In fishponds, where this association has the center of its development in this country, it shows great variability and occurs in several forms adapted not only to a certain sequence of ecophases but also to their topic position combined with the character of the soil: (a) subass. *oenanthetosum aquaticae:* with increased participation of *Oenanthe aquatica* (dif), which is frequently dominant; in the deepest parts of ponds and in centers of summer-drained fishponds with a long limosal ecophase, in deep organic soils (A-horizon at 0.5–0.7 m). The soil splits into large polygons only after prolonged drying, but the cracks cut across the whole A-horizon. (b) subass. *typicum:* with full participation of the dominant species, but with a certain sequence of phenophases, e.g., with a spring aspect of *Ranunculus sceleratus* and a summer aspect of *Rumex maritimus*. Mainly in central parts of fishponds in the sedimentary zone with a moderately deep A-horizon (0.4–0.5 m) and with an even duration of the limosal and terrestrial ecophases: *Ranunculus sceleratus* flowers during the limosal and *Rumex maritimus* during the terrestrial ecophase. (c) subass. *chenopodietosum rubri:* with abundant to dominant *Chenopodium rubrum* (dif), in the sedimentary zone between the central part of a fishpond and its inner littoral, with a relatively short limosal ecophase in rapidly drying-out soil. This community may also penetrate into the central part of the fishpond in very dry years. (d) subass. *atriplicetosum patulae:* with a high participation not only of *Atriplex patula* (dif) but also of the species of the orders *Polygono-Chenopodietalia* and *Agrostietalia stoloniferae*, occurring in the asedimentary zone in rapidly decomposing litter of reedswamp dominants (förna).

Distr.: Cz: all regions; Eu: all regions.

Polygono-Bidentetum tripartiti (Koch 1926) Sissingh 1946

Indicating group of species: *Polygonum hydropiper* (ch, dom), *P. lapathifolium* (ch, dom), *Bidens tripartitus* (ch), *B. radiatus* (loc), *Polygonum mite* (loc)

Habitats: Usually mineral soils which are waterlogged for a long or moderately long period in early summer and with a moderately deep organic layer. The very poor stands occur not only in small village ponds and along the edges of village ditches without drainage but also in dung waters, here in contact with the *Chenopodietum glauco-rubri*.

Distr.: Cz: all regions; Eu: all regions (?).

Bidentetum cernui Slavnić 1951

Indicating group of species: *Polygonum hydropiper* (ch, dom), *Bidens cernuus* (ch, dom), *Leersia oryzoides* (ch), *Epilobium palustre* (ch)

Habitats: In the sublittorals and eulittorals of summer-drained ponds, usually in ecotopes supplied with spring water, or in bays, in soils waterlogged with running water. A long limosal ecophase or a soil otherwise waterlogged for a very long time represent a fundamental condition for the development of this community type, which also occurs (but not regularly) in bare soils in islets of drifted materials. In more eutrophicated waters, *Epilobium hirsutum* prevails as the ecological dominant.

Distr.: Cz: SB; Eu: W, C, N.

Pulicario vulgaris-Bidentetum (Ambrož 1939) Hejný h.l.

Indicating group of species: *Pulicaria vulgaris* (ch, dom), *Bidens tripartitus* (ch), *B. cernuus* (ch), *Agrostis stolonifera* (loc), *Anthemis cotula* (dif)

Habitats: In littorals and eulittorals on shores of small village ponds or in larger ponds in contact with villages. Very probably an anthropogenous derivative of the previous association, but usually concentrated in waterlogged soils in early summer. Never occurs at the edges of springs in ponds remote from villages or in exposed belts of drifted materials.

The following relevé documents the association:

Date and location: August 10, 1964; Horní Malovický fishpond (SB), shore below a village. Area: 10 m^2. Cover degree: 90%.

Pulicaria vulgaris 4
Chenopodium glaucum 4
Polygonum lapathifolium 3
Rumex maritimus 3
Potentilla supina 2
Epilobium hirsutum 1
Bidens tripartitus 1

Distr.: Cz: SB, SM, WS; Eu: C.

Polygono brittingeri-Chenopodietum rubri Lohmeyer 1950 occurs only in gravel soils in the subripal of running waters in summer, and quite exceptionally in fishponds.

Of the other types of vegetation of bare emerged bottoms, only the most important and most widely distributed associations are introduced here. A complete list of all the vegetation units is being worked out; the classification of these short-lived communities will be confined to narrowly defined units which correspond better to the basic ecological niches of the constituent species. Broadly collected material is required for the elaboration of a correct concept for the classification of these vegetation units.

— Foederatio *Cyperion flavescentis* Koch 1926
(including *Elatini-Eleocharition ovatae* and *Radiolion linoidis*)

Relatively species-poor communities of the plants belonging to the ecological group of pelochthophyta. They occur both in mineral and organogenous soils emerging periodically after previous flooding. The constituent plants have a very short life cycle lasting three to four months. The soils are fully water-saturated at the beginning of the development of juvenile populations of the species and are quite dry at the time of their senescence. They require, therefore, a long limosal to terrestrial ecophase during their growing season. The communities occur (a) in organo-mineral soils in sedimentary zones as synusia of the associations of *Bidention tripartiti*; (b) as independent units in transitional biotopes, and (c) independently in erosion biotopes in mineral soils such as sands, gravel or peaty sands. The following associations are listed here:

Eleocharito ovatae-Caricetum bohemicae (Klika 1935) Pietsch in Pietsch et Müller-Stoll 1965

Indicating group of species: *Eleocharis ovata* (ch, dom), *Carex bohemica* (ch), *Plantago maior* ssp. *pleiosperma* (ch)

Habitats: On emerged bottoms, typical of the transitional biotopes of fishponds, i.e., on muddy soils with a humus horizon to a depth of 0.1–0.2 m and with sand or expanding clay in the subsoil. The following forms may be observed: (a) subass. *typicum*, with *Carex bohemica* prevailing over the short *Eleocharis ovata*, and with *Gnaphalium uliginosum*. It occurs in ecotopes with a rapidly sinking ground water level, i.e., with a short limosal ecophase. (b) subass. *eleocharietosum ovatae*, with tall and vigorous forms of *Eleocharis ovata* prevailing, with *Alisma plantago-aquatica* and intensely regenerating juvenile plants of *Typha latifolia*. The subassociation becomes established during a long limosal ecophase, at spring sites and in bays of fishponds. The humus horizon is usually up to 0.4 m deep.

In both forms, this is the most widely distributed association of emerged bottoms in SB and in the Bohemian-Moravian upland; it does not occur in SM.

Distr.: Cz: SB; Eu: C, W, N.

Coleantho-Spergularietum echinospermae (Vicherek 1972) Hejný h.l.

Indicating group of species: *Coleanthus subtilis* (ch, dom), *Spergularia echinosperma* (ch), *Limosella aquatica* (ch)

Habitats: In the mostly eroded or rarely transitional areas of emerged shores of fishponds, in soils usually possessing a highly specific structure: a pellicle of mud (2–3 cm thick) coarse-sand to 10 cm and clay subsoil. Optimum development in spring, from May to June in colline regions, to July in submontane regions. The communities of this association develop not only in drained ponds but also on emerged shores.

Distr.: Cz: SB (diffuse); Eu: C, W, N.

Dichostyli-Gnaphalietum uliginosi Horvatić 1931

Indicating group of species: *Dichostylis micheliana* (ch, dom), *Gnaphalium uliginosum* (ch), *Cyperus fuscus* (ch)

Habitats: Very similar to those of the ass. *Eleocharito ovatae-Caricetum bohemicae* developing in the transition and sedimentary zones of emerged bottoms in mid-summer, but differing in its affinity either to calcareous or alkaline soils.

Distr.: Cz: SB (rare), SM; Eu: C, S.

Gypsophilo muralis-Potentilletum norvegicae (Ambrož 1939) Hejný h.l.

Indicating group of species: *Gypsophila muralis* (ch), *Potentilla norvegica* (ch), *Juncus bufonius* (ch), *Gnaphalium luteo-album* (ch)

Habitats: Emerged fishpond shores, typical development in the asedimentary areas on sandy to gravelly-sandy soils with the muddy pellicle very thin or absent. This pellicle is rapidly mineralized during summer drainage. The community is frequently in contact with the *Eleocharieto ovatae-Caricetum bohemicae.* The usually short limosal and long terrestrial ecophases determine the development of the communities.

Distr.: Cz: SB; Eu: W, C.

Stellario uliginosae-Isolepetum setaceae (Koch 1926) Moor 1936

Indicating group of species: *Stellaria uliginosa* (ch), *Isolepis setacea* (ch)

Habitats: Wet road edges, spring sites and springs in the asedimentary areas of fishponds, emerged fishpond bottoms or shores with a long limosal ecophase, in gravelly-sandy soils with a marked humus horizon (1–6 cm deep). In poorer form these communities are formed after scraping off the shallow fishponds littorals. On the shores, they are often in contact with the ass. *Ranunculo-Juncetum bulbosi.*

Distr.: Cz: SB, WB; Eu: C, N, W.

Junco tenageiae-Radioletum linoidis Pietsch 1961

Indicating group of species: *Juncus tenageia* (ch), *Radiola linoides* (ch), *Lycopodiella inundata* (ch)

Habitats: Emerged bottoms and shores; development typical in asedimentary areas of fishponds in peaty sands, occurring mainly in ancient lake basins.

Distr.: Cz: SB (spurious); Eu: C, W, N.

Hyperico humifusi-Spergularietum rubrae Wojczik 1968

Indicating group of species: *Hypericum humifusum* (ch), *Spergularia rubra* (ch), *Centunculus minimus* (ch)

Habitats: Emerged shores in the erosion areas of ponds, frequent occurrence also after scraping off of the littoral belts on a sandy bottom or near the deponia of scraped-off material. The community is typical not only of fishponds, but also of gravel-pits and waterlogged fields.

Distr.: Cz: SB; Eu: C, W, N.

Cypero fusci-Samoletum Müller-Stoll et Pietsch 1965

Indicating group of species: *Samolus valerandi* (ch, dom), *Cyperus fuscus* (ch, dom), *Plantago maior* ssp. *winteri* (ch)

Habitats: On emerged shores in calm bays with optimum development along the shoreline (at −2 to +5 cm). Communities typical of sites with an increased content of sulfate and chloride both in soil and water.

Distr.: Cz: SM; Eu: C, S, E; A: C.

— Foederatio *Cypero-Spergularion salinae* Slavnić 1948

Species-poor communities of annual halophytes, both obligatory and facultative, belonging to the life form of pelochthophyta, colonizing slightly to strongly saline (alkaline) soils of emerged bottoms and shores of periodic waters and fishponds. The following associations are listed here:

Heleochloetum schoenoidis (Soó 1933) Topa 1939

Indicating group of species: *Heleochloa schoenoides* (ch), *H. alopecuroides* (ch), *Juncus ambiguus* (ch), *Spergularia marginata* (ch), *Plantago major* ssp. *winteri* (ch), *Veronica anagalloides* (ch)

Habitats: Species-poor communities on bottoms of periodic waters or on emerged shores of village fishponds in heavy clay which frequently splits into polygons with conspicuous salt effiorescences on the surface.

Distr.: Cz: SM, WS; Eu: C, E, S.

Crypsidetum aculeatae (Bojko 1932 n.n.) E. Topa 1939

Indicating group of species: *Crypsis aculeata* (ch, dom), *Spergularia salina* (ch), *Cyperus fuscus* (ch), *Puccinellia distans* ssp. *limosa* (ch), *Chenopodium glaucum* (ch)

Habitats: Species-poor communities of emerged bottoms or shores of periodic waters and fishponds, in rapidly drying saline loamy sands forming small polygons but deep cracks. The community requires a moderately long limosal ecophase in early summer followed by a long terrestrial ecophase. Prostrate forms of the species prevail.

Distr.: Cz: SM, WS; Eu: C, E, S.

1.2.3 Floristic and Phytogeographical Characteristics of the Opatovický and Nesyt Fishponds

The plant species variability of the Opatovický and Nesyt fishponds reflects the physicogeographical conditions of the Třeboň basin and Lednice region, as described in Section 1.1. Complete lists of vascular plant species found within the cadastral areas of the two fishponds contain altogether 373 and 579 species which have been ascertained at the Opatovický and Nesyt fishponds respectively. The relatively small number of species ascertained at Opatovický gives only an insufficient picture of the plant species diversity encountered in South Bohemian fishponds: most of them are richer in submerged vascular plants and in those colonizing emerged bottoms. Opatovický, as a highly managed and economically impor-

Table 2. Vicariant vascular plant species at the Opatovický and Nesyt fishponds

Opatovický	Nesyt
Achillea ptarmica	*A. aspleniifolia*
Agropyron repens	*A. r.* var. *halophila*
Alisma plantago-aquatica	*A. lanceolatum*
Bolboschoenus maritimus ssp. *maritimus*	*B. m.* ssp. *compactus*
Carex vulpina	*C. otrubae*
Cirsium palustre	*C. canum*
Eleocharis acicularis, E. ovata	*E. uniglumis*
Festuca ovina	*F. pseudovina*
Juncus bufonius	*J. ambiguus*
Lotus uliginosus	*L. tenuis*
Lycopus europaeus	*L. exaltatus*
Plantago major ssp. *intermedia*	*P. m.* ssp. *winteri*
Schoenoplectus lacustris	*S. tabernaemontani*
Scorzonera humilis	*S. parviflora*
Senecio aquaticus (syn. *barbareifolius*)	*S. erraticus*
Spergularia echinosperma, S. rubra	*S. marina, S. media*
Taraxum officinale	*T. bessarabicum*
Trifolium arvense, T. dubium, T. spadiceum	*T. fragiferum* ssp. *fragiferum*, ssp. *bonannii*
Utricularia neglecta	*U. vulgaris*
Veronica scutellata	*V. anagalloides, V. catenata*
Zannichelia palustris ssp. *palustris*	*Z. p.* ssp. *pedicellata*

tant fishpond, is relatively poor in species belonging to such species-rich genera as *Batrachium*, *Callitriche*, *Carex*, *Elatine*, *Juncus*, and *Potamogeton*, and in nymphaeids. The shores of South Moravian fishponds host numerous terrestrial species, among them many halophytes and anthropophytes. Table 2 lists ecological vicariants within species or genera occurring at the two fishponds; their selection and differentiation represents a result of long-term development of the two respective regional floras. Some of these species occur solely in one of the regions compared, others markedly prefer one of them. In Table 3, the species most characteristic of the two fishponds are grouped according to their properties as ecological indicators. The differences between the vascular plant-species outfits of the two fishponds stand out clearly from this table.

From the phytogeographical point of view, South Bohemia and South Moravia belong to two different phytochorions as defined by Dostál (1960). South Bohemia (including the Opatovický fishpond) belongs to the Hercynicum, the region of Central European forest flora, while South Moravia (including the Nesyt fishpond) belongs to the Pannonicum, which is the region of Central and East European thermophilous flora. A more detailed phytogeographical classification of each territory is evident from the following scheme:

	South Bohemia (Opatovický fishpond)	South Moravia (Nesyt fishpond)
Region	Hercynicum	Pannonicum
Sub-region	Sub-Hercynicum	—
District	Boreo-Hercynicum	Eupannonicum
Territory	South Bohemian fishpond basins	Lower Morava basin

Table 3. Comparison between species sets of ecological indicators at the Opatovický and Nesyt fishponds

Opatovický	Nesyt
1. Acidophilous species	
Avenella flexuosa, Calluna vulgaris, Carex canescens, C. echinata, C. pallescens, C. pilulifera, Cicuta virosa, Coleanthus subtilis, Comarum palustre, Genista germanica, Juncus filiformis, Naumburgia thyrsiflora, Molinia coerulea, Nardus stricta, Pedicularis palustris, P. silvatica, Peucedanum palustre, Polygala vulgaris, Selinum carvifolia, Succisa pratensis, Vaccinium myrtillus, Viola palustris	none
2. Calcicole and thermophilous species	
none	*Artemisia absinthium, Botriochloa ischaemum, Bryonia alba, Conium maculatum, Eryngium campestre, Juncus inflexus, Loranthus europaeus, Marrubium vulgare, Podospermum laciniatum, Reseda luteola*
3. Subhalophytes and halophytes	
none	*Althaea officinalis, Aster tripolium ssp. pannonicus, Atriplex hastata agg., Batrachium baudotii, B. rionii, Bupleurum tenuissimum, Carex distans, C. hordeistichos, C. secalina, Chenopodium botryoides, Crypsis aculeata, Glaux maritima, Heleochloa alopecuroides, H. schoenoides, Hibiscus trionum, Juncus gerardii, Melilotus dentatus, Phragmites communis* var. *salsa, Plantago maritima, Puccinellia limosa, Rumex stenophyllus, Salicornia prostrata, Samolus valerandi, Suaeda pannonica, Triglochin maritimum*
4. Anthropophyte species	
Aesculus hippocastanum, Capsella bursa-pastoris, Chenopodium album, Ch. ficifolium, Pyrethrum parthenium, Quercus palustris, Salix matsudana cv. *Tortuosa, S. sepulcralis*	*Acer negundo, Amaranthus* (several species), *Amorpha fruticosa, Aster novi-belgi, Atriplex* (several species), *Calendula officinalis, Cannabis sativa* ssp. *ruderalis, Chenopodium* (several species), *Celtis occidentalis, Citrullus lanatus, Cnicus benedictus, Corylus colurna, Cosmos bipinnatus, Euphorbia marginata, Juglans nigra, J. regia, Kochia trichophylla, Phaseolus vulgaris, Portulaca oleracea, Robinia pseudoacacia, Solanum lycopersicum, S. tuberosum*

1.2.4 Phytocenological Comparison of the Opatovický and Nesyt Fishponds

The two fishponds in which our IBP studies were made, were chosen as typical representatives of the two phytogeographical and biogeographical regions. The differences and similarities between them thus acquire validity in a broader European or, at least, Central European context. Since the plant communities are the structurally best evaluated functional parts of the ecosystems studied, their comparison can give an overall picture of the differences and similarities between our two study areas. The phytocenological situation and comparison between the two fishponds are analyzed by Hejný (1973a). It is therefore not necessary to repeat details; only the presence or absence of individual associations can be compared from a more general viewpoint here.

1. Aquatic plant communities. Opatovický fishpond is characterized by an absence of macrophyte communities over most of its open-water area. The communities of the foed. *Lemnion*, *Utricularion*, *Potamogetonion pusilli* and *Oenanthion aquaticae* are developed only in bays or in contact with the reed belt. But even the bays are only partly colonized by these communities.

In Nesyt, on the other hand, most of the open water is colonized by aquatic plant communities. The ass. *Potamogetoneto-Zannichellietum pedicellatae* and *Bolboschoenetum maritimi continentale* occur in ecoperiods with changing water level, which take place regularly in Nesyt, but only exceptionally in Opatovický.

It is clear from this basic ecological background how difficult the comparison is. The greater frequency of littoral ecoperiods in Nesyt also brings about a full development of varied plant communities in its bays (ass. *Batrachietum rionii*, *Bolboschoenetum maritimi continentale* and *Hippuridetum vulgaris*). Some particular comparisons will now follow.

(a) The communities of the foed. *Lemnion* are concentrated in bays in both ponds as independent units or as a mosaic in open spaces, "lagoons", within the reed belts or as synusia in the littoral communities. In all instances, the communities belong to the ass. *Lemnetum gibbae*.

(b) The communities of the foed. *Utricularion vulgaris* show much the same features in both ponds; they differ only at the subassociation level. *Lemno-Utricularietum utricularietosum neglectae* has a more stabilized (but still periodical) position in Opatovický, appearing in years following an unstable water level.

(c) The communities of the foed. *Potamogetonion pusilli* exhibit a distinct diversity in distribution and frequency in both fishponds. Only some diversity exists in their species composition. Both associations, one in each pond, are very poor in species. In Opatovický, it is because of their fragmentary development, in Nesyt, the alkaline water precludes greater variety in water-plant species.

(d) The communities of the foed. *Batrachion aquatilis* are represented only by the ass. *Batrachietum rionii* in Nesyt, occurring relatively frequently in its bays in years following limoso-terrestrial ecoperiods. In Opatovický, no associations of this foederatio have ever been recorded.

(e) The communities of the foed. *Bolboschoenion maritimi* have a similar distribution in Nesyt as other water-plant communities favored by a changing water level. The communities of the foed. *Oenanthion aquaticae* are represented only in bays of the two ponds, the ass. *Hippuridetum vulgaris* is known only from Nesyt.

2. Reedswamp communities. In contrast to aquatic plant communities, the reedswamp communities are relatively copious in both ponds. They acquire diverse forms and vary both in space and time. Communities maintaining the same structure and distribution for a long time are found only rarely in either pond. On extreme sites, the reedswamp stands are also rare, even as initial stages. It is very interesting and surprising that these enduring perennial communities are only stable during long littoral ecoperiods. During or after ecoperiods with a changing position of the water level, their structure is altered and the communities may even be damaged or destroyed. The only really stable macrophyte plant communities are those of the littoral tall sedges.—Some remarks on particular differences:

(a) The ass. *Phragmitetum communis*: great variability and regeneration capability in Nesyt.

(b) The same association occurs in two subassociations in Opatovický but communities of the ass. *Typhetum angustifoliae* possess a greater regeneration capability here.

(c) *Glycerietum maximae* covers large areas in Opatovický but in Nesyt it occurs only in its western and southeastern bays in hypertrophic conditions, near inflows of waste water. In Opatovický, these communities are also developed best in hypertrophic habitats affected by waste water. Two differing forms of the *Glycerietum maximae* communities may be distinguished in general in the littoral of Opatovický:

subass. *typicum*: this form consists of tall and usually fertile shoots in dense and productive stands, largely occurring in calm bays;

subass. *utricularietosum:* this form consists of short and usually sterile shoots with thin culms. Stands of this form very often combine with synusia of *Utricularia* and *Lemna.*

(d) *Typhetum latifoliae* builds patches or strips of short-term duration in both ponds.

(e) The ass. *Phalaridetum arundinaceae* is very rare in Nesyt and *Sparganietum erecti* is rare in Opatovický. *Acoretum calami* has only been observed in Opatovický in fragments.

3. Communities of tall sedges. A certain difference between the two ponds has been observed in the occurrence of the communities of tall sedges. They are physiognomically conspicuous in the outer littoral of both fishponds. The main differences may be summarized as follows:

(a) In Opatovický, the zonation of perennial plant communities *(Pragmition-Caricion rostratae-Caricion gracilis)* is continuous without any gaps. *Caricetum gracilis* therefore covers large areas in the littoral; its communities are rich developed and belong to two subassociations.

(b) In Nesyt, on the other hand, the zonation of perennial plant communities is partly discontinuous. The south shores of Nesyt have been disturbed by ploughing, and wet meadows have been turned into fields. But the littoral of Nesyt is also affected by frequent drying in the growing season. This brings about a development of communities of the foed. *Agropyro-Rumicion crispi.* The present relative distributions of communities of the ass. *Caricetum ripariae* and *Meliloto-Caricetum otrubae* are a result of long littoral and terrestrial ecophases.

(c) The peaty parts of Opatovický (east shore) provide habitats not only for communities of the ass. *Caricetum gracilis* but also for *Caricetum rostratae* and *Calamagrostietum canescentis*. The occurrence of *Calamagrostis canescens* as ecological dominant in gradually drying-out organic soils on terrestrialized sites is a significant symptom of changing conditions, from wet to relatively dry.

4. The relative representation of communities colonizing emerged bare soils is difficult to compare. The very name of the Nesyt fishpond (= insatiable = never saturated) indicates the great frequency of its incomplete filling with water from a relatively small catchment area in a climatically rather dry region. Opatovický, on the contrary, is mostly filled up and a low water level is only exceptional here. The long ecocycles make this fishpond somewhat more similar to a lake than to a fishpond. Hence Nesyt is rich in species and communities of emerged bare soils whereas Opatovický is very poor in them; these communities develop here only in fragments.

The two fishponds clearly differ in many aspects, but their common feature is a rich development of reedswamp communities in various forms. It was therefore of interest to investigate in some detail the ecology of these communities in the two fishponds differing in most ecological parameters as well as in the history, development and species outfit of their biotic communities. Their approximate distribution in the two ponds is shown in the maps presented in Section 1.1.

References see pp. 93—95.

1.3 Algal Populations Related to Different Macrophyte Communities

P. MARVAN and J. KOMÁREK

The pelagic communities of algae from the South Bohemian fishponds are discussed in Section 4 (see Table 1 in Sect. 4.1). In this Section, we give brief characteristics and comparisons of the species composition in individual microphyte synusia associated with principal macrophyte communities. The basic data originate from the Opatovický fishpond (Komárek et al., 1973) and are completed by further observations on Nesyt fishpond (Marvan and Ettl, in press). The macrophyte communities are classified as in Section 1.2.

The following indices were used in comparisons and evaluations of the species composition of algal communities:

(a) Index I_S

$$I_S = \frac{\Sigma\,|d_j|}{N_u},$$

where d_j – difference between the values of abundance of the jth species in a given pair of algal sets compared, $N_u (= N_a + N_b - N_c) =$ total number of species represented in at least one set. Index I_S is used for expressing the differences in species composition between two algal communities; its values are based on those of abundance quoted by Komárek et al. (1973, Tables III to IX) with only a few corrections.

(b) Index of common occurrence I_C

$$I_C = \frac{S_c}{S_u},$$

where S_c denotes the number of samples containing both species compared and S_u denotes the number of samples containing at least one of the compared species.

(c) Sørensen's (1948) index of homotoneity H (see Moravec, 1971; Komárková and Komárek, 1975)

$$H = 100\,\frac{2N_c}{N_a + N_b}$$

where N_a = number of species in community A, N_b = number of species in community B, N_c = number of species common to both communities.

(d) Coefficient of stability K_S (Komárek, 1973)

$$K_S = \frac{N_S}{N_I}$$

where N_S is the number of species in a community, performing their complete vegetation cycle in the locality under consideration, and N_I is the number of incidental accessory species.

(e) Diversity of the community D

$$D = -\sum_{i=1}^{N} (M_i/M)\log_2(M_i/M)$$

where M = total number of individuals in a sample, M_i = number of individuals of the ith species in the sample, and N = total number of species in the community.

1. *Lemnion minoris*. Only few algal species and individuals occur in typically developed communities of *Lemnetum gibbae* in both fishponds, Opatovický and Nesyt. The characteristic diatom *Achnanthes hungarica* accompanies these communities in both fishponds. Other algal species have penetrated into the *Lemnion* communities from the surroundings. In Opatovický, the set of algal species is most similar to that associated with communities dominated by *Bolboschoenus maritimus* (see Table 1).

In Nesyt, *Lemna gibba* and filamentous algae often cover relatively large areas of open water beyond the reed-belt, forming a varied carpet above the submerged vegetation. Algal filaments here provide mechanical support to *Lemna*. The algae thus act as an ecological dominant. Within the reed belt, however, *Lemna gibba* tends to suppress the filamentous algae.

2. *Hydrocharition morsus-ranae*. Communities of the ass. *Lemno-Utricularietum utricularietosum neglectae* which occur in the littoral of the Opatovický fishpond host some 230 species of algae; this is the highest number of all communities followed. Among these species, over 50% may be regarded as constant ($K_S > 1$); the species composition of this algal synusium therefore markedly differs from that accompanying other types of fishpond higher plant communities (Komárek, 1973). Planktonic algae prevail in this specific community, whereas sessile epiphytes constitute only 12% of it. Characteristic seasonal changes take place in this community (Fig. 2 in Sect. 4.1). During a short period in spring, loosely attached filamentous algae, particularly *Tribonema* and *Microspora* spp., develop rapidly—they frequently grow on sprouting turions. In the summer season numerous species of chlorococcal algae predominate, with accessory desmids and filamentous algae (*Spirogyra*, *Mougeotia*). During August, diatoms begin to develop and predominate; later, their brown mucilage covers the dying plants almost completely (with dominant *Fragilaria capucina*).

Table 1. Indexes I_S of algal communities in different monospecific helophyte stands in the Opatovický fishpond (explanations of three numbers see in Sect. 4.1, p. 300)

		Phalaris arundinacea	*Phragmites communis*	*Schoenoplectus lacustris*	*Typha angustifolia*	*Phragmites communis*	*Schoenoplectus lacustris*	*Typha latifolia*	*Glyceria maxima* (littoral stands)	*Glyceria maxima* (floating)	*Bolboschoenus maritimus*	*Persicaria amphibia*	*Carex gracilis*
Erosion littoral	*Phalaris arundinacea*												
Erosion littoral	*Phragmites communis*	1.000 1.000 1.048											
Erosion littoral	*Schoenoplectus lacustris*	1.182 1.106 1.071	0.806 0.681 0.597										
Erosion littoral	*Typha angustifolia*	1.239 1.169 1.130	0.974 0.861 0.845	0.953 0.814 0.836									
Accumulation littoral	*Phragmites communis*	1.227 1.187 1.175	0.933 0.815 0.874	1.038 0.889 0.909	0.942 0.863 0.871								
Accumulation littoral	*Schoenoplectus lacustris*	1.213 1.158 1.149	0.947 0.825 0.890	0.955 0.795 0.836	0.944 0.845 0.856	0.788 0.704 0.703							
Accumulation littoral	*Typha latifolia*	1.159 1.112 1.104	0.891 0.769 0.839	0.966 0.804 0.844	0.908 0.826 0.849	0.796 0.723 0.700	0.912 0.772 0.767						
Accumulation littoral	*Glyceria maxima* (littoral stands)	1.208 1.143 1.138	0.935 0.822 0.922	0.984 0.826 0.886	0.985 0.898 0.899	1.029 0.919 0.924	1.033 0.901 0.897	0.910 0.800 0.815					
Accumulation littoral	*Glyceria maxima* (floating)	1.179 1.139 1.109	0.875 0.776 0.770	1.011 0.845 0.788	1.074 0.966 0.941	0.862 0.766 0.788	0.938 0.816 0.825	0.874 0.756 0.763	0.905 0.815 0.845				
Accumulation littoral	*Bolboschoenus maritimus*					0.964 0.846 0.865	0.929 0.782 0.814	0.862 0.723 0.766	0.929 0.807 0.859	0.890 0.746 0.700			
Accumulation littoral	*Persicaria amphibia*					0.845 0.770 0.788	0.814 0.710 0.733	0.760 0.652 0.664	0.883 0.802 0.813	0.725 0.654 0.684	0.971 0.829 0.853		
Accumulation littoral	*Carex gracilis*					1.050 0.933 0.926	1.079 0.955 0.946	1.070 0.949 0.932	1.163 1.038 1.028	0.952 0.839 0.776	0.964 0.900 0.867	1.047 0.936 0.937	
Accumulation littoral	*Amblystegium riparium*					1.242 1.155 1.146	1.275 1.175 1.165	1.051 0.990 0.991	1.290 1.165 1.160	1.264 1.162 1.129	1.101 1.021 1.014	1.214 1.124 1.118	1.275 1.180 1.135

In Nesyt, where the foederatio is represented by sparse patches of *Lemno-Utricularietum utricularietosum vulgaris*, the accompanying microvegetation was not followed.

3. *Potamogetonion pusilli.* In the Opatovický fishpond, *Persicaria amphibia* is relatively most important among the edifying species of the association *Parvopotamogetoneto-Zannichelietum palustris.* The composition of the algal synusium is substantially species-poorer here than in the previous foederatio. Planktonic elements retreat in favor of epibionts whose share increases to some 33% (Fig.4 in Sect.4.1). The occurrence of *Zygnema* clusters is characteristic of the summer season. The species composition of the microvegetation resembles that in *Typhetum latifoliae* and *Glycerietum maximae*, and differs little from that in communities of *Phragmites* and *Schoenoplectus.*

In Nesyt, *Persicaria amphibia* is rare, and its companion algal vegetation was not studied. In the extensive communities of *Potamogeton pectinatus* and *Zannichelia pedicellata*, representing the *Potamogetoneto-Zannichelietum pedicellatae*, the most important components of the microvegetation are filamentous algae such as *Fragilaria* spp., *Cladophora fracta*, *Spirogyra* spp., *Oedogonium* spp., *Enteromorpha intestinalis*, loosely attached to the higher plants. These algae, at the peak of their development, often form continuous covers on the macrophyte leaves and stems. The epiphyton proper is only poorly developed in Nesyt, contrary to its rich development on *Persicaria amphibia* in Opatovický. This difference is apparently due to the different character of the two fishponds, namely with respect to the bottom substrate.

4. *Bolboschoenion maritimi* (in South Moravia) and *Glycerio fluitantis-Oenanthetum aquaticae subass. bolboschoenetosum* (in South Bohemia). In both specimen fishponds, *Bolboschoenus maritimus*, the edifying species of these communities, forms loose stands in which relatively large amounts of incoming radiation penetrate below the water surface. Typically epiphytic algae are relatively poorly represented, and accessory planktonic elements prevail. The usual principal component of the algal vegetation colonizing *Bolboschoenus* in Nesyt is clusters of filamentous algae, of which large amounts become accumulated and later decompose on the spot. The poor differentiation of this microvegetation from the other types is reflected in its rather low values of the index I_S, particularly against the algal synusia in communities dominated by *Typha latifolia* or by floating *Glyceria maxima* (Table 1).

5. *Oenanthion aquaticae.* The algal vegetation colonizing the *Hippuridetum vulgaris* occurring in the shallow western part of the Nesyt fishpond has been studied occasionally. Typical is the mass development of *Spirogyra* in the warmed up water of the "lagoons" as well as the formation of epipelic covers on the bottom of pools by benthic strata of blue-green algae (*Oscillatoria* spp.) and diatoms (*Cylindrotheca gracilis*, *Caloneis amphisbaena* and others). The high share of *Euglena*, *Phacus* and other saprophilous species in the algal synusium indicates an increased nutrient level in this microhabitat. The species composition of the whole algal community is rich and reflects relative isolation from other littoral macrophyte communities.

6. *Phragmition communis.* The conditions affecting the development of microphytes in these communities are rather different from those in other community types. Within the foederatio, the dominant helophyte and the habitat characteristics determine the differences in algal microvegetation existing between individual associations. Table 1 demonstrates the floristic similarity between the algal synusia in various littoral communities in the Opatovický fishpond. Relatively closest to each other are the algal synusia accompanying the *Phragmitetum* and *Schoenoplectetum* occurring both in the erosion and accumulation types of habitats. In those of the former type, the microvegetation accompanying the *Typhetum angustifoliae* is also closely related to that of *Phragmitetum.* Differences, however, exist in the course of seasonal development of the algal synusium in dense *Phragmites* stands on the one hand, and in loose stands of *Schoenoplectus* on the other (Komárek, 1973). Sessile diatoms (especially of the genera *Gomphonema, Cymbella, Synedra, Achnanthes minutissima* and others) and chlorococcal algae *(Characium ensiforme)*, blue-greens *(Homoeothrix stagnalis)* and green filamentous algae (*Stigeoclonium* spp.) represent important constituents of the reed belt algal vegetation, particularly in the communities of *Typhetum angustifoliae, Phragmitetum communis, Schoenoplectetum lacustris* and in floating *Glyceria maxima.* Filamentous non-sessile algae such as *Fragilaria construens, F. capucina,* and *F. vaucheriae* are also seasonally prominent here. On the other hand, planktonic algae are relatively less represented in the *Phragmition* communities (Fig. 4 in Sect. 4.1).

The character of the algal communities of the *Glycerietum maximae* in the Opatovický fishpond differs according to habitat: (1) Loose *Glyceria* stands occurring in accumulation biotopes along the waterward edge of the reed-belt are colonized by a vegetation similar to that found in stands of *Bolboschoenus*, with much the same course of seasonal development. (2) Floating *Glyceria* colonizing the outer edge of the inner littoral hosts an algal vegetation analogous to that found in *Phragmitetum, Typhetum angustifoliae* or *Schoenoplectetum.* (3) Dense *Glyceria* stands occurring in a shallow littoral or limosal ecophase possess an algal synusium resembling that of *Typhetum latifoliae.* When the *Glyceria* canopy becomes quite dense in summer, the colonization by algae is severely suppressed.

These differences correspond to the plant-sociological classification of the above *Glyceria* communities by some authors, with (2) being regarded as a variant of *Phragmitetum communis*, and (3) as a true *Glycerietum maximae.* The communities of *Phalaridetum arundinaceae*, also belonging to the *Phragmition* foederatio, are distinguished by a species-poor microvegetation; hence the high index I_S between this microvegetation and that forming the algal synusium in other *Phragmition* communities.

7. *Caricion gracilis.* The spring maxima of a filamentous algal synusium dominated by *Tribonema* spp. characterize the two associations representing this foederatio at the Opatovický and Nesyt fishponds respectively: *Caricetum gracilis* and *Caricetum ripariae* (see p. 52). After the termination of about a month of their mass development in March to April, the algae become heavily reduced. This is the result both of the heavy shading by the macrovegetation and of the regularly occurring fall of water level to or below soil surface.

The zone of sedge communities is characterized by the presence of certain algal indicators of more or less oligotrophic and acid conditions. They are present even in Nesyt but they are represented more abundantly in Opatovický. In Nesyt, these elements are confined to the landward edge of the littoral, where they occur together with abundant halophilous to mesohalic diatoms. The presence of the latter reflects the increased salt content (especially that of sulphate) in the water and soil, which is enforced by the fall of water level in summer. The algal synusium accompanying the *Caricetum ripariae*, as well as the landward edge of the *Phragmition* communities in Nesyt, thus markedly differs in species composition from the algal synusia of the fishpond's other communities. The landward strip is characterized by the presence of such algae as *Tribonema* spp., *Ophiocytium* spp., *Eunotia curvata*, *Navicula salinarum*, *N. simplex*, *Nitzschia commutata*, *Hantzschia vitrea*, *Synedra minuscula*, and by the absence of species abundant in the waterward part of the reed belt such as *Enteromorpha intestinalis*, *Cladophora fracta*, *Amphora veneta*, *Synedra fasciculata* and *S. pulchella*. Mechanical barriers and reduced mixing of the water masses contribute to the establishment of these sharp boundaries; but the principal reason is the markedly differing habitat conditions in the outer and inner zones of the Nesyt littoral. Within the central and inner littoral, the algal vegetation comprises no typically mesohalic elements, and only some halophilous algae are present.

Similar but less pronounced differences are also observed in the Opatovický fishpond (Fig. 4 in Sect. 2.1.2; Fig. 6 in Sect. 4.1). Here, even the communities of *Caricetum gracilis* host some typically planktonic algae while clusters of *Tribonema*, acidophilous desmids, or *Eunotia curvata* are encountered even in the fishpond's open water. The reason for this poorer differentiation is the relatively narrow reed belt and less heterogeneous habitat conditions within the Opatovický littoral. Nevertheless, even here the specific character of the algal vegetation of the inner littoral (closely related to the pelagial vegetation) is clearly reflected in the high value of the I_S coefficient (cf. Table 1).

References see pp. 93—95.

1.4 Aerial Photography and Mapping of Fishpond Vegetation

P. ŠMÍD

The mapping of the fishpond vegetation in the Czechoslovak IBP Wetland project served a twofold purpose: (a) estimation of the areas covered by stands of different vegetation types, which might also serve as production types; (b) documentation of different stages of development of the littoral vegetation and of its changes under various kinds of influence. The aerial photographs permit an examination of the long-term spreading of the stands, of their succession, as well as of their seasonal development. For example, the mass occurrence of filamentous algae, or of communities dominated by *Bolboschoenus* of short duration in the Nesyt fishpond could be mapped out from the photographs. So could the formation of numerous "lagoons" (patches of open water) within the reed-belt communities in the Opatovický fishpond, resulting from decay of the central parts of *Glyceria* colonies after long-term excessive flooding. Such a documentation provides valuable information for making decisions on management and conservation.

The application of any standard topographical survey techniques to wetlands is complicated by the difficult access to, and poor visibility within, the tall and dense vegetation. The variable pattern of the vegetation also represents a complicating factor.

For the aerial photography of details of the vegetation as well as of whole sections of fishpond littorals, a low-flying model aircraft was used. This was equipped with a short-wave radio transmitter for control of both the camera and the flight parameters. The method was worked out at the Institute of Geography, Czechoslovak Academy of Sciences, Brno (Stehlík, 1969). Aerial photography from such a model aircraft is relatively cheap and can be accomplished on the most suitable days and at times of day chosen by the investigator. The main disadvantage of the technique is an insufficiently precise control of flight height and of the angle of the model aircraft. The transfer of the photographed situation on to a map is therefore less easy than in standard aerial photography and the resulting map is somewhat less accurate. A precise outline map with contours and sometimes also a precise topographical location of certain conspicuous points in the littoral are inevitably required for producing an accurate map.

At the Nesyt fishpond, an outline map of 1:5000 was used as a baseline map, and the aerial photographs were taken on the dates and in the ways indicated:

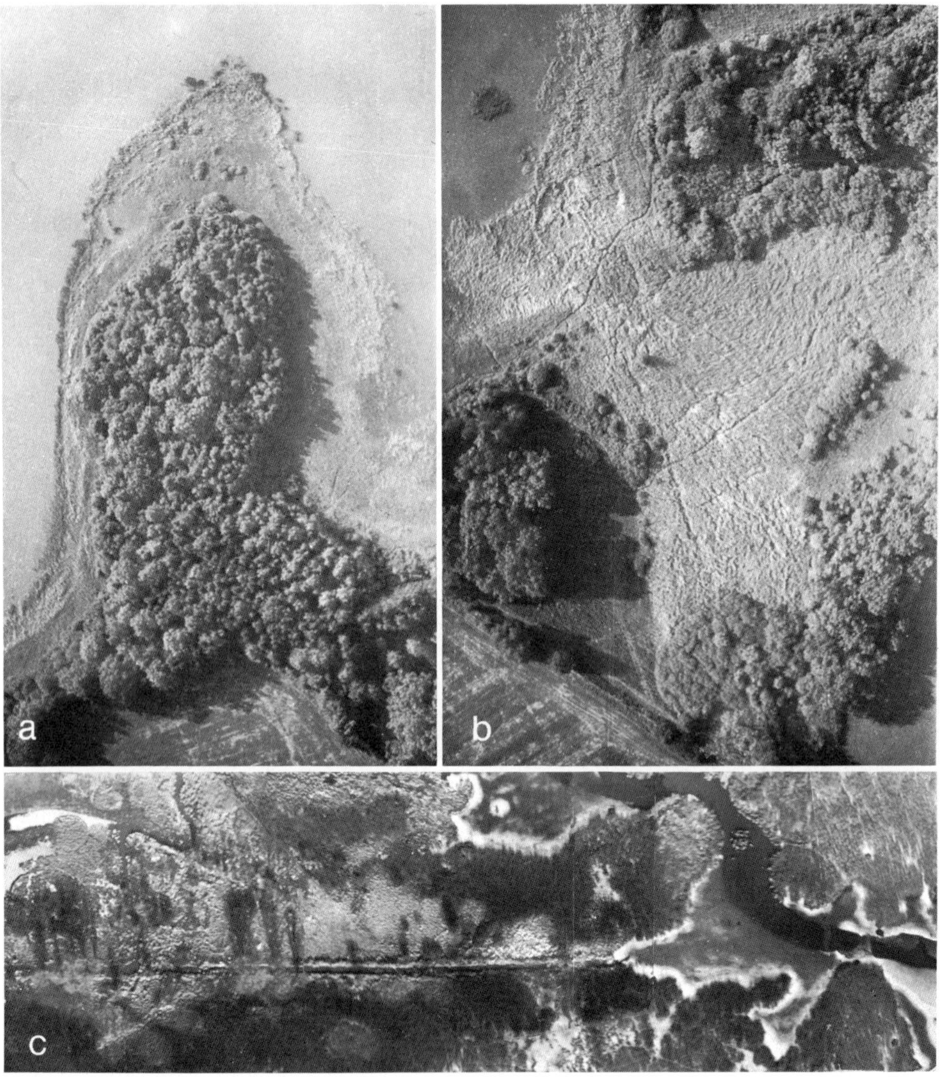

Fig. 1a–c. Aerial photography. (a) S.E. shore of Opatovický fishpond with littoral belts of *Schoenoplectus lacustris* in bay at left; (b) Terrestrialized bay biotope S at E shore of Opatovický fishpond, colonized by luxuriant stands of *Phragmites communis* and *Glyceria maxima*. In the littoral *(top left)* mixed stands of *Glyceria maxima* and *Phragmites communis* are seen, and a young colony of *Typha angustifolia* in the sublittoral; (c) W. bay of frozen Nesyt fishpond, with ice-free inflow. (Photos: Geographical Institute of the Czechoslovak Academy of Sciences)

1: January 1971; black and white 35 mm film was used. In the winter photographs, the small patches of frozen open water ("lagoons") are conspicuous, the muskrat houses are also easily recognizable (Fig. 1c). The stands of *Phragmitetum communis* are distinguishable from the *Typheta*, the latter showing a typical coarse-grained structure. Indeed, *Typha angustifolia* forms looser stands and its remaining dead leaves are thick and usually broken. The communities of *Glyceria*, *Phalaris* and tall sedges (*Carex riparia* and others) with almost horizontally oriented leaves appear much brighter and are distinguishable from the *Phragmiteta* and *Typheta*, which appear darker owing to their vertically oriented stems.

2, 3: July 1971, using ORWO-Color inverse color 35 mm film, and September 1971, using the same film material as in July.

In the color photographs taken in summer, *Phragmitetum* appears bright bluish-green, *Typheta* are deep brownish-green. The significant difference in color is probably caused by differences in leaf inclination and leaf aging. In the fall photographs, the groups of *Typha latifolia* can be distinguished from those of *T. angustifolia*: the shoots of the latter are already dying off and becoming dry, showing a bright brown color. *T. latifolia*, on the other hand, remains green till late fall.

4: July 1972, ORWO-Chrom inverse color film, 6 cm × 6 cm. In that year, the water level was low in the Nesyt fishpond and the photographs showed the *Bolboschoenus* stands (yellowish green) and other plant communities of the emerged fishpond bottom. The submerged vegetation with dominant *Potamogeton pectinatus* was also evident.

Aerial photography for mapping of the Opatovický fishpond was accomplished in August 1971 by the same technique (Fig. 1 a, b). In this case, a cadastral map of 1:5000 was used as a baseline map. The much more patchy vegetation cover, especially along the eastern shore, made it quite difficult to recognize and map out the normal fishpond shoreline. In the color photographs, the pale green stands of *Glyceria* were easily distinguishable from the *Phragmiteta* and other reedswamp communities. *Schoenoplectetum lacustris* and *Typhetum angustifoliae* appear green. In the whole littoral area, the lagoons and areas of dead *Glyceria* stands represented a typical component during the ecoperiod of raised water level, which was characteristic of the 1971 growing season (Fig. 3 in Sect. 2.2).

The resulting vegetation maps of the Opatovický and Nesyt fishponds are shown in Figures 4 and 5 in Section 1.1. The total areas occupied by each vegetation type as mapped out are presented in Table 3 in Section 1.1. These vegetation types correspond to the most important community types, or groups thereof, as outlined in Section 1.2, but some communities of short-term duration or occupying only small areas could not be mapped out.

References see pp. 93—95.

1.5 Animal Populations in Fishpond Littorals

J. PELIKÁN, K. HUDEC, and K. ŠŤASTNÝ

The grazing–predatory food chain in fishpond littorals depends, above all, on the primary production in the littoral plant communities. The participation of herbivores in the further fate of organic matter produced by the plants cannot be ignored. They may destroy (but not consume!) up to one fifth of the primary production (for a detailed evaluation see Sect. 3.1.7). Herbivorous macrofauna is represented by molluscs (Sect. 6.5), by a great number of insect species such as flies of the genus *Lipara* on reed, mining dipterous larvae (Sect. 6.4), noctuid larvae of *Archanaria (Nonagria)*, *Leucania*, and *Tapinostola*, larvae of *Tineidae*, *Pyralidae*, and *Tortricidae* in *Typha* flowers; by aphids (mostly *Hyalopterus pruni* sucking on reed leaves (Sect. 6.4), by thrips *(Chirothrips, Haplothrips, Iridothrips)* and by phytophagous *Coleoptera* (mostly *Chrysomelidae*, *Curculionidae*, see Sect. 6.3).

Important direct grazers of aquatic macrophytes are certain birds, especially greylag geese *(Anser anser)* and phytophagous ducks (Sect. 6.2). For nest-building, these birds mainly use dead plant material, but populations of certain species show preferences for peculiar types of littoral plant communities (see Table 1). Out of the small mammals, the muskrat *(Ondatra zibethica)* and the water vole *(Arvicola terrestris)* are locally important as destructors and consumers of the central and inner littoral *Phragmites*, *Typha*, *Schoenoplectus*, and *Glyceria*, especially of their young sprouts. The common vole *(Microtus arvalis)* and, locally, the bank vole *(Clethrionomys glareolus)* act in a similar way in terrestrial ecophase of reedswamps in the outer littoral (see Sect. 6.1). Figure 1 presents the distribution of populations of small mammals in the terrestrial, limosal, and littoral ecophases of the reed-belt in the Nesyt fishpond in relation to the position of water level and vegetation type.

Carnivores are represented by numerous populations, frequently occurring at high densities. Spiders compete with predatory beetles, *Coleoptera*: *(Carabidae, Staphylinidae, Dytiscidae)* and with bugs *(Hemiptera)*. A relatively large set of specific dragon flies *(Odonata)* is associated with the fishpond communities. About six species of frogs and two species of lizards prey on insects, and so do several species of reed warblers *(Acrocephalus, Locustella)* and four species of shrews (of the genera *Sorex*, *Neomys*, and *Crocidura*). An important predominantly predatory element is the black-headed gull *(Larus ridibundus)*, living in large colonies in helophyte stands of numerous fishponds. The waterfowl and other birds associated with fishponds may, however, be mostly regarded as omni-

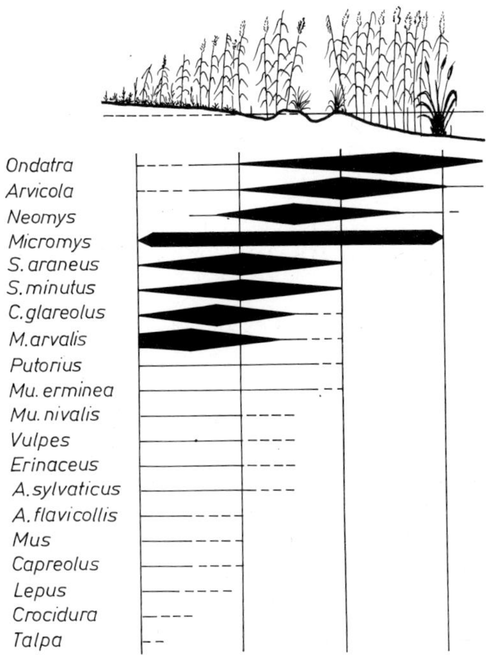

Fig. 1. Distribution of populations of small mammals in the Nesyt fishpond reed-belt. *Left to right:* terrestrial, limosal, and littoral zones. (From Pelikán, 1975)

Table 1. Situation of nests of more frequent species in littoral plant communities (K. Hudec, 1975)

Species	Type of stand				Age of stand		Density of stand		Distance from open water		
	P	T	C	M	Old	New	Dense	Sparse	0	5	10 m
Podiceps cristatus	35	11	—	—	16	14	1	13	7	12	8
P. nigricollis	10	2	—	1	7	+	3	1	1	—	1
P. ruficollis	26	7	1	—	8	7	3	7	18	2	2
Ixobrychus minutus	23	3	—	—	17	2	4	1	4	2	—
Aythya ferina	16	1	11	6	10	3	3	3	4	1	6
Gallinula chloropus	8	7	7	4	5	4	1	11	3	1	1
Fulica atra	159	67	1	9	89	37	10	98	79	23	16
Acrocephalus arundinaceus	123	14	—	1	46	37	9	5	33	14	4
A. scirpaceus	108	38	1	11	29	41	8	14	30	16	6

P, *Phragmitetum communis;* T, *Typhetum angustifoliae*; C, *Phragmitetum* with *Carex* undergrowth; M, mixed community.

vorous, though individual species differ in their food requirements. Trampling of the plants and eutrophication by bird excrements are important secondary effects of the rich bird life in fishponds, especially in waterfowl reserves and sanctuaries. Grazing greylag geese *(Anser anser)* also locally affect the species composition of littoral helophyte communities (Květ and Hudec, 1971), and so do mute swans

Fig. 2. Night heron *(Nycticorax nycticorax)* nesting in the Velký and Malý Tisý nature reserve. (Photo: J. Formánek)

(Cygnus olor) breeding in several pairs in numerous fishponds in both South Bohemia and South Moravia. Plant communities of the Malý Tisý nature reserve near Třeboň (see Sect. 8) are locally affected by the guano of the night heron *(Nycticorax nycticorax)*, breeding there in large numbers (Fig. 2). Much of the birds' food, however, originates from outside the fishponds. Certain bird populations feeding partly or, sometimes, predominantly in surrounding agricultural areas thus import appreciable amounts of materials, especially minerals, into a fishpond and its littoral. A quantitative estimation of this import has not yet been attempted.

Table 2. Seasonal variation (in densities per 10 ha) of bird populations identified in biotopes of dams along the Nová Řeka canal in the Třeboň basin. (Data by K. Šťastný)

Species	Spring	Summer	Fall	Winter
Accipiter nisus		0.14		
Acrocephalus palustris	0.14			
Actitis hypoleucos	0.14	0.14		
Aegithalos caudatus	3.86	1.25	11.68	25.14
Alcedo atthis			0.64	0.14
Anthus trivialis	0.29	0.54		
Ardea cinerea	0.14			
Bucephala clangula	0.43			
Carduelis spinus			19.32	43.96
Certhia brachydactyla } *Certhia familaris*	16.86	12.07	9.11	6.04

Table 2 (continued)

Species	Spring	Summer	Fall	Winter
Coccothraustes coccothr.			0.82	
Columba palumbus		0.29	1.61	
Cuculus canorus	1.43			
Dendrocopos major	2.29	4.25	3.75	1.25
Dendrocopos medius	0.57	0.96	1.46	0.29
Dendrocopos minor		0.48		0.29
Dryocopus martius	0.14	0.14	0.18	
Emberiza citrinella	1.29	0.82		
Emberiza schoeniclus		0.29		
Erithacus rubecula	0.86	10.29	1.14	
Ficedula albicollis	13.43	0.29		
Ficedula hypoleuca	0.29			
Fringilla coelebs	19.29	6.46	2.29	
Garrulus glandarius	1.86	1.78	25.18	2.32
Hirundo rustica		0.96		
Chloris chloris	1.14	0.29	0.18	
Jynx torquilla	0.14			
Locustella fluviatilis	0.29	0.14		
Motacilla alba	1.14	0.14		
Muscicapa striata	13.86	2.61		
Oriolus oriolus	0.29	0.54		
Parus ater			1.29	
Parus caeruleus	20.14	32.14	16.07	22.54
Parus major	8.43	16.36	15.25	1.64
Parus palustris } *Parus atricapillus* }	5.71	13.32	12.18	7.57
Passer montanus			1.61	
Phasianus colchicus		0.68	0.18	0.14
Phylloscopus collybita } *Phylloscopus trochilus* }	5.29	4.68	0.82	
Picus viridis		0.14		
Prunella modularis	0.57	2.07		
Pyrrhula pyrrhula	0.43	1.64	1.46	0.14
Regulus regulus	0.29		4.71	5.64
Serinus serinus	0.14			
Sitta europaea	10.14	16.21	18.50	11.82
Streptopelia turtur	0.14			
Strix aluco	0.14			
Sturnus vulgaris	47.42			
Sylvia atricapilla	5.00	10.29		
Sylvia borin	12.57	10.04		
Sylvia communis	3.14	2.21		
Sylvia curruca	0.29	0.29		
Troglodytes troglodytes		0.68	0.32	
Turdus philomelos	2.57	0.82	0.32	
Turdus merula	3.71	3.29	0.50	
Turdus iliacus			0.18	
Turdus pilaris				0.14
Turdus viscivorus		0.43		
Total	206.29	160.16	150.75	129.06

Also recorded: *Ciconia ciconia, Picus canus.*

Rather varied biotopes frequently coexist in a fishpond and its surroundings. This is particularly true of large fishpond systems with their mighty dams and interconnecting canals and ditches. The variety of biotopes is reflected in a corresponding variety of animal populations living together in a relatively small area but occupying different niches. A two years' survey of the set of bird populations living on and along the dams fringing the flood plain of the ancient Nová Řeka (New River) canal passing through a wooded and fishpond-rich part of the Třeboň basin has detected altogether 60 species populations (Table 2).

Duck farms are situated on the shores of a great many ponds. Each farm is stocked with several hundreds or thousands of both young and adult domestic ducks rotated several times each year. The ducks produce excrements passing directly into the fishpond water, and also affect the fishpond plant communities mechanically (for more details see Sect. 7.3.2).

Top carnivores are represented by the grass-snake *(Natrix natrix)*, commonly living along the fishpond shores, by the grey and, occasionally, purple heron *(Ardea cinerea, A. purpurea)*, by the common tern *(Sterna hirundo)*, the marsh harrier *(Circus aeruginosus)* and by several mammals such as the fox *(Vulpes)*, the polecat *(Putorius)*, feral cats *(Felis)*, the stoat and the weasel *(Mustela erminea* and *M. nivalis)*, mostly searching for food in the fishpond littorals. Carnivorous ducks and many other predatory birds such as the white-tailed eagle *(Haliaeetus albicilla)* are present only during their migrations.

At least four fifths of the energy flow beyond the primary producers' level passes through the detritus food chain. Some dead plant material is consumed by numerous populations of saprophagous larvae of flies *(Diptera)* such as *Tipulidae* or *Chironomidae*. The dead organic matter also supports, together with small planktonic and benthic organisms, great numbers of mosquitoes (larvae of *Culicidae*, Sect. 6.5). Imaginal stages of all these insects are preyed upon by reed warblers and other birds such as the bearded tit *(Panurus biarmicus)*, which is common in the South Moravian reedswamps. By this pathway, a part of the energy originally contained in the dead plant material is transferred to the predatory food chain. The relations between the saprophagous and predatory evertebrate communities are discussed in Section 6.5. The total biomass of the evertebrate macrofauna as well as its species composition are largely controlled by the feeding activities of the fish. The accessibility of the outer littoral to relatively big fish, especially to carp *(Cyprinus carpio)*, depends on the position of water level and on the density of the littoral helophyte communities. The fish usually have easy access into the inner littoral overgrown by *Typha* or tall and frequently relatively loose *Phragmites* or *Schoenoplectus*.

The carp and other cultivated fish species are, of course, the principal terminal producers in fishponds. The fishpond littoral, usually densely overgrown by helophyte stands, is mostly regarded as unimportant for the food production in the trophic chains leading to fish production (Straškraba et al., 1967). But this need not be true during certain critical periods such as early spring, when the relatively warm water in accessible parts of the reed-belt and in its lagoons is more productive than the open water in the fishpond pelagial (see Sects. 2.1 and 4.3). Communities of submerged macrophytes as well as loose stands of emergent macrophytes acquire, however, great importance for the cultivated fish, both as shelter and as sites of abundant production of fish food (see, e.g., Kořínková, 1971). The value of the higher vegetation in fishponds for fish farming has been briefly assessed by Putschögl (1973).

The principal results of ecological surveys of animal populations and communities in South Moravian fishponds are given by Pelikán (1975) for mammals, by Hudec (1975) for birds, and by Obrtel (1972) for certain *Coleoptera*. A hydrobiological survey of the Lednice fishponds has been accomplished by Losos and Heteša (1971). For South Bohemian fishponds, many faunistic papers concerning animal life have been published by numerous authors, but only in local Czech journals, for avifauna see Šťastný (1973) and Řezníčk (1973).

References see pp. 93—95.

1.6 Decomposers in the Fishpond Littoral Ecosystem

B. ÚLEHLOVÁ

Microorganisms form an important biotic component in every ecosystem. The share of their biomass in an ecosystem may be important during favorable conditions in certain periods of the year. Their rich and manifold enzymatic abilities break down all sorts of organic matter. Thus the microorganisms by means of their metabolic activities support the energy flux through the ecosystem and the cycling of mineral elements. In natural ecosystems, microbial activities control the recycling and availability of mineral nutrients as well as the plant productivity.

Wetland ecosystems are typical detritus-based systems (Odum, 1971). The microorganisms settle on dead plant remnants which are characterized by wide C:N ratios, they use them as a source of energy, enrich them with nitrogen, ferment them and make them available and more palatable for organisms of further trophic levels. The microorganisms thus form the base of a trophic pyramid and initiate a highly diversified and complicated detritus food web.

Let us consider more closely the spatial distribution of microorganisms and microbial processes in wetland ecosystems. The microorganisms are present there in the following environments: air, water, soil, detritus or mud, and transition areas between these substrates. Further, they may be associated with either living or dead plants and animals. All the substrates may occur in different combinations, thus providing various microhabitats supporting diverse microbial populations. Their complete enumeration and description is impossible, as wetlands comprise a multitude of such microhabitats, each forming a peculiar biological-chemical-physical unit supporting a specific microbial community (Day et al., 1973).

Because of their ubiquitous character, many taxonomical groups of microorganisms may participate in decomposition processes in the fishpond littoral: bacteria, actinomycetes, fungi, algae, and protozoa.

Several ecological factors, besides temperature and humidity, may be important and decisive for the selection and establishment of microbial populations. Pugh (1958), Pugh and Williams (1968), Taligoola (1969), and Pugh and Mulder (1971) report, for example, on fungal communities associated with individual developmental stages of different wetland plants such as *Phragmites communis*, *Typha latifolia*, or *Salsola kali*. Further factors of importance are the acidity and redox properties of the environment: fungi prefer an acid environment while bacterial populations thrive better in neutral conditions. Anaerobic organisms are more active under low oxygen tension, and humification processes prevail; under high oxygen tension, high respiration of aerobic populations accompanies the full mineralization of organic substrates.

The ecological relationships between microbiota in littoral ecosystems are only little known. They are extremely complicated and unstable, changing in time; moreover, relationships ranging from symbiosis to parasitism become established between organisms belonging to different taxonomic groups.

In the course of the ecosystem studies of the Nesyt fishpond in South Moravia, observations and experiments were performed, whose aim was an evaluation of the numbers and composition of microflora associated with different stages of decomposition of dead plant remnants, particularly of *Phragmites communis* and *Typha angustifolia.*

Material and Methods. During 1971 and 1972, the plant material and soil samples were collected along a transect across the entire reed-belt of the Nesyt fishpond, at monthly intervals. The transect comprised four different vegetational zones: (1) limoso-terrestrial mixed community with *Phragmites communis, Typha angustifolia,* and *Carex riparia (Phragmitetum communis caricetosum ripariae);* (2) littoral and nearly pure community of *Phragmites communis (Phragmitetum communis phragmitetosum;* (3) lagoon without emergent vegetation and nearly void of macrophytes other than *Lemna gibba;* (4) littoral pure *Typha angustifolia* community with some *Lemna gibba* undergrowth *(Typhetum angustifoliae).* In each zone, the respective numbers of the microorganisms were estimated in the standing dead plant material, litter, and soil or mud (sapropel).

The plant or soil material to be analyzed was placed in sterile bottles and flooded with sterile water. The bottle was shaken vigorously for 10 min in order to suspend the microorganisms in the water. A series of dilutions was prepared and plated out on to the following media: meat-pepton agar to count aerobic bacteria; pepton agar for bacteria digesting relatively stable plant products; cellulose agar for aerobic cellulolytic bacteria; Ashby agar for nitrogen fixers; cellulose agar for cellulolytic fungi; and agar with glucose-yeast extract and casein agar for actinomycetes. Aaronson (1970) gives full recipes for all the agar media used.

Microbial analyses of the water were performed with water samples from all four vegetational zones during 1971. The following agar media were used for the isolation and cultivation of microorganisms: meat-pepton agar, to obtain a count of aerobic bacteria; Ashby agar supporting nitrogen fixers; medium I for nitrifiers taking part in the first step of nitrification; medium II for nitrifiers oxidizing NO_2^- to NO_3^-; and a medium for microorganisms carrying out desulfurification. All the agar media used are described by Rodina (1965).

Concurrently with microbial analyses, the analyses of microbial N and P present in the water samples were performed. The total P and total N were estimated in water samples filtered (1) through a normal metallic filter (pore diameter 45 µm) and (2) through a bacterial filter (porosity 0.4 µm), using the methods described by Golterman and Clymo (1969). Microbial P and N were calculated from the differences between these two types of analyses (Dvořák, 1973).

Microbial Population Densities in Ecosystem Structures. Annual averages of counts of microorganisms in standing dead materials of the different plants tested are shown in Table 1. Standing dead *Typha angustifolia* is much more densely colonized then standing dead *Phragmites* (on dry weight basis). Both kinds of dead material collected in aquatic habitats yield higher counts of microflora than those from terrestrial habitats. The differences between terrestrial and aquatic habitats are particularly pronounced in the case of *Typha.* Approximately 10^6

Table 1. Annual averages in counts of microorganisms in 1 g dry standing dead material of *Phragmites communis* and *Typha angustifolia* in the Nesyt fishpond littoral

	Moisture[a]	PR	MPA	Akt. 2	Ashby	Akt. 1	CB	CF
		10^6	10^6	10^6	10^6	10^6	10^6	10^4
Phragmites, terrestrial	8.29	7.8	7.1	3.0	1.9	2.0	8.8	1.1
Phragmites, aquatic	8.69	26.2	4.2	3.2	1.8	1.1	2.8	1.7
Typha, terrestrial	12.93	69.9	49.4	87.5	3.3	8.1	96.9	2.0
Typha, aquatic	60.60	332.6	357.6	512.0	139.8	60.1	584.6	2.6

[a] Given in % fresh weight.
PR: agar medium for microflora living on plant remnants, MPA: meat pepton agar, Akt. 2: agar medium for actinomycetes, Ashby: agar medium for diazotrophs, Akt. 1: agar medium for actinomycetes, CB: agar medium for cellulolytic bacteria, CF: agar medium for cellulolytic fungi.

Table 2. Annual averages in counts of microorganisms in 1 g dry litter in terrestrial and aquatic biotopes of the Nesyt fishpond littoral

	Moisture[a]	PR[b]	MPA	Akt. 2	Ashby	Akt. 1	CB	CF
		10^8	10^8	10^8	10^8	10^7	10^8	10^5
Terrestrial:								
Phragmitetum communis, with *Carex riparia*, surface litter layer 0–8 cm	32.36	2.4	1.5	4.9	3.8	4.0	2.4	2.4
Phragmitetum communis, with *Carex riparia*, subsurface litter layer 8–15 cm	77.58	3.4	1.7	3.8	2.8	4.7	4.0	2.0
Aquatic:								
Phragmitetum communis, surface litter bottom layer 0–3 cm	70.02	7.3	3.6	6.7	3.3	9.6	9.0	1.5
Phragmitetum communis, subsurface litter bottom layer 3–10 cm	83.48	7.3	3.1	6.7	4.2	8.1	7.1	1.5
Typhetum angustifoliae floating on water surface	90.72	19.5	3.6	18.5	1.9	10.0	7.5	1.6

[a] Given in % fresh weight. — [b] Abbreviations as in Table 1.

individual microorganisms exist in 1 g of dry standing dead material of littoral helophytes. The moisture content of the standing dead material may differ considerably with habitat, which may control the counts of microorganisms.

Table 2 shows the annual average counts of microorganisms in the terrestrial and aquatic litter from the Nesyt fishpond littoral. Here, the substrate humidity again substantially affects the density of microorganisms, as in the standing dead

Table 3. Annual averages in counts of microorganisms im 1 ml water in different biotopes of the Nesyt fishpond littoral

	MPA	Nitr. 1	Nitr. 2	Ashby	Desulf.
	10^3	10^3	10^3	10^3	10^3
Phragmitetum communis with *Carex riparia*	5.1	5.2	3.5	7.7	11.9
Phragmitetum communis	9.9	6.2	3.7	10.6	10.7
Lagoon	6.4	4.4	2.9	6.8	7.9
Typhetum angustifoliae	4.8	5.8	2.7	6.4	8.1

MPA: meat pepton agar, Nitr. 1: agar for nitrifying organisms (NH_4 to NO_2), Nitr. 2: agar for nitrifying organisms (NO_2 to NO_3), Ashby: agar for diazotrophs, Desulf.: agar for desulfurizing organisms.

Table 4. Annual averages of microorganism counts in 1 g sapropel in the two principal helophyte communities of the Nesyt fishpond

	Moisture[a]	PR[b]	MPA	Akt. 2	Ashby	Akt. 1	CB	CF
		10^7	10^7	10^7	10^7	10^6	10^7	10^4
Phragmietum communis	88.8	31.1	0.01	0.01	7.6	2.1	8.0	3.3
Typhetum angustifoliae	83.07	7.3	1.2	1.4	1.6	1.0	1.7	5.4

[a] Given in % fresh weight. — [b] Abbreviations as in Table 1.

material. Mean values of the microbial counts are 10^8 individuals per 1 g of dry material; this is much more than the counts obtained in standing dead material. The reasons for this phenomenon may be (1) much larger surface of the litter; (2) more developed detritus food chain in the litter; (3) higher nutritive value of the litter for the microorganisms.

The annual mean counts of microorganisms in water samples collected in the four vegetational zones of the Nesyt fishpond littoral in 1971 are presented in Table 3. The data demonstrate that the numbers of microorganisms tend to increase from open water towards the shore. The water sampled in the zone occupied by pure *Phragmites (Phragmitetum communis)* is distinguished by the markedly highest counts of all the groups tested. This zone is rather isolated from the open water by both the lagoon and the zone of *Typhetum angustifoliae*. The concentration of substance released from decomposing plant materials is high here, and so is the supply of energy resources required by the microorganisms, while the export and dilution of organic substances are poor (see Sect. 2.3). A different situation seems to prevail in the zone of *Typhetum angustifoliae* in the inner littoral. The water-containing substances released during the decomposition of plant material are permanently diluted by the relatively pure water from the open water area of the fishpond.

Of interest are the data pertinent to nitrifiers: the counts of organisms using NH_4 (nitrifiers I) are twice as high as the counts of organisms using NO_2 as

Table 5. Minimum and maximum microbial counts in different materials during their decomposition in littoral of the Nesyt fishpond

			PR [a]	MPA	Ashby
Standing dead air	10^7	min.	0.8	0.4	0.2
		max.	33.3	35.8	14.0
Litter—terrestrial environment soil surface	10^8	min.	2.3	1.5	2.8
		max.	3.3	1.7	3.8
Litter—aquatic environment bottom	10^8	min.	7.3	3.1	1.9
		max.	19.5	3.6	4.2
Mud	10^7	min.	7.3	0.01	1.6
		max.	31.1	1.2	7.6
Water	10^3	min.	—	4.8	6.4
		max.	—	9.9	10.6

[a] Abbreviations as in Table 1.

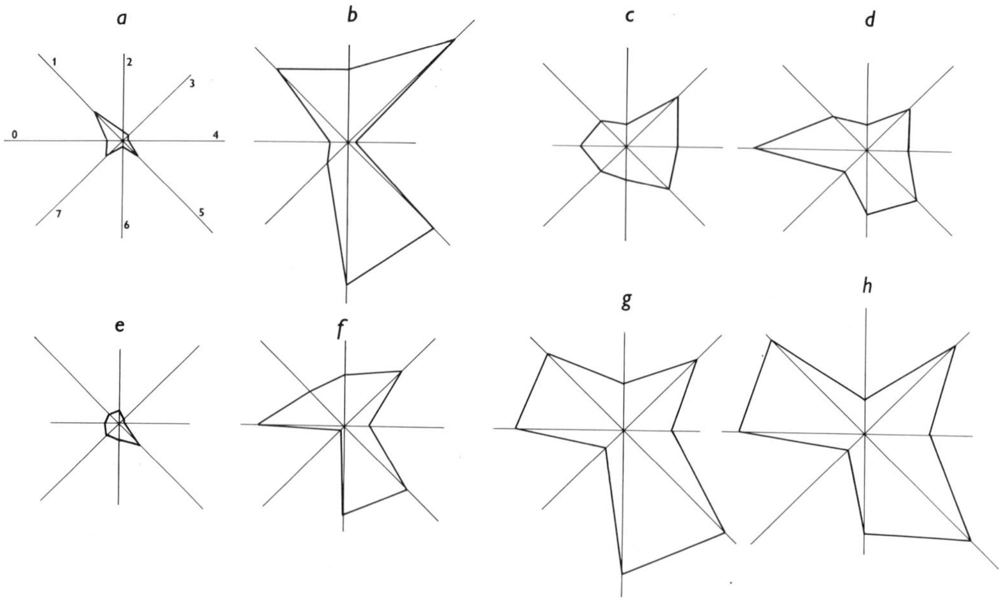

Fig. 1a–h. Microbial populations on different substrates in Nesyt fishpond littoral. Counts of microorganisms from terrestrial ecophase: (a) Standing dead *Phragmites communis.* (b) Standing dead *Typha angustifolia.* (c) Litter in the mixed community of *Phragmites communis*, *Carex riparia*,and *Typha angustifolia.* (d) Upper soil layer in the same community. Littoral ecophase: (e) Standing dead *Phragmites communis.* (f) Standing dead *Typha angustifolia.* (g) Litter in pure *Phragmites communis.* (h) Detritus in pure *Phragmites communis.* Connotation of axes (except 0 and 7 in 10^6 counts): *0*, soil moisture in air-dried sample (% of fresh weight); *1*, organisms on pepton agar; *2*, organisms on meat—pepton agar; *3*, organisms on glucose yeast extract agar; *4*, organisms on Ashby agar; *5*, organisms on casein agar; *6*, organisms on agar for cellulolytic bacteria; *7*, 10^4 counts of organisms on agar for cellulolytic fungi. In (a), (b), (e) 1 cm corresponds to $12.5 \cdot 10^6$ germs; in (c), (d), (f), (g), (h), 1 cm corresponds to $12.5 \cdot 10^7$ germs

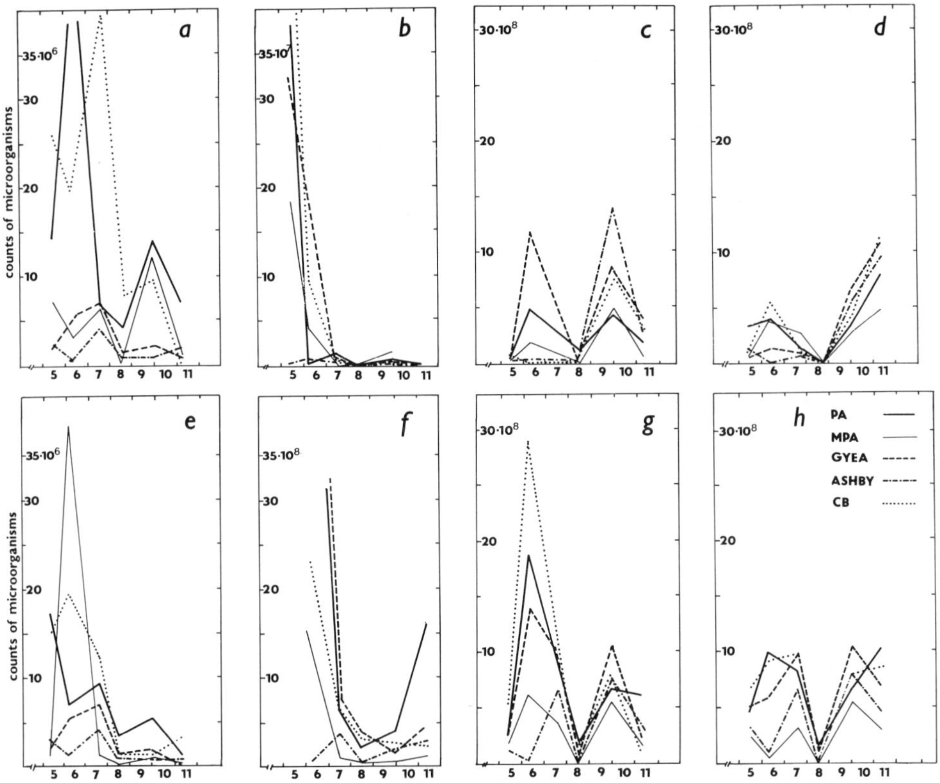

Fig. 2. Seasonal patterns (*abscissa:* time in months) of microbial counts in standing dead material, litter and soil or mud in Nesyt fishpond littoral with (a) to (h) the same as in Figure 1

nitrogen source. Only a part of NH_4–N thus appears to be nitrified to NO_3, while the remainder follows the organic nitrogen metabolic pathways.

The extremely low counts of microorganisms in the water result most probably from the poor supply of energy resources and other substances indispensable for microbial growth.

Table 4 presents the counts of microorganisms in the (dry) sapropel originating from the Nesyt fishpond. The counts are again lower in comparison with the litter. Sapropels represent relatively stable final products of decomposition processes in wetlands. Their colonization by microorganisms is therefore poorer than in the litter. Table 5 summarizes the microbial data discussed, giving the approximate ranges for plant standing dead material, litter, mud, and water in the littoral of the Nesyt fishpond. The greatest differences have been found in the standing dead material, which is most exposed to the variable weather. The fact that higher numbers of organisms were found on the poor medium (PR) than on the rich agar (MPA) is also remarkable.

Similarity and Diversity in Microbial Populations. The degree of similarity between microbial populations colonizing different selective media, can be derived from the diagrams in Figure 1. Plant materials having reached different

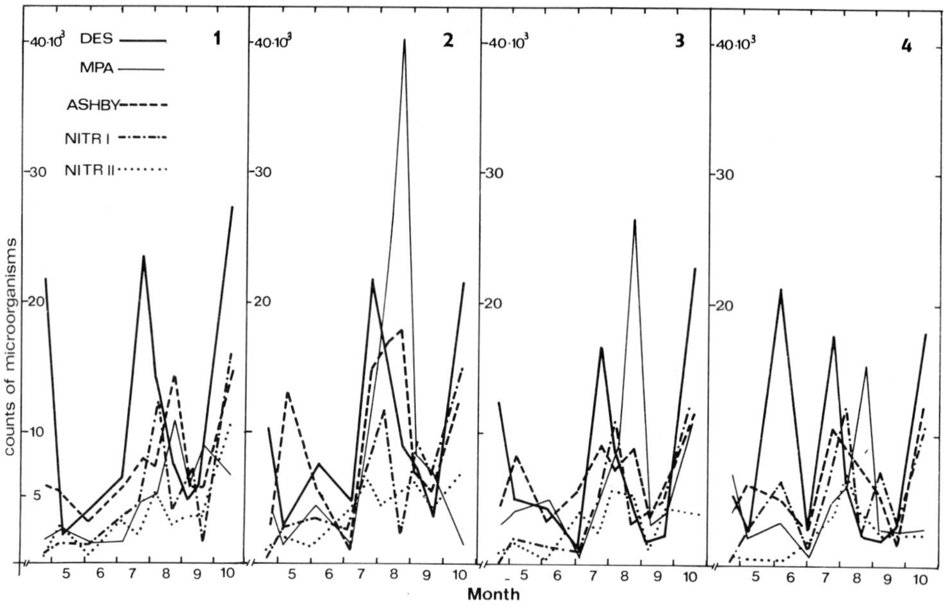

Fig. 3. Seasonal patterns of microbial counts in the water from four vegetational zones of the Nesyt fishpond littoral. *1*, mixed community of *Phragmites communis*, *Carex riparia*, and *Typha angustifolia*; *2*, pure *Phragmites communis* stand; *3*, lagoon without emergent vegetation; *4*, pure *Typha angustifolia* stand

stages of decomposition, i.e., standing dead material, litter and detritus, host rather similar microbial populations in the aquatic environment. The populations differ in materials exposed to terrestrial conditions. Interesting are the differences between the kinds of microflora colonizing the standing dead material of the two principal reed-belt dominants *Phragmites communis* and *Typha angustifolia*. Differences exist not only in the total counts but also in the relative importance of diverse physiological groups, and in species diversity, as revealed by direct observation. Microflora of the standing dead *Phragmites* is much the same in both terrestrial and aquatic habitats. The same is true for *Typha angustifolia*. Beijerink's rule that "the material chooses the microflora" is fully confirmed by these results.

Seasonal Patterns of Microbial Populations. Figure 2 illustrates the seasonal patterns of microbial populations studied in the same materials as described above. The populations are very abundant on standing dead material in spring, and become reduced later during the growing season. Later, only small numbers of microorganisms survive the unfavorable conditions: The microbial populations colonizing the litter and soil or mud exhibit two marked booms—one in spring, and the other in autumn. A depression in counts of all groups of microorganisms is evident in August when the senescence of *Phragmites* shoots becomes apparent (Mason and Bryant, 1975; Husák, 1971, etc.). This month represents an important "turning point" in the life of a littoral ecosystem. During the first half of the growing season, microbial booms take place in the terrestrial soil (in May and early June); and then in the mud (in June and July), where the temperature increases more slowly than in the soil.

Figure 3 presents the seasonal changes in microbial counts in water sampled in the four vegetational zones. The curves document the population dynamics and succesion of different microbial groups. The highest counts of desulfurizing bacteria were found in all the water samples at the end of July. The boom of microorganisms favored by MPA took place only toward the end of August, probably after the onset of senescence of the *Phragmites* and *Typha* shoots. The highest population densities of microorganisms were recorded in the water from the *Phragmitetum*, from where they decreased in either direction of the transect. The seasonal patterns of the aquatic microbial populations are much the same in both *Phragmitetum* and *Typhetum angustifoliae*. Desulfurizing bacteria and nitrifiers show several concurrent peaks.

At different stages of their decomposition, the plant remnants are evidently colonized by microbial populations of different density. The microbial populations display more or less specific kinds of succession and specific seasonal patterns in each kind of material. This is in agreement with the results of other sporadic studies on the microflora in wetlands (Furusaka et al., 1969; Takeda and Furusaka, 1970; Sato and Furusaka, 1972; Day, 1972).

References see pp. 93—95.

1.7 Development of Wetland and Aquatic Vegetation in the Třeboň Basin Since the Late Glacial Period

V. Jankovská

The development of wetland and aquatic plant communities in the South Bohemian pond basins is documented by paleogeobotanical studies of basal layers of mires (Jankovská, 1970). Macroscopic and pollen analyses of peat-bog profiles dated by the C^{14} method (Jankovská, unpublished) show that shallow waters occurred in the Třeboň basin as early as the late glacial period. The reconstruction of past mire communities was followed in two peat-bog profiles situated one in the southeastern part of the basin, near České Velenice (Velanská cesta), some 30 km S.E. of Třeboň, and the other in its northern part near Veselí nad Lužnicí (Švarcenberk)[1].

In the first profile, in a filled small water body the oldest finds, dated back to the older Dryas period (Dryas II, Dryas I), are green planktonic algae such as *Pediastrum, Scenedesmus, Tetraedron, Botryococcus;* the first vascular aquatic species appeared later *(Myriophyllum alterniflorum* and *Sparganium* cf. *minimum).* At the beginning of Alleröd (10,000 B.C.), floating-leaved plants appeared, especially some species of the genus *Potamogeton: Potamogeton* cf. *natans, P.* cf. *gramineus, P.* cf. *obtusifolius, P.* cf. *praelongus, Nuphar pumila* and *Batrachium* sp. This community resembled the present relic subarctic plant association of *Nupharetum pumili* Oberd. 1957. Mixed communities of *Equisetum limosum, Menyanthes trifoliata, Comarum palustre,* and some mosses, e.g., *Scorpidium scorpioides* and *Calliergon giganteum,* growing on the shores, contributed to the filling of the shallow water body. This phase ended between the younger Dryas (Dryas III) and Preboreal periods (about 8150 B.C.). Subsequently, the marsh was overgrown by species of the present *Scheuchzerio-Caricetea fuscae* Nordh. 1936 ex Tx. 1937 and began to be colonized by *Phragmites communis.* The reed stands prevailed here at the beginning of the Boreal period (6800 B.C.), but oligotrophic to dystrophic communities prevailed at its end. At present, the site is overgrown by a forest of *Pinus silvestris* with an undergrowth of *Vacciniaceae.*

The other mire profile is situated at the Švarcenberk fishpond north of the center of the Třeboň basin. The mire has replaced a late glacial shallow water

[1] The mire Velanská cesta appears, according to conclusions of a new paper (Jankovská, unpublished), older than reported in Jankovská (1970). This is documented by an amended pollen diagram (Fig. 1).

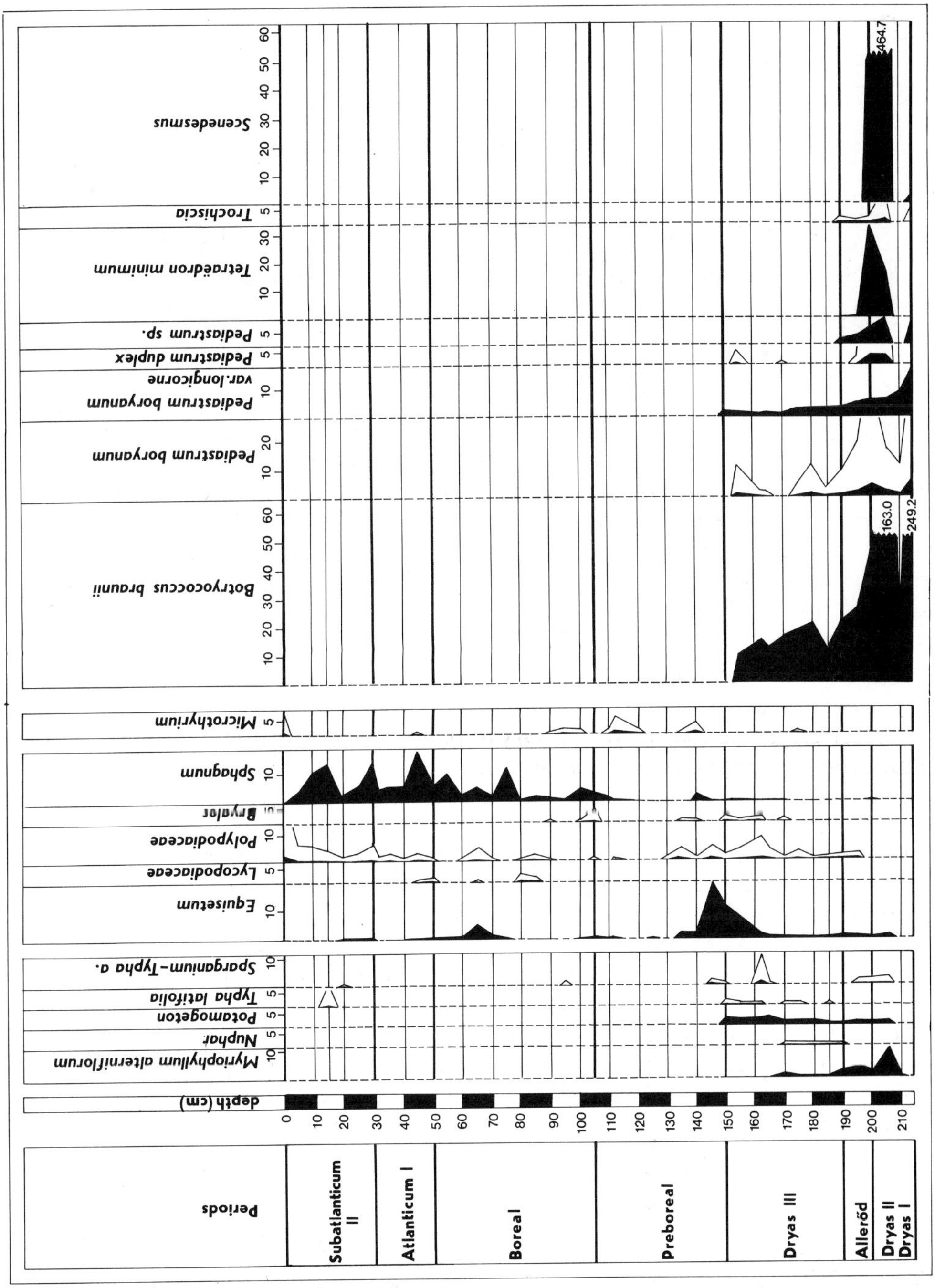

Fig. 1. Pollen diagram of macrophyta and algae in peat-bog profiles of Třeboň basin

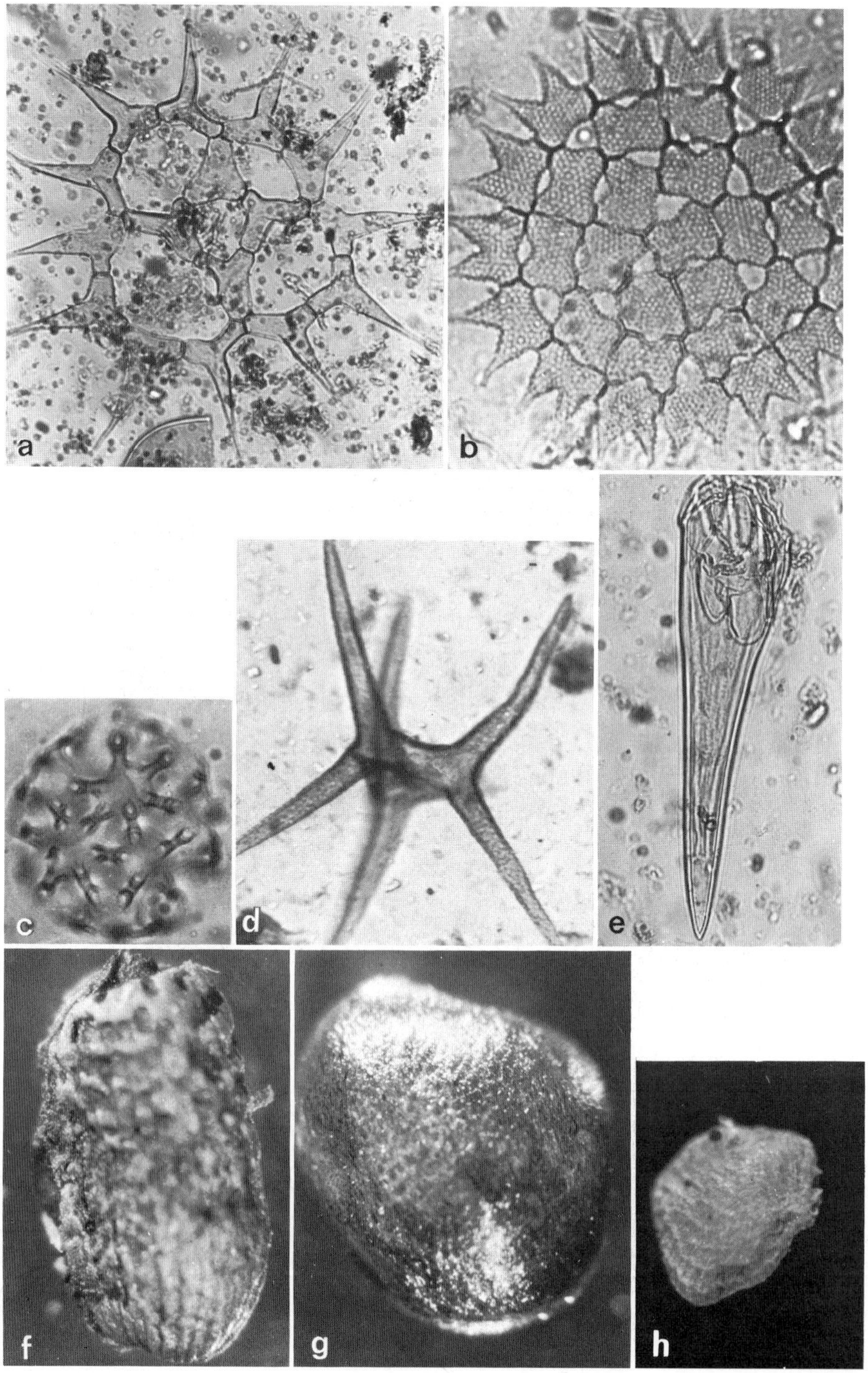

a
b
c
d
e
f
g
h

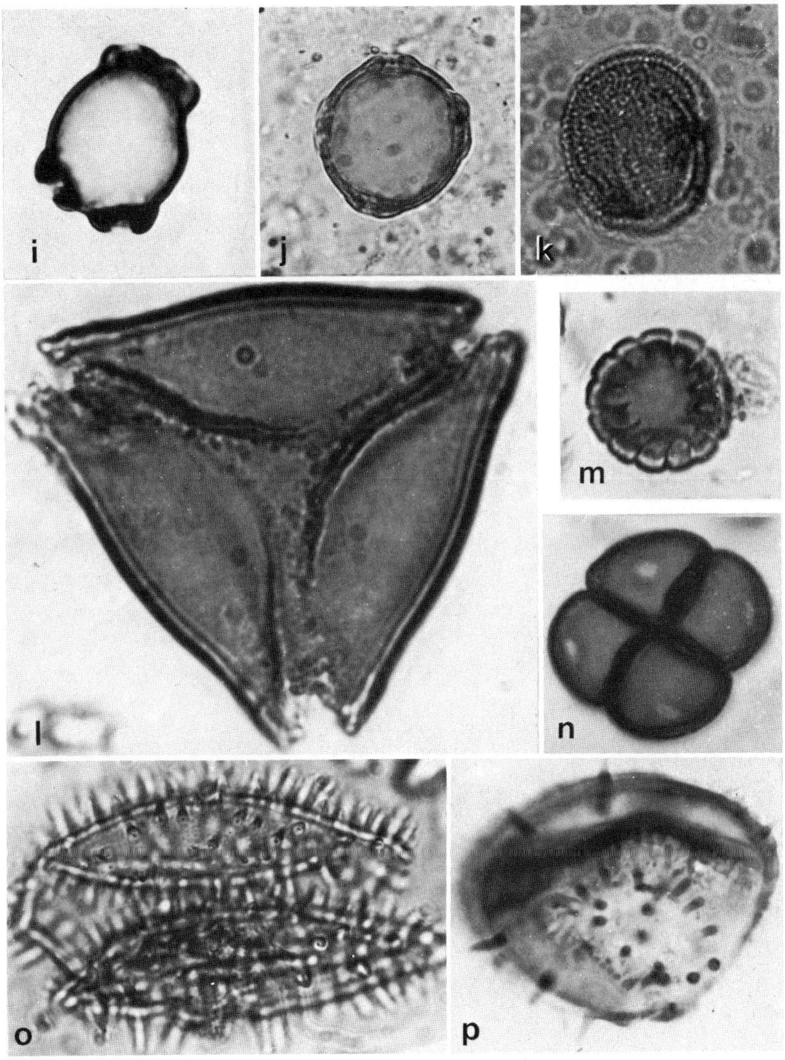

← Fig. 2a–p. Microphotos of algae, trichomes, seeds, and pollen grains of higher plants found in peat profiles in Třeboň basin. I, locality Velanská cesta; II, locality Švarcenberk. (a) *Pediastrum simplex*, Alleröd, depth 370 cm, II, 1500 × ; (b) *Pediastrum boryanum* var. *cornutum*, Boreal, 270 cm, II, 1000 × ; (c) *Coelastrum reticulatum*, Boreal, 250 cm, II, 1000 × ; (d) *Nuphar* cf. *pumila* (trichome, astrosclereid), younger Dryas, 165 cm, I, 500 × ; (e) *Ceratophyllum demersum* (trichome), younger Dryas, 365 cm, II, 500 × ; (f) *Calla palustris* (seed), younger Atlanticum, 130 cm, II, 25 × ; (g) *Menyanthes trifoliata* (seed), younger Atlanticum, 75 cm, II, 25 × ; (h) *Batrachium* sp. (seed), Boreal, 275 cm, II, 25 × . Pollen grains, 1000 × : (i) *Myriophyllum alterniflorum*, younger Dryas, 155 cm, I; (j) *Myriophyllum spicatum*, younger Dryas, 360 cm, II; (k) *Potamogeton* sp., Alleröd, 375 cm, II; (1) *Trapa natans*, Boreal, 250 cm, II; (m) *Utricullaria* sp., younger Atlanticum, 70 cm, II; (n) *Typha latifolia*, Preboreal, 280 cm, II; (o) *Nuphar* cf. *luteum*, Boreal, 270 cm, II; (p) *Nuphar pumila*, younger Dryas, 175 cm, I. (Photo: V. Jankovská)

body. The initial plant community was poor, with predominant algae. The communities of floating-leaved plants appeared later; they were closely related to the present foederationes *Potamion* W. Koch 1926 and *Nymphaeion* Oberd. 1957. The following plants prevailed: *Potamogeton* cf. *natans*, *P.* cf. *obtusifolius*, *P.* cf. *gramineus*, *Batrachium* sp., *Myriophyllum spicatum*, *M. verticillatum*, *Utricularia* sp., *Nymphaea* sp., *Nuphar* cf. *luteum; Trapa natans* also occurred here during the first phases of the Holocene. At the beginning of the Boreal period, plant communities of the foederationes *Phragmition communis* W. Koch 1926 and *Magnocaricion elatae* W. Koch 1926 had filled the depression completely. *Phragmites communis* and *Equisetum limosum* were dominant during the whole Boreal and Early Atlantic periods; *Lythrum salicaria*, *Lycopus europaeus*, *Typha latifolia*, *Scirpus* sp. and several *Carex* species also occurred here. The subsequent invasion by *Alnus glutinosa*, and later by *Picea* was followed, in the Late Atlantic, by an invasion of *Pinus silvestris* while *Alnus* gradually retreated.

From the results of these two as well as of other analyses of peat-bog profiles, it may be concluded that aquatic and wetland communities related to the present foederationes *Potamion* and *Nymphaeion* have been present in the Třeboň basin since the late glacial period. *Typha latifolia* and *Phragmites communis* appeared here probably during the Pre-boreal period and reedswamp communities developed fully during the Boreal period. Alder swamp forest (*Alnetea glutinosae* Br.-Bl. et Tx. 1943) has presumably been present here since the Early Atlantic.

The results of pollen analyses suggest a sporadic occurrence of oligotrophic communities *Oxycocco-Sphagnetea* Br.-Bl. et Tx. 1943 in the Třeboň basin as early as the end of the Pre-boreal period. Full development of these communities, however, begins at the end of the Boreal period and in the Atlantic period. In the Late Atlantic period, the communities of Pino-Ledion Tx. 1955 penetrated into the larger mires. More extensive stands of *Pinus mugo* ssp. *uncinata* var. *rotundata* f. *pyramidata* probably appeared in the Třeboň basin at the end of the Late Atlantic and in the Sub-boreal period.

Similar results were also obtained in the neighboring České Budějovice basin (Rybníčková et al., 1975).

References see pp. 93—95.

References

Aaronson, S.: Experimental Microbial Ecology. New York and London: Academic Press, 1970

Ambrož, J.: Floristic notes to South Bohemian peat bogs. (In Czech) Čas. Nár. Musea, **109**, 30–38 (1935)

Ambrož, J.: Flora of the drained fishpond soils in the Třeboň basin. (In Czech) Sbor. Přírodověd. klubu v Jihlavě, Jihlava, 1–83 (1939)

Březina, P.: Woodlands of the Třeboň Basin, Czechoslovakia. (In Czech). Praha: Rozpravy ČSAV, 85 (10). Academia (1975)

Chábera, S., Šabatová, E.: Survey of the Hydrography of South Bohemia. (In Czech). České Budějovice: Pedagogická čtení 1–72 (1965)

Czudek, T. (ed.): Geomorfologické členění ČSR (Geomorphological division of the Czech Socialist Republic). Studia Geographica (Brno) **23**, 1–137 (1972)

Day, W. J.: Community structure and carbon budget of a salt marsh and shallow bay estuarine system in Louisiana. Center for Wetland Resources Publ. No. LSU-S6-72-04, Baton Rouge, 1973

Demek, J., Balatka, B., Czudek, T., Láznička, Z., Linhart, Y., Loučková, H., Panoš, V., Raušer, J., Seichterová, H., Sládek, J., Stehlík, O., Štelcl, O., Vlček, V.: Geomorphology of Bohemia and Moravia. (In Czech). Praha, 1965

Dostál, J.: The phytogeographical regional distribution of the Czechoslovak flora. Sb. Česk. Spol. zeměp. **65**, 193–202 (1960)

Dvořák, J.: Nitrogen and phosphorus compounds in the water of the littoral of the Nesyt fishpond. In: Littoral of the Nesyt fishpond. Květ. J. (ed.). Studie ČSAV, 1973/15:139–141. Praha Academia, 1973

Furusaka, C., Hattoei, T., Sato, K., Yamagishi, H., Hattori, R., Nioh, I., Nishio, M.: Microbiological, chemical and physico-chemical surveys of the paddy-field soil.—Rep. Inst. Agric. Res. (= Sci. Rep. Res. Insts. D). Tôhoku, Univ. **20**, 89–101 (1969)

Golterman, H. L., Clymo, R. S.: Methods for Chemical Analysis of Fresh Waters. IBP Handbook No. 8. Oxford: Blackwell, 1969

Hejný, S.: Ein Beitrag zur ökologischen Gliederung der Makrophyten der tschechoslowakischen Niederungsgewässer. Preslia **29**, 349–368 (1957)

Hejný, S.: Ökologische Charakteristik der Wasser- und Sumpfpflanzen in den Slowakischen Tiefebenen (Donau- und Theissgebiet). Bratislava, 1960

Hejný, S.: Über die Bedeutung der Schwankungen des Wasserspiegels für die Charakteristik der Makrophytengesellschaften in den mitteleuropäischen Gewässern. Preslia **34**, 359–367 (1962)

Hejný, S.: The dynamic characteristics of littoral vegetation with respect to changes of water level. Hidrobiologia **12**, 71–85 (1971)

Hejný, S. (ed.): Ecosystem Study on Wetland Biome in Czechoslovakia. Czechosl. IBP/PT-PP Rep. No. **3**, Třeboň, 1973a

Hejný, S.: Plant communities and their typical features in the littoral of Opatovický fishpond. In: Ecosystem Study on Wetland Biome in Czechoslovakia. Hejný, S. (ed.): Czechosl. IBP/PT-PP Rep. No. **3**, Třeboň, 1973b, pp. 29–37

Heteša, J., Hudec, K., Husák, Š.: Lednice fishponds and their investigation. In: Littoral of the Nesyt Fishpond. Květ, J. (ed.). Studie ČSAV 15: 11–15. Praha: Academia 1973.

Holub,J., Hejný,S., Moravec,J., Neuhäusl,R.: Übersicht der höheren Vegetationseinheiten der Tschechoslowakei. Rozpravy ČSAV **77**, (3), 1–75 (1967)

Hudec,K.: Density and breeding of birds in the reedswamps of Southern Moravian ponds. Acta Sci. Nat. (Brno) **6**, 1–41 (1975)

Husák,Š.: Productivity and structure of intact and cut stands of *Phragmites communis* and *Typha angustifolia* in the Nesyt fishpond. (In Czech). Thesis, Purkyně University, Brno, 1971

Jankovská,V.: Ergebnisse der Pollen- und Großrestanalyse des Moors "Velanská cesta" in Südböhmen. Folia Geobot. Phytotaxon. Bohemoslov. **5**, 43–60 (1970)

Jeník,J.: Geobotanical map of the Třeboň region: second approximation. (In Czech). Quaest. Geobiologicae **14**, 7–32 (1974)

Jeník,J., Větvička,V.: Ecology and structure in stands of *Salix* spp. in the Třeboň basin. In: Ecosystem Study on Wetland Bicome in Czechoslovakia. Hejný,S. (ed.). Czechosl. IBP/PT-PP Rep. No. **3**, Třeboň 1973, pp. 39–46

Komárek,J.: Seasonal changes in the algal microflora of Opatovický fishpond (South Bohemia). In: Ecosystem Study on Wetland Biome in Czechoslovakia. Hejný,S. (ed.). Czechosl. IBP/PT-PP Rep. No. **3**, Třeboň, 1973, pp. 185–196

Komárek,J., Ettl,H., Marvan,P.: A review of algae in Opatovický fishpond (South Bohemia) in 1971–1972. In: Ecosystem Study of Wetland Biome in Czechoslovakia. Hejný,S. (ed.). Czechosl. IBP/PT-PP Rep. No. **3**, Třeboň, 1973, pp. 175–178

Komárková,J., Komárek,J.: Comparison of pelagial and littoral primary production in a South Bohemian fishpond/Czechoslovakia/. In: Limnology of Shallow Waters. Salánki,J. and Ponyi,J. E. (ed.). Symposia Biol. Hung. **15**, Akad. Kiadó, Budapest, 1975, 77–95

Kořínková,J.: Quantitative relations between submerged macrophytes and populations of invertebrates in carp ponds. Hidrobiologia **12**, 377–382 (1971)

Květ,J., Hudec,K.: Effects of grazing by grey-lag geese on reedswamp plant communities. Hidrobiologia **12**, 15–40 (1971)

Lange,L.de: Gibbosity in the complex *Lemna gibba*—*Lemna minor*: literature survey and ecological aspects./Aquatic Botany **1**, 327–332 (1975)

Losos,B., Heteša,J.: Hydrobiological studies on the Lednické rybníky ponds. Acta Sci. Nat. Brno **5**, (10) 1–54 (1971)

Luther,H., Rzóska,J.: Project Aqua. IBP Handbook No 21. Oxford: Blackwell, 1971

Marvan,P., Ettl,H.: Algae of the Nesyt Fishpond. Praha: Academia, in press

Mason,C.F., Bryant,R.J.: Production, nutrient content and decomposition of *Phragmites communis* Trin. and *Typha angustifolia* L.J. Ecol. **63**, 71–95 (1975)

Moravec,J.: A simple method for estimating homotoneity of sets of phytosociological relevés. Folia Geobot. Phytotaxon. Bohemoslov. **6**, 141–170 (1971)

Nekvasilová,H.: Vegetation of the Řežabinec Fishpond Near Ražice. (In Czech). Thesis, Charles University, Praha, 1973

Novotný,J.: Die Teiche in Südböhmen (In Czech). Sb. Česk. Spol. zeměp. **77**, 37–51 (1972)

Oberdorfer,E.: Süddeutsche Pflanzengesellschaften. Jena, 1957

Obrtel,R.: Soil surface *Coleoptera* in a reedswamp. Acta Sci. Nat. (Brno) **6** (6), 1–35 (1972)

Odum,E.: Fundamentals of Ecology. Philadelphia: W.B.Saunders, 1971

Pelikán,J.: Mammals of Nesyt fishpond, their ecology and production. Acta Sci. Nat. (Brno) **9**, (12) 1–45 (1975)

Pelíšek,J.: Principal types of soil-moisture dynamics in alluvial regions of Czechoslovakia. (In Czech). In: Vegetační problémy při budování vodních děl. Ježdník,T. (ed.). Praha: Academia, 1964

Pugh,G.J.F.: Leaf litter fungi found on *Carex paniculata* L. Trans. Br. Mycol. Soc. **41**, 185–196 (1958)

Pugh,G.J.F., Williams,G.M.: Fungi associated with *Salsola kali*. Trans. Br. Mycol. Soc. **51**, 389–396 (1968)

Pugh,G.J.F., Mulder,J.L.: Mycoflora associated with *Typha latifolia*. Trans. Br. Mycol. Soc. **57**, 273–282 (1971)

Putschögl,V.: Role of macrophyte plant communities in the management of the Nesyt fishpond. In: Littoral of the Nesyt fishpond. Květ,J. (ed.). Studie ČSAV 15: 153–155, Praha: Academia, 1973.

Řezníček,J.: The populations of waterfowl on the Opatovický fishpond in the summer periods of 1970, 1971 and 1972. In: Ecosystem Study on Wetland Biome in Czechoslovakia. Hejný,S. (ed.). Czechosl. IBP/PT-PP Rep. No. **3**, Třeboň, 1973, 221–223

Rodina,A.G.: Methods of Water Microbiology. (In Russian). Moscow, Leningrad: Nauka, 1965

Rybníček,K.: A comparison of the present and past mire communities of Central Europe. In: Quaternary Plant Ecology. Birks,B.J.H., West,R.G. (eds.). 14th Symposium of the British Ecological Society. Oxford: Blackwell, 1973

Sato,K., Furusaka,C.: Bacteriological studies on the percolated soils. Rep. Inst. Agric. Res. (= Si. Rep. Res. Insts. Series D) Tôhoku Univ. **23**, 1–15, 1972

Spirhanzl,J., Káš,V.: Agronomical and Pedological Study of Soils Formed on South Bohemian Terciary Sediments. (In Czech). Sb. výzk Úst. Zeměd. ČSR **147**, 1–114 (1936)

Šťastný,K.: The quantity of birds frequenting the reed-beds of the Opatovický pond. In: Ecosystem Study on Wetland Biome in Czechoslovakia. Hejný,S. (ed.). Czechosl. IBP/PT-PP, Rep. No. **3**, Třeboň, 1973, pp. 225–229

Stehlík,O.: Contribution aux méthodes de l'investigation de l'erosion du sol. Travaux du Symposium international de geomorphologie appliqueé, Bucarest, Mai 1967. Bucarest, 1969

Straškraba,M., Kořínková,J., Poštolková,M.: Contribution to the productivity of the littoral region of ponds and pools. Rozpr. Česk. Akad. Věd., Řada Mat. Přír. Věd. **77**, 1–80 (1967)

Takeda,K., Furusaka,C.: On the bacteria isolated anaerobically from paddy-field soil succession of facultative anaerobes and strict anaerobes. Soil Sci. **44**, 343–348 (1970)

Taligoola,H.K.: Studies in the colonization of *Phragmites communis* Trin. by microfungi. Ph. D. Thesis, University of Nottingham, 1969

Zapletal,K.: The Present State of the Geology of the Carpathians and Depressions in Moravia and Silesia. (In Czech) Brno, 1950

Fig. I. Resistance thermometers installed in a littoral *Phragmites* stand. (Photo: J.P. Ondok)

Section 2
Environmental Factors in Fishpond Littorals

After setting a macroclimatological framework—with due attention to the radiation climate—for our investigations, this section characterizes the peculiarities of the microclimate in various fishpond habitats and plant communities, with emphasis on the littoral reed-belt communities. Their rather high primary production is assumed to result from an efficient interception and use of the incoming radiant energy. The radiation microclimate of the reedswamp plant communities thus receives special attention. A sufficient supply of mineral nutrients from the water, soil and bottom sediments is also essential for attaining a high level of primary production. By taking the example of the two specimen fishponds, the basic characteristics and dynamics of mineral nutrient content are shown for the water, as well as bottom sediments and soils in various fishpond littoral habitats, particularly within the reed belt (Fig. I).

2.1 Climatic Conditions

2.1.1 Fundamental Climatological Characteristics

K. PŘIBÁŇ and P. ŠMÍD

The climate of the Třeboň basin is influenced by the adjacent Šumava highlands, by the flat surface relief of the basin itself, and by the thermal properties of the vegetation cover alternating with water bodies (Švec et al., 1967, 1969). In the climatic division of Czechoslovakia, the Třeboň region is classified as moderately warm (Quitt, 1971). Despite the oceanic character of the basin, which was originally a wetland, the distribution of precipitation is typical of Central Europe with abundant rainfall in summer (see the climatic diagram in Fig. 1 and Table 1). The air temperatures are positively affected by the heat capacity of the larger fishponds, whose water accumulates heat. Temperature inversions occur fairly frequently, being a product of cloudless nights combined with lack of wind in the rather flat landscape configuration of the Třeboň basin. The subalpine föhn blowing from the southwest sometimes affects the Třeboň region as well. Local thunderstorms are frequent in summer.

In the region of Lednice the local subcontinental and relatively warm and dry climate is reflected in a wider amplitude of both temperature and rainfall during a year (see climate diagram in Fig. 1 and Table 1). In summer, rainfall frequently does not compensate for evaporation. Easterly wind directions prevail. The average annual distribution of precipitation shows two maxima: June–July and October–November. The intense evaporation from the South Moravian alluvia, including the Lednice fishponds, results in a partly independent local water cycling with frequent local thunderstorms and rains (Quitt, 1971; Husák, 1971).

In both regions the IBP investigations of fishpond littoral ecosystems were accompanied by measurements of both the local macroclimate and of the microclimate of the littoral vegetation stands. This section presents the fundamental macroclimatic characteristics. Results of the micrometeorological measurements are reported in Section 2.1.2. The measurements were started in 1964 and continued until 1974. A meteorological station was set up 2 km from the Opatovický fishpond in South Bohemia. Data from the meteorological station 6 km from Lednice characterize the macroclimate at the Nesyt fishpond in South Moravia. At Opatovický, the air temperature and humidity were registered with thermohygrographs during the whole year, and the incident global radiation was measured

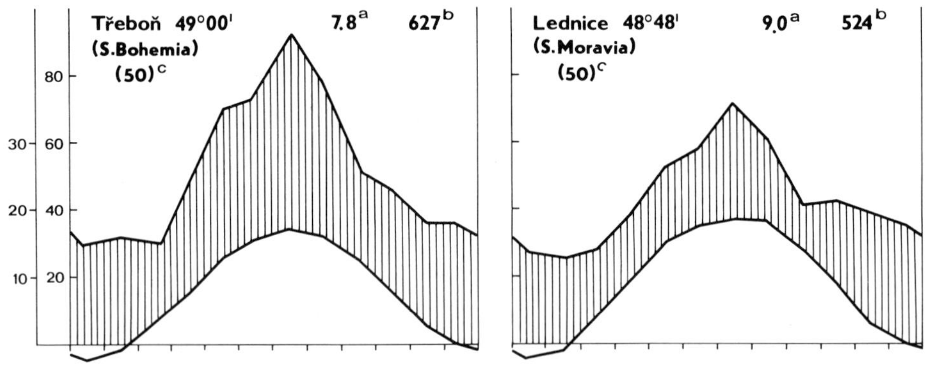

Fig. 1. Climatic diagrams of Třeboň and Lednice. *Abscissa:* Mean monthly temperature (*upper line*, °C). and precipitation (*lower line*, mm); [a] Annual mean temperature (°C). [b] Total precipitation (mm). [c] Years of observation

Table 1. Location and climatic characteristics (means of 1901–50) of Třeboň and Lednice

	Třeboň	Lednice
Latitude	49° 00′ N	48° 48′ N
Longitude	14° 46′ E	16° 48′ E
Altitude (m)	433	164
Type of climate	subatlantic	subcontinental
Growing season	April to September	April to October 10
Average temperatures (°C)		
annual	7.8	9.0
April to September	14.0	15.4
coldest month (January)	− 2.2	− 1.7
warmest month (July)	17.7	19.0
Average precipitation totals (mm)		
annual	627	524
April to September	415	323
wettest month (July)	94	70
driest month (January)	30	26
Average relative air humidity (%)		
annual	79	—
April to September	74	—
Mean annual sunshine duration (h) (1965–1969)	1,641	1,713

with Kipp and Zonen solarimeters during each growing season from 1964 to 1974. At Lednice, a Bellani pyranometer was installed for this whole period, while a Kipp and Zonen solarimeter was in use from 1967 to 1974. The missing solarimeter data were calculated from the pyranometer readings using a regression equation constructed by Palát (unpublished). The measurements were completed by data recorded at the standard stations of the Hydrometeorological Institute at Třeboň and Lednice. The long-term data of 1901–1950 on air temperature and precipitation were also obtained at these stations. For more details see Šmíd

Table 2. Maxima, minima and mean air temperatures (°C) per month, growing season, and year from 1964 to 1974

	Jan.	Feb.	March	April	May	June	July	Aug.	Sept.	Oct.	Nov.	Dec.	Apr.–Sept.	Year
Třeboň: Opatovický fishpond														
Mean	−3.8	−1.1	2.1	7.5	12.4	15.8	17.2	16.3	12.1	7.2	2.9	−0.6	13.5	7.4
Max.	0.1	1.0	9.4	4.6	15.1	17.0	18.7	18.2	13.4	11.7	5.4	3.3	14.3	8.1
Min.	−6.0	−4.5	−1.0	6.1	10.9	13.9	15.8	15.0	9.3	3.8	0.3	−5.0	12.7	6.8
Lednice: meteorological station														
Mean	−2.3	1.7	4.2	9.8	14.4	17.6	19.1	18.2	14.2	8.7	4.2	0.1	15.6	9.2
Max.	1.5	4.9	7.9	11.4	17.1	19.6	20.8	20.7	15.6	13.0	6.6	3.3	16.1	9.8
Min.	−5.4	−1.6	1.2	7.9	12.4	15.6	17.6	16.4	12.2	6.1	1.1	−4.2	14.5	8.4

Table 3. Absolute minima and maxima of air temperature (°C) recorded at Třeboň and Lednice during the growing season from 1964 to 1974 and means for all 11 years

	April		May		June		July		August		September	
	min.	max.	min.	max.	min.	max.	min.	max.	min.	max.	min.	max.
Třeboň: Opatovický fishpond												
Absolute values	−6.2	22.3	−3.7	31.9	0.5	33.2	3.0	33.8	2.0	35.3	−3.7	31.2
Mean	−4.4	17.3	0.0	26.0	4.2	28.9	5.5	30.6	4.4	31.5	−0.1	27.1
Lednice: meteorological station												
Absolute values	−4.4	27.9	−1.5	30.2	3.8	32.2	5.6	35.2	5.0	34.6	−3.3	30.0
Mean	−2.3	23.2	2.0	27.1	6.4	30.2	7.8	32.1	6.9	31.6	0.8	27.0

Table 4. Monthly means of incident global radiation at Třeboň and Lednice from 1964 to 1974 (MJ cm^{-2})

	April	May	June	July	August	September
Třeboň						
Mean	381	491	526	551	483	293
Max.	501	567	601	676	618	401
Min.	262	421	460	488	399	232
Lednice						
Mean	419	532	550	555	487	340
Max.	493	619	654	671	553	427
Min.	279	430	435	434	460	287

Average daily sums during the growing season as defined for each year and station above (MJ $cm^{-2} d^{-1}$):

Třeboň: Opatovický fishpond										
1964	1965	1966	1967	1968	1969	1970	1971	1972	1973	1974
18.1	14.6	14.0	16.5	14.5	17.7	13.9	15.7	15.0	17.2	15.0
Lednice: meteorological station[a]										
—	—	—	17.3	17.3	16.6	15.5	16.1	13.7	15.5	13.6

[a] For 1964 to 1966, the mean daily sums could not be calculated from the Bellani pyranometer readings.

(1973) and Přibáň (1973a). The averaged long-term data are summarized in the climate diagrams (after Walter and Lieth, 1960) given in Figure 1, and in Table 1.

The long-term average of air temperature in the Lednice region is higher by 1.2° C. The maxima of monthly mean air temperatures are also higher and the ranges of their variation in individual years are greater at Lednice than at Třeboň. The climate of the Třeboň basin thus appears relatively more oceanic than that of the Lednice region. The subatlantic character of the climate in the Třeboň basin is also evident from the greater amounts of precipitation.

The results of ten years' measurements (1964–1974) specify the temperature conditions at the Opatovický and Nesyt fishponds during our IBP investigations; the averaged data are summarized in Table 2.

During the growing season the monthly mean temperatures naturally vary for each month in individual years. At Opatovický the maximum difference was 5° C, the lowest monthly mean air temperature during the growing season (April to September) was recorded in April, 1970 (6.1° C), and the highest was in July, 1967 (18.7° C). At Nesyt, the maximum difference was 6.5° C. The lowest monthly mean air temperature recorded was in April, 1965 (8.4° C) and the highest was also in July, 1967 (20.8° C). These and other data from Table 2 illustrate the warmer character of the growing season in the Lednice region as compared with the Třeboň basin. The absolute monthly minima and maxima of air temperatures during the growing season differ less, as they usually occur during meteorological situations affecting large areas in Central Europe. The absolute temperature minima and maxima recorded at Třeboň and Lednice in individual months during

Table 5. Average dates of the onset and termination of characteristic daily mean air temperatures and their average duration (in days) at Třeboň and Lednice. Average data of 1901–50

Daily mean temperatures higher than	0° C			5° C			10° C			15° C		
	from	to	duration	from	to	duration	from	to	duration	from	to	duration
Třeboň: Opathovický fishpond	21 Feb.	8 Dec.	291	29 March	1 Nov.	218	29 Apr.	2 Oct.	157	5 June	30 Aug.	87
Lednice: meterological station	17 Feb.	15 Dec.	302	20 March	8 Nov.	234	19 Apr.	10 Oct.	175	20 May	8 Sept.	112

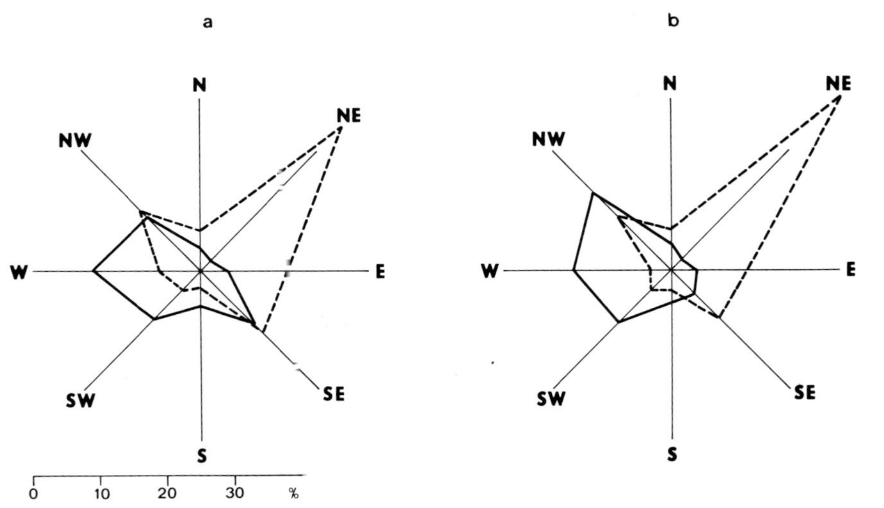

Fig. 2 a and b. Average distribution of wind directions: percentage of all observations at Třeboň (—) and at Lednice (---) for the whole year (a) and for June to August (b)

the ten years of IBP investigations are given in Table 3. In both the Třeboň basin and Lednice region night frosts occur practically every year in April; in May and September, they are relatively more frequent in the Třeboň basin. On the other hand, day temperatures over 20° C occur in the Lednice region every year in April, while they do rarely in the Třeboň basin. During summer, the main difference between the two regions is in the night temperatures, which tend to be lower in the Třeboň basin. Table 4 gives data characterizing the input of solar radiation at Třeboň and Lednice. It is evidently somewhat greater in the Lednice region.

Table 5 provides additional data illustrating the average course of temperatures during a year and growing season in either region. The more favorable conditions and the longer effective growing season at Lednice are again evident.

Data on the distribution of wind directions in the two regions are presented in Figure 2. Westerly winds clearly prevail in the suboceanic Třeboň basin but they retreat in favor of easterly winds (particularly from the northeast and southeast) in the subconinental Lednice region. Calm days occur more frequently in the Třeboň basin: 37.6% and 39.3% of all days during the year and in June to August respectively—as compared with 16.4% and 14.8% at Lednice.

References see pp. 153—155.

2.1.2 Microclimate in Fishpond Littoral Ecosystems

P. ŠMÍD and K. PŘIBÁŇ

The reed-belt is a complicated system where three physical phases are usually present, i.e., the solid (bottom soil and plant mass), liquid (water) and gaseous (air) media. According to the position of the water level, three basic ecophases may be distinguished in the littoral zone (see Sect. 1.2) with different vegetation as well as microclimatic conditions.

The energy budget approach represents the most suitable method of describing the microclimatic conditions of any active surface. It is based on the equation:

$$R_n = G + H + LE$$

with R_n for net radiation flux, G for ground heat flux, H and LE for the sensible and latent heat fluxes into the atmosphere respectively (L is the latent heat of vaporization of water—about 590 cal g^{-1}—and E is the rate of evaporation). A great disadvantage of the direct energy-balance measurements is the requirement for a large homogeneous area with an active surface of one type; moreover, rather elaborate instruments are needed. For these two reasons, much simpler techniques were used for microclimatic measurements in the IBP investigations of fishpond littorals. Nevertheless, some valuable results have been obtained, providing at least the possibility of estimating the components of the heat budget for the most important types of fishpond littoral vegetation dominated by helophytes.

Table 1. Ground temperatures in various zones of the Nesyt fishpond. (Data by Šmíd)

Date	Zone and depth	Temperature °C						Energy used to heat ground	
		At water surface			At −0.2 m depth				
		max.	min.	range	max.	min.	range	MJ m^{-2}	kcal m^{-2}
2. Aug. 1975	Open water −0.6 m	24.1	21.5	2.6	24.0	21.5	2.5	6.3	1,505.7
30. Aug. 1972	Open water −1.0 m	18.5	16.9	1.6	18.2	16.6	1.6	6.7	1,601.3
30. Aug. 1972	Littoral *Typhetum angustifolia*, −0.6 m	18.4	16.3	2.1	18.2	16.4	1.8	5.0	1,195.0
30. May 1973	Littoral *Phragmitetum*, −0.3 m	19.0	14.5	4.5	15.6	14.3	1.3	3.9	932.1
30. May 1973	Limosal *Phragmitetum* with *Carex riparia* undergrowth	16.3	11.7	4.6	11.1	10.6	0.5	1.3	310.7

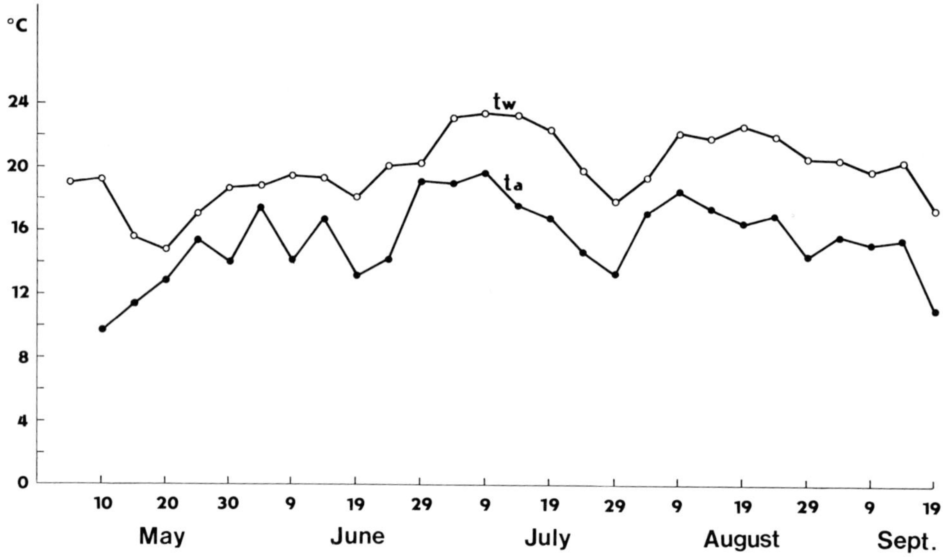

Fig. 1. Five days' averages of water temperature at –0.2 m *(tw)* in the Opatovický fishpond and of air temperature *(ta)* measured in a nearby standard meteorological screen. (After Přibáň, 1973b)

Table 2. Spring temperature conditions in a fishpond littoral (Hlohovecký fishpond near Lednice, 15 April 1971. (After Šmíd, 1973)

Habitat		Temperature °C		
		min.	max.	av.
Air (at 2 m height) above the mown meadow		−0.3	11.9	6.3
Water (−0.1 m depth) in *Phragmites* stand:	cut:	6.8	19.3	12.7
	uncut:	5.0	10.4	7.9

Continuous measurements of air, water, and soil temperatures and relative air humidity were taken with platinum resistance thermometers, (Fig. 1 in Introduction to this Sect.) pairs of the sensors, wet and dry, for air-humidity measurements, attached to automatic six-point data recorders. Short-term measurements of the same parameters were also taken with mercury-in-glass thermometers and Assmann-type ventilated psychrometers respectively. The accuracy of the two kinds of temperature measurements was 0.5 and 0.1° C respectively. Thermohygrographs exposed in standard meteorological screens provided a rough check on the air-temperature and humidity data, where required.

Microclimate of Shallow Water. Some basic remarks on the thermal properties of shallow open water are necessary for understanding the microclimatic conditions in helophyte stands. The following are the main characteristics of water (as compared with soil): (1) great heat capacity; (2) small reflexivity; the albedo of

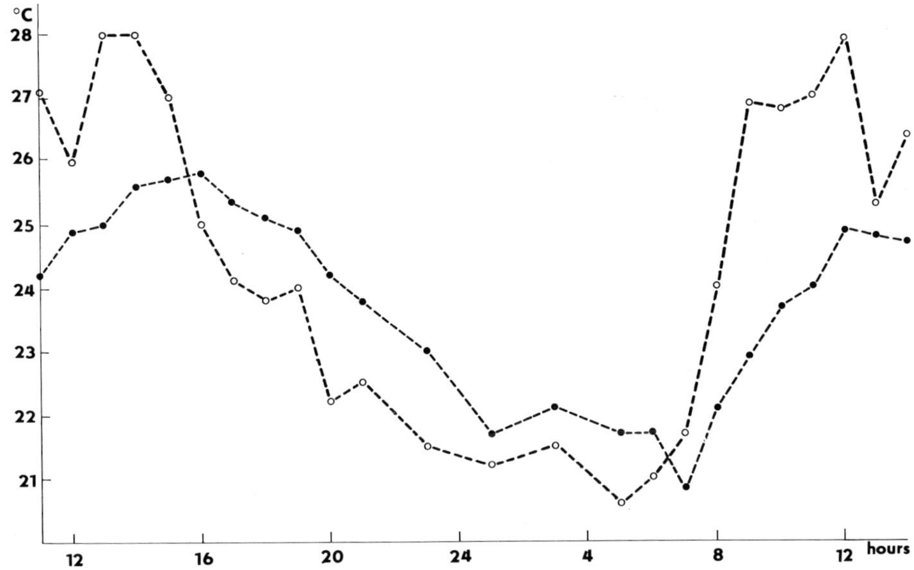

Fig. 2. Daily course of water temperature in a duckweed community (○--○) *(Lemnetum gibbae)* and in open water (●--●) of the Podzámecký fishpond, Central Bohemia, on a bright and cloudless summer day. (After Rejmánková, 1973)

water strongly depends on the solar angle, and is only about 10% at a sun elevation of 40°: this is less than the albedo of other natural surfaces; (3) penetration of incident radiation deep into the water.

All these factors result in an increased ground heat flux in shallow waters (see Table 1), which leads to a rapid increase in water temperature in spring (Fig. 1). This feature is probably strengthened by small heat losses by night eradiation, which tend to be prevented by an increased air humidity and fog formation under stable atmospheric conditions. The increased water temperature allows a rapid development of algal vegetation in spring, mainly in the lagoons (patches of open water within the reed belt) as well as in littoral reed stands where the standing dead stems have been removed by winter cutting (Table 2).

Thermal stratification usually does not develop in relatively large bodies of shallow water where the mixing effect of wind and waves in considerable (Fig. 3). In small and sheltered shallow waters, quite marked thermal stratification may develop (Geiger, 1961; Martin, 1972).

Submerged vegetation of varying density and/or a cover of floating higher plants and filamentous algae are often present in these shallow waters. Microclimatic conditions in such habitats were studied by Úlehlová (1970) in the Netherlands, and by Rejmánková (1973) in South Moravia. These investigations have confirmed that the surface layers of water may become overheated and a steeper temperature gradient may occur below at daytime. This is indirect evidence that the submerged vegetation reduces the ground heat flux. These facts, as well as a few direct measurements of evaporative water loss (Gavenčiak, 1972), indicate

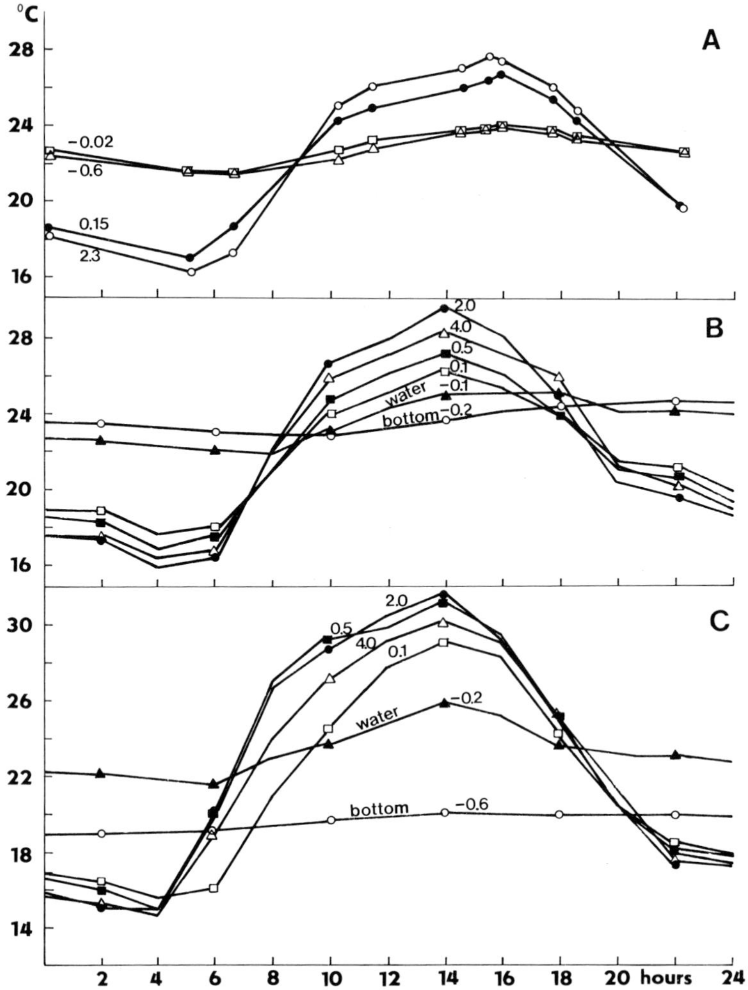

Fig. 3 A–E. Daily course of water or soil and air temperature in various zones of fishpond littorals. (A) Open water, depth –0.6 m, Nesyt, 2 Aug. 1973; (B) littoral *Phragmitetum communis*, E. shore of Opatovický. (C) *Typhetum angustifoliae*, W. shore of Opatovický, both measurements on 21 July 1972; (D) stand of *Eleocharis acicularis* on bare bottom of Sádky pond near Třeboň, 10. July 1974 (Velásquez, 1975); (E) terrestrial stand of *Phragmitetum communis* in spring, Nesyt, 21 March 1974. t_a and t_s are temperatures of the air and soil surface, resp. Numbers at the lines indicate heights of measuring sensors in *m*

that the latent heat flux and evaporation are greater in aquatic vegetation than in open water.

The great heat capacity of water also considerably affects the thermal stratification in the lower air layers. In contrast to a solid surface (soil or plant canopy), a conspicuous temperature inversion occurs above the water surface at daytime (mainly before noon), but unstable stratification (i.e. negative temperature gradients = less than adiabatic = with temperature decreasing with height faster

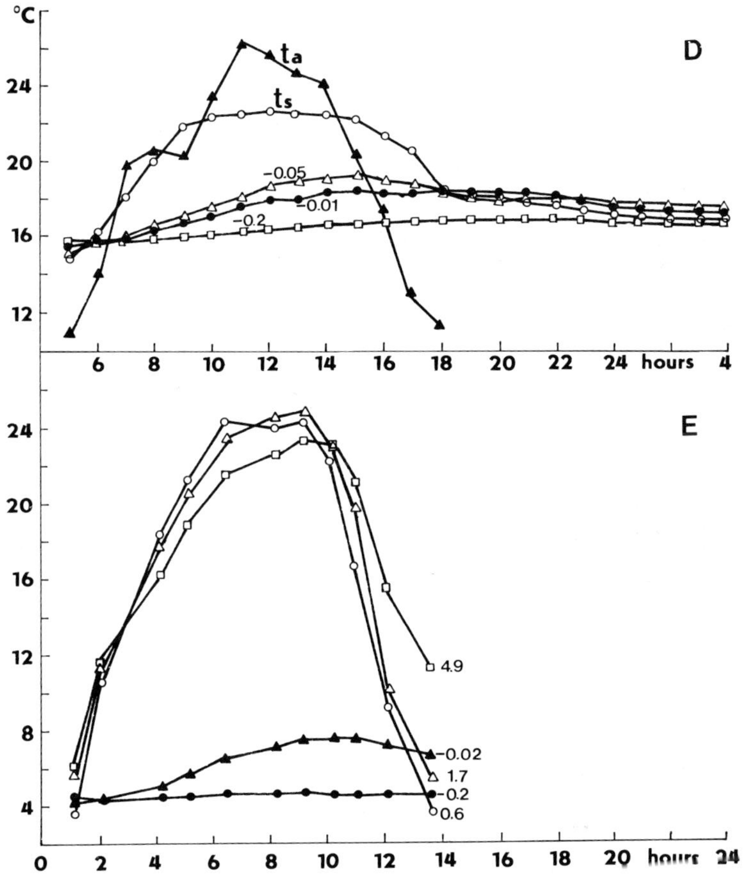

than by 0.01 K m^{-1}) develops at night (Fig. 3), and the water-vapor pressure gradient also remains negative. This feature causes considerable evaporation from the open water surface at night, but due to the generally stable air stratification in the lowest several hundred meters of air, this usually only leads to the formation of fog or heavy dew on the shore or in the littoral stands. Temperature differences between the pond and shore may cause local winds (Szumiec, 1973).

Table 3 illustrates a typical zonation of temperature conditions in a naturally vegetated fishpond at the peak of the growing season. The temperature-modifying effect of the floating vegetation depends on its density and/or on the size of the leaves. (The *Trapa* colony was small and loose.) The overheating of the air within the helophyte stands at midday is evident and so are the milder air temperatures. Most extreme is the microclimate in the outer littoral swamps.

Microclimate in Fishpond Helophyte Communities. The main factor influencing the microclimatic conditions in these communities is the presence of large amounts (usually about 1–3 kg m^{-2}) of living and/or dead plant mass, with the winter-cut stands in spring forming an exception. A thick layer of litter usually

Table 3. Differentiation of air and water temperatures (°C) in a natural zonation of aquatic and littoral vegetation on a sunny day and cloudness night in summer; Záhorský fishpond, South Bohemia, 8–9 July 1964. (Data by J. Květ)

	Air at 2 m[a]			Air at 0.1 m			Water at −0.1 m		
	max.	min.	range	max.	min.	range	max.	min.	range
Open water	25.6	10.7	14.9	25.5	12.2	13.3	21.7	19.6	2.1
Floating vegetation dominated by:									
Potamogeton natans	23.5	9.8	13.7	25.2	10.9	14.3	22.1	18.3	3.8
Nymphaea candida	24.6	10.7	13.9	26.3	11.9	14.4	22.1	18.9	3.2
Trapa natans	24.8	11.7	13.1	24.3	12.2	12.1	22.7	19.5	3.2
Helophyte vegetation, monospecific stands of:									
Schoenoplectus lacustris	24.5	10.6	14.1	22.7	11.9	10.9	21.7	19.5	2.2
Typha angustifolia	26.5	9.9	16.6	25.1	11.7	13.4	22.2	18.5	3.7
Littoral swamp communities:									
Caricetum elatae	27.3	10.2	17.1	26.9	9.3	17.6	24.7	12.9	11.8
Caricion fuscae	26.9	9.9	17.0	26.8	8.1	18.7	26.7	12.2	14.5

[a] All levels of measurement related to actual water level.

insulates thermally the deeper-situated soil of bottom. This insulating effect is more pronounced in the terrestrial and limosal ecophases, when the litter is not flooded and is only partly saturated with water. A dense undergrowth of sedges and grasses provides additional insulation. In early spring, ice was found to persist in limosal stands of *Phragmitetum communis* with *Carex riparia* undergrowth for a longer time than anywhere else in the whole Nesyt fishpond. The relatively lowest bottom temperatures have been observed here for most of the year (Fig. 4). More favorable water and bottom temperatures occur in the *Phragmitetum communis* or *Typheta* in littoral ecophase with no dense undergrowth, with the litter layer water-saturated and thus possessing a greater thermal conductivity, and—mainly in deeper parts of the littoral usually occupied by *Typhetum angustifoliae*—with the inner littoral water mixing considerably with that from the open central part of the fishpond. Despite this mixing, the reduction of incoming radiation by a fully developed canopy of the helophyte vegetation (LAI up to 5–7, see Sect. 2.2) usually makes the water cool in the reed belt in summer. These cool conditions will reduce the rate of organic matter decomposition. In spring, the littoral microclimate appears to be relatively the most favorable of the whole fishpond (Fig. 4). The increased water and bottom temperatures promote rapid plant development. Winter cutting of the reeds even improves this favorable situation by removing the shading and insulating dead material (Table 2). After winter cutting, the new reed shoots usually grow well and uniformly. The absence of dead shoots, however, also increases the risk of frost damage to the developing young shoots (Květ at Nesyt and Van der Toorn and Mook, the Netherlands, personal communications).

The emerged bottoms of drained ponds, where the shading and insulating effect of the canopies of plant communities is usually much smaller than in

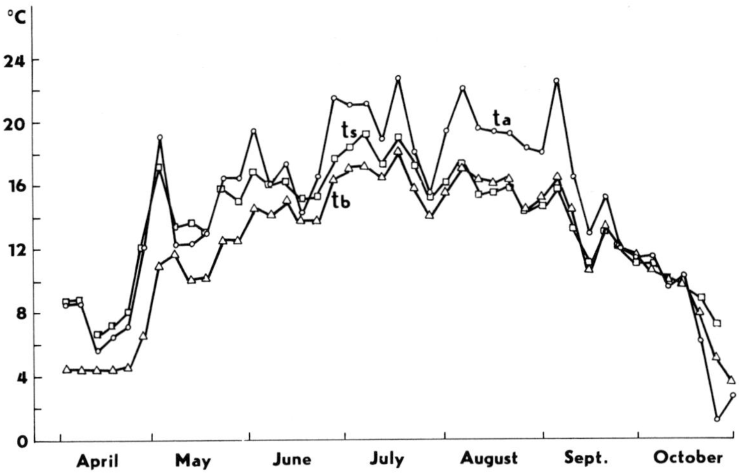

Fig. 4. Five days' averages of water temperature *(ts)* in middle of water column (ranging from 30 to 5 cm) in a littoral *Phragmites* stand; temperature in surface layer of mud (below bottom surface *tb* −2 cm) in a limosal stand of *Phragmites* with dense undergrowth of *Carex riparia*, both at Nesyt; air temperatures *(ta)* measured in standard meteorological screen at Lednice

helophyte communities, warm up rapidly (see the great daily amplitude of temperature, reacting to a considerable depth; Fig. 3). These suitable thermal conditions, combined with a better aeration, promote the decomposition of organic material in empty ponds.

The microclimate within the helophyte canopies depends primarily on the ratio of dead to living plant material, which changes in the course of a year.

From fall to spring, most of the incoming radiation is absorbed by the dry stems and converted into sensible heat. As the turbulent flux of this heat into the atmosphere is reduced, a considerable overheating of the reeds, as compared with the air above, may occur at daytime. Some data on wind reduction by a reed canopy are given by Willer (1949), other measurements were taken by P. Šmíd (unpublished) at the Nesyt fishpond. At night, steep vertical gradients of air temperature develop within the dead canopy. A marked temperature inversion often develops within a reed stand as a result of rapid heat loss from the dead material. Yet the calm and relatively warm conditions in helophyte stands provide shelter to numerous animals during winter and early spring.

In summer, the microclimate of the fully developed canopies is quite different. The temperature gradients are usually very small in helophyte stands in littoral ecophase, only rarely exceeding 1–2° C. Sometimes, especially in the afternoon, positive gradients develop, with the canopy air becoming colder than the air above (Šmíd, 1975; Přibáň, 1973b). On the other hand, in limosal and terrestrial stands of *Phragmites* with an undergrowth of sedges or grasses, the air temperatures may be higher immediately above this undergrowth on clear days than in the *Phragmites* canopy itself.

At night, when a stable air stratification occurs, a significant temperature inversion develops within the canopy. Rich and heavy dew is also common in

Table 4. Average values of the Bowen ratio (mean ±2 standard error) in the air above vigorously growing littoral *Phragmites* for the interval between 10 and 15 h on 5 sets of bright days, during the growing season; Nesyt fishpond. (After Šmíd, 1975)

Time interval	No. of days of observation	Bowen ratio
10–13 May	3	0.48 ± 0.20
18 May–1 June	7	0.21 ± 0.04
26 June–17 July	7	−0.03 ± 0.02
6–12 Aug.	6	0.06 ± 0.02
16 Sept.–6 Oct.	4	0.39 ± 0.08

fishpond reed belts; it is probably formed as the relatively warm and moist air from the open water area encounters the colder leaf surfaces of reed and other helophytes.

As the gradients of air temperature are rather small at daytime and those of absolute air humidity (water-vapor pressure) become considerable, the Bowen ratio (proportional to the ratio between the temperature and humidity gradients recorded between two chosen levels above the canopy) tends to be small or even negative during summer in helophyte communities. As this ratio is also equal to the ratio between the sensible and latent heat fluxes, most of the incoming radiant energy appears to be converted into latent heat and used for evapotranspiration (Šmíd, 1975) (Table 4).

Below a *Phragmites* canopy, the air remains colder and moister than within the canopy during a whole day. The increased air humidity within the canopy is due to an intense transpiration of the leaves (see Sect. 3.3). Similar results have been obtained by Willer (1949) and by Dykyjová and Hradecká (1973, 1976).

References see pp. 153—155.

2.2 Radiation Climate in Fishpond Littoral Plant Communities

J. P. Ondok

The results of detailed measurements of both global and photosynthetically active radiation in littoral communities have been reported by Ondok (1973a, b, c, 1975; in press, 1978). The most detailed description of methods used is given in Ondok (in press, 1978). This paper summarizes the characteristics of the radiation regime in different helophyte stands colonizing fishpond shores. This section will frequently refer to the above papers for more detailed descriptions of some of the methods and theoretical considerations including the literature used.

As evident from Section 1, the helophytes dominant in fishpond littoral plant communities frequently form nearly monospecific communities with a poor undergrowth. The dominance of one species is somewhat less pronounced only in some communities, e.g., those dominated by *Carex* species. The zonation of emergent macrophyte communities in fishpond and its dependence on the habitat zonation are described in Section 1. Within one community type, the zonation is reflected in gradients both of shoot density, and of the spatial distribution of clusters (pattern) of the dominant. These gradients more or less follow the slope of the bottom. Nearly monospecific invasion stands of the following dominant species were selected for investigations of the radiation regime: *Phragmites communis, Glyceria maxima, Typha angustifolia, Typha latifolia, Sparganium erectum, Acorus calamus*, and *Schoenoplectus lacustris*. All stands were more or less flooded with water, i.e., at littoral ecophase. The only mixed community, dominated by *Carex gracilis*, belonged to the association *Caricetum gracilis naumburgietosum thyrsiflorae*.

Two layers were distinguished: the canopy layer above the water surface and the water layer with submerged parts of the macrophytes, also containing some floating and submerged macrovegetation (e.g., *Lemna, Riccia*) and algae. In fully developed stands with a dense canopy, even less than 1% of incident global radiation reaches the water surface. The fraction of radiation penetrating into the water layer is therefore significant only at the beginning of the growing season.

Measurement and Evaluation Techniques. Global radiation is divided into three main components when penetrating a stand: (1) radiation intercepted by the canopy; (2) that intercepted by the water layer; (3) that reflected back to the atmosphere. For obtaining a budget of global radiation within a stand, three

a

b

Fig. 1a–d. Hemispherical photographs of stands of *Typha latifolia* (a) *Carex* (sp) (b) *Pragmites communis* (c) and *Schoenoplectus lacustris* (d). Camera was placed horizontally in (a) and vertically in (b) to (d). (Photo: J. P. Ondok)

c

d

radiation fluxes are considered: (a) direct sun radiation *(S)*; (b) diffuse sky radiation *(D)*; and (c) radiation scattered due to transmission and reflection of the leaves and due to reflexion from the water surface *(C)*. In our measurements, however, other radiation components were easier to measure: (a′) irradiance due to the downward flux of diffuse sky and scattered radiation *(U)*; (b′) irradiance due to the upward flux of scattered radiation *(V)*; and (c′) irradiance due to the direct and downward flux of diffuse sky and scattered radiation *(Z)*. The first two components were measured on shaded plots, the third on sunlit plots.

The components *S*, *D*, and *C* were than calculated as follows:

$$S = Z - U; \quad D = U - V; \quad C = V.$$

A device for recording the relative sunfleck area from the ratio between the sunlit points and the total number of points arranged in a grid was used for measuring *Z* (Ondok, 1973a). At the same time, the method of hemispherical photography was adapted and applied to the littoral helophyte stands (see Fig. 1). More details on its use are given by Ondok (in press, 1978). The diffuse sky and scattered radiation were estimated from measurements with phytoactinometers furnished with a selenium photocell (VEB Carl Zeiss, Jena, GDR) and with the following filters: BG 29, BG 33, and 17 (Schott, Jena, GDR). The daily amounts of PhAR in the stands were also recorded with Kubín-Hládek integrators (see Ondok, 1973a and Fig. 2).

Incident global radiation was measured above the stand with two Kipp and Zonen solarimeters, one of them equipped with a shading ring excluding the direct radiation component. The records of radiation above the stand combined with the evaluation of hemispherical photographs (see Fig. 1) permitted an assessment of the penetration of both direct and diffuse sky radiation into the stand. However, if using this method, the scattered radiation had to be evaluated in another way. The Kahlsico photometer (Model 268 WA 310) was used for measurements of irradiance at different levels in the water. For further details see Ondok (in press, 1978).

Interception of Radiation. Under clear sky conditions, most of the incoming radiant energy belongs to the component of direct sun radiation (70–75%), the diffuse sky radiation amounts to 15–20%, and scattered radiation ranges between 1 and 5%. The variation of these values is primarily given by the sun elevation. Under overcast sky conditions the first two components may acquire different values but the third component remains much the same.

The penetration of radiation into a stand depends on a number of factors, among which stand geometry, canopy density and wind conditions are the most important for the extinction of both direct and diffuse radiation. The flux of scattered radiation is mainly affected by optical properties of the leaves, i.e., their transmission and reflexion.

The stand geometry has been examined in *Phragmites* stands because of the easy measurement of leaf inclination and orientation. Young reed leaves are inclined more or less vertically (erectophile foliage; de Wit, 1965) and become more horizontal in the course of their development. Hence the leaf inclination prevailing in the stand is vertical early in the spring and becomes horizontal

Fig. 2a and b. Integrators of the Kubín-Hládek type (1963) recording daily amounts of PhAR 370–720 nm in (a) a young *Phragmites* stand, May; (b) a young *Typha angustifolia* stand, early June. (Photo: D. Dykyjová)

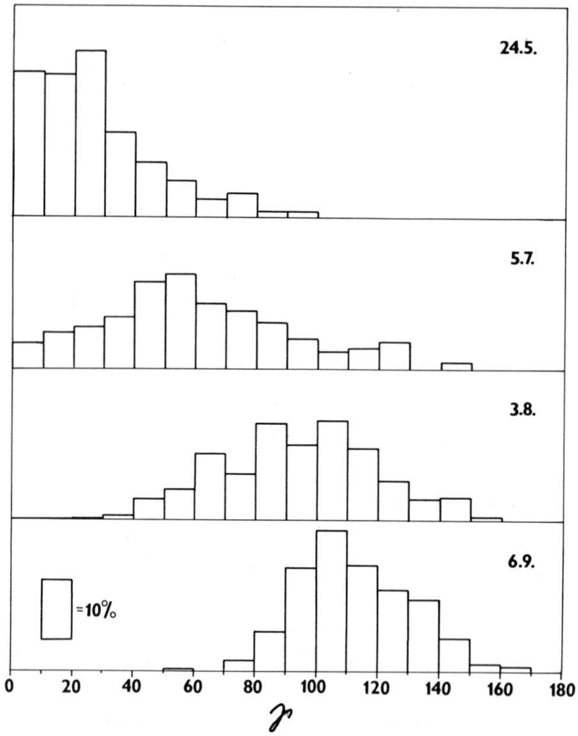

Fig. 3. Histogram of class frequencies (%) of leaf inclination (γ) in a reed stand on different dates. γ is measured from the vertical

(planophile foliage) by the end of the growing season. This development is documented in Figure 3. Most of the helophytes examined such as *Typha angustifolia*, *T. latifolia*, *Acorus calamus*, and *Sparganium erectum*, possess a more or less erectophile foliage, which enables the radiation to penetrate quite deep into their stands. Most of their young and physiologically active assimilatory tissues are located in the lower canopy layers and the deeply penetrating radiation is used for their photosynthesis. In *Phragmites* and *Glyceria maxima*, the youngest leaves are concentrated in the higher canopy layers. Their successive leaves grow on the stems and aquire, on an average, a near-horizontal position.

Leaf orientation is frequently anisotropic in reed-bed stands. This anisotropy mainly corresponds to the prevailing wind direction and increases during the growing season (Ondok, 1973a).

The influence of leaf arrangement on the penetration of radiation into a *Phragmites* stand has been examined by simulation techniques (Ondok, in press, 1978). The most favorable penetration of radiation may be expected with a prevailingly south-oriented erectophile foliage. This conclusion is valid for the latitude of 49° N.

The wind acts as another factor influencing the penetration of radiation particularly into littoral helophyte stands. The different rates of heating of the air above the water surface and above the adjacent land give rise to breezes even in

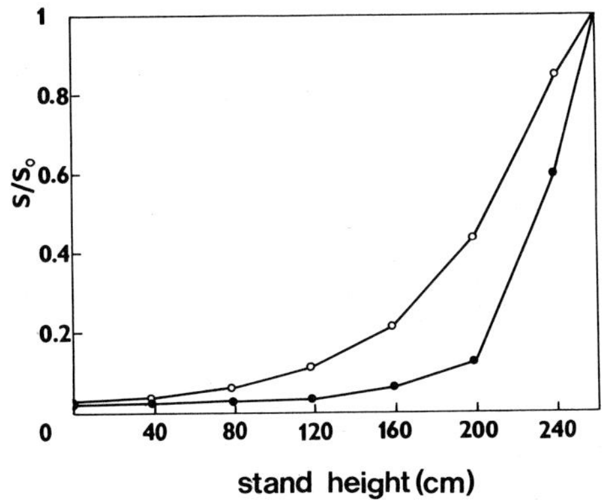

Fig. 4. Extinction of direct sun radiation (*S*) in a *Phragmites* stand under windless (● - - - ●) and windy (○ - - - ○) conditions. Date of measurement: 23 July

otherwise calm weather. The resulting leaf movements bring about an alternating irradiance of the leaves, which may affect favorably the photosynthetic efficiency of the leaves. For each leaf, frequency of changes in irradiance by direct radiation is positively correlated with both the wind speed and its level of insertion. Applied to whole leaf-canopy layers, this phenomenon is shown in Figure 4.

The influence of horizontal variation in shoot density (and thus also in foliage density) on the extinction of incoming radiation is difficult to evaluate as the sunrays mostly come in at low angles and penetrate the stand along paths cutting across patches differing in canopy density; this circumstance reduces the importance of this variability. However, great differences still exist in light extinction between stands markedly differing in average shoot density. Figure 5 illustrates such differences. The extinction curve for direct radiation pertains to the sun elevation at noon on 21 June, and thus represents a seasonal minimum value.

The optical properties of the leaves determine the flux of scattered radiation in a stand. In the helophytes studied, reflexion plays a more important role than transmission because their leaves are usually thick and opaque. The average transmission mostly does not exceed 4%, while the reflexion ranges between 10 and 15%. Both the reflection and transmission vary in the course of a growing season and depend on the average age of the foliage and on sun elevation. The average values for the individual species studied given in Table 1 are expressed in terms of percentage energy contents in the spectral region of 300–700 nm (PhAR) of the incident radiation.

Daily and Seasonal Radiation Régime. The portioning of incoming radiant energy into the components intercepted by either the canopy or the water layer, and the albedo, depends on the factors reported above. If we know the daily course, of sun elevation and of both direct sun and diffuse sky radiation, we may calculate the daily régime of either global or photosynthetically active radiation. The selected examples in Table 2 indicate the daily sums of radiation which are intercepted by the canopy or water layer and reflected to the atmosphere in fully developed littoral helophyte stands. Despite their great variation, the values given in Table 2 illustrate the order of individual littoral helophyte stands as regards

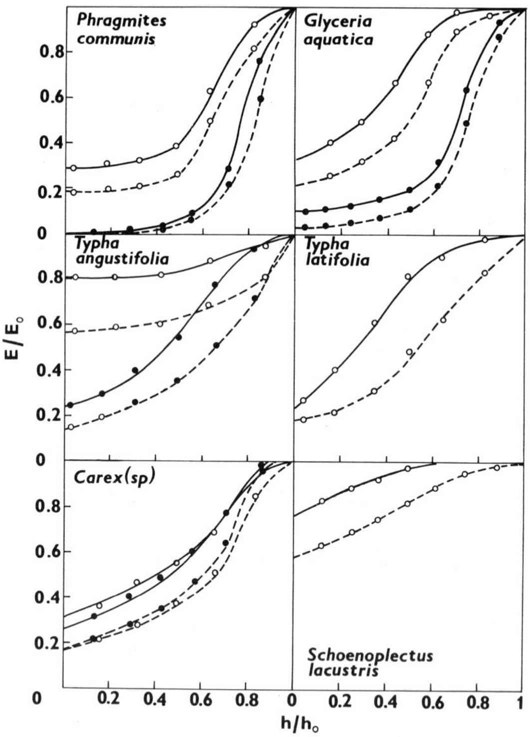

Fig. 5. Extinction of direct sun radiation at noon sun elevation (———) and of diffuse sky radiation (- - - -) in loose (—○—) and dense (—●—) helophyte stands. Height scale (coordinate) expressed in fractions from total stand height

Table 1. Average values of reflection, transmission, and absorption of PhAR by leaves of several helophyte species

Species	Transmission (%)	Reflection (%)	Absorption (%)
Phragmites communis	3.43	10.83	86.64
Typha latifolia	2.12	10.27	87.61
Typha angustifolia	1.66	10.65	88.69
Glyceria maxima	4.23	12.51	83.26
Sparganium erectum	3.27	10.83	85.90
Acorus calamus	2.73	15.10	82.16

their transparency for incoming radiant energy. The values pertain to PhAR (300–700 nm) but the order will remain the same for global radiation. Although the interception of PhAR by the canopy and water layers varies considerably with the type of stand, only little variation exists in the albedo.

Evaluation of the seasonal course of the radiation régime requires a complicated computation procedure and may be followed with advantage by means of a

Table 2. Daily amounts of PhAR (300–700 nm) intercepted by canopy (X_1), by water layer (X_2), and reflected back to atmosphere (X_3) as percentage of incident PhAR[a] above the stand of several littoral helophytes

Species	Date	X_1 %	X_2 %	X_3 %	Density, Water level (cm above bottom)
Phragmites communis	17. July	73.3	23.1	4.6	low 30–40
Phragmites communis	29. Aug.	94.1	1.2	4.7	high 40–50
Glyceria maxima	25. July	88.3	7.1	4.6	high 0–10
Glyceria maxima	26. Aug.	64.4	30.9	4.7	low 80–90
Typha angustifolia	13. Aug.	56.8	38.6	4.6	low 30–40
Typha angustifolia	5. Sept.	80.0	15.2	4.8	high 70–90
Carex (sp.)	17. Aug.	63.7	32.1	4.8	mean 0–10
Schoenoplectus lacustris	26. Aug.	22.9	72.6	4.5	low 50–60

[a] PhAR calculated as 0.45 of total radiation.

model which simplifies, to some extent, the real situation (Ondok, 1975, in press, 1978).

The model has two inputs: direct sun and diffuse sky radiation given as hourly sums of recorded values or generated as stochastic values from density functions derived from recorded data. One hour is thus the time-scale unit.

The daily course of hourly sums of global radiation under clear sky conditions may be expressed by the function:

$$E^* = 930[\sin\delta \sin\varphi + \cos\delta \cos\varphi \cos(n\pi/12)] \text{ (Wm}^{-2}) \quad (1)$$

where E^* is the hourly sum of global radiation for the nth hour (the computation of n starts from noon), δ is sun declination and φ is latitude. E is reduced by a coefficient k which simulates the variation in sky cloudiness and is generated by the function:

$$k_{n+1} = k_n + e \quad (2)$$

where e is a normal variate: $N(0, 0.118)$ and k_o is generated for each day by means of the density function:

$$P(x < k_0) = 0.008\, e^{3.9k} \quad (3)$$

The formulae (2) and (3) are derived from empirical distributions of (E/E^*) and $(E/E^*)_{n+1} - (E/E^*)_n$. Then

$$E = kE^* \quad (4)$$

The direct sun and diffuse sky radiation components are calculated by means of the empirical formulae:

$$S = (1-h)E \qquad D = hE \quad (5)$$

derived on the basis of a negative correlation between the reduction of global radiation by cloudiness as simulated by k and the concurrent relative increase in

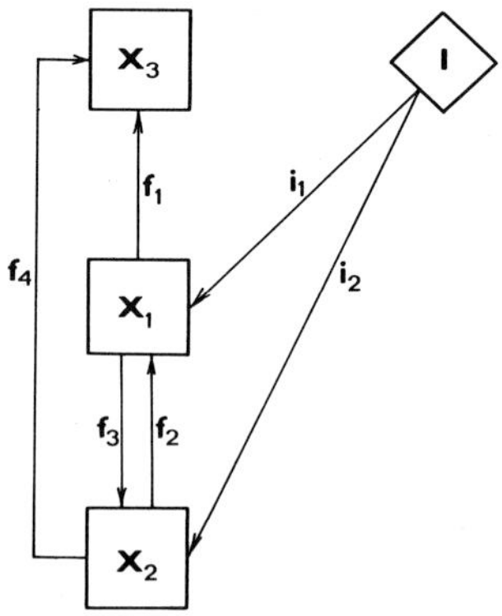

Fig. 6. Structural scheme for both radiation components. I, input; X_1, X_2, X_3, radiation intercepted by leaves, water, and albedo, respectively. Transfer functions relate to radiation intercepted by leaves (i_1), by water layer (i_2), reflected by canopy (f_1), reflected by water and than intercepted by canopy (f_2), transmitted by leaves to the water layer (f_3), and reflected by water table to the atmosphere (f_4)

Table 3. List of transfer functions in the model of PhAR regime in a *Phragmites* stand

Model I (direct radiation)	Model II (diffuse sky radiation)
$i_1 = 1 - B$	$i_1 = 1 - A$
$i_2 = B$	$i_2 = A$
$f_1 = \frac{r}{1+\sin\gamma_2}\frac{1-AB}{1-B}$	$f_1 = \frac{r}{2}(1+A)$
$f_2 = (1-A)$	$f_2 = (1-A)$
$f_3 = \frac{t}{1-\sin\gamma_2}\frac{A-B}{1-B}$	$f_3 = \frac{2tL_o}{\pi}\frac{A}{1-A}$
$f_4 = \varrho A$	$f_4 = \varrho A$

t, r, transmission and reflection coefficients of leaves; ϱ, reflection coefficient of water surface; t, r, ϱ expressed as fractions of PhAR; A, $\exp(-2L_o/\pi)$; B, $\exp(-L_o/\pi \sin\gamma_2)$; L_o, leaf area index; γ_2, sun elevation.

diffuse sky radiation. For reducing S and D to the PhAR spectral region, the respective coefficients of 0.42 and 0.6 are used (Tooming and Niilisk, 1967).

The models of portioning of direct sun and diffuse sky radiant energy between the canopy, water, and atmosphere have an identical structural scheme, illustrated in Figure 6, but the transfer functions must be defined separately for each model. The formulae for the transfer functions are listed in Table 3, and their derivations are by Ondok (1975). On the basis of this list, the following system of equation has been constructed:

$$\begin{aligned} dX_1/d\gamma_2 &= (1-f_3-f_1)[(1+f_1f_2)i_1+f_2i_2]I \\ dX_2/d\gamma_2 &= (1-f_2-f_4)[(1+f_1f_2)i_2+f_3i_1]I \\ dX_3/d\gamma_2 &= [(f_1+f_3f_4)i_1+(f_4+f_2f_1)i_2]I \end{aligned} \qquad (6)$$

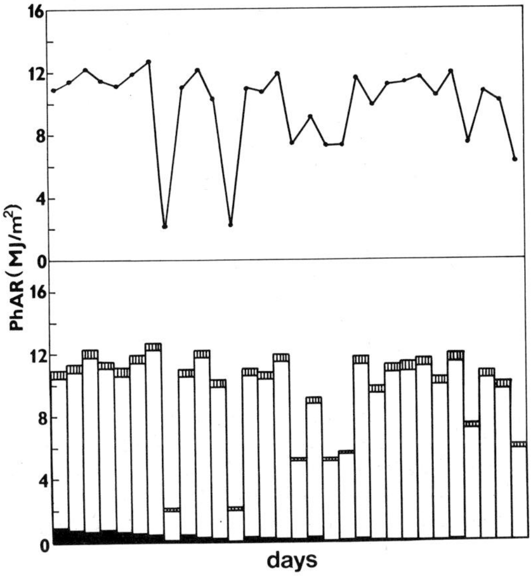

Fig. 7. Example of a simulated course of PhAR regime in a reed stand during a month (July). *Top*: daily sums of PhAR above the stand; *bottom*: daily sums of PhAR intercepted by canopy (□), by water (■), and reflected back to the atmosphere (▤)

Table 4. Seasonal totals of PhAR intercepted by the canopy (X_1), water (X_2), and reflected to the atmosphere (X_3), obtained from the model of radiation régime in a *Phragmites* stand (May to Sept.)

	Wh cm^{-2}			
	Direct sun radiation	Diffuse sky radiation	Total	Total %
X_1	408.7	517.3	926.0	70.6
X_2	121.5	205.1	326.6	24.9
X_3	27.3	31.1	58.6	5.5
Total	557.7	753.5	1,311.2	
Total (%)	42.5	57.5	—	100.0

In the equations, appropriate proper values have to be substituted for the input and transfer functions in either model. The daily course of PhAR régime in a reed stand results from an integration of system (6) for both radiation components; for details see Ondok (1975). By introducing the seasonal course of δ and L_o, the whole seasonal course of the radiation régime is obtained. An example of such seasonal course during one month is given in Figure 7. Table 4 contains the total sums of PhAR intercepted by the canopy and water layer, and reflected to the atmosphere as obtained from the seasonal course.

The model proposed may serve for investigating the radiation régime under simulated conditions (e.g., variable shoot density, various types of distribution of

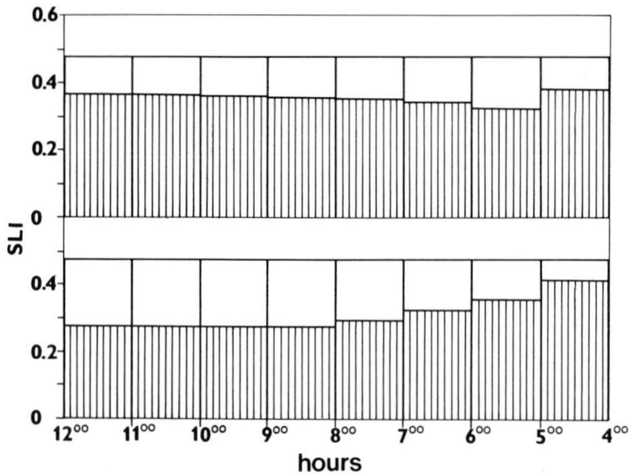

Fig. 8. Daily course of SLI (*clear area*) and SDI (*hatched area*) in reed stand (6 July, $L_o=4.76$) with a planophile (*top*) and erectophile (*bottom*) foliage

leaf inclination and orientation) and for estimating the importance of individual factors for the radiation régime. The 1-h scale used may be replaced by another one with shorter time units without changing the structural scheme of the model. Only the functions for generating the stochastic variate k have to be replaced by others derived from empirical distributions corresponding to the new time units.

Radiation Régime and Photosynthesis. When using radiation data for calculating photosynthesis, the distribution of irradiance on the leaf surfaces has to be evaluated for sunlit and shaded leaves separately. One method of calculating such distributions is given by Ondok (1973c). The relative amount of direct radiation projected on to a leaf surface within a stand layer is equal to the difference between relative sunfleck area (a) immediately above and below this layer:

$$d_k = a_l - a_{l-1} \tag{7}$$

The projection depends on leaf position and sun elevation:

$$d'_k = \frac{d_k}{(|\sin\gamma_1 - \cos\gamma_1 \operatorname{cotg}\gamma_2 \cos\zeta|)} \tag{8}$$

where d'_k is the projected area, γ_1 is leaf inclination, γ_2 is sun elevation and ζ represents the azimuth distance between sun and leaf orientation. The projected area corresponding to any of ($m \times n$) classes of leaf inclinations and orientations is:

$$\mathrm{SLI}_{ij} = \begin{cases} d'_k f(\gamma_{1i}, \varphi_{1i}) & \text{if } \dfrac{d'_k}{L_k} \leq 1 \\ L_k f(\gamma_{1i}, \gamma_{1i}) & \text{in other instances} \end{cases} \tag{9}$$

where L_k is leaf area index in the kth layer. SLI means sunlit area index for the class considered, and $f(\gamma_1, \varphi_1)$ is the frequency function of leaf inclination and

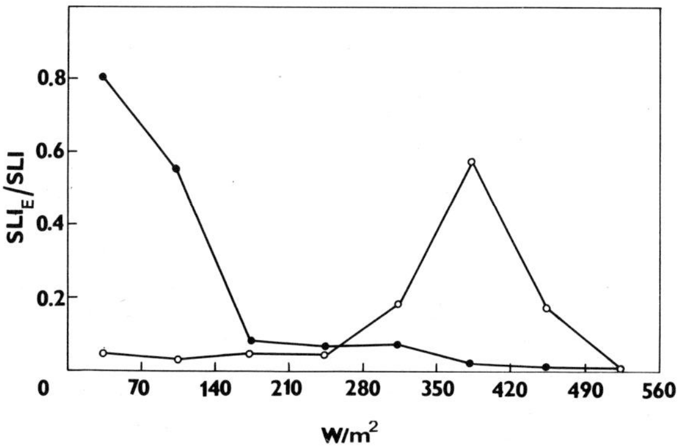

Fig. 9. Distribution of irradiance on leaf surfaces in a planophile foliage of a reed stand with $L_o = 4.76$ on 6 July. Ordinate: density expressed as fraction of SLI; coordinate: class intervals of E. Sun elevations: 17° (—○—) and 64° (—●—)

orientation. Each of $(m \times n)$ classes is coordinated with a certain value of irradiance (E_{ij}):

$$E_{ij} = E_0(|\sin\gamma_1 - \cos\gamma_1 \operatorname{cotg}\gamma_2 \cos\zeta|) \tag{10}$$

where E_0 is irradiance with direct sunlight above the stand. The density function is obtained with E_{ij} arranged in classes, the class density being expressed by fractions of foliage area sunlit by the irradiance lying within each considered class interval of E_{ij}. This density function varies with foliage density, with patterns of its geometrical arrangement and with sun position. Erectophile foliage usually has higher values of SLI than planophile foliage. This is documented by Figure 8. Only one example of variation in the distribution of E as dependent on leaf position is given in Figure 9 for the planophile foliage of a reed stand; for more data see Ondok (in press, 1978). His results prove that a more uniform distribution of leaf inclination and orientation is correlated with a more uniform distribution of E. Both erectophile and planophile foliage show marked peaks in the frequency histograms of irradiance.

The computation of distribution of irradiance by diffuse and scattered radiation on shaded leaves is performed easily as it may be assumed that the irradiance on leaf surface does not depend on the direction of incoming radiation in this case. It is equal to the irradiance measured on a horizontal plane. The distribution of E on SLI and SDI serves as input a photosynthetic model which is described in Section 3.5.

References see pp. 153—155.

2.3 Water Chemistry in the Fishpond Littorals

B. ÚLEHLOVÁ and S. PŘIBIL

Water chemistry is one of the important characteristics of a water body: it depends on the geology of the area and on the chemistry of the inflowing water, and it may be influenced by the biotic components of the water body. On the other hand, it affects both the composition and the species diversity of the indigenous biota. Water chemistry is therefore sometimes used as a basis for the classification of water bodies, freshwater and wetland ecosystems (Thienemann, 1926; Naumann, 1932; Björk, 1967).

Extensive littorals of such shallow water reservoirs as fishponds represent highly differentiated biotopes with respect to both the stability and variation in the physical and chemical properties of their aquatic environment. These properties are determined, in addition to the basic geological and climatic factors, by the quality of inflowing water (influenced by the character of the catchment area and by its agricultural, silvicultural, and other management) as well as by the management of the fishpond itself (including fertilizer application, additional feeding of the fish stock, and other management measures). A specific characteristic of large areas in the fishpond littoral is the accumulation of gradually decomposing dead organic matter as well as a penetration, in both space and time, of open-water or terrestrial biotic communities into the littoral. Heat exchange between air — water — land as well as hydrological factors contribute to the variation in physical and chemical properties of the littoral habitats. Water movements, driven predominantly by air movements, result in either displacement or sedimentation of both inorganic and organic matter in the littoral. Detritus benches deposited on certain sites separate, frequently almost entirely, certain areas of the littoral, especially in its landward outer zone, from the rest of the fishpond. The spatial pattern of emergent macrophyte communities, reflecting the pattern of sedimentation and erosion, induces further differentiation of environmental conditions in the fishpond littoral. Individual littoral zones and biotopes are thus distinguished not only by their specific hydrochemical characteristics but also by the structure and productivity of their plant and animal communities and by the rates of decomposition of dead organic matter (see Sect. 5).

2.3.1 Nesyt Fishpond

The water chemistry of the Nesyt fishpond depends above all on the water quality of its main water source, the brook "Mikulovský Potok", which drains a highly productive agricultural catchment area about 150 km^2 in size. Above the inflow into the pond, the brook passes through a highly saline area. Zimmerman (1916) suggested that the high content of mineral salts reflected the presence of the sea during the Miocene, and that the high concentration of dissolved mineral nutrients in the water of Nesyt originated from the local soils and subsoils. Today it is impossible to neglect the contribution of mineral nutrients leached from the intensely cropped and fertilized agricultural land.

Many authors (Zapletálek, 1932; Zimmerman, 1916; Fischer, 1920; Bayer and Bajkov, 1929; Soudek, 1929; Bílý, 1932, and others) discussed the specific character of the Nesyt fishpond from the biological point of view and suggested that it is affected to a large extent by water chemistry. The first analyses of the Nesyt water were performed by Rzehak and Kornauth at the beginning of this century, and several authors used the result of these analyses.

Jírovec (1936) analyzed waters sampled across the north shoreline of the Nesyt fishpond in the years of 1933, 1934 and twice in 1935: in spring and in fall. It was the first systematic investigation of the water chemistry of the pond.

The aim of the present chemical study was to obtain data on the spatial variation, and also on possible changes during the last forty years, of the chemical characteristics of the Nesyt water. Such data should also provide the basis for considering the effects of fishpond management on the water chemistry.

2.3.1.1 Material and Methods

Water samples were analyzed within 24 h after sampling during 1969 and 1970, from April to October at monthly intervals. The contents of Ca^{2+}, Mg^{2+}, Na^+, K^+, NH_4^+, HCO_3^-, SO_4^{2-}, Cl^-, NO_3^-, PO_4^{3-}, and pH were estimated according to methods recommended by the Committee for Standard Methods of Water Analyses (Hofman et al., 1965).

Five sites were chosen for the investigation to characterize the diversity of local conditions in the Nesyt fishpond (see also map in Fig. 2 in Sect. 1.2). Three sites are situated on its north shore. Site 1 is just at the mouth of the Mikulovský Potok brook in the western bay of Nesyt, in a dense stand of *Hippuridetum vulgaris*. The bottom is covered with a thick (about 20 cm) layer of black, fine-textured sapropelic mud smelling of hydrogen sulfide. The subsoil is also a black, smelling mud with a loam-clay texture.

Site 2 is situated at the north shore, in a stand of *Phragmitetum communis* whose primary production had been studied for several years (Květ et al., 1969; Hradecká and Květ, 1973). The layer of semidecomposed plant litter is about 8–10 cm thick and covers a transient layer with mud and sand, and a sandy loam subsoil enriched with black humic substances.

Site 3 is in open water near the fishpond dam in front of the outlet from the fishpond, where the water quality integrates all kinds of influence acting during its passage through the fishpond. The site is practically void of organic sediments, the soil is sandy clay loam enriched with humic substances.

The two other habitats investigated are located near the south shore of the Nesyt fishpond. Site 4 is in its accumulation zone, in its southwest part, at the borderline between an extensive *Phragmitetum* and open water. The site has a

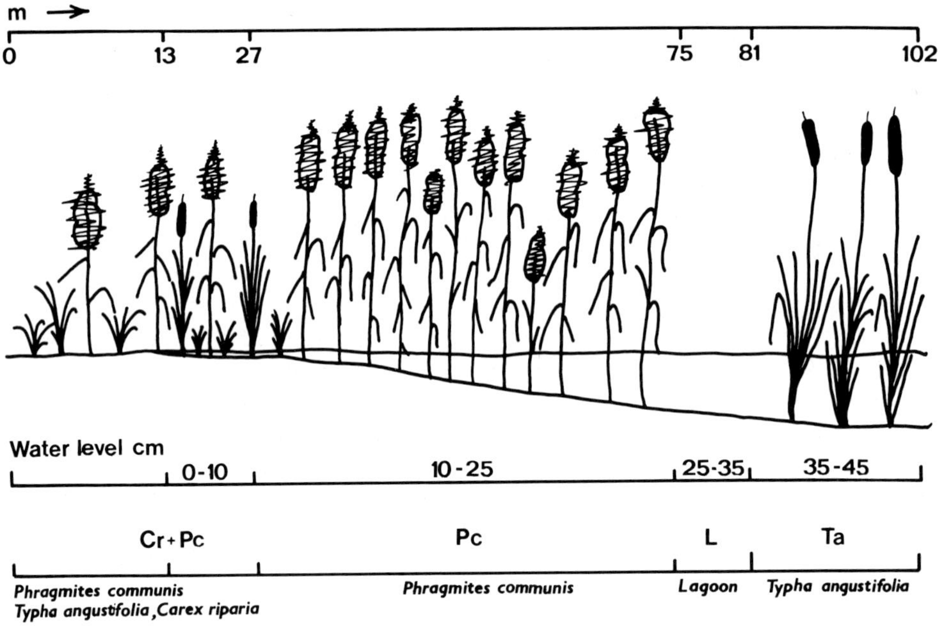

Fig. 1. Littoral transect in the Nesyt fishpond

black smelling sapropel (mud) layer about 20 cm thick, the subsoil is clay loam highly enriched with sapropel. The site is significantly influenced by waterfowl.

Site 5 is situated in the southwest bay of Nesyt, on the borderline between the reed-belt and open water. There is a layer of black fine sediment about 10 cm thick; the sandy subsoil is highly enriched with black mud and contains clay and loam inclusions.

During 1971, chemical analyses were performed, at two-week intervals, of water samples collected along a transect across the littoral, crossing four different vegetational zones, at the west shore of the Nesyt (Fig. 1). The four zones included a mixed stand of *Phragmitetum communis* with *Typha angustifolia* and *Carex riparia* (Cr + Pc); a pure stand of *Phragmites communis* (Pc); a lagoon free of helophyte vegetation (L); and a pure stand of *Typhetum angustifoliae* (Ta), in the inner littoral adjoining the pelagial of the fishpond. Water samples were analyzed for important forms of nitrogen and phosphorus as separated by filtration through filters with pore sizes of 45 and 0.4 μm.

The total and phosphate phosphorus, the ammonia-, nitrite-, and nitrate-nitrogen, as well as the Kjeldahl nitrogen were estimated according to the methods given by Golterman and Clymo (1969); for further details see also Dvořák (1973). The results of analyses were used for calculating the concentrations of bacterial and dissolved organic phosphorus and nitrogen from the differences between data on total P and N in filtrate A (45 μm) and data on total P and N in filtrate B (0.4 μm), and between data on total P and N in filtrate B and on mineral P and N forms in the same filtrate.

2.3.1.2 Spatial and Seasonal Variations of the Water Chemistry

Table 1 presents average data on cation and anion contents of the water samples from the five different sites of the Nesyt fishpond, as estimated during 1969 and 1970.

The values resulting from the analysis of variance of the chemical data are given in Table 1. All chemical characteristics vary significantly with time but only some of them vary significantly according to the site of sampling.

The water samples from sites 1 and 4 are alike, and they differ from the remaining samples. This difference appears to result from the import of considerable amounts of allochthonous organic matter, the sources of which are village wastes at site 1, and bird guano at site 4. Apparent also are differences between the years. They reflect most probably the effects of fishpond management. In 1969, the fish crop was harvested in the autumn and shallow parts of the Nesyt littoral remained drained till the following spring. During the winter drawdown of 1969/1970 mineralization and oxidation of organic matter took place in the freely exposed parts of the fishpond bottom. The pond was refilled with water during the early spring of 1970, and subsequently restocked with young fish.

Figures 2–5 provide information on the seasonal changes in water chemistry. Certain cations and anions show typical patterns of seasonal variations despite the differences in management between 1969 and 1970 mentioned above. So nitrates and ammonia are relatively very high in early spring and low during the rest of the year. Potassium, bicarbonate and phosphate tend to increase more or less steadily from spring to fall. Sulfate attains high cocentrations in late spring but decreases strongly in fall. The same seasonal pattern of sulfate variation was recorded in Nesyt by Jírovec (1936).

Table 1. Water chemistry in the Nesyt fishpond in 1969 and 1970. Annual mean concentrations of ions in mg/l and their spatial and seasonal variations

Site (see text)	HCO_3^-	SO_4^{2-}	Cl^-	PO_4^{3-}	NO_3^-	NH_4^+	K^+	Na^+	Ca^{2+}	Mg^{2+}	pH
1969											
1	507	118	56	3.10	0.63	—	34.6	70.3	91.8	28.0	6.90
2	514	223	61	2.78	0.33	—	36.4	77.4	92.9	37.9	7.08
3	528	213	62	2.50	0.32	—	37.0	81.1	89.3	34.9	7.03
4	481	160	56	3.16	0.60	—	40.2	77.2	82.9	34.4	7.08
5	505	230	59	2.44	0.40	—	32.6	80.4	91.8	43.4	7.00
1970											
1	403	179	57.9	2.88	4.70	0.89	27.8	56.9	98.8	51.2	7.2
2	293	226	54.0	1.96	2.07	0.70	23.2	54.2	64.7	62.3	7.6
3	295	216	53.8	2.56	2.52	0.35	24.2	53.8	57.3	63.3	7.6
4	376	201	56.2	3.19	4.99	1.02	26.5	56.2	83.9	56.2	7.6
5	273	256	49.5	2.03	1.97	0.51	20.7	59.3	60.5	61.7	7.7
Spatial variation	++	–	–	–	–	—	++	+	++	++	++
Seasonal variation	++	++	+	++	++	++	++	++	++	++	++

– Not significant; + significant at 0.05 level; ++ highly significant at 0.01 level.

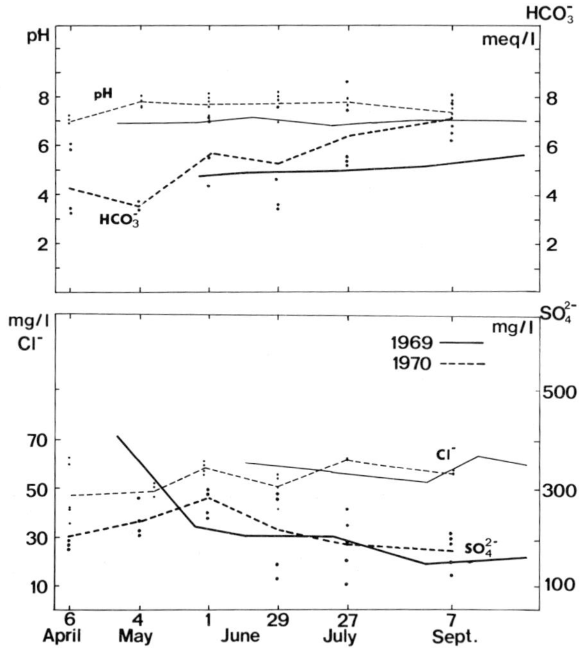

Fig. 2. Temporal variations of pH and HCO_3^-, Cl^-, and SO_4^{2-} in water of Nesyt fishpond

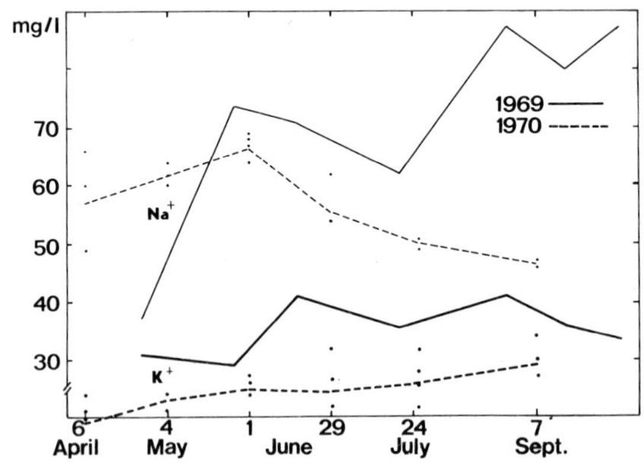

Fig. 3. Temporal variations of K^+ and Na^+ in water of Nesyt fishpond

The overall chemical differences between 1969 and 1970, which can be attributed largely to management, can be seen by comparing the annual means of the respective chemical water characteristics. Nitrate and sulfate increase in the second year after the oxidation of the bottom sediments during the exposure of fishpond bottom. An increased level of magnesium and the decreased contents of bicarbonates, calcium, sodium, and potassium in the second year may be explained by the complete renewal of the fishpond water body. This resulted in the establishment of new dynamic equilibria between the biotic and abiotic components of the ecosystem. Nonetheless, with the advancing season, the observed

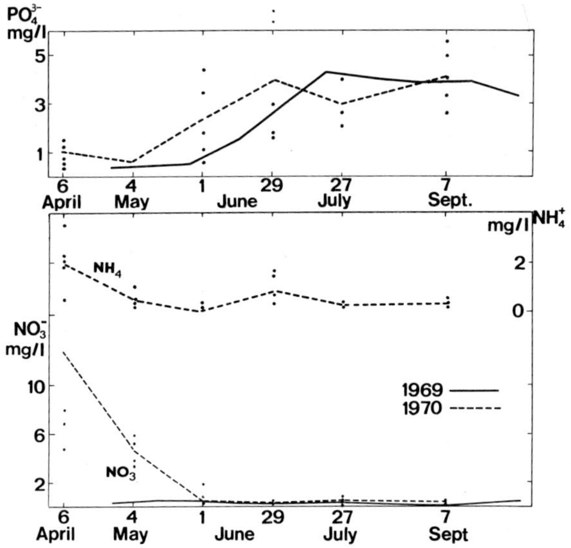

Fig. 4. Temporal variations of PO_4^{3-}, NH_4^+, and NO_3^- in water of Nesyt fishpond

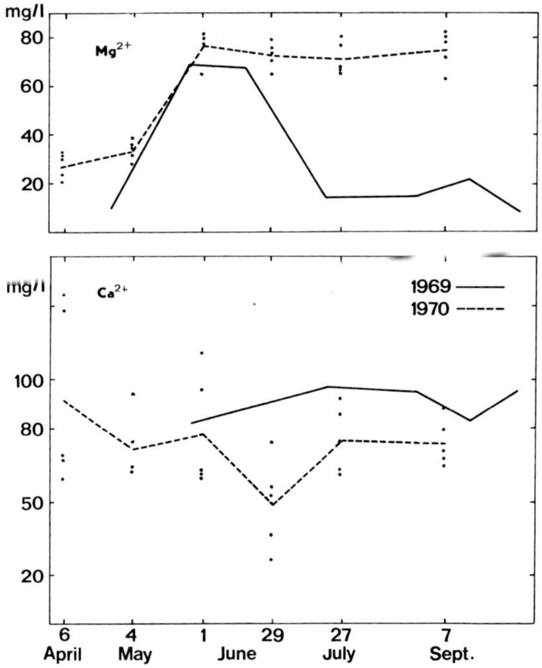

Fig. 5. Temporal variations of Ca^{2+} and Mg^{2+} in water of Nesyt fishpond

concentrations tended to approach the values found in the preceding year (Table 2).

A comparison of the present data on the chemistry of Nesyt fishpond with those of Rzehak and Kornauth (1910, cited in Jírovec, 1936) and Jírovec (1936) indicates that the content of sulfates was higher and that of chlorides lower at the

Table 2. Water chemistry in stands of *Phragmitetum communis* in Nesyt (annual averages)

Year	HCO_3^-	SO_4^{2-}	Cl^-	NO_3^-	PO_4^{3-}	NH_4^+	K^+	Na^+	Ca^{2+}	Mg^{2+}	pH
	mg/l										
1969	507	189	59.0	0.46	2.80	—	36.16	77.3	89.74	39.56	7.02
1970	328	215	54.2	3.25	2.52	0.69	24.50	56.1	73.00	58.90	7.50
Average	417	202	55.6	1.85	2.66	0.69	30.33	66.7	81.37	49.23	7.26

Table 3. Cl^- and SO_4^{2-} contents in water of the Nesyt fishpond estimated at different periods of the 20th century

	Cl^-	SO_4^{2-}
	ml g^{-1}	
Rzehak and Kornauth, 1910 (cited in Jírovec, 1936)	37.6	342.5
Jírovec, 1936	50.0	283.0
Úlehlová and Rejthar, 1973	55.6	202.0

Table 4. Comparison of water chemistry in selected water bodies in Norway, Poland, and Czechoslovakia. Concentrations of all ions in mg l^{-1}

Ions	Norway	Poland	Czechoslovakia	
			Nesyt	Opatovický
NH_4^+	0.020–0.040	0.07 – 0.47	0.03 – 5.80	0.103– 1.50
NO_3^-	0.010–0.027	0.029– 0.240	0.030– 4.90	0.025– 1.31
PO_4^{---}	0.002–0.021	0.011– 0.063	0.14 – 3.10	0.011– 0.17
SO_4^{--}	2.3 –6.44	0.07 –21.78	118 –256	20 –120
Cl^-	0.43 –0.60	1.45 –16.56	49.5 – 62.00	8.4 – 30
K^+	0.40 –0.50	1.07 – 5.53	20.70 – 40.20	6.4 – 18.2
Na^+	0.70 –1.29	0.85 –12.40	53.80 – 81.10	5.2 – 7.4
Ca^{2+}	0.96 –8.05	0.40 –50.2	57.3 – 92.2	19.9 – 29.4
Mg^{2+}	0.51 –0.73	0.87 –18.40	28.00 – 63.30	4.3 – 6.5

beginning of the century (Table 3). Should this hypothesis prove correct, it would be necessary to regard these changes as evidence of adverse effects of intense agriculture, whereby the redundant chlorides from mineral fertilizers tend to accumulate as waste. Sulfates, unlike chlorides, can enter biological cycles and their decreasing content can be related to the increasing eutrophication of the pond.

A comparison of the water chemistry in the Nesyt and Opatovický fishponds with that in three freshwater Norwegian lakes (Gröterud, 1972) and in seven Mazurian lakes (Poland) (Korycka, 1969) is presented in Table 4. The analytical data show distinctly the highly polluted waters in Czechoslovak fishponds and a semisaline character of the Nesyt fishpond.

2.3.1.3 Transect Study of Variation in Phosphorus and Nitrogen Water Content

The seasonal and spatial patterns of various phosphorus and nitrogen forms in the water along a transect across the littoral of Nesyt, as described above, are shown in Figures 6–10. Phosphate phosphorus plays a decisive role in the seasonal changes of phosphorus content in the water. Each increase in phosphate content was followed by an increase in bacterial phosphorus. Dissolved organic phosphorus was low in nearly all samples. Two main peaks of phosphorus compounds appear on the curves for water samples from different sites, one in late May and early June, the other in late July to early August. In these periods, phosphate phosphorus was intensely released either from dying-off biotic populations (autolysis?) or from decaying organic materials. Water from the L and Ta sites, which is in permanent exchange with water from the fishpond pelagial, displays similar seasonal dynamics of phosphorus compounds.

The sites CrPc and Pc attained a limosal character in October during water discharge from the fishpond. Consequently, the phosphate phosphorus increased to high levels in comparison with the phosphate phosphorus contents concurrently observed in sites L and Ta remaining inundated at that time. The most conspicuous increase in phosphate phosphorus occurred at the Pc site and was accompanied by a similar increase in microbial counts.

The situation appears more complicated as concerns the nitrogen transformations. There exists again a likeness of the two sites L and Ta adjoining the pelagial of the fishpond, and also a marked increase in contents of all nitrogen forms on sites CrPc and Pc because of the water drawdown in the fall. Dissolved organic nitrogen (DON) is an important constituent among the water-contained nitrogen compounds, as contrasting with dissolved organic phosphorus. The annual course shows only small changes around the level of about 10 mg l^{-1}. The trans

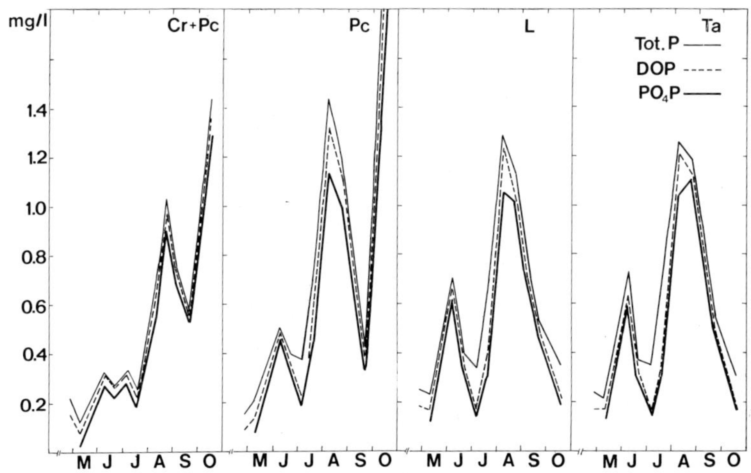

Fig. 6. Spatial and seasonal variations of total P, dissolved organic P, and PO_4-P, in water of littoral transect (cf. Fig. 1) in Nesyt fishpond

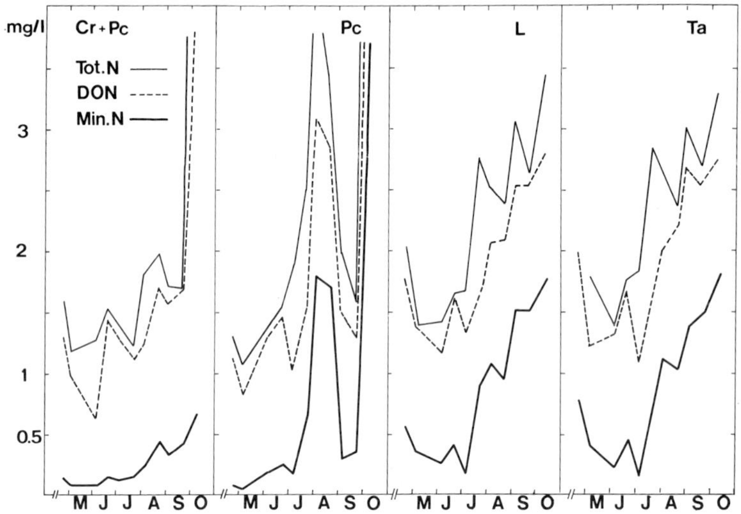

Fig. 7. Spatial and seasonal variations of total, dissolved organic and mineral nitrogen, in water of littoral transect (cf. Fig. 1) in Nesyt fishpond

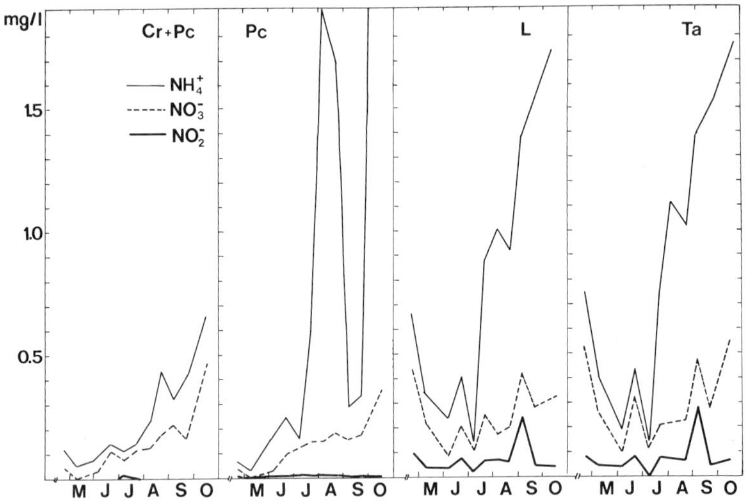

Fig. 8. Spatial and seasonal variations of mineral nitrogen forms in water of littoral transect in Nesyt fishpond

formations and seasonal patterns of mineral nitrogen forms appear more complicated (Fig. 8). The ammonia and nitrate peaks occur at all sites simultaneously during the early part of the year, while in the later part the peaks of nitrates lag behind those of ammonia. The bacterial nitrogen usually increases after the peaks of mineral or organic forms of nitrogen. The mineral nitrogen forms remain mostly at low levels in the water of the CrPc and Pc sites during the first phase of

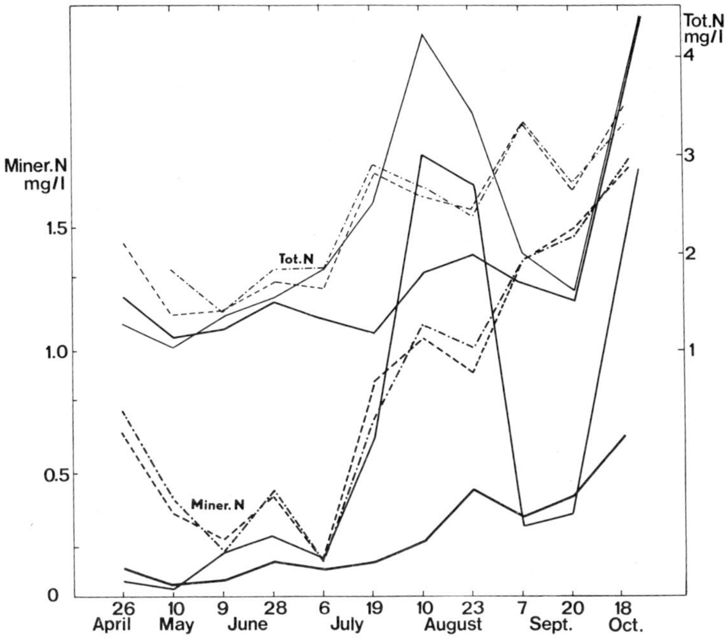

Fig. 9. Spatial and seasonal variations of total and mineral nitrogen in water of littoral transect in Nesyt fishpond

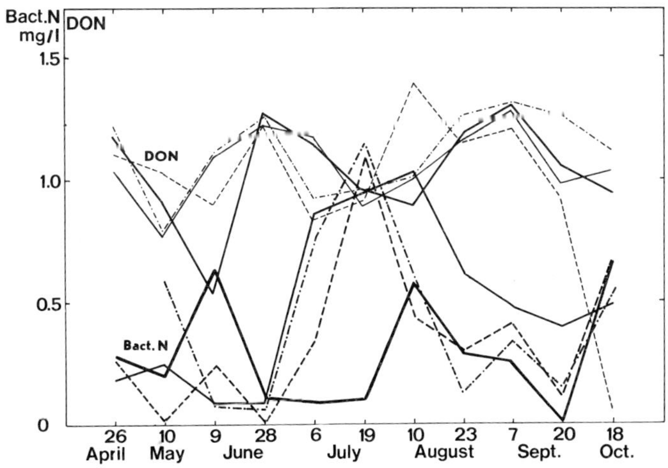

Fig. 10. Spatial and seasonal variations of dissolved organic and bacterial nitrogen in water of littoral transect in Nesyt fishpond

the growing season, when intense macrophyte growth is taking place; the nitrite contents are very low or nil here throughout the whole vegetation period. Figures 9–10 demonstrate the yearly courses of all nitrogen forms studied. The curves reveal a similarity between the L and T sites as well as the uniqeness of the Pc site.

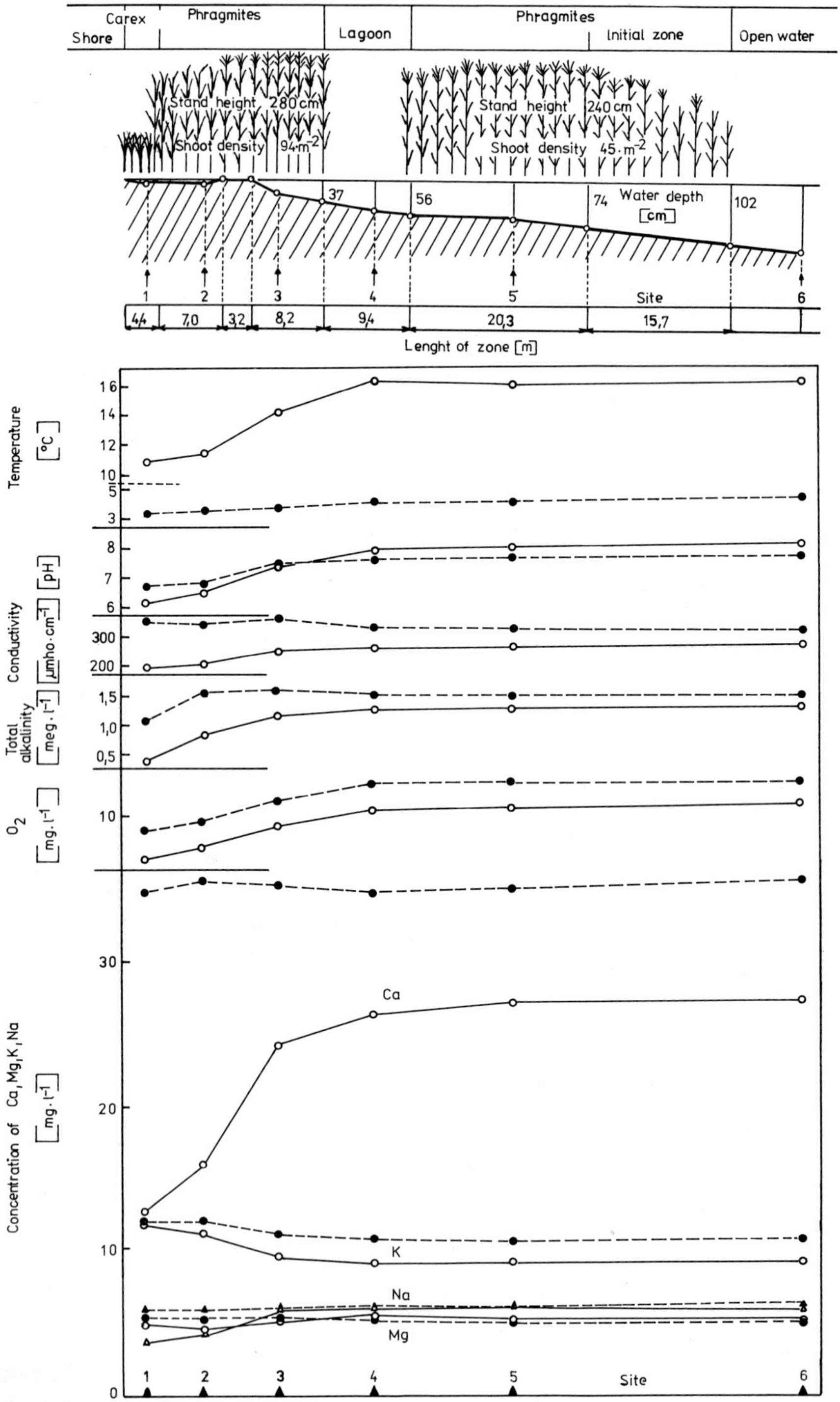

Fig. 11. Opatovický fishpond. Changes in the physical and chemical parameters of the aquatic environment along a transect from shore to open water in the stand of *Phragmitetum communis*. ○——○ vegetation period (1. April–30. Sept.); ●- - - - -● period of vegetation rest (1. Oct.–31. March)

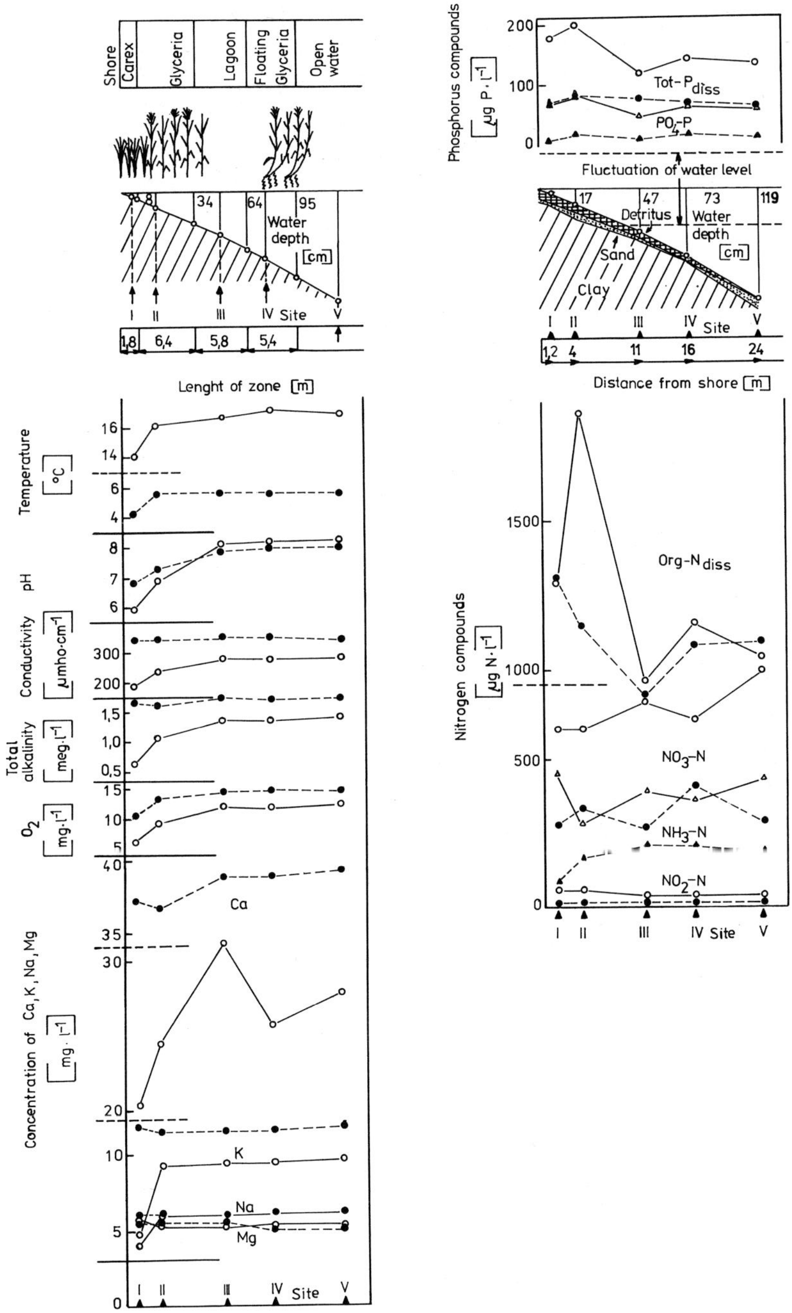

Fig. 12. Opatovický fishpond. Changes in the physical and chemical parameters of the aquatic environment along a transect from shore to open water in the stand of *Glycerietum maximae*. ○——○ vegetation period (1. April–30. Sept.); ●- - - -● period of vegetation rest (1. Oct.–31. March)

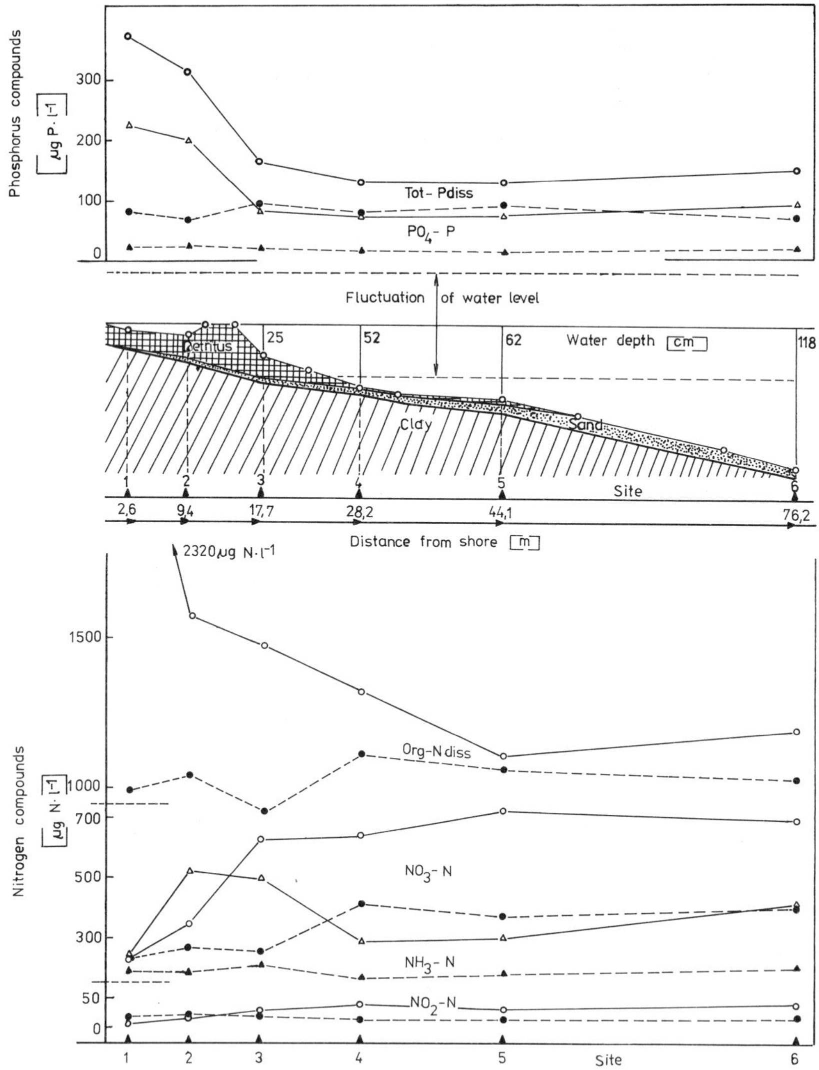

Fig. 13. Spatial variation of selected chemical parameters in littoral *Phragmitetum communis*. Sites 3, ○——○ (the densest stand); 6, ●- - - - -● (open water; see Fig. 11)

2.3.2 Opatovický Fishpond

A three-year study was made (1973–75) of the principal physical and chemical parameters of the aquatic environment in the two principal monodominant reed-belt communities of the Opatovický fishpond, namely in stands of *Phragmites communis* ("V"-biotope of the erosion—accumulation type on the eastern shore, ass. *Phragmitetum communis*) and of *Glyceria maxima* (accumulation biotope on the S.W. shore, ass. *Glycerietum maximae*). The study has revealed how the variation in the parameters studied develops within the stands in both space, i.e., along

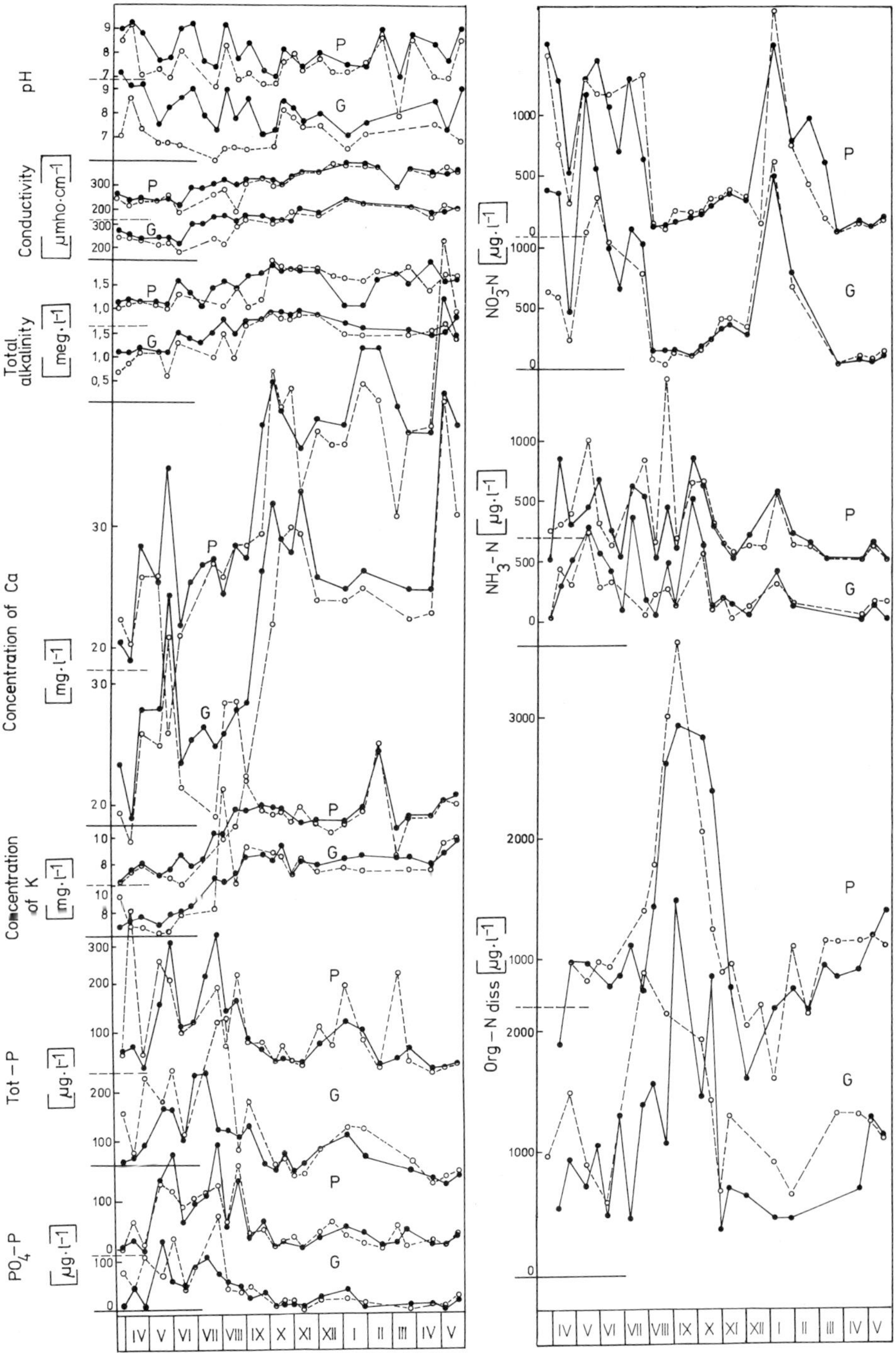

Fig. 14. Temporal variation of selected chemical parameters in *Phragmitetum communis* and *Glycerietum maximae*. Sites No. II, ○---○ (the densest stand); V, ●—● (open water; see Fig. 12)

transects laid down across the reed-belt (Figs. 11 and 12), and time, i.e., on different sampling dates during a year (Figs. 13 and 14). The sampling sites were located along the transects in individual visually distinguishable zones of the two reed-belt communities. Wooden foot-bridges situated above the water level facilitated the sampling and prevented disturbance to the biotopes. The samples were taken at two-week intervals. Average values of several of the parameters recorded permitted the identification, in each community, of two to three zones, clearly differentiated even during the period of vegetation rest. Particularly great differences were observed between the densest parts of the reed-belt plant communities and open water.

The two reed-belt communities were also intensely studied, for their primary production and mineral nutrient uptake, by Dykyjová and Hradecká (1976)—see also Sections 3.1.1 and 3.4; the characteristics of their bottom soils are presented in Section 2.4. By contrast to the naturally eutrophic South Moravian Nesyt fishpond, the Opatovický fishpond is filled with naturally oligotrophic to dystrophic water originating from an extensive woodland area with numerous peat-bogs. The eutrophication of Opatovický is only of recent date, which is also evident from the chemical parameters of its water. Large amounts of fertilizers and manure added repeatedly during the growing season, disturb the uniform character of the fishpond water to such an extent that the specific seasonal fluctuations of its chemical characteristics cannot be distinguished even in the littoral zone.

Nevertheless, the characteristic gradients of many physical parameters such as water temperature, conductivity, total alkalinity, and oxygen content, have been observed, especially in the outer littoral (Fig. 13). This pertains to both transects followed across the *Phragmitetum* and *Glycerietum* respectively; in both of them, the outer littoral is rather isolated from the rest of the fishpond by a "threshold" of detritus drifted ashore. The characteristic decrease in total dissolved P and total N (Fig. 14) as well as the increase in Ca^{2+} concentration have been observed along both transects. For the relationship between the changes in water chemistry and the growth and production of phytoplankton and periphyton see Section 4.3.

References see pp. 153—155.

2.4 Structure and Chemistry of the Fishpond Bottom

D. DYKYJOVÁ and B. ÚLEHLOVÁ

As briefly described in Section 1.1, the character of submerged soils on fishpond bottoms is different in many respects in the two regions investigated, South Bohemia and South Moravia.

In the Třeboň basin, unsaturated freshwater marsh soils occur on acid sand, clay or peat (pH 3.5–5) and in some places on transitional types to more base-saturated subsoils (Neuhäusl, 1965). The soils resemble those described in detail by Pearsall (1938) and Misra (1938) in the English Lake District. The shallow littoral overgrown by helophyte plant communities always accumulates larger amounts of organic matter in the sediments than the deeper bottom below open water. The main source of plant mineral nutrients is the interstitial water of these sediments, but the rooted helophytes take up a substantial part of their nutrients from the subsoil; they function as a "nutrient pump" between the subsoil, sediments and littoral water (Björk, 1967).

Owing to an intense decomposition (see Sect. 5) and mineralization of organic material, the water is nutrient richer in the vegetated littoral than in the open fishpond area (pelagial) (see also Bernatowicz and Zachwieja, 1966; Planter, 1970, 1973; Pieczyńska, 1972). The bottom and water chemistry of the reed-belt depend on the shore configuration and wave motion, so that two essentially differing kinds of fishpond habitats or zones occur: the accumulation and the erosion zone (see Fig. I in the Introduction). During the IBP investigations of primary production in relation to the quality of fishpond bottom sediments, marked gradients of several characteristics of bottom fertility have been found. These occurred along transects laid down across the reed-belt perpendicular to the shoreline.

In the Opatovický fishpond, attention was mainly focused on the relationship between bottom configuration, sediment fertility and primary production. Accumulation of mineral nutrients in the plant material was also assessed as it reflects the nutrient level both in the water and in the soil, and characterizes the feedback in nutrient enrichment of the biotope by the vegetation (see Sect. 3.4).

Two main types of submerged soils occur in the Opatovický fishpond: (1) unsaturated sandy soils of mineral origin with a blue-grey impermeable clay layer in the subsoil (a permanently reduced G horizon); (2) pseudogleys or semigleys prevailing on clay subsoil, especially in relatively frequently drained parts of the littoral. Typical rusty dots and veins of precipitated Fe^{3+} occur in the gleyic

total nutrient pool in the soil. But the growth of the helophytes, as in terrestrial plants, is related to available nutrients and not to their total content in the soil. This extraction method provides data useful for comparing the chemical composition of the total nutrient pool in different soil profiles (Table 1 and Fig. 1 and 2). Carbon, i.e., 50% of combustible organic matter and nitrogen were assessed by the same methods in extracts obtained by either technique.

Boyd (1968) has criticized the application of extraction methods used in agricultural soil testing to subhydric soils. Unfortunately, adequate extraction procedures applicable to submerged soils and mineral nutrition of rooted aquatic plants have not been elaborated. However in comparisons between different littoral habitats of the accumulation or erosion zones, relative differences in nutrient uptake by the helophyte stands can only be correlated with relative nutrient levels on a particular site (see Sect. 3.4.2). Tables 1–3 present examples of the total and "available" nutrient supply in different erosion or accumulation biotopes colonized by stands of one helophyte species. Deeper soil profiles of the limosal accumulation biotopes colonized by *Phragmites* or *Glyceria* betray a former fishpond bay, richly eutrophicated and gradually filled with gyttja under anerobic conditions (Fig. 1 b, c). The secondary peaks of N, C, and total P found at the depth of 0.85 m in a thick layer of petrified black sediments, below pure blue-grey clay, prove this assumption. In Table 4, biotope 2, the total contents of K, Ca, and Mg also show secondary peaks at the same depth.

Another example is the mouth of the Branná brook into the Opatovický fishpond (see Fig. 4 a–c) which is filled with plenty of both organic and inorganic

Fig. 4a–c. Branná brook. (a) Silty and muddy terraces at mouth at autumnal drawdown of Opatovický fishpond; (b) stream-bed with sedimented decomposing remnants of *Glyceria maxima* (Photo: J.P. Ondok); (c) inclined aerial photography of inflow into fishpond at high water level, lagoons replacing dead parts of *Glyceria* stands. (×) (Photo: Geographical Institute of Czechoslovak Academy of Sciences)

Fig. 5. Vertical profile of the rhizosphere and subsoil in *Phragmites communis* biotope in Nesyt fishpond. Successive layers (downwards) of förna (0–7 cm), humic matter (to 15 cm), a prevailingly mineral horizon (to 45 cm), all on top of a sandy clay gleyic subsoil. (After Husák, 1971; photo Š. Husák)

material brought in from the surrounding fields (cf. Fig. 3a–c). The detritus of the *Glycerietum maximae* and *Caricetum gracilis* accumulated in the surface layers adds new organic matter and mineral nutrients to the humus-rich A horizon of the profile.

Correlations between the nutrient pools in the habitats and the biomass of different helophyte communities have been found in the Opatovický fishpond (see Sects. 3.4.1 and 3.4.2). Analogous habitats occur in the littorals of other South

Bohemian and South Moravian fishponds and show much the same differences in helophyte production as related to the nutrient status in the soil.

The part of South Moravia where the Lednice fishponds are situated is characterized by numerous localities of steppe character, reflecting the rather dry climate of the region, with recurrent periods of drought. The geological structure of this area is formed by granite with superimposed dolomite and limestone, whose outcrops occur in several places in the proximity of the Lednice fishponds. The upper layers are formed by loess sands, loams, and clays of Tertiary and Diluvial origin. The shores and bottom of the Nesyt fishpond are covered with sand, loam, and organic sediments.

Deep layers of muddy, black, humic sediments (sapropels) occur in shallow accumulation parts of the pond which are free of rooted vegetation, as well as in some places with running water. Förna tends to accumulate under the plant cover (Fig. 5).

Gleyic layers with typical oxidation—reduction horizons are always found at different depths under the upper layers of sedimented loam or sand. During 1969 and 1970, the sediments and soils were analyzed, originating from five different sites located in both the accumulation and erosion biotopes of the Nesyt fishpond. The samples were taken at monthly intervals. The sampling sites were identical with those selected for the study of water chemistry (see Sect. 2.3.1). The aim of this study was to prove (i) the seasonal dynamics of available nutrients in the sapropels and bottom soils; and (ii) the chemical diversity of the sites investigated. A further objective was to obtain basic data for calculating the budget of mineral nutrients in the littoral vegetation. Differences between the two years of investigation may provide information on the changes resulting from fishpond management.

Figures 6–8 present the results of analyses of dry weight content in air-dried samples, and of the carbon, nitrogen, "exchangeable" phosphate, potassium, sodium, calcium, magnesium, ammonium, and nitrate contents in the bottom sediments and soils, carried out during 1969 and 1970. Most of the mineral constituents are present in the soils at lower concentrations than in the bottom sediments. This is due to the less developed sorption complex (exchange capacity), smaller content of organic matter, and high content of skeleton in the soils. Only magnesium represents an exception: its content in the soil surpassed that in the bottom sediment on several sampling dates.

While the carbon and nitrogen contents remained essentially the same during the period of our investigations, most of the remaining constituents showed more or less pronounced fluctuations. The most pronounced changes occurred in the contents of exchangeable magnesium and calcium. These changes result most probably from the absorption and release of these nutrients by diverse populations of green plants during their development. Some populations may act as temporary stores, others as "nutrient pumps" according to the theory of Björk (1967).

The average content of carbon, nitrogen, and mineral constituents in the bottom sediments and soils is presented in Table 3, separately for each of the five localities and for the years 1969 and 1970. The data illustrate the great diversity of the sampling sites. Sites 1 and 2, both situated in accumulation biotopes, are relatively stable and show a similarity in the composition of their bottom sedi-

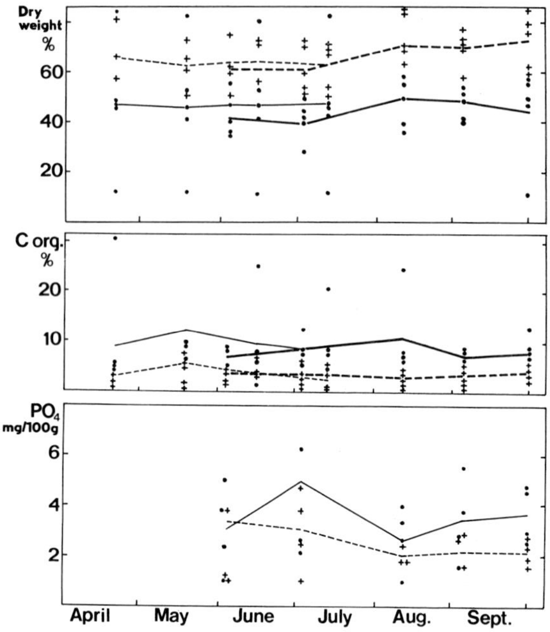

Fig. 6. Time course of changes in dry weight percentage contents, carbon and phosphate contents in bottom sediments of Nesyt fishpond. Sediments: —— 1970, —— 1969; soils- - - 1970, - - - 1969

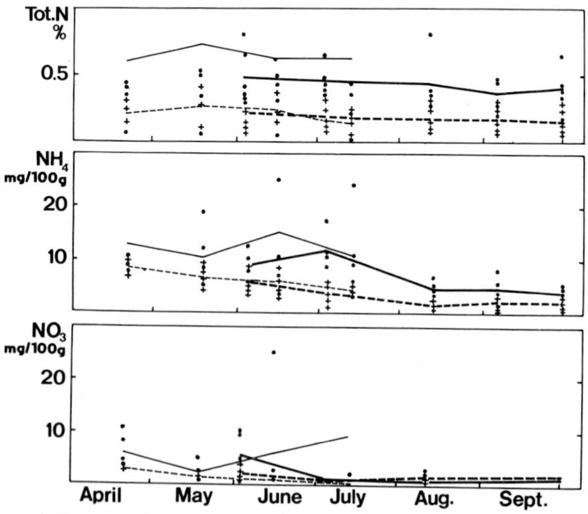

Fig. 7. Time course of changes in contents of total nitrogen and its mineral forms in bottom sediments of Nesyt fishpond (cf. Fig. 6)

ments and soils. Substantially different sediments and soils are found at site 4 in the erosion part of the pond.

The results document, in agreement with a number of other studies (Gorham, 1953; Entz et al., 1963; Rybak, 1969; Pieczyńska, 1972), that a distinct both horizontal and vertical differentiation exists in the chemical properties of sediments and soils in littoral parts of ponds and lakes, parallel with differentiation in

Table 3. Spatial variation of sediment (0–5 cm) and soil (5–15 cm) chemistry in different biotopes in the Nesyt fishpond

Site[c]	Season	Sediments								
		C[a]	N[a]	PO_4^{3-} [b]	NH_4^+ [b]	NO_3^- [b]	K^+ [b]	Ca^{2+} [b]	Mg^{2+} [b]	Na^+ [b]
1 *Hippuridetum vulgaris* accumulation biotope	1969	7.28	0.36	5.6	9.2	1.9	45	482	54	15
	1970	7.53	0.47	—	11.1	4.9	49	339	70	20
2 *Phragmitetum communis* accumulation biotope	1969	14.06	0.59	—	—	—	20	543	70	25
	1970	27.75	2.01	—	23.8	18.1	53	479	49	25
3 Near the dam, erosion biotope	1969	6.66	0.28	4.4	5.0	1.5	24	287	49	12
	1970	—	—	—	5.1	1.6	54	—	42	—
4 *Phragmitetum communis* erosion biotope	1969	8.15	0.41	2.5	6.5	2.7	41	475	61	13
	1970	7.30	0.42	—	7.9	1.9	32	342	50	15
5 *Phragmitetum communis* erosion-accumulation biotope	1969	7.59	0.45	1.8	6.6	3.1	41	388	71	17
	1970	5.06	0.30	—	7.1	2.8	23	279	76	13
		Soils								
1	1969	4.60	0.24	2.5	3.7	1.6	44	304	33	7
	1970	4.43	0.22	—	6.6	1.3	20	288	67	13
2	1969	2.66	0.10	—	2.6	1.4	10	248	55	6
	1970	3.16	0.16	—	7.3	1.4	9	162	54	13
3	1969	2.91	0.10	4.0	1.8	2.0	10	247	56	5
	1970	0.85	0.08	—	—	1.6	6	221	42	5
4	1969	5.52	0.30	1.7	3.4	2.0	33	393	62	10
	1970	5.72	0.30	—	6.1	1.3	33	308	96	8
5	1969	2.64	0.14	2.0	3.6	1.7	27	345	70	13
	1970	1.45	0.12	—	5.0	1.2	18	269	44	11

[a] Average annual contents in % dry w.
[b] Average annual contents in mg per 100 g dry w.
[c] Biotopes described in text.

the vegetation and in water chemistry (Neuhäusl, 1965; Zachwieja, 1965; Bernatowicz and Zachwieja, 1966; Björk, 1967; Planter, 1970, 1973).

The differences between the data of 1969 and 1970 (see Figs. 6–8 and Table 3) pertain above all to the chemical changes affected by fishpond management. The autumnal water discharge from the fishpond in 1969, after which the shallower sites remained drained for several months, had pronounced effects mainly on the bottom sediments and, to some extent, on the soils. The exposure and aeration of the bottom was accompanied by mineralization of organic substances in the bottom sediments: the average C and N contents increased somewhat in 1970 and a considerable increase was observed for all mineral forms of nitrogen.

The characteristics of certain sites may change radically, as illustrated by data for site 3 in Table 3. The organic layer present on site 3 in 1969 disappeared completely and was buried below sedimented sand in 1970.

Our results confirm the findings of Pieczyńska (1972) that the bottom characteristics of eroded littoral biotopes may undergo pronounced changes during short time periods.

The time courses of the contents of mineral substances, especially of sodium and potassium (Fig. 8), indicate that new dynamic equilibria develop between

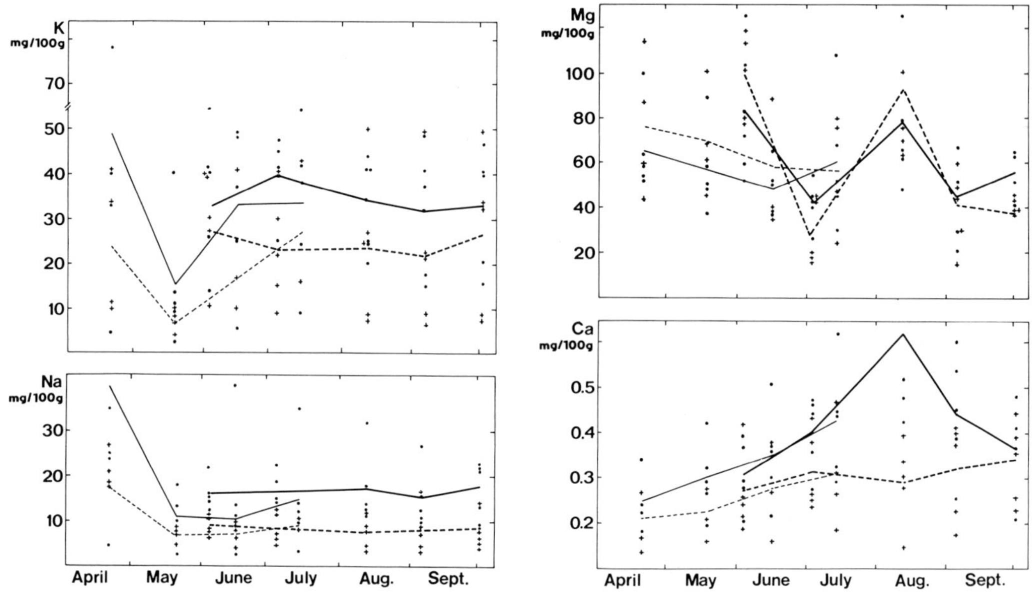

Fig. 8. Time course of the exchangeable K, Na, Mg, and Ca contents in bottom sediments of Nesyt fishpond. Curves as in Figure 6

Table 4. Water, sapropel, and bottom soil analyses of halotrophic (Nesyt fishpond) and eutrophic (Opatovický fishpond) sites of *Phragmitetum communis.* Summarized according to Úlehlová et al. (1973), Přibil (unpublished data), and Dykyjová and Hradecká (1976)

Contents	Water (seasonal variations)		Sapropel (0–25 cm) (spatial variations)		Bottom soil (25–40 cm) (spatial variations)	
	Nesyt	Opatovický	Nesyt	Opatovický	Nesyt	Opatovicky
			% dry w.		% dry w.	
Organic matter			5.06– 27.75	36.6 – 45.6	1.45– 5.72	2.40– 2.46
N total			0.30– 2.01	0.17– 2.80	0.12– 0.30	0.01– 0.25
	mg/l		mg/100 g dry weight		mg/100 g dry weight	
N total	—	0.01 – 3.62				
NH_4-N	0.35– 0.89	0.103– 1.50	5.1– 23.8	—	3.0– 7.3	—
NO_3-N	0.46– 4.99	0.025– 1.31	1.6– 18.1	—	1.2– 1.7	—
PO_4-P	1.96– 3.19	0.011– 0.17	3.6	22.0–100.1	2.5	0 – 6.1
Tot. P. diss.	—	0.065– 0.35				
K	20.7 –36.16	6.4 –18.2	5.4– 53.5	10.1– 18.6	8.7– 33.5	0.8– 3.8
Na	53.8 –77.3	5.2 – 7.4	3.1– 24.9	5.2– 8.1	8.2– 12.6	1.5– 1.8
Ca	57.3 –98.8	19.9 –29.4	221 –472	222 –355	162 –308	55 –68
Mg	39.56–63.3	4.3 – 6.5	42 – 76	17 – 77	44 – 96	0 –14

different phases (compartments) of the ecosystem subsequent to the drainage and refilling of the fishpond.

A general outline of the differences in trophic conditions of both fishpond ecosystems is presented in Table 4.

References see pp. 153—155.

References

Bayer, E., Bajkov, A.: Hydrobiological study of Lednice fishponds I. (In Czech). Sb. Vys. Šk. Zeměd. Brno **14**, 1–165 (1929)

Bernatowicz, S., Zachwieja, J.: Types of littoral found in the lakes of the Masurian and Suwalki lakelands. Ekol. Pol. Ser. A **14**, 519–547 (1966)

Bílý, J.: New or less known Moravian diatoms. (In Czech). Sb. Klubu Přír. Brno, 16–20, 1932

Björk, S.: Ecological investigations of *Phragmites communis*. Studies in theoretical and applied limnology. Folia Limnol. Scand. **14** (1967)

Boyd, C. E.: Some aspects of aquatic plant ecology. In: Reservoir Fishery Resources Symposium. Athens, Georgia, April 5–7, 1968

Dvořák, J.: Nitrogen and phosphorus compounds in the water of the littoral of the Nesyt fishpond. In: Littoral of the Nesyt fishpond. Květ, J. (ed.): Studie ČSAV Praha **15**, 1973, pp. 139–141

Dykyjová, D., Hradecká, D.: Productivity of reed-bed stands in relation to the ecotype, microclimate and trophic conditions of the habitats. Pol. Arch. Hydrobiol. **20**, 111–119 (1973)

Dykyjová, D., Hradecká, D.: Production ecology of *Phragmites communis* Trin. I. Relations of two ecotypes to microclimate and nutrient conditions of habitat. Folia Geobot. Phytotax. Bohemoslovaca **11**, 23–61 (1976)

Dykyjová, S., Květ, J.: Comparison of biomass production in reedswamp communities growing in South Bohemia and South Moravia. In: Productivity of Terrestral Ecosystems. Production Processes. Dykyjová, D. (ed.): Czechosl. IBP/PT-PP Rep. No. **1**, Praha, 1970, pp. 71–76

Entz, B., Ponyi, J. E., Tamas, G.: Sedimentuntersuchungen im südwestlichen Teil des Balaton, in der Bucht von Keszthely in 1962. Ann. Inst. Biol. Tihany **30**, 103–125 (1963)

Fischer, R.: Die Algen Mährens und ihre Verbreitung. Verh. Naturforsch. Ver. Brünn **52**, 1–94 (1920)

Gavenčiak, Š.: Research on water loss by evapotranspiration from water plants. (In Slovak, with German and Russian summary). Vodohospodářský Časopis (Bratislava) **20**, 16–32 (1972)

Geiger, R.: Das Klima der bodennahen Luftschicht. Braunschweig: Friedr. Vieweg und Sohn, 1961

Golterman, H. L., Clymo, R. S.: Methods for Chemical Analyses of Fresh Waters. (IBP Handbook No. 8). Oxford: Blackwell Sci Publ., 1969

Gorham, E.: Chemical studies on the soils and vegetation of waterlogged habitats in the English Lake District. J. Ecol. **41**, 345–360 (1953)

Gröterud, O.: Ice analyses. Data from three Norwegian Lakes. Hydrobiologia **40**, 371–391 (1972)

Hofman, P., Havránek, M., Čuta, J., Chalupa, J., Maděra, V., Hamáčková, J., Kohout, M.: Official Methods of Chemical Water Analyses. (In Czech). Praha: SNTL, 1965

Hradecká, D., Květ, J.: Morphological and production characteristics of three clones of Phragmites communis Trin. from the Nesyt area. In: Littoral of the Nesyt Fishpond. Květ J. (ed.). Studie ČSAV **15**, 1973, pp. 97–102

Husák, Š.: Productivity and structure of intact and cut invasion stands of *Phragmites communis* Trin. and *Typha augustifolia* L at the Nesyt fishpond. (In Czech) Thesis, Botany, Dep., Fac. of Science, Purkyně Univ. Brno (1971)

Jírovec, O.: Water chemistry of the Lednice fishponds. (In Czech). Věst. Král. Čes. Spol. Nauk Tř. II, 1–19 (1936)
Korycka, A.: Seasonal changes in water chemical composition in seven lakes. Pol. Arch. Hydrobiol. **16** (29), 1–29 (1969)
Květ, J., Svoboda, J., Fiala, K.: Canopy development in stands of *Typha latifolia* L. and *Phragmites communis* Trin. in South Moravia. Hidrobiologia **10**, 63—75 (1969)
Martin, N. A.: Temperature fluctuations within English lowland ponds. Hydrobiologia **40**, 455–469 (1972)
Misra, R. A.: Edaphic factors in the distribution of aquatic plants in the English lakes. J. Ecol. **26**, 411 (1938)
Naumann, E.: Grundzüge der regionalen Limnologie. Die Binnengewässer, Vol. XI. Stuttgart: Schweitzerbart 1932
Neuhäusl, R.: Vegetation der Röhrichte und der sublitoralen Magnocariceten in Wittingauer Becken. Vegetace ČSSR s. A., Vol. 1. Praha: Academia, 1965, pp. 11–117
Ondok, J. P.: Photosynthetically active radiation in a stand of *Phragmites communis* Trin. I. Distribution of irradiance and foliage structure. Photosynthetica **7**, 8–11 (1973a)
Ondok, J. P.: II. Model of light extinction in the stand. Photosynthetica **7**, 50–57 (1973b)
Ondok, J. P.: III. Distribution of irradiance on sunlit foliage area. Photosynthetica **7**, 311–319 (1973c)
Ondok, J. P.: IV. Stochastic model. Photosynthetica **9**, 201–210 (1975)
Ondok, J. P.: Régime of global and photosynthetically active radiation in helophyte stands. Studie ČSAV **10**, 1–112, Praha: Academia, 1977
Pearsall, W. H.: The soil complex in relation to plant communities. J. Ecol. **26**, 180–194 (1938)
Pieczyńska, E.: Production and decomposition in the eulittoral zone of lakes. In: Proc. IBP-UNESCO Symp. on Productivity Problems of Freshwaters. Kajak, Z., Hillbricht-Ilkowska, A. (eds). Warsaw, Cracow: PWN, 1972b
Planter, M.: Physicochemical properties of the water of reed-belts in Mikołajskie, Tałtowisko and Sniardwy lakes. Pol. Arch. Hydrobiol. **17**, 337–356 (1970b)
Planter, M.: Physical and chemical conditions in the helophyte zone of the lake littoral. Pol. Arch. Hydrobiol. **20**, 1–7 (1973)
Přibáň, K.: Climatological characteristic of the Třeboň basin and of Opatovický fishpond area. In: Ecosystem Study on Wetland Biome in Czechoslovakia. Hejný, S. (ed.). Czechosl. IBP/PT-PP Rep. No. **3**, Třeboň, pp. 11–14, 1973a
Přibáň, K.: Microclimatic measurements of temperatures in pure reed stand. In: Ecosystem Study on Wetland Biome in Czechoslovakia. Hejný, S. (ed.). Czechosl. IBP/PT-PP Rep. No. **3**, pp. 65–70, Třeboň, 1973b
Quitt, E.: Klimatické oblasti Československa (Climatic areas of Czechoslovakia; in Czech). Studia Geographica (Brno) **16**, 1–73 (1971)
Rejmánková, E.: Seasonal changes in the growth rate of duckweed community *(Lemnetum gibbae)*. Folia Geobot. Phytotaxon. Bohemoslov. **8**, 1–13 (1973)
Rybak, J. I.: Bottom sediments of the lakes of various trophic types. Ekol. Pol., Ser. A **17**, 611–662 (1969)
Šmíd, P.: Fundamental climatological and hydrological characteristics of grassland ecosystems in the Lanžhot area. In: Ecosystem Study on Grassland Biome in Czechoslovakia. Rychnovská, M. (ed.). Czechosl. IBP/PT-PP Rep. No. **2**, Brno, 1972, pp. 11–16
Šmíd, P.: Microclimatological characteristic of reed swamps at the Nesyt fishpond. In: Littoral of the Nesyt Fishpond. Květ, J. (ed.). Studie ČSAV **15**, 29–38 (1973)
Šmíd, P.: Evaporation from a reedswamp. J. Ecol. **63**, 299–309 (1975)
Šmíd, P., Palát, M.: Climate of the area of the Lednice Fishponds. In: Littoral of the Nesyt Fishpond. Květ, J. (ed.). Studie ČSAV **15**. Praha: Academia, 1973, pp. 25–27
Soudek, Š.: Contribution to the study of zooplankton of Lednice fishponds. (In Czech). Sb. Masarykovy Akad. Práce **3**, 39–79 (1929)
Švec, R., Nekovář, F., Vojtěch, S.: Outline of the Geography of South Bohemia. I–III. (In Czech.) Rozpravy Pedag. Fak., Ř. Přírod. Věd. 1–67, 1–123, 1–65, České Budějovice 1967, 1968, 1969

Szumiec, M.: Zeiträumige Temperaturänderungen über Laichteichen und über dem Boden in den niedrigsten Luftschichten. Verh. Int. Ver. Theor. Angew. Limnol. IX, Fishponds **16**, 1311–1317 (1966)

Szumiec, M.: Determination of water temperature in shallow water bodies. Acta Hydrobiol. **15**, 247–257 (1973)

Thienemann, A.: Die Binnengewässer Mitteleuropas. Die Binnengewässer, Vol. I. Stuttgart: Schweitzerbart, 1926

Tooming, K., Niilisk, H.: Transition coefficients from integrated radiation to photosynthetically active radiation. (In Russian). In: Fitoaktinometričeskie Issledovaniya Rastitelnogo Pokrova. Valgus, Tallin: 1967

Úlehlová, B.: An ecological study of aquatic habitats in north-west Overijssel, The Netherlands. Acta Bot. Neerl. **19**, 830–858 (1970)

Úlehlová, B., Husák, Š., Dvořák, J.: Mineral cycles in reed stands of the Nesyt fishpond in southern Moravia. Pol. Arch. Hydrobiol., Warszawa **20**, 121–129 (1973)

Úlehlová, B., Rejthar, L.: Water chemistry in the Nesyt fishpond. In: Littoral of the Nesyt Fishpond. Květ, J. (ed.). Studie ČSAV **15**. Praha: Academia, 1973, pp. 39–43

Velásquez, J.: Communities and production ecology of *Eleocharis acicularis* (L.) R. et Sch. CSc. Thesis, Institute of Botany, Czechosl. Acad. Sci. (Ms). Průhonice, 1975

Walter, H., Lieth, H.: Klimadiagram. In: Weltatlas. Jena: VEB Fischer-Verlag 1960

Willer, A.: Kleinklimatische Untersuchungen in *Phragmites*-Gelege. Verh. Int. Ver. Theor. Angew. Limnol. **10**, 566–574 (1949)

Wit, C. T. de: Potential photosynthesis of crop surfaces. Agric. Sci. **7**, 141–149 (1965)

Zachwieja, J.: Daily variation of temperature, O_2, CO_2, pH, and alkalinity in the littoral zone of the Mamry Lake. (In Polish). Pol. Arch. Hydrobiol. **13**, 5–27 (1965)

Zapletálek, J.: Hydrobiological study of the Lednice fishponds. (In Czech). Sb. Vys. Šk. Zeměd. Brno **24**, 1–70 (1932)

Zimmermann, F.: Die Fauna und Flora der Grenzteiche bei Eisgrub. I. Gastropoda et Acephala.-Verh. Naturforsch. Ver. Brünn **54**, 1–26 (1916)

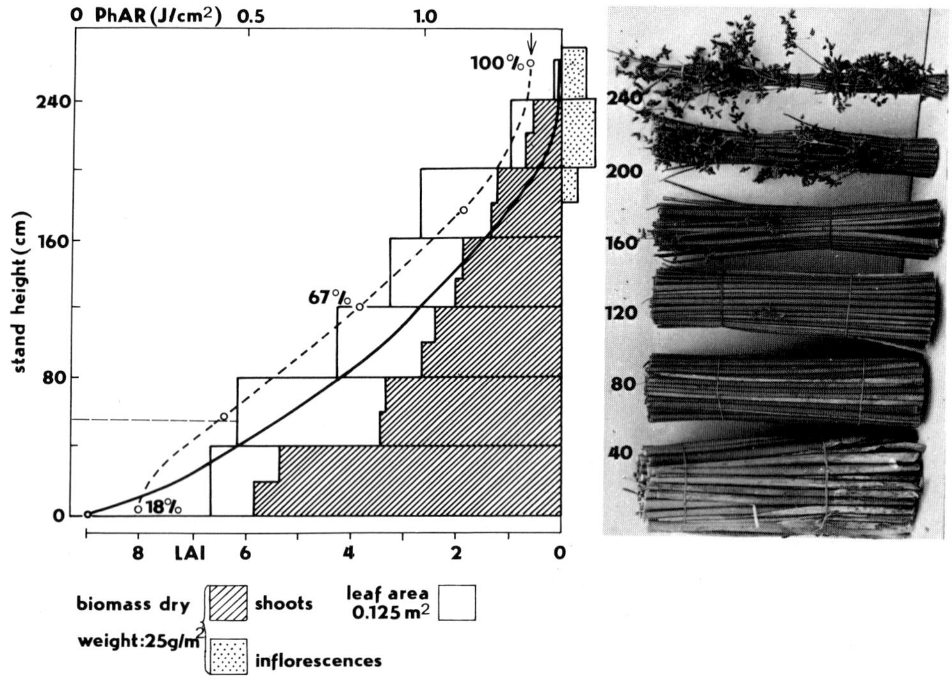

Fig. I. Diagram of vertical structure of a *Schoenoplectus lacustris* stand, harvested in a stratified manner from 1 m^2 of stand area. Photo: shoots, divided horizontally into 40 cm high segments. The cumulative leaf area index *(full line)* follows similar curve to radiation profile; *dashed line;* instantaneous percentage values of irradiance of incident PhAR in open space, J/cm^2, recorded in same biotope. (Photo and drawing by D. Dykyjová)

Section 3
Primary Production and Production Processes in Littoral Plant Communities

In the IBP research project, the primary production by higher vegetation and certain production processes occurring in plants were studied most thoroughly in fishpond reed-belts. This section presents the principal results of these studies. The intraspecific variability and its role in determining plant production are acknowledged. However, in a natural vegetation, the validity of any production estimate depends much on the method and sampling technique used. The various ways—both direct and indirect—of assessing the growth, development and production both of above- and belowground plant parts have therefore been evaluated. Examples are shown of the growth analyses and dry-matter economy of reedswamp as well as other fishpond littoral plant communities, and of the assessment of their vertical structure by stratified harvesting. Selected data on biomass and estimates of annual net production are then given for a number of fishpond littoral higher-plant communities. Illustrated also is the storage of energy in their biomass and their efficiency of solar energy conversion into net primary production. Of the production processes, the photosynthetic production by common reed communities was ascertained in two ways: from growth-analytical data and by measurements of gas exchange; finally, a model of the photosynthetic production has been proposed.

Water relations have also been studied mainly in common reed, while the nutrient uptake and its relation to net production have been followed in all ecological dominants of the fishpond reed-belts. The results of cultivation experiments in hydroponics prove the findings made in the field. As duckweeds frequently form an important component of the fishpond littoral ecosystems, a brief survey is presented of the growth, production and mineral nutrient uptake by duckweed communities in ponds.

3.1 Plant Growth and Estimates of Production

3.1.1 Intraspecific and Clonal Variability and Its Importance for Production Estimates

D. Dykyjová

The primary production of natural communities is controlled not only by the climatic and edaphic characteristics of the habitat, but also by the genome structure of the populations of local ecotypes in the range of their geographical area (McNaughton, 1966). Littoral stands of emergent macrophytes, especially reed-bed stands, present relatively simple natural communities in which the population of the ecotype is the dominant component of the community.

Phragmites communis is the most polymorphous and widespread species, growing in uniform clonal populations in the variable littoral habitats. In its pure stands, it was possible to measure the variability of biomass production of one *Phragmites* clone, extended over an area within which there is large variation in edaphic and microclimatic conditions. Sometimes, however, clones of different reed ecotypes characterized by different chromosome idiograms may grow side by side under similar environmental conditions (Björk, 1967). Björk (1963, 1967) and van der Toorn (1971, 1972) have reported that tetraploid reed forms ($2n = 48$) are found in highly diverse habitats and show distinct differences in morphology and production. In comparative studies on the morphological parameters, biomass and production of reed stands in the littoral of the Opatovický fishpond, the biometric and production characteristics of two different reed clones (possibly ecotypes), growing close to each other in two different habitats within the same fishpond littoral, were investigated for seven years. Both clones, those on the littoral (V) and limosal (S) sites, retained their different morphological and production parameters during all seven growing seasons (Dykyjová, 1971a). Both clones are tetraploids (Pazourková, 1973). The clone V represents a typical littoral reed growing in the whole erosion littoral zone of the fishpond with the water level varying between 0.1 and 1 m above the bottom and exposed to the prevailing westerly wind blowing across the fishpond. The other S clone grows in a limosal terrestrialized bay never actually flooded by the fishpond water, but auxotrophicated by occasional flooding from a waste-water channel from the nearby pig farm. This habitat is sheltered from wind by tall stands of spruce and oaks. The microclimate is therefore milder here than in the open fishpond littoral (see Sect.

Table 1. Biomass (dry wt. m^{-2}) and some other production characteristics of littoral and limosal reed stands (Dykyjová and Hradecká, 1973)

Number of shoots		Shoot height (cm)		Dry wt. ($g\,m^{-2}$)				Leaf area $m^2\,m^{-2}$
				Shoots		Underground		
$\bar{x}$	max.	$\bar{x}$	max.	$\bar{x}$	max.	$\bar{x}$	max.	
Littoral ecotype								
80–127	90–144	200–260	260–320	1,110–1,920	1,380–2,200	5,890	8,560	3.8– 9.5
Limosal ecotype								
72–142	81–154	230–280	270–320	1,520–3,000	1,650–3,250	3,360	5,930	5.5–11.5

Mean ($\bar{x}$) and maximum values in seasons 1964–1970.

2.1.1). Morphological and biometrical parameters of leaves, stems, and inflorescences are larger in the auxotrophicated limosal ecotype (Fig. 1a). Each season its net production was therefore greater than in the littoral clone. The biomass (dry weight) followed in both reed clones for seven successive seasons (1964–1970), is presented in Table 1 (Dykyjová, 1971a; Dykyjová and Hradecká, 1973).

Hradecká (1973) describes, in her comparative morphological study of the common reed inflorescences sampled in South Bohemia, South Moravia, and South Slovakia, a large variability in the development of the *Phragmites* panicles. The subpannonian region of South Moravia, and even more of South Slovakia, with a warmer climate and a longer growing season, are characterized by a greater variability of reed clones and ecotypes, according to increasing diversity of forms and species with the decreasing geographical latitude and increasing temperature. According to Hradecká (1973), *Phragmites* forms with markedly green edges of the paleal nerves (in green or flavescent panicles) are only typical of South Moravia and South Slovakia (subpannonian area) whereas in South Bohemia (subatlantic area) they are found only exceptionally along communications and rivers. Similarly, the reeds with purple or purplish-brown glumes and lemmae (i.e., with purplish-brown and dark panicles) are typical of South Bohemia; in South Moravia they are found only along rivers. The observations made so far seem to imply that the differences in the morphology of *Phragmites* from different regions could partly be explained on plant-geographical grounds. The migration, hybridization, and adaptation of different reed types to various habitats seem to be taking place even at present. For illustration of the diversity of *Phragmites* forms in the pannonian region, diagrams (contact copies) of extreme types are presented: leaves and panicles of the largest terrestrial form growing on loess soils near the Nesyt fishpond; growth diagram of internodes, leaves and panicles of the shortest form from a saline habitat at the Nesyt fishpond; tall form with large leaf area and purplish-black panicles growing along the banks of the Dyje river near the Šakvický fishpond in South Moravia (Fig. 1b–d).

Pazourková (1973), investigating the chromosome numbers in different *Phragmites* forms growing in Czechoslovakia or introduced from foreign habitats, has found an aneuploid chromosome number (2n–50) with two extra chromosomes

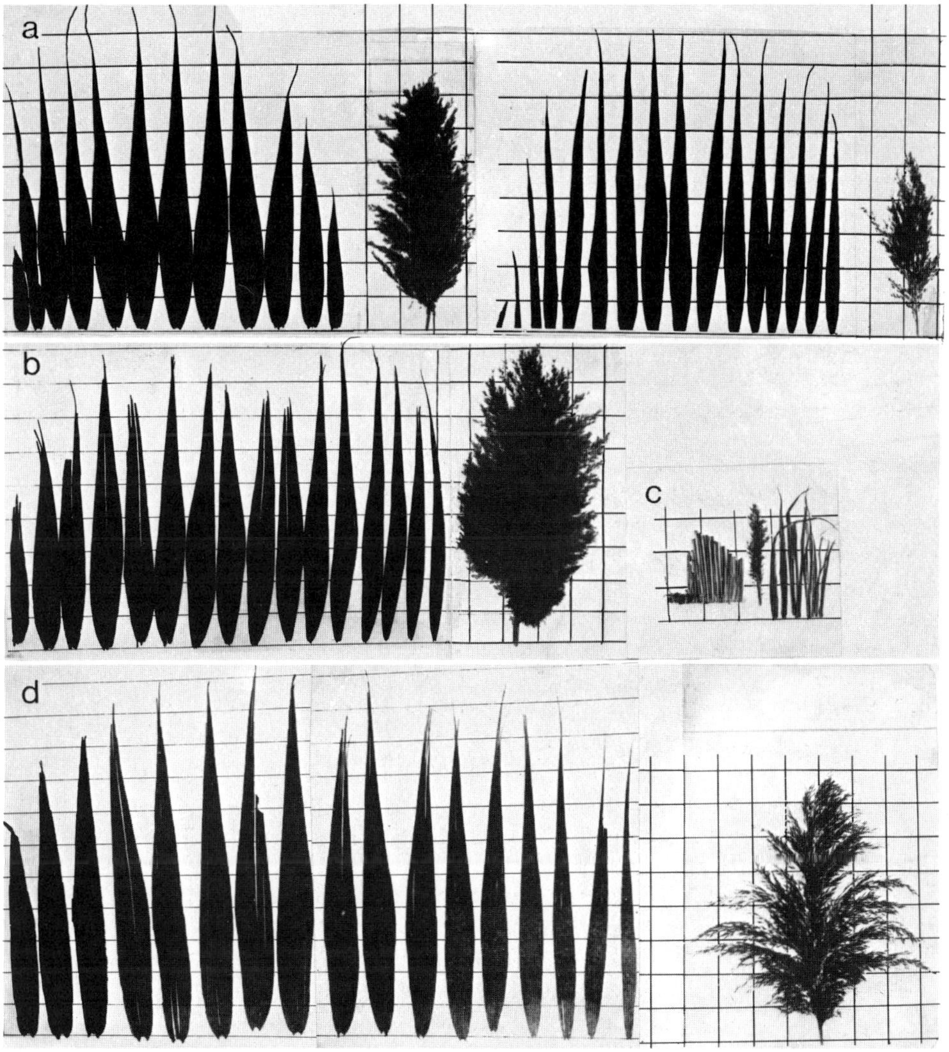

Fig. 1a–d. Growth diagrams of leaf blades, in ascending order from base to apex, and panicles of (a) Limosal *(S, left)* and littoral *(V, right)* *Phragmites* clones from the Opatovický fishpond. (b) Terrestrial *Phragmites* form growing on South Moravian loess near Nesyt fishpond. (c) Internodia, leaf sheaths, leaf blades, and panicle of a short *Phragmites* form from a saline habitat in littoral of Nesyt fishpond. (d) *Phragmites* form from the Šakvický fishpond, South Moravia, with a large leaf area. Scale in (a) to (d) 5 by 5 cm

in the caryotype of the short reed from the saline habitats at the Nesyt fishpond. In the caryotype of *Phragmites* from the South Moravian Hlohovecký fishpond, one pair of chromosomes with satellites was present.

Even more widely varying forms may be found in more southern countries, for example, the floating "plaur" reed form of 4–6 m in height, in the Danube Delta (Romania) (Fig. 2a). This giant form is octoploid (chromosome number 2 n = 96,

Fig. 2a and b. Growth diagram of leaf blades, and panicle of the (a) floating ("plaur") reed form growing in the Danube Delta (Romania). (b) Contact copies of a dwarf *Phragmites* ecotype in comparison with two panicles of a littoral *(left)* and terrestrial *(right) Phragmites* form growing in neighboring freshwater habitats. *On the right*: sterile creeping diploid shoots (for details see text). Scale in (a) and (b) 5 by 5 cm

Raicu et al., 1972) and denoted as var. *gigantissima*. Raicu et al. quote a stem height of 416 cm (August 30) and a leaf area of 4000 cm^2 per shoot. In measurements of a stem sampled on September 3, 1970, the stem height was 520 cm and the leaf area of 22 leaves per stem was 3320 cm^2. In the mediterranean habitats in highly saline and alkaline soils in the Bouches du Rhône, La Camargue nature reserve, a dwarf form of *Phragmites* grows no taller than 15–20 cm (Fig. 2b). It has developed sterile creeping shoots (Dykyjová, 1971a) with a diploid chromosome number $2n = 24$ (Pazourková, 1973), after transplantation into a freshwater nutrient solution under Central European climatic conditions.

All examples mentioned show how necessary it is to respect the forms or ecotypes of local populations in the examination of productivity as controlled by climatic or edaphic factors in a larger geographical area.

The other dominant reed-belt species appear as variable as the common reed. *Typha angustifolia*, displaying in many cases a very high production, inclines to hybridization with *T. latifolia*. It is possible that the highly productive stands found in our measurements, composed predominantly of sterile polycormones, were hybrids exhibiting a heterosis effect. Similar cases are described by Bray et al. (1959) for highly productive *Typha latifolia* populations. According to S. G. Smith (personal communication), many stands observed in South Bohemia and South Moravia might be formed by hybrid *Typha* populations.

Two stands of *T. angustifolia*, growing side by side in the western littoral of the Opatovický fishpond, differ in average shoot height: 3.30 m and 2.60 m, respectively, but their shoot density is the same.

The highly variable species *Bolboschoenus maritimus* occurs in South Bohemia only in its acidotrophic subspecies: *B. maritimus* ssp. *maritimus* and its productivity is controlled by the nutrient content in the fishpond bottom and by the water level. In South Moravia, many populations of the alkalitrophic *B. maritimus* ssp. *compactus* (Drobov) Hejný occur in mixed stands or side by side with the acidotrophic subspecies. Hybrid swards probably occur here as well.

References see pp. 285—291.

3.1.2 Selection of Sampling Areas in Assessment of Production

J. P. ONDOK and J. KVĚT

Most of our estimates of macrophyte production and biomass have been made in nearly monospecific stands which frequently occur in the littorals of South Bohemian and South Moravian fishponds. Communities formed by mixed stands were taken into consideration only exceptionally, for example, when mapping fishpond littorals and estimating the plant biomass of relatively large areas (Květ and Ondok, 1973). Selection of the monospecific stands has revealed differences and peculiarities in the production of individual helophyte populations. Mixed communities dominated by *Carex* species often occur in the epilittoral where they occupy extensive areas. The marginal waterlogged terrestrial communities also contain a large number of species (see Sect. 1.2).

In a monospecific stand, the selection of sampling plots and methods depends on the aim of the investigation. In this case, three different techniques were applied:

(a) The biomass was measured in plots with the highest shoot density, selected intentionally within a stand. This technique provides an estimate of the maximum production capacity of a stand; it also serves in interspecific comparisons of production at a defined level.

(b) The average biomass present in the whole area occupied by a helophyte stand was estimated by various methods. The shoot density, shoot height, and their spatial distribution are greatly variable in such stands. Hence the problem of selecting the most appropriate sampling method had to be solved first. The methods employed will be briefly described in the next section.

(c) The seasonal changes in shoot biomass have been followed by both direct (harvesting) and indirect methods which provided the basis for growth analysis.

All three techniques yielded data on the aerial biomass. The underground biomass was assessed only in some instances; here, however, the old underground parts could not be distinguished from those produced during the current season. This problem is discussed in Sections 3.1.3 and 3.1.4.

3.1.2.1 Destructive Methods

Individual shoots were harvested from a certain number of selected sampling plots and the biomass (dry weight) of individual aerial plant parts (stems, leaves, inflorescences) as well as leaf area were determined separately in each plot. Sometimes, the harvest was made in a stratified manner in order to assess the vertical distribution of the biomass and leaf area (see Sect. 3.1.6). The selection of the size of the sampling plots depended on the type of the stand and on the spatial distribution of the shoots. Most of the biomass estimates are based on sampling plots of 1 m × 1 m or 0.5 m × 1 m in area.

Helophytes often form clusters varying in size and determined by vegetative shoot spreading (see Sect. 3.1.3 and Fig. 1) or by variation in the bottom substrate. A sampling plot therefore has to be larger than the cluster size. The dimensions of the clusters were determined by "cluster analysis" (Greig-Smith, 1964; Kershaw, 1964), which was applied to the principal types of monospecific helophyte stands (for more details see Ondok, 1970b, 1971b). Figure 2 shows the results of such an analysis using the "mean square" parameter for some selected helophyte stands. Table 1 summarizes the minimum sampling areas for these stands.

The selection of sampling plots inside a stand may be accomplished in several ways:

(a) Harvesting from randomly located sample plots. In littoral helophyte stands the random sampling requires a large number of replicates if the error of an estimate is to be kept within a narrow range. This is due to the high variation in shoot density and height, which, in addition, is complicated by a pronounced density gradient from the shore toward open water. An example of such a gradient is shown in Figure 3. This sampling technique becomes unmanageable if many replicates are needed.

Fig. 1. Remnants of clusters of *Typha angustifolia* on the shore of the drained Opatovický fishpond after stands had been destroyed by grazing muskrats. (Photo: J. P. Ondok)

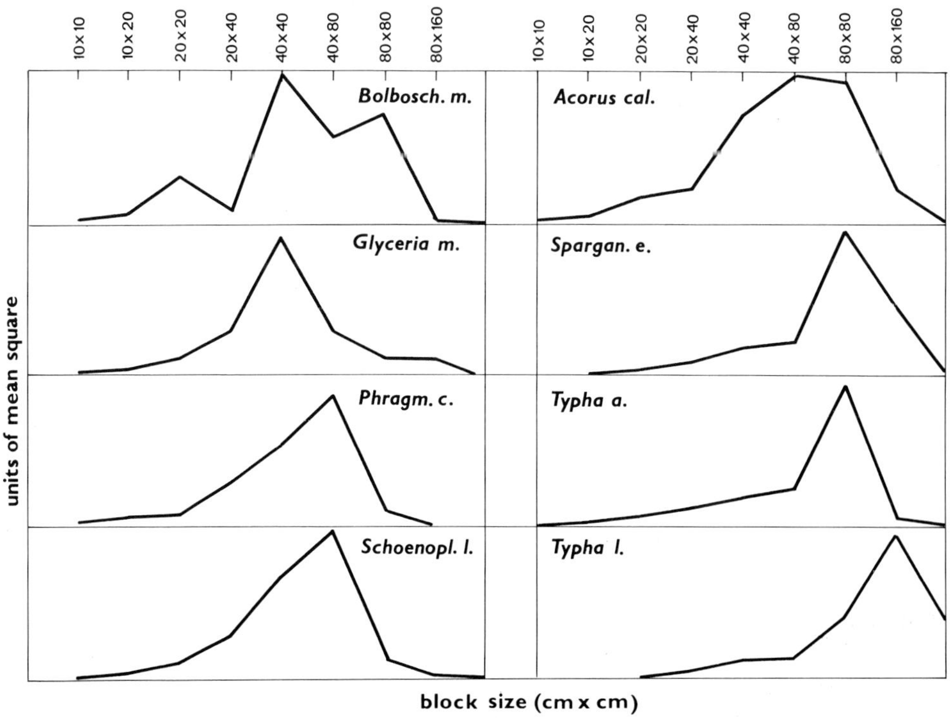

Fig. 2. Cluster analysis using mean square method for some helophyte stands. Peaks show maximum variation revealing occurrence of clusters at the dimensions (horizontal scale) belonging to these peaks

Table 1. Minimum sampling area (a_{min}) determined by means of cluster analysis

Species	a_{min} (m × m)
Bolboschoenus maritimus	(0.4 × 0.4)
Glyceria maxima	(0.4 × 0.6)
Phragmites communis	(0.4 × 0.8)
Schoenoplectus lacustris	(0.8 × 0.8)
Acorus calamus	(0.8 × 0.8)
Sparganium erectum	(0.8 × 0.8)
Typha angustifolia	(0.8 × 0.8)
Typha latifolia	(1.6 × 1.6)

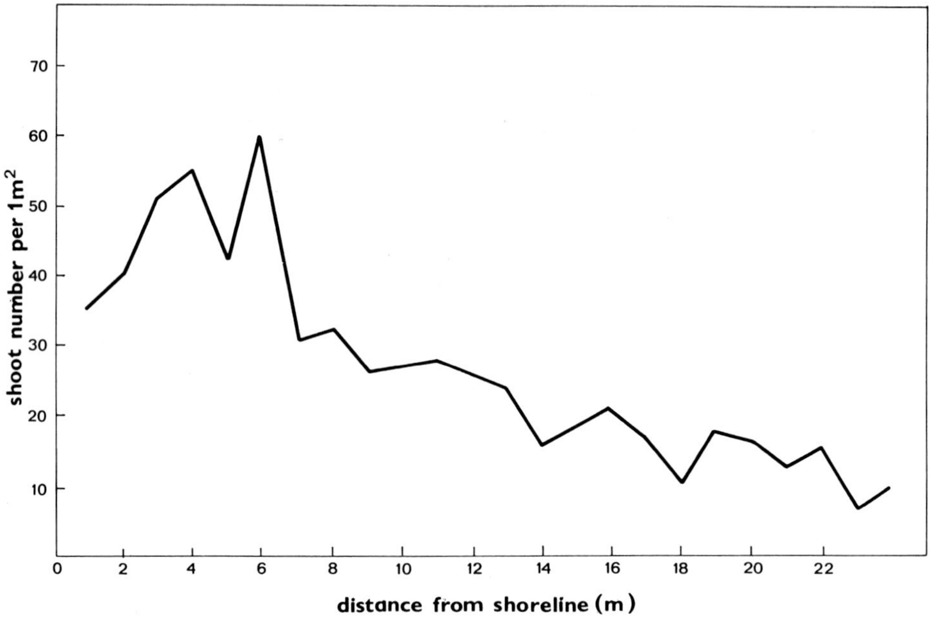

Fig. 3. Gradient of shoot density per 1 m² from open water toward the shore in a *Typha angustifolia* stand

(b) Harvesting from sample plots located along a transect. The sample plots are located along a line transect cutting across the stand. Individual sample plots may be contiguous or, in more extensive areas, located at selected distances along the transect. Ondok (1971 b) has discovered that the most advantageous location of a transect is in the direction of the maximum variation in shoot density within a stand. In littoral stands, this direction is usually perpendicular to the shoreline. Harvesting along a transect provides a relatively good estimate of average production in the whole stand.

3.1.2.2 Indirect Methods

Several variants of semidestructive methods have been applied to the littoral helophyte stands. For these methods in general, see Ondok (1971 c).

Table 2. Mean shoot density per m² in some helophyte stands determined by the nearest distance method according to Clark and Evans (1954)

Stands	Site, date	Water level (m)	Mean density	Confidence interval (P=0.05)
Glyceria maxima	Of, 25 May	0 –0.5	217	189–251
Bolboschoenus maritimus	Of, 8 June	0.05–0.10	87	67–119
Phragmites communis	Of, 20 Aug.	0 –0.4	65	51– 85
Schoenoplectus lacustris	Of, 13 July	0 –0.4	157	120–214
Schoenoplectus lacustris	Vdf, 23 July	0.5 –0.6	88	67–119
Typha angustifolia	Vdf, 12 July	0.3 –0.4	33	25– 44
Typha angustifolia	Vyf, 30 July	0.6 –0.7	51	40– 68

Of, Opatovický fishpond; Vdf, Vdovec fishpond; Vyf, Vyšehrad fishpond.

(a) Estimation of biomass from average stand density and average shoot weight. The average shoot density is determined in a nondestructive way, the average shoot weight is determined by destructive sampling. Average shoot density (n) is determined by counting shoots in contiguous quadrats sized 0.5 m × 0.5 m along a transect or by "plotless" methods based, e.g., on measurements of distances between individual regularly spaced points (e.g., along a line) and the nearest shoots (see, e.g., Clark and Evans, 1954; Cottam and Curtis, 1956). The average weight of one shoot (w) is determined by harvesting individual shoots nearest to the regularly spaced points. The average shoot biomass per unit stand area is then obtained by multiplying w by n. If the standard errors of w and n are known the confidence interval of the calculated stand biomass is estimated on the basis of both the mean and variance in the following way:

$$m_{h\cdot w} = m_h \; m_w \tag{1}$$

$$s^2_{h\cdot w} = m_h s^2_w + m_w s^2_h + 2\, m_h m_w s_{h,w} \tag{2}$$

where $m_n(m_w)$, $s^2_h(s^2_w)$ and s_{hw} are the estimates of the mean, variance or covariance of $h(w)$. This method is especially suited for large stands or for stands with widely varying shoot density. An example of the evaluation of the mean shoot density of some helophyte stands by means of the nearest distance method is given in Table 2.

(b) Method of standard permanent quadrats. This method enables us to follow the seasonal dynamics of stand growth. Certain parameters (e.g., shoot length, shoot density, number of leaves) are measured in all shoots inside a selected permanent quadrat at selected time intervals, harvesting simultaneously other shoots with the same parameters as those of the plants measured inside the quadrat. These shoots are taken from outside the permanent quadrats on areas with similar shoot density. In this way, "paired" samples are obtained, which approximate the actual biomass and other parameters of the shoots inside the permanent quadrat.

Another variant of this technique employs regression equations which permit an estimation of the biomass or other plant parameters from measurements on intact shoots. The shoots needed for establishing the regression equations can be harvested independently of measurements within the permanent quadrat. More details are given by Ondok (1971c).

3.1.2.3 Average Shoot Biomass

The indirect methods have been used for estimating the average production in monospecific helophyte stands in the littoral of the Opatovický fishpond (Ondok, 1973a) and for estimating the seasonal growth of aerial parts and leaf area in a *Phragmites* stand (see Sect. 3.1.5).

The average weight per shoot was estimated from a sample of 100 shoots, obtained by the nearest distance sampling method. A thin pole, on which points were indicated at 1 m intervals, was moved across the stand investigated. The shoots nearest to the points indicated were those to be harvested. In some instances, especially where the stand formed rather small patches or where counting the shoots was difficult (e.g., in *Glyceria* stands), the shoots were sampled from individual 0.5 m × 0.5 m quadrats along the transect. The results obtained in this way (b) are quite comparable with those obtained by using the nearest distance sampling method (a). The average shoot biomass data obtained by either the (a) or (b) method are presented in Table 3 along with other stand and habitat characteristics.

These estimates are subject to error, the principal source of which is the difficult definition of boundaries between individual stand types as they occur in nature, rather than the procedure of averaging itself. These considerations set certain limits on the interpretation of the data presented in Table 3.

3.1.2.4 Relationship Between Shoot Biomass and Production

In helophyte communities, the seasonal maximum shoot biomass mostly closely approaches their annual net aboveground production if the stands are healthy and unaffected by insect or fungus pests or by large-scale grazing (Westlake, 1965). The losses may amount, locally, to some 10% of the annual net production in the case of muskrat *(Ondatra)* grazing on *Typha latifolia* (Pelikán et al., 1970), insect damage to *Phragmites* destroys up to 10 to 20% (see Skuhravý in Sect. 6.4), and the damage by fungus diseases has not yet been evaluated. Všetička (unpublished data) has listed available references on fungi growing on *Phragmites* and *Typha* spp. and about 10% of them may have a pronounced unfavorable effect on the production of these helophytes. Grazing by grey-lag geese *(Anser anser)* may locally suppress the shoot production almost completely in *Phalaris arundinacea* and *Glyceria maxima; Schoenoplectus*, *Bolboschoenus* and *Phragmites* are also heavily damaged on certain sites while *Typha* remains unaffected (Květ and Hudec, 1971). Birds mostly seem to clip off and use for nest-building only negligible amounts of live material (Kožená-Toušková, 1973) but the mechanical damage caused by large birds such as swans *(Cygnus olor)* may be locally quite substantial. Mechanical damage by wind was found to be appreciable in a young stand of *Phragmitetum communis* following a period of rapid

Table 3. Average aerial shoot biomass (dry weight in $g\,m^{-2}$) and other characteristics of some helophyte stands measured in the littoral of the Opatovický fishpond. (After Ondok, 1973a)

Stand type	Date of sampling 1972	Sampling method (see text)	Height of water level (m)	Shoot biomass
Phragmites communis	July–Aug.	a	0.00–0.90	578–1,210
Glyceria maxima	12 July	b	0.00–0.90	659–1,122
Typha angustifolia	25 July	b	0.80–0.90	1,781
Bolboschoenus maritimus	24 July	b	0.70–0.90	167

extension growth in the culms during extraordinarily warm weather in May. The development of firm mechanical tissues had not kept pace with the extension growth and the culms broke in the nodes at a height of about 1 m. Frost damage may destroy young shoots emerging in the spring, especially in winter-cut reed stands in limosal or terrestrial ecophase (see Šmíd, 1973 and Sect. 2.1). Most of this damage to shoot production is extremely difficult to evaluate quantitatively in view of the subsequent regeneration of the shoots and mobilization of rhizome reserves for their growth and for that of new tillers or offshoots. Simulation of the damage by cutting or in some other way (see Sects. 5.4 and 7.4), might help in quantifying these production losses.

Květ (1971) has estimated, from changes in vertical canopy structure, the losses of net production due to leaf shedding during the seasonal development of a healthy *Phragmites* stand from spring to the maximum of shoot biomass. These losses amounted to some 6% of seasonal maximum shoot biomass and to 29% of maximum leaf biomass. This, of course, does not include the net losses due to self-thinning after the shoot biomass has attained its maximum. It is estimated that in *Phragmites* stands with less than 5% of infested or damaged shoots, the total losses in shoot production do not exceed 10% of seasonal maximum shoot biomass. Periodically taken samples of litter confirmed the very small losses due to leaf shedding and shoot mortality before seasonal maximum biomass is attained (Husák, 1971) in a vigorously growing littoral invasion *Phragmites* stand at the Nesyt fishpond, where these losses ranged between 1.8 and 3.3% in three plots within the stand. Even if twice as much litter should disappear or be decomposed during the approximately three-week intervals, it is justified to assume, in general, very small losses in *Phragmites*. Its seasonal maximum biomass therefore, indeed, approximates fairly closely to its net annual aboveground production.

The situation is similar in stands of *Schoenoplectetum lacustris* as long as not too many stems are broken and destroyed by wave action, and in ungrazed stands of *Bolboschoenus maritimus* (losses of 6–12%). Here, however, an exact timing of the sampling for the seasonal maximum biomass is important because the shoots die off rather suddenly and rapidly. The same precautions must be observed when sampling the seasonal maximum biomass in communities of tall sedges such as *Caricetum elatae* or *C. gracilis* where, moreover, delayed tillering has to be observed afterwards and the biomass of the newly formed tillers included in the estimate of net aboveground production.

In *Typha angustifolia*, also at Nesyt, the dry weight of the dead leaves shed before attainment of the seasonal maximum shoot biomass (assessed relatively early in the season, on July 29), together with that of dead stems and leaf sheaths and leaf bases, amounted to 5.5 to 12.4% (Husák, 1971). Dykyjová (unpublished data) quotes a range of 8 to 10%. This difference from *Phragmites* is due to the relatively earlier senescence and death of leaves in the fertile shoots of *Typha*. But the growth curves of the aboveground biomass in stands of both *Typha* species are distinguished by a plateau lasting for about two months when the biomass is at a steady state and near its seasonal maximum: the losses by leaf shedding and others are compensated by the formation of new leaves and new offshoots (see Sect. 3.1.3). The dry weight of the lost plant parts must naturally be included in an estimate of annual production. The total weight of all litter collected in a *Typha angustifolia* stand whose shoot biomass was about 2430 gm^{-2} on five occasions between June 23 and September 29, amounted to some 330 gm^{-2}, i.e., about 13.6%. If losses of litter due to decomposition are considered, the real losses will be up to 20%. In *Typhetum latifoliae* in the area of former Lake of Kobylí, South Moravia, the losses estimated in much the same way also amounted to some 20% of the seasonal maximum biomass. In *Sparganium erectum*, whose growth pattern is similar to that of *Typha latifolia*, a slightly smaller estimate of losses has been made by Dykyjová, 8.1–13.3% (unpublished data). In *Glyceria maxima*, the actual shoot biomass at any one time is known to be considerably less than the annual shoot production (one half to three-quarters of it at the maximum in South England, according to Westlake, 1975). The inherent tendency of *Glyceria maxima* to grow and propagate vegetatively during the whole year is, however, curtailed by the relatively low winter temperatures in Central Europe. Hence the most vigorous tillers can grow and develop fully during a growing season and the weaker and later-formed tillers are checked in their growth by the onset of the cold season. The growth curve of a *Glyceria* stand thus approaches in shape the growth curves of *Typha* stands (see Sect. 3.1.5, and Fig. 1 in Sect. 3.1.3). Nonetheless, it may be extrapolated from Jakrlová's (1974) data on wet *Glyceria* meadows in South Moravia that the dry weight of litter represented about 30% of the shoot biomass when the stand of *Glycerietum maximae* had just reached its maximum in June. Some of the litter may have originated from the winter or previous fall but other litter was added to that amount later during the season. Independently, Dykyjová (unpublished data) has arrived at the value of some 20–30% as indicating the difference between the annual net production and biomass of aerial shoots in *Glycerietum maximae*. The losses are greater in dense than in loose stands.

The following conclusions have been drawn for estimates of net annual shoot production as related to those of seasonal maximum biomass in fishpond littoral helophyte communities. Two groups of their dominant species may be distinguished: in one, including *Phragmites communis*, *Schoenoplectus lacustris* and possibly *Bolboschoenus maritimus*, *Acorus calamus*, and littoral tall sedges (*Carex* spp.), no serious mistake is made if the seasonal maximum biomass of their aboveground parts is identified with the annual net aboveground production. The associated error is definitely smaller than both the sampling error and that due to the fact of the initial spring growth of the aerial shoots making use of the reserves from the rhizomes. In the other group, including *Typha angustifolia*, *T. latifolia*,

Sparganium erectum, *Glyceria maxima*, a certain correction factor, ranging between 1.1 and 1.3, has to be applied in order to arrive at a reasonable estimate of net aboveground production from seasonal maximum shoot biomass. Even in these species, however, seasonal maximum biomass represents a sufficiently exact relative measure of shoot production in comparative studies, e.g., between biotopes or clones, provided the biomass is sampled in a comparable way and at the same stage of stand development.

All these considerations only apply to stands which are not seriously affected by insect or fungus pests and which are not visibly grazed.

3.1.2.5 Estimation of Underground Biomass and Net Production

In most herbaceous plant communities, the estimation of underground biomass and of its seasonal changes is far more difficult than the estimation of aboveground biomass. This statement applies to all fishpond littoral communities, especially if they are flooded. Relatively easy is the estimation of underground biomass only in certain shallow-rooted plants and communities of emerged fishpond bottoms and shores. In most developed perennial helophyte communities in fishponds, the recovery of complete root and rhizome systems is only possible during periods of drawdown of the water level for fish cropping (mostly in fall) or during winter or summer drainage of a pond. A short-term drainage, however, precludes the study of seasonal changes in underground biomass. An equally complicating factor is the highly variable pattern of horizontal distribution of the underground organs in practically all helophyte stands. For example, the coefficients of variation between Fiala's (1973a) monoliths from eight stands of *Phragmitetum communis* varied, with five replicates in each stand, between 3.5 and 74.9% (sic!) for root biomass, 7.5 and 23.0% for rhizome biomass, and 7.6 and 27.5% for total underground biomass. Each monolith had been excavated below an area of 0.25 m by 0.5 m to a depth of −0.8 to −1.2 m, according to the penetration of the rhizomes. The root biomass had been assessed only to the depth of −0.2 m as this layer of soil contained 70–80% of all root biomass.

The classical technique of monolith excavation thus yields data of limited value which, however, may increase if the samples are taken so as to show differences along a certain gradient in habitat characteristics and/or community structure. The excavation of monoliths has thus remained the most widely used sampling technique for underground biomass, because it has yielded relatively comparable data, however variable they may be.

Particularly suitable is the excavation or recovery of the rhizome and root systems along transects. The data reported by Fiala and Květ (1971) from transects across the boundaries between a pure *Bolboschoenetum maritimi* and a *Typhaetum latifoliae*, and between the latter community and a *Phragmitetum communis*, elucidated the changes in the vertical distribution of underground organs and in underground biomass with advancing succession in the former Lake of Kobylí in South Moravia.

The great differences in underground biomass between various stands of *Phragmitetum communis* occurring in the southern regions of Bohemia, Moravia and Slovakia have been analyzed by Fiala (1973a, 1976). He has found a regression of the mean rhizome diameter (y, cm) on the total number (x) of rhizome sections (transversal, longitudinal or tangential) encountered in a profile wall 1 m wide:

$y = 0.00046\,x^2 - 0.0678\,x + 3.91$, with $r = 0.869$

and another regression of rhizome biomass (W_r, $g \cdot m^{-2}$) on x:

$W_r = 0.00069\,x^2 + 0.101\,x + 1.339$, with $r = 0.864$

It should, therefore, be possible to estimate the rhizome biomass of *Phragmites* from either the mean rhizome diameter or from the number of rhizome sections occurring per unit width of a soil-profile wall. But this approach does not solve the estimation of root biomass which may vary even more widely than that of the rhizomes. In the eight stands of *Phragmitetum communis* investigated by Fiala (1973a, 1976), the biomass ratios of roots / shoots, rhizomes / shoots and total underground organs / shoots varied within the following respective limits: 0.07–4.4, 1.5–5.5, and 1.8–9.9. All three highest values were found in the limosal biotope "S" of the Opatovický fishpond (see Sect. 2.4). The corresponding values for Opatovický, "V" biotope, and Nesyt, northern shore, were as follows: 0.07 and 1.1, 4.0 and 3.0, 4.1 and 4.1. The rhizome / shoot biomass ratio seems to be small in biotopes with a fluctuating water level, which become limosal to terrestrial in late summer to fall. But this ratio becomes three times higher in biotopes with a stabilized hydrological regime. The dependence of the root / shoot biomass ratio on habitat seems far less clear. The total underground biomass, however, seems to be positively related to the supply of mineral nutrients from the soil, especially to that of nitrogen (regression coefficients of 0.787 and 0.884 for rhizome and root biomass, respectively); the root biomass was positively correlated with the content of total carbon in the soil.

The highest underground biomass may thus be expected to occur in eutrophic organogenous soils characteristic of accumulation biotopes in managed fishponds. These conclusions, valid for the stands of *Phragmitetum communis*, remain to be verified for the other reedswamp helophyte communities. Their underground biomass has also been assessed by the monolith-excavation technique (Dykyjová and Květ, 1970; Husák, 1971, and other papers), but in a smaller number of stands and habitats.

On flooded sites, the monolith-excavation technique is useless. An attempt has been made here to use the coring technique for an assessment of seasonal changes in underground biomass. In reedswamp stands, it is technically feasible to sample cylindrical cores, some 0.2 m in diameter, of the bottom material containing rhizome sections and roots. Many replicate cores may be sampled from one site. Yet, a statistical analysis of such a procedure applied to a stand of *Phragmitetum communis* at Nesyt, on three occasions during a year, proved a weakness of this technique. The diameter of the cores was evidently too small in relation to the degree of clustering in the horizontal distribution of the rhizomes (Fiala et al., 1968).

In reedswamp helophytes, the only method applicable to the study of seasonal changes in underground biomass thus seems to be an analysis of the growth and development of their polycormones either in natural habitats or in cultivation experiments. These two techniques are described in the Sections 3.1.3 and

3.1.4, respectively. An advantage of the experimentation with young and healthy polycormones is that their losses of rhizome biomass by death and decay are negligible. Hence all the increment in rhizome biomass may be identified with their net production over a given period.

In mature stands, however, the mere difference between the seasonal minimum and maximum biomass represents no estimate of either rhizome or root annual net production. With rhizomes, the calculation is complicated, apart from their uneven spatial distribution, by their different life span not only in different species but also in the same species growing in different habitats; this is particularly true of *Phragmites*, which may occur in rather diverse habitats. In the South Moravian Šakvický and Nesyt fishponds, most of its rhizomes have been found to live for four years. This was indicated by the age dependence of their carbohydrate content and bulk density (see Sect. 3.1.3) and by the poor growth and mortality of aerial shoots growing from the four-year-old vertical rhizomes. The long-term average net production of rhizomes can be approximated by dividing their total biomass by their average age (four years in the case of the two above stands of *Phragmites communis*). The average annual net production of the rhizomes would thus amount to 940 and 550 gm^{-2} of dry weight in the invasion littoral stands of the Nesyt and Šakvický fishponds, respectively (Fiala, 1973). For Šakvický, the value represents some 30% of the seasonal maximum aboveground biomass reported by Dykyjová and Květ (1970), while for Nesyt it represents some 52% of the maximum aboveground biomass attained in littoral ecophase during a favorable season (Květ et al., 1969), which corresponds to about 48% of the net annual production (see Sect. 3.2.3); with respect to the smaller amount of shoot biomass produced in the same stand during a limosoterrestrial ecoperiod, the quoted average rhizome production would then represent between 60 and 75%.

Estimation of the annual net production of roots is still less exact and more complicated than that of rhizome production, because of the difficult separation of live and dead roots and the mostly unknown life span of the roots. According to Westlake (1965), this is probably less than one year, on an average, in most helophytes; Schierup (1970) claims a turnover time of five years for *Phragmites* roots. A similar discrepancy, between two to four years, and about one year, exists in the literature as to the average life span of roots in grasslands. Further studies are needed to elucidate the root production in different species and communities of rooted perennial macrophytes occurring in different habitats. Very few dead roots or their remnants were, however, found even in three year-old outdoor hydroponic cultures of helophytes by Dykyjová et al. (1971). (This text on mature stands has been mostly adapted from Fiala, 1976, where more details and numerous other references may be found.)

In the IBP studies on mature invasion reedswamp stands growing in favorable littoral or limosal biotopes, indirect estimates of total underground net production per year were relied on, regarding it equal to a certain fraction of the seasonal maximum aboveground biomass. This assumption appears justified on the basis of: (1) Fiala's findings reported above; (2) his studies on underground production in two-year-old polycormones of *Typha;* and (3) the development of the total underground/aboveground (R/S) ratio in one- to three-year-old hydroponic sand cultures of various fishpond littoral helophytes (see Sect. 3.1.4). The respective fractions applied were: 0.5–1.0 for *Phragmites;* 0.4 for *Typha latifolia;* 0.8–0.9 for *T. angustifolia;* 1–1.5 for *Schoenoplectus lacustris;* 0.50–0.90 for *Acorus calamus;* and 1–1.90 for *Bolboschoenus maritimus* ssp. *maritimus.*

The estimated fraction of 0.5 used for *Phragmites* is also justified on the basis of an indirect calculation of its underground production described by Ondok in Section 3.1.5.

References see pp. 285—291.

3.1.3 Seasonal Development of Helophyte Polycormones and Relationship Between Underground and Aboveground Organs

K. FIALA

Helophytes dominant in fishpond reed-belts are perennial plants with extensive systems of underground organs: rhizomes, tillers, and roots (Fig. 1). Their stands arise from originally single colonies, polycormones, each developing from one vegetative diaspore or from seed. Polycormones (Pénzes, 1960) extend each year occupying larger areas of fishpond bottom or shore. In order to recognize the rate of the polycormone development and to assess the relation between the production of aboveground and underground biomass, some field experiments were made.

During 1969 and 1970, small polycormones of *Typha angustifolia* and *Typha latifolia* were cultivated in a small shallow fishpond near Lednice in South Moravia. Before planting out the seedlings from which the small polycormones were started, silon-mesh nets (supported by polythene sheets) had been buried in the mud 10–20 cm below the bottom level. This enabled the extraction of the underground organs from the mud. During the first growing season, five consecutive destructive harvests were taken and three during the next season (Fiala, 1971a, 1973a). The rhythm of offshooting and growth in the aerial shoots was followed in those polycormones to be harvested last each year as follows: the emerging shoots were successively dated and labeled, and the rhizome system was horizontally recovered and mapped out in detail at the end of the growing season. The changes in the content of reserve materials were followed in underground organs of *Phragmites communis* Trin. forming a stand situated at the southwestern shore of the Nesyt fishpond. The upper parts (0.4–0.5 m long) of vertical rhizomes of equal age carrying terminal aerial shoots, and all new rhizomes (i.e., rhizomes growing from these one-year-old vertical rhizomes) were anlyzed. For determination of the total carbohydrate content, the underground organs were hydrolyzed for 3 h with 2% HCl and the photometric method of Nelson (1944) was used. The volume of the rhizomes was determined in a calibrated glass cylinder and the bulk density of the rhizomes (dry wt./fresh vol) was then calculated (Fiala, 1973b, c, 1976).

The seasonal development of *Typha latifolia* and *T. angustifolia* polycormones is characterized by a high rate of offshoot formation during the whole growing

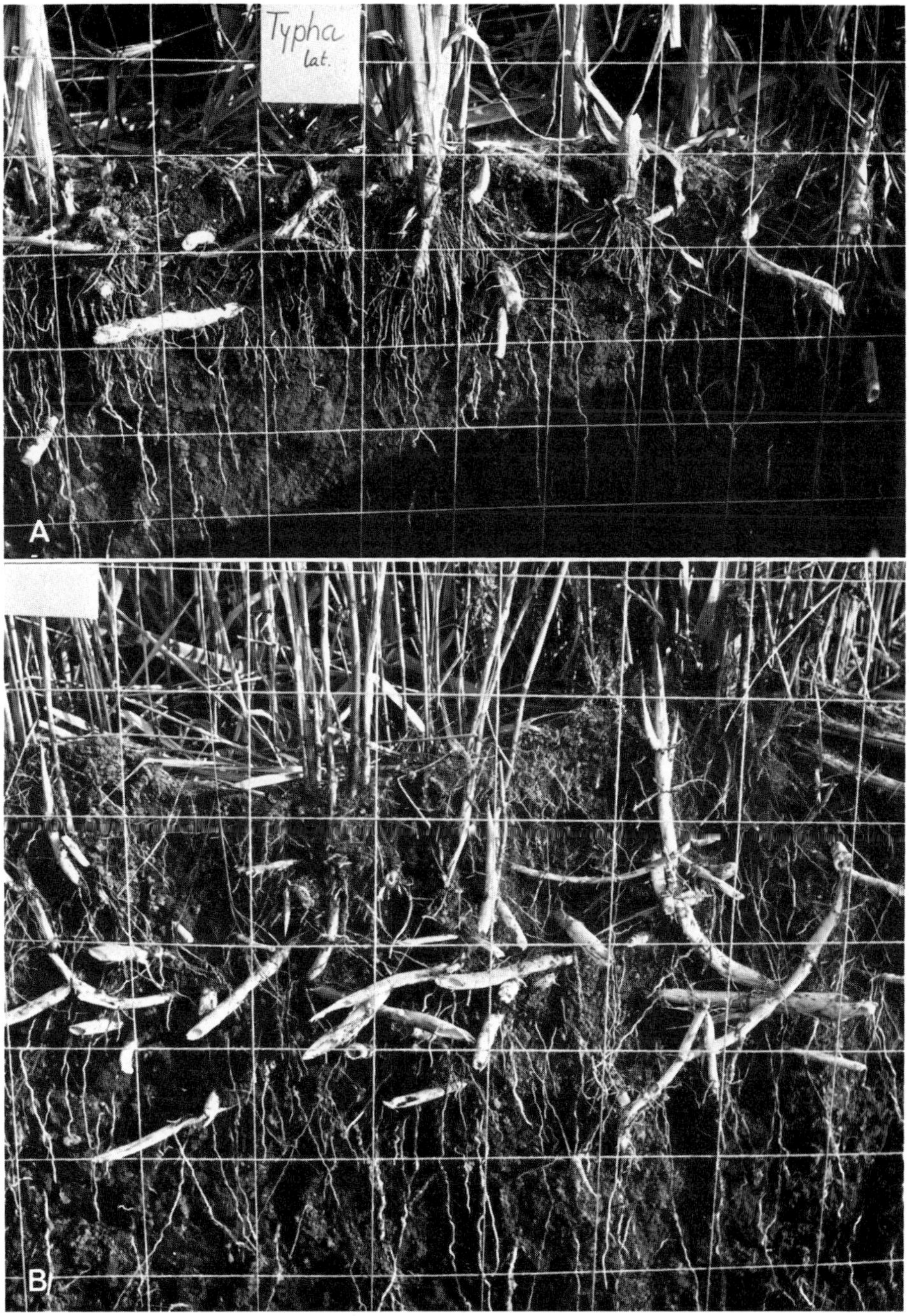

Fig. 1A and B. Underground organs in situ of *Typha latifolia* (A) and *Phragmites communis* (B) with deeper penetrating rhizomes grid: 10 cm by 10 cm. (Photo: J. Svoboda)

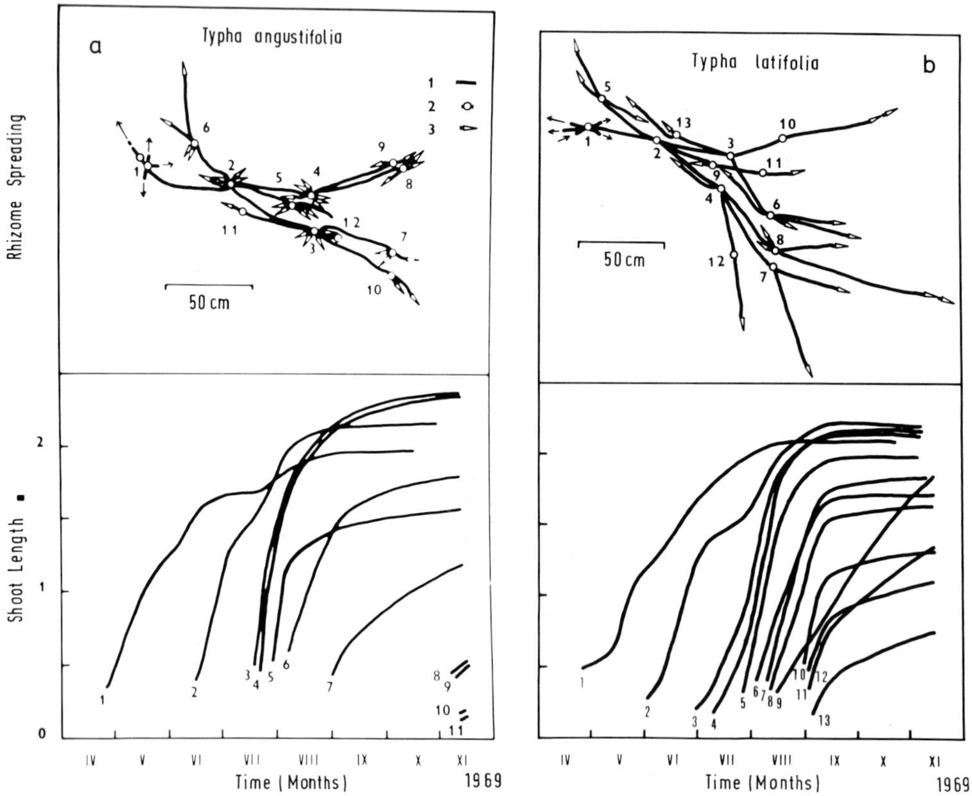

Fig. 2. (a) *Above:* Spread of rhizomes in a one-year-old *Typha angustifolia* polycormone (horizontal projection of part of polycormone); *1*, rhizomes formed in the course of 1969; *2*, shoots which emerged in the course of 1969; *3*, hibernating buds; *Below:* Emergence of individual shoots (labeled successively *1* to *12*) and their growth in length during growing season of 1969. (b) *Above:* As in (a) for a one-year-old *Typha latifolia* polycormone; *Below:* As in (a) for *T. latifolia*

season (Figs. 2, 3). New rhizomes, each about 0.4–1 m in length, grow from the lateral buds formed at the bases of aboveground *Typha* shoots. These rhizomes do not branch, but bear new shoots; from the base of each such shoot, further rhizomes follow about the same direction as those of the previous order. All parts are connected and the plants form ramified polycormones. Most new *Typha* shoots emerge during the period of most favorable climatic conditions (end of July to beginning of August). In the first growing season, 2.9 new shoots of *Typha latifolia* were formed per day from 22 July to 7 Aug. (mean value) and 9.5 shoots in the second year from 29 July to 28 Aug. (Fig. 2). During the same periods, *T. angustifolia* produced only 0.9 and 6.1 shoots per day respectively. The situation was reversed in the fall when *Typha angustifolia* produced only 2.4 shoots per day at the end of the first growing season (from 3 Sept. to 12 Nov.), and 16.2 shoots at the end of the second growing season, while *T. latifolia* produced only 1.6 and 7.5 shoots respectively. Offshoot formation is increased in fall, when the growth of

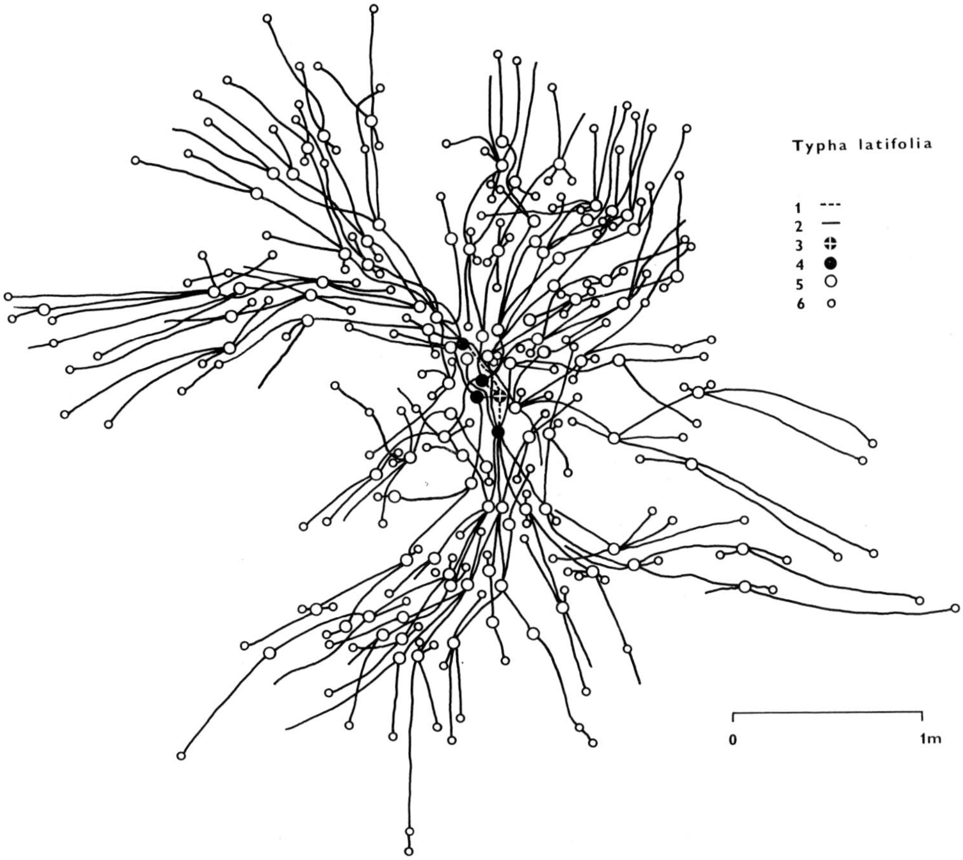

Fig. 3. Vegetative spreading of the rhizomes in a one-year-old *Typha latifolia* polycormone during growing season of 1969, horizontal projection of whole polycormone; *1*, rhizomes formed during preceding year; *2*, rhizomes formed in the course of 1969; *3*, remnants of dead shoots from preceding year; *4*, shoots which emerged at end of preceding year; *5*, shoots which emerged in course of 1969; *6*, hibernating buds

aboveground organs is evidently suppressed by low night temperatures. The autumnal differences between the rates of offshoot formation in the two *Typha* species are mainly due to their different morphological structure. *T. angustifolia* produced more new shoots (hibernating buds) attached to the mother shoots by short (5–10 cm long) and thick rhizomes (see Fig. 2a). The total number of shoots produced during the first year was much the same in both *Typha* species (212 in *T. angustifolia*, 220 in *T. latifolia*); by the end of the second year, it was greater in *Typha angustifolia* (1390 shoots) than in *T. latifolia* (1082 shoots).

The growth curves of young *Typha* shoots (Fig. 2) indicate that each of these shoots is probably able to supply assimilates required for growth to two consecutive offshoots (mainly rhizomes; see also Ondok, 1972).

The high rate of both rhizome and shoot net production determines the high rate of vegetative spreading of *Typha* polycormones. At the end of the first year, the approximate size of the area occupied by each *Typha latifolia* polycormone

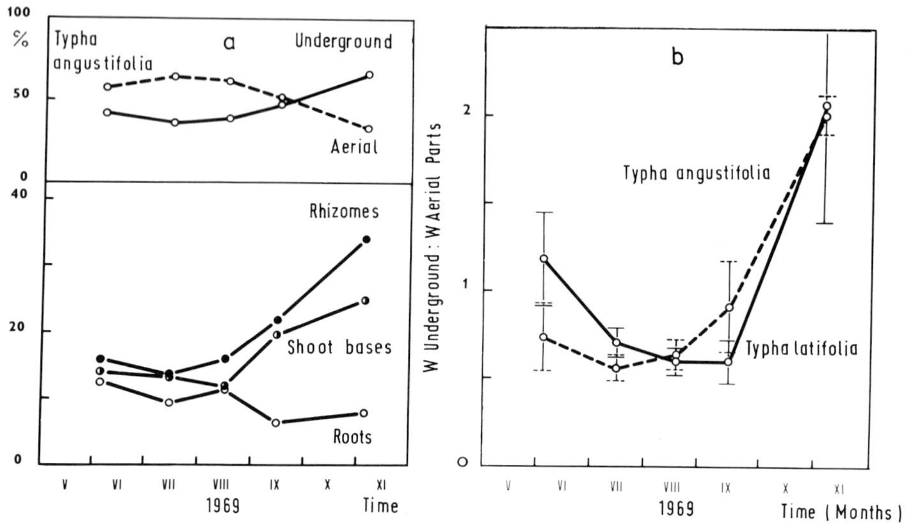

Fig. 4. (a) Seasonal changes in distribution of dry weight in rhizomes, roots, and shoot bases of one-year-old (1969) *Typha angustifolia* polycormones; (b) seasonal changes in dry weight ratio between underground and aerial parts (R/S ratio) in *Typha angustifolia* and *T. latifolia* polycormones during 1969 growing season. (Vertical bars, ± twice standard error)

was 10–16 m^2, with about 120 m of new rhizomes (see Fig. 3). Each *T. angustifolia* polycormone occupied 4–9 m^2 (with about 50 m of rhizomes). These areas increased to 50–58 m^2 (total rhizome length 480 m) and to 37–40 m^2 (380 m of rhizomes) respectively by the end of the next growing season (Fiala, 1973c, and in press). The crop growth rate (CGR) of all aerial organs reached its maximum in both *Typha* species in summer, and two periods are highly important for the production of their underground organs: one in late July to mid-August, the other in the fall (late September to October). In *Typha angustifolia* it is mainly the rhizomes that participate considerably in the daily increments in all underground organs, while in *Typha latifolia* it is both the rhizomes and shoot bases. The average daily increment in rhizome length reached a peak only in summer. In mid-July, the average increment in length of new rhizomes was 9.9 and 6.5 cm day^{-1} per shoot in *Typha angustifolia* and *T. latifolia* respectively. In both *Typha* species, the participation of the growth rate of all underground organs (among them, above all of rhizomes) in the total growth rate of whole polycormones increased linearly with time. The participation of the growth rate of roots fell from high values at the beginning of the year to minimum values at the end of August to mid-September, and again rose to its highest values in fall (Fiala, 1973c, 1976).

Seasonal changes in the distribution of newly formed biomass into different parts of the *Typha* polycormones were expressed in terms of percentages of total polycormone dry weight (Fig. 4a). During both growing seasons, a larger portion of the biomass mostly appeared to be present in the aboveground organs. In the fall of the first growing season, the rhizomes constituted 30–35% of dry weight of the whole polycormones. In two-year-old polycormones, the dry weight of flowering shoots was particularly high in *T. latifolia*, whereas such shoots were absent in

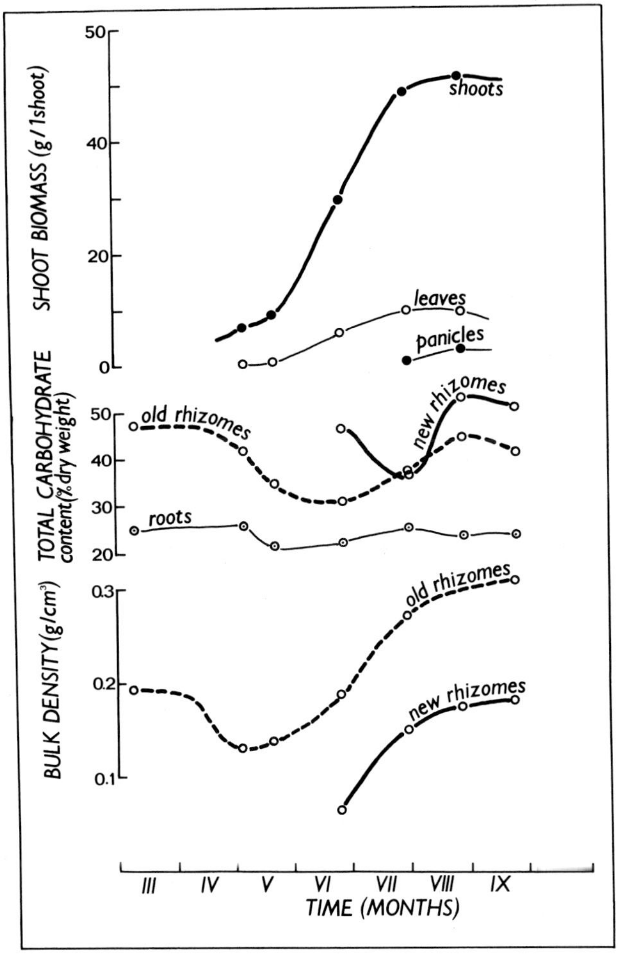

Fig. 5. Seasonal changes in total shoot, leaf, and inflorescence dry weight, total carbohydrate content, and bulk density of underground organs in *Phragmites communis* during 1971 growing season

one-year-old polycormones. This explains the smaller participation of the underground organs in total polycormone biomass at the end of the second growing season.

The share of the aerial parts in total biomass increases during the first half of the growing season. In summer (first half of Aug.), the dry-weight ratio of the underground to aboveground organs (R/S ratio) attained about 0.6 in both one- and two-year-old polycormones of both *Typha* species. At the end of the first growing season, R/S$\simeq$2 (Fig. 4b); in two-year-old polycormones the autumnal R/S ratio was 0.55 and 0.99 in *T. latifolia* and *T. angustifolia* respectively.

The high increments in the underground biomass of *Typha* recorded at the end of the growing seasons were evidently due to an accumulation of large amounts of reserves in the rhizomes. Considerable seasonal changes in the reserve content in

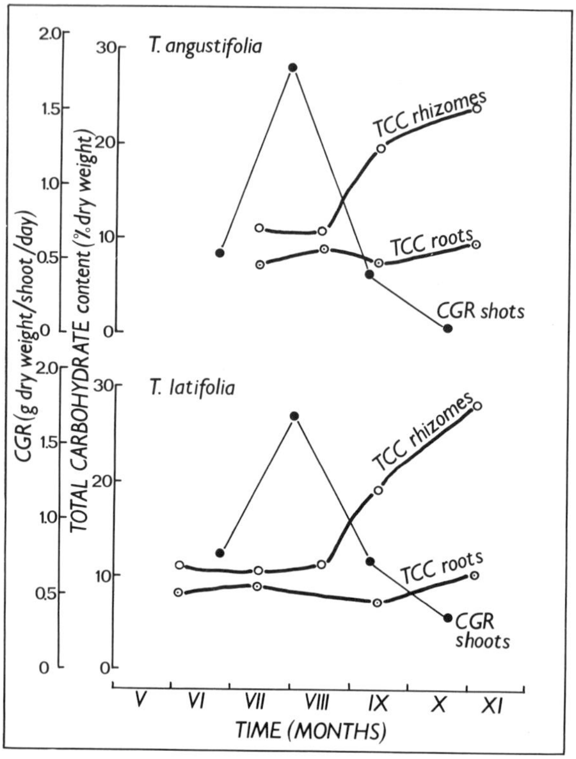

Fig. 6. Seasonal changes in rate of dry matter production (CGR) in aerial organs and course of accumulation of total carbohydrate in underground organs of *Typha angustifolia* and *T. latifolia* (TCC) in 1969

underground organs are also typical of the other helophytes with large rhizome systems, such as *Phragmites*. The relationship between the seasonal course of growth in aerial shoots of *Typha* and *Phragmites*, and the seasonal course of total carbohydrate content in their rhizomes and roots are compared in Figures 5 and 6. In *Phragmites*, considerable amounts of reserve material were supplied to the aboveground organs from the rhizomes during late April and the first half of May. This subsidy obviously determined the initial rapid shoot growth. During the spring, the total carbohydrate content decreased by about 40% in the old *Phragmites* rhizomes and their bulk density also decreased by about 30%. A similar decrease in bulk density was recorded in old *Typha* rhizomes in spring: by some 60% and 40% in *Typha angustifolia* and *T. latifolia* respectively.

On the other hand, during July and August, carbohydrates accumulated not only in the new *Phragmites* rhizomes, most of which had obviously been formed during July, but also in its older live rhizomes. Unlike the carbohydrate content, the bulk density began to increase at the end of May; at the end of July it was twice that recorded after the spring exhaustion of reserves, and its fall values surpassed 0.3 g cm^{-3} (Fig. 5). The formation of mechanical tissues thus probably precedes the accumulation of reserve materials in one-year-old rhizomes.

In both *Typha* species, reserve material apparently accumulates later (mostly in Sept. and Oct.) than in *Phragmites*. Both the total carbohydrate content and bulk density recorded in the young *Typha* rhizomes (above all of *T. latifolia*) in fall were two to three times higher than those recorded in summer: the autumnal total carbohydrate content varied between 20 and 30% of dry weight. In fall, the photosynthetic activity seems to decrease earlier in *Typha angustifolia* than in *T. latifolia*. The latter species, therefore, accumulates more reserve material.

The total carbohydrate content per unit dry weight found in young (formed in the previous year) *Phragmites* rhizomes in spring exceeds the content assessed in the same rhizomes at the end of the same growing season. A similar difference was also recorded in old *Typha* rhizomes: more reserve material is apparently taken out of old rhizomes for shoot growth than is later returned to them from the shoots. The hibernating buds and obviously also the growth of *Phragmites* roots in fall probably draw on the reserve material accumulated in rhizomes (for details see Fiala, 1973c, 1976). The rates of mobilization and accumulation of reserve material in underground organs of helophytes deserve more attention and the data reported here permit a first orientation in the problem.

Our understanding of the seasonal development of *Typha* polycormones may be summarized in the following way: the growing season may be approximately divided up into four successive periods: (1) end of April to May: mobilization of reserve materials and development of the photosynthetic apparatus linked with a rapidly decreasing R/S ratio; (2) June: termination of, or a slow-down in the uptake of reserve substances from the rhizomes, continuing increase in aboveground biomass, appearance of the flowering shoots, the first offshooting and growth of new underground organs; persistence of a marked decrease in the R/S ratio; (3) July to August: maximum development of the photosynthetic apparatus, formation of numerous offshoots, high increase in underground biomass, particularly in the rhizomes; a relative stabilization of R/S ratio; (4) September to November: termination of the development of the photosynthetic apparatus, decrease in aboveground biomass, intense formation of offshoots and hibernating buds, accumulation of reserve substances in underground organs linked with an increase in the R/S ratio.

References see pp. 285—291.

3.1.4 Experimental Hydroponic Cultivation of Helophytes

D. Dykyjová and K. Véber

Assessment of the annual production of underground organs is complicated in most emergent macrophytes not only by their perennial character but also by the great spatial variation in their underground biomass and by the difficult access to their underground organs. Their annual increments were therefore followed, in successive seasons, in polycormones started from young seedlings or cuttings in hydroponic cultures for three years. At the end of each season, the one-to three-year-old polycormones were harvested and their increments evaluated. Most of the species investigated possess large aboveground shoots and extensive underground rhizome systems. A successful cultivation of such plants is therefore determined by the space, area, and depth of the cultivation tanks. Since high production of aerial shoots is obtained only from stands at least several years old, it was necessary to produce "mature" polycormones for assessments of underground production. The winter survival of the underground organs in the outdoor cultivation units remained to be tested.

3.1.4.1 Technique of Hydroponic Cultures

The hydroponic equipment for both the summer (outdoor) and winter (in greenhouses) cultivation of helophytes as well as the first results are described by Dykyjová et al. (1971, 1972). Only a brief description of the principles of cultivation will therefore be given here.

The outdoor hydroponic equipment (Fig. 1a and b) consisted of six identical units, each provided with independent circulation of the nutrient solution; six different régimes of mineral nutrition could be applied simultaneously. Each unit consisted of three pairs of south-facing cultivation tanks 1 m^2 in area and 0.5 m in depth. Each pair formed a tier in a cascade-like set-up. The solid and inert substrates, mostly coarse diatomaceous earth (kieselguhr), fine washed river sand, or a mixture of both materials, were packed on top of a drainage layer (inert gravel covering a perforated false bottom). The nutrient solution circulated from a reservoir (placed at the highest level) to the tanks by regularly repeated portions controlled by a simple siphon; from the lowest tank it flowed into containers placed below ground level, from where it was continuously repumped to the reservoir. At one-to three-week intervals, according to the growth rate of the plants, the nutrient solution was completely changed.

In the initial experiments, a slightly modified nutrient solution devised for the mass cultures of algae was used (Šetlík and Simmer, see Šetlík et al., 1970). For the higher plants, the concentration of Ca^{2+} was increased by replacing NH_4NO_3 with an equivalent amount of $Ca(NO_3)_2$ and the total concentration of micronutrients was slightly reduced. This nutrient solution, which has proved most suitable for the cultivation of helophytes, has the following composition:

Composition of the nutrient solution. (After Šetlík and Simmer; modified)

Compound	mg l^{-1}	Compound	mg l^{-1}
$CO(NH_2)_2$	600.0	H_3BO_3	6.18
$Ca(NO_3)_2 \cdot 4H_2O$	800.0	$MnSO_4 \cdot H_2O$	2.23
$MgSO_4$	963.0	$CuSO_4 \cdot 5H_2O$	2.49
KH_2PO_4	680.0	$ZnSO_4 \cdot 7H_2O$	2.87
$Fe_2(SO_4)_3 \cdot 9H_2O$	28.1	$CoCO_3 \cdot 7H_2O$	2,81
EDTA complex	88.62	$(NH_4)_2MoO_4$	1.96

In the first year of experiments (1967), either seedlings or rhizome cuttings with one either terminal or lateral bud were planted out at a density of five plants per tank. This applied to *Phragmites*, both *Typha* species and *Acorus calamus*, while with *Schoenoplectus lacustris* and *Bolboschoenus maritimus* nine plants per tank were planted out. However, these numbers became too high for all the species in the subsequent growing season. Hence only one plant per tank was then

Fig. 1a and b. General view of the outdoor cascade-like hydroponic equipment in spring (a), and a single cascade unit (b) with young *Schoenoplectus lacustris* polycormones in summer. (Photo: J.P. Ondok)

left. In the fall, the shoots were harvested and the tanks were drained. The solid substrate with the underground organs remained covered only with the litter containing mostly dead leaves and some dead stems during winter, but the substrate temperature remained sufficiently high to ensure the survival of the underground organs. In spring, the shoots of all the species started to grow at about the same time as those in the natural stands. During the growing season, temperatures and relative air humidity were recorded within the stands in the tanks. The incoming global radiation was also measured (see Sect. 2.1). For calculating the coefficient of solar energy conversion in the plants cultivated, the daily totals of global radiation were converted to PhAR (= 0.45 times global radiation).

Winter hydroponics were installed in the greenhouses of the large dual-purpose unit for algal mass cultures (Šetlík et al., 1970). The plants were cultivated in large but shallow troughs 10 m long, 1 m wide and 0.15 m deep, during winter season, when the glass roof of the greenhouse was not in use for cultivation of algae. The solid substrate, a mixture of river sand with diatomaceus earth, was periodically subirrigated with a nutrient solution of the composition described above. The cultivation area of the trays was much larger than that of the summer hydroponic units (see Fig. 2). It was thus possible to harvest a relatively large number of experimental plants at several sampling intervals during the cultivation season. In this way, the dynamics of the initial development of both the aboveground and underground organs and the R/S ratio could be followed. The greenhouse was only partially air-conditioned and divided into two compartments. Hence only two different temperature regimes were set up, with daily mean temperatures of 12° C and ±20° C.

3.1.4.2 Thermoperiodism of Helophytes

The plants were cultivated from lateral winter buds, cuttings or seedlings planted out at appropriate distances. The planting material must be grown from the autumnal buds formed in the season preceding the cultivation. If the material is taken in the course of the growing season, the new plants will be feeble and will eventually die off. It was therefore necessary to learn when the winter dormancy of the buds is accomplished. In the first experiments, the cuttings sampled from wild-growing plants in fall were stored at different ranges of temperature for one month. The best subsequent development took place in plants grown from the material stored at +1 to +3° C. Three different planting times were tested, i.e., November, December, and January, to verify the effect of bud dormancy. The growth in both the shoot and underground biomass, the leaf area, and the R/S ratio recorded at successive sampling intervals were followed in the plants, developed from the cuttings planted out on different planting dates (Fig. 3). In some instances the biomass and R/S ratio were determined only at the end of the cultivation, after five to six months.

The winter cultures were not illuminated artificially. The incoming solar radiation was measured with integrating recorders of PhAR (370–730 nm) of the Kubín and Hládek type, see Kubín (1973), placed both inside and above the greenhouse roof. The density of the radiation flux transmitted through the glass roof increased with increasing sun elevation during late winter and early spring

Fig. 2a and b. Winter hydroponic cultures of *Acorus calamus* and *Bolboschoenus maritimus* (2-month-old polycormones, March 1967) (a) in a cold greenhouse; (b) 5-month-old culture of *Schoenoplectus lacustris* in a warm greenhouse (Dykyjová et al., 1972)

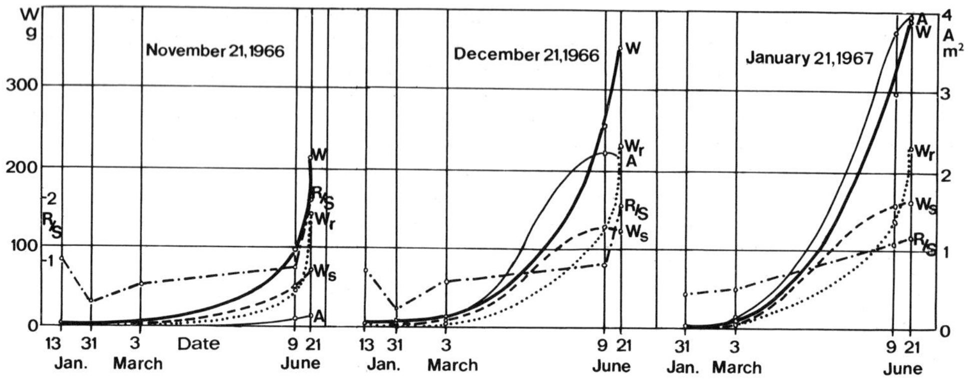

Fig. 3. Course of biomass (dry weight) in the shoots (W_s) and underground organs (W_r), and of total biomass (W), of the leaf area (A) and of the R/S ratio in *Typha latifolia* planted on 3 different dates in a warm greenhouse. Abscissa: sampling dates; ordinate: dry weight and leaf area as mean values of ten polycormones. (After Dykyjová et al., 1972)

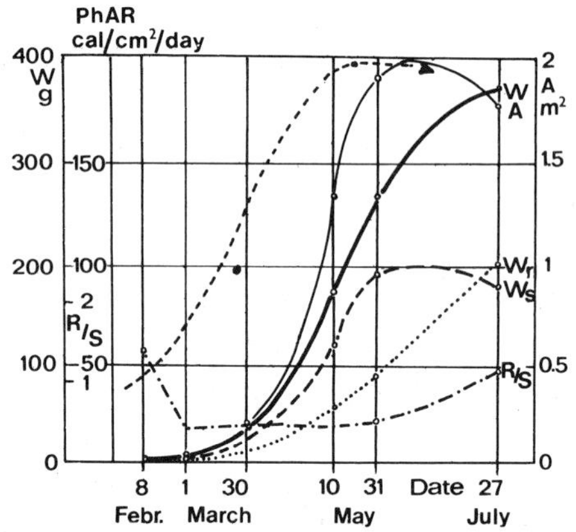

Fig. 4. Course of dry weight of shoots (W_s), underground organs (W_r) and total biomass (W) and of the R/S ratio in *Typha latifolia* planted in January in a warm greenhouse as one-year-old seedlings (mean values of ten polycormones), ----, course of transmitted PhAR in greenhouse in cal cm^{-2} day^{-1} as monthly means from January to June (Dykyjová et al., 1972)

(see Fig. 4). The air and substrate temperatures and relative air humidity were recorded continuously.

Examples of Results. In the winter hydroponics, in the experiments with different planting dates, outstanding symptoms of thermoperiodism and dormancy were found in all species investigated. All plants which developed from cuttings planted out in November were small as a result of an uncompleted bud-break period. This effect was more pronounced in the warmer greenhouse conditions (20° C) and did not disappear even during the later spring. At this time, the higher irradiance brought about a better growth in the plants started later in winter, in December or January. The greater growth rate was maintained in the experimen-

Table 1. Development of aboveground and underground biomass dry weight (in g), of leaf area, and R/S ratio in *Schoenoplectus lacustris* planted 21 Nov. 1966, in a warm greenhouse (mean values of ten plants) (Dykyjová et al., 1972)

Date of harvest 1967	Max. stem height cm	Wg above-ground	Wg under-ground	R/S	Number of shoots	Leaf area dm^2
12 Jan.	89	0.92	1.88	2.06	2	1.16
2 Feb.	128	1.32	0.26	0.20	2	1.92
1 March	149	3.14	0.72	0.23	3	3.20
30 March	162	10.60	3.75	0.35	6	5.95
5 May	213	41.00	21.70	0.53	17	31.70
22 June	210	53.00	52.24	0.95	25	31.90
2 July	205	42.00	69.50	1.65	30	18.60

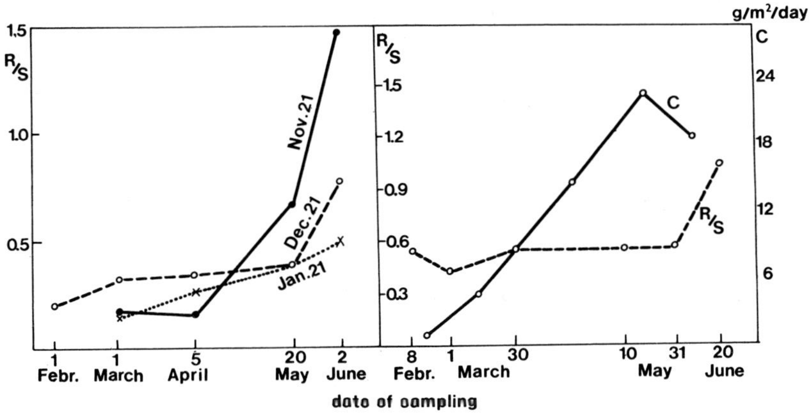

Fig. 5. *Left*: Development of R/S ratio in *Schoenoplectus lacustris*. *Right*: Rate of shoot dry matter production (*C*) and R/S ratio in *Typha latifolia*

tal plants of later sowing dates for the rest of the cultivation season (Fig. 4). After two months of cultivation, *Typha latifolia* shoots started from cuttings stored at low temperatures and planted out in January attained the same shoot height and density as those in the natural stands; *Phragmites* plants started as one-year-old seedlings attained a high density during the spring months as well. *Typha angustifolia* requires a very high irradiance. It developed very slowly in the winter cultures as compared with the summer outdoor hydroponics, and produced only few lateral shoots even after termination of its bud break. Nevertheless, these plants started in January developed several larger clusters, each bearing a few flowering shoots. *Schoenoplectus lacustris* and *Acorus calamus* developed better in the cooler greenhouse at 13–15° C, but they became easily etiolated at a low irradiance in warmer conditions. The formation of large shoots began in them only in late spring (April to May, see Table 1 and Fig. 2b).

The biomass and leaf area increased gradually in all species and so did the R/S ratio. The later the plants were harvested, the more underground organs had been

Fig. 6a and b. One-month-old cutting of *Typha latifolia* (a) and its two-year-old fertile polycormone (b) in summer hydroponic cultures. (Photo: J. P. Ondok)

formed. *Schoenoplectus lacustris*, for instance, grown in the warm greenhouse and harvested in July, when the stems were starting to die off, attained a higher R/S ratio than the plants in the natural biotopes (Fig. 5).

The species compared differ somewhat in their response. *Typha angustifolia* in the vegetative growth phase may develop a ratio of $R/S \simeq 2$, while the aerial parts prevail when fertile shoots are formed. *Phragmites* produced approximately the same amount of underground and aboveground biomass ($R/S \simeq 1$) in mature polycormones, while in older ones the aboveground organs prevailed. In the other species the R/S ratio was mostly less than 1. (Fig. 4 in Sect. 3.1.3). The R/S ratio thus appears to be a relatively constant quantitative species characteristic, but only for a given growth phase of the polycormone.

In the outdoor hydroponics, the experimental plants were started from seedlings (*Phragmites* and *Typha latifolia* in some instances) or from cuttings (most of *Typha latifolia* and all the other reed-belt species) or from tubers (*Bolboschoenus maritimus*). The small seedlings must be planted out quite early, and successively transplanted into bigger pots, since otherwise their growth and development may be interrupted. The vegetative propagation of the planting material makes it possible to obtain rapidly large plants possessing the genetic uniformity of a clone

Fig. 7. One- *(below)* and two- *(above)* year-old polycormones of *Phragmites communis* in the outdoor hydroponic cultivation unit. (Photo: J. P. Ondok)

(Fig. 6). But even in the outdoor summer cultivation, the period of dormancy and the thermoinduction of bud break must be respected. Only hibernating buds thus represent suitable material for propagation. Sufficiently large plants are produced only if the buds are either taken from natural stands in late fall and stored outdoors in peat for the winter, or sampled in natural biotopes during the winter months. The initial biomass of the planted material was about 2.5—10 g fresh weight, corresponding to 0.2—2 g of dry weight.

In view of the long vegetation cycle of the helophytes investigated, a single experiment could only be carried out with each species; owing to the small number of replicates, no successive sampling took place during the growing season, and only the final biomass of both aerial and underground organs was assessed. From the final harvests of 1967, 1968, and 1969, the annual net production by one- to three-year-old cultures was compared (Fig. 7) with natural stands and the coefficients of solar energy conversion were estimated (see Sect. 3.1.8 and Table 2).

The total seasonal production of aerial organs and the one- to three-year production of underground parts are employed in estimates of the annual underground production in natural biotopes. The annual underground increment was found to be 0.5–1.2 (Fig. 8) times the annual production of shoot biomass. Similar values have frequently been found in natural stands at the initial stage of polycormone development, as in both *Typha* species, *Schoenoplectus lacustris*, *Acorus calamus* or *Bolboschoenus maritimus*. These results correspond to the findings by Fiala as well as by other authors (see Sect. 3.1.3).

Table 2. Comparison of cultivated and natural stands of *Phragmites communis* and *Typha angustifolia*

Species	Age of culture	Date of		Growing season	No. of shoots	Biomass ($g\ m^{-2}$)[a]			R/S	*CGR* ($g\ m^{-2}\ day^{-1}$)[b]		η %[c]	
		planting	harvest	days	per m^2	shoots	underg. organs	total		shoots	total	shoots	total
Phragmites communis	1 year	Apr. 1969	Sept. 1969	158	80	482	493	975	1.02	3.0	6.1	0.63	1.26
	1 year hibernating	Sept. 1967	Sept. 1968	153	275	2,409	1,160	3,569	0.48	15.7		3.96	
	2 years	May 1967	Aug. 1968	105	1,435	5,726	2,730	8,456	0.46	54.5		12.41	
	3 years	May 1967	Sept. 1969	158	1,210	3,401	8,886	12,287	2.61	21.5		4.44	
Maximum yields in natural stands	Opatovický fishpond		Sept. 1966	130	144	2,960	—	—	—	22.7		5.75	
Typha angustifolia	1 year	Apr. 1969	Sept. 1969	154	9	405	310	715	0.76	2.6	4.6	0.53	0.94
	1 year hibernating	Sept. 1966	Sept. 1967	128	31	1,229	1,281	2,510	1.04	9.6		2.03	
	2 years	May 1967	Aug. 1968	100	177	3,039	2,070	5,109	0.68	30.4		6.71	
Maximum yields in natural stands	Spálený fishpond		Aug. 1964	112	72	2,934	—	—	—	26.2		5.84	
	Opatovický fishpond		Aug. 1966	103	92	3,710	—	—	—	36.0		8.40	

[a] In cultivated stands: per m^2 cultivation tank area; in natural perennial stands: per m^2 ground area.
[b] Average daily rate of dry matter production per season.
[c] Coefficient of solar energy conversion (% PhAR) per season.

Fig. 8. (a) Rhizomes and roots of one-year-old polycormone of *Phragmites communis* harvested from the outdoor hydroponic culture; (b) Rhizomes and roots of two-year-old polycormone of *Acorus calamus*. (Photo: J. P. Ondok)

The annual net production as well as the coefficients of solar energy conversion appeared exceptionally high in most experiments (Table 2). The mean rate of total dry matter production per day (*C*), from shoot emergence to the attainment of seasonal maximum shoot biomass depends, like the coefficient η, on the harvest date. The values of $C = 30$ to 60 g m^{-2} day^{-1} recorded in the one-year-old cultures have otherwise been found only in most intense crop cultures. Three-year-old *Phragmites communis* formed flowering shoots of 2.6 m in height; its maximum shoot density was more than 1360 culms per 1 m^2. The highest shoot biomass in *Phragmites* was attained in two-year-old polycormones (5.72 kg m^{-2} of dry weight). Two-year-old polycormones of *Typha latifolia* also produced more biomass than its natural stands, 5.24 kg m^{-2} and 6.50 kg m^{-2}, of the shoot and total biomass respectively, but the R/S ratio was only 0.38, as a large part of the assimilates had been apparently translocated to the flowering spadices. Similarly, two-year-old fertile *Typha angustifolia* produced 3.47 kg m^{-2} of shoots and 6.50 kg m^{-2} of total biomass, but its R/S ratio was higher, 0.87. The age dependence of the distribution of the produced assimilate was very clear in this species. Its one-year-old polycormones, with only vegetative shoots, stored equal or even larger amounts of dry matter in the underground parts than in the shoots: the R/S ratio ranged from 1.0 to 1.45. The biomass of two-year-old polycormones, with a large number of both sterile and fertile shoots, approached the biomass of the natural stands, and the R/S ratio declined. In cultivated *Schoenoplectus lacustris*, the biomass was not as high as in natural biotopes, apparently because this species is adapted to deeper flooding, which was not possible in the shallow cultivation tanks.

Large branched polycormones developed from single buds of rhizome cuttings in cultivated *Acorus calamus* (Fig. 8b). Its shoot production was considerably higher than in natural biotopes. The rapidly vegetatively propagating *Bolboschoenus maritimus* formed very dense stands in outdoor hydroponics, with the total biomass of 10 kg m^{-2} of dry weight greatly exceeding the biomass found in natural habitats; the R/S ratio was as high as 2.6 in this instance.

These first experimental data cannot be generally valid and have to be checked in more detailed experiments under more exact control of the conditions of cultivation. In comparisons of the production in several years old outdoor hydroponic cultures with that in natural stands, some specific factors have to be considered:

a. The mechanical barrier of the tank walls causes a considerable accumulation of underground organs and an exceedingly dense crowding of the shoots on a small area.

b. In each tank, the stand surface was typically hemispherical like the crowns of solitary deciduous trees. The interception of incident solar radiation must therefore be different from that in natural closed canopies. However, the high total net production, frequently exceeding that in natural biotopes, is to be ascribed in the first place to the favorable mineral nutrition; this proves the great potential production of the fishpond littoral helophyte communities (see also Sect. 3.4.3).

References see pp. 285—291.

3.1.5 Estimation of Seasonal Growth of Underground Biomass

J. P. Ondok

Seasonal growth of the aboveground parts of the helophytes can be determined directly by the harvest method. But it is difficult to assess the growth of the underground organs in this way (see Sects. 3.1.3 and 3.1.4). The photosynthetic model described in Section 3.2.3 was used to estimate the growth of underground biomass in the reed stand.

3.1.5.1 Growth of Aboveground Stand Biomass

The seasonal development of the aerial parts in a reed stand was investigated by the indirect method (Ondok, 1971c, 1973a). This method permits the estimation, with sufficient accuracy, of the growth of aboveground biomass on a selected plot within a stand.

In a littoral *Phragmites* stand (Opatovický fishpond), four quadrats of 0.5 m^2 in size were selected early in the spring. During the growing season both the number and lengths of the shoots were measured inside each quadrat at short intervals; each time, equivalent individual shoots were harvested elsewhere in the stand. Four parallel samples were thus obtained, each composed of a corresponding number of shoots of the same lengths as in one of the permanent quadrats. The coefficient of variation between the replicates was usually less than 5% for all primary data: leaf area and dry weight of both fresh and senescent leaves, stems and inflorescences per quadrat. The measurements were carried out at weekly intervals, only in the fully developed stand were the intervals longer. Figure 1 presents the results of the seasonal measurements taken in one permanent quadrat.

The early growth of the stems submerged in water (the water level fluctuated between 0.4 and 0.6 m during the growing season) is mainly determined by assimilate translocation from the underground storage organs. The delayed leaf development (as compared with stem growth) implies that this translocation endures until the shoots become self-supporting. Later on, surplus assimilates are translocated to the rhizomes and roots. In addition, assimilates are translocated from the dying-off leaves and other aerial organs. The rate of leaf senescence and dying seems to be enhanced in dense stands. The dry dead leaves remain on the stems till late fall and only then fall off. The dry stems remain in the stand for the next one or two growing seasons and only then fall down. All this has been observed in the reed-belts of both South Bohemian and South Moravian fishponds.

3.1.5.2 Growth of Underground Biomass and Translocation

If we know the growth of all aerial plant parts, their photosynthesis and dark respiration, we may estimate the translocation of assimilates from the underground organs into the aerial parts in the spring and the growth of underground biomass from surplus assimilates in summer and fall. The block scheme of com-

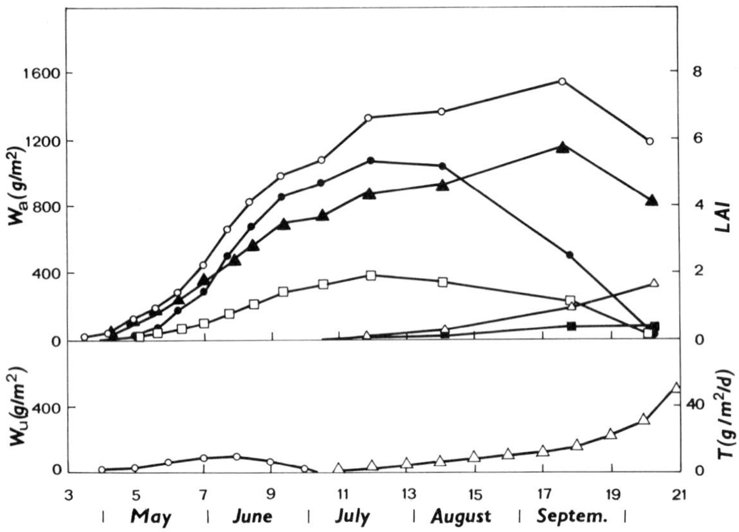

Fig. 1. *Top:* Seasonal changes in leaf area$_{LAI}$ (●—●), and in dry weight W_g of stems (▲—▲), fresh leaves (□—□), inflorescences (■—■), dead leaves (△—△), total aboveground biomass (○—○) in a reed stand determined by indirect nondestructive method; *Bottom:* Growth of underground biomass W_u (△—△) and inflow rate T of assimilates from the reserve to aerial organs (○—○) in a reed stand determined by means of a photosynthetic model

putation is given in Figure 2. The time scale is divided into ten-day units. The net photosynthesis calculated from the photosynthetic model (see Sect. 3.2.3) was converted into dry-weight increments (in g/m^2/d^1), by regarding 1543 mg CO_2 as equivalent to 1 g biomass (Šesták et al., 1971). The photosynthetic function was established from data computed in the model for consecutive time intervals under conditions of average irradiance and air temperature within these intervals, using the growth data on leaf area and aerial biomass from Figure 4:

$$F = \exp(4.376 - 0.328\ T + 0.00456\ T^2) \quad (\text{g m}^{-2}\ \text{d}^{-1}) \tag{1}$$

where F is the photosynthetic function and T is time in ten-day intervals ($2 < T < 22$, T_0 corresponds with March 21). The dark respiration of aboveground biomass is subtracted from the amount of daily net photosynthesis W_f. The respiration is regarded as dependent on the mean night temperature (t_a):

$$r_a = \exp(0.308\ t_a - 11.57) \quad (\text{g g}^{-1}\ \text{d}^{-1}) \tag{2}$$

where r_a is the dark respiration of aboveground biomass (see Fig. 4). The seasonal course of the mean night temperature was approximated by a polynome in T fitted to measured data (see Fig. 3):

$$t_a = 10.46 - 3.514\ T + 0.912\ T^2 - 0.0643\ T^3 + 0.00135\ T^4 \tag{3}$$

Now, the ΔW_f values are compared with corresponding daily increments of aboveground biomass ΔW_a. The difference $\Delta W_u = \Delta W_f - \Delta W_a$ represents the trans-

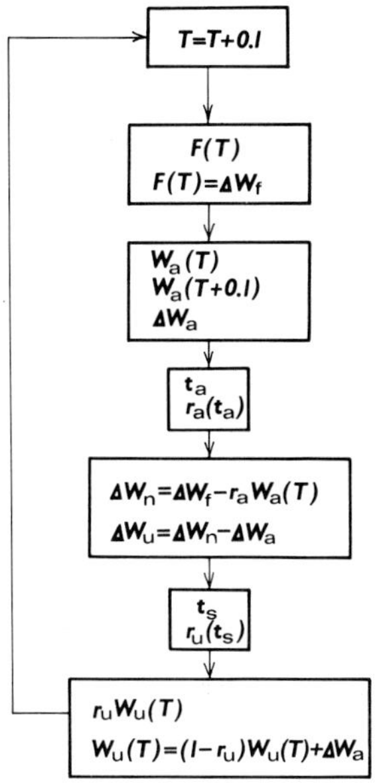

Fig. 2. Block scheme for computation of translocation from the underground reserves and of seasonal growth of underground biomass (for explanation see text)

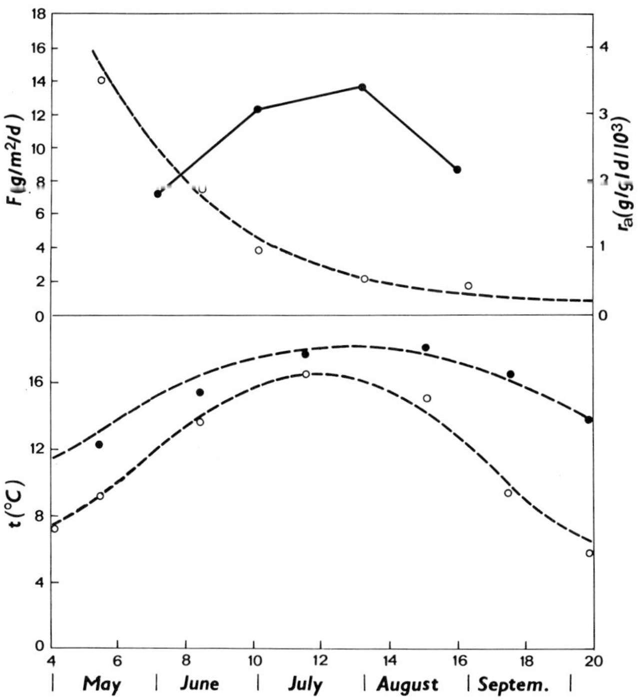

Fig. 3. *Top:* Seasonal course of assimilation F (○—○) and respiration r_a (●—●) rates in a reed stand; *Bottom:* Seasonal course of night mean temperature of air (●—●) and of diurnal mean soil temperature (○—○) at a depth of 0.2 m. *Dashed line* marks fitted course

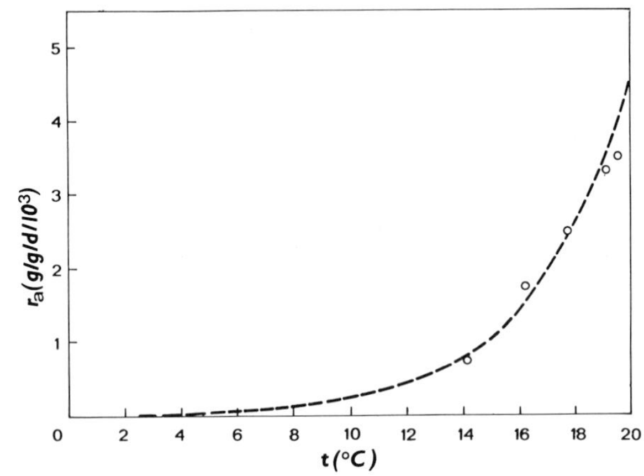

Fig. 4. Dependence of respiration of aboveground biomass on night mean air temperature

fer of assimilates to the underground organs (mainly to rhizomes). When the respiration of actual underground biomass formed during the current year, r_u (older underground biomass is not considered) is subtracted from W_u an estimate of the net underground production at any chosen phase of seasonal stand development is obtained. The value of r_u is given as a function of the daily average temperature of the bottom substrate at the depth of 0.2 m (t_s). (All the temperature data were measured by Přibáň.) Of course, r_u is not known. In the function:

$$r_u = \exp(at_s + b) \tag{4}$$

the parameters a, b were changed to obtain a respiration either higher or lower than that of the aboveground biomass. It has been found that even a tenfold reduction or increase in r_u with respect to r_a affects the final value of net underground production to a very small extent (ca. 3%). The seasonal course of t_s was again approximated by a polynome in T:

$$t_s = 3.379 + 2.312\,T - 0.09055\,T^2 \tag{5}$$

where $2 < T < 22$ (see Fig. 3).

For the seasonal course of leaf area index (L) and aboveground biomass (W_a), individual points from the empirical curves (see Fig. 1) were interpolated in order to obtain the data on L, W_a, and ΔW_a for all days of the growing season.

The results are shown in Figure 1 which illustrates both the storage rate of underground biomass and the course of translocation of assimilates from the underground reserves. The total inflow is 410 g of dry weight per season. The final ratio of underground / aboveground biomass was 0.44 at the end of the growing season. But if this final underground biomass is related to the seasonal maximum aboveground biomass a smaller ratio of 0.34 will result.

Dykyjová et al. (1971) have reported on this ratio in outdoor hydroponic cultures of reed after one, two and three years of cultivation. Their data are compared with the above results in Table 1. As only a negligible portion of dead rhizomes and roots was present even in three-year-old cultures, the underground /

Table 1. Ratio of underground to aboveground biomass (R/S) per season obtained in hydroponic cultures of *Phragmites communis*. (After Dykyjová et al., 1971)

Cultivation interval	Date of		R/S per season
	planting	harvest	
1 year	1 April 1969	29 Sept. 1969	0.79–1.60
1 year with many young hibernating shoots	16 Sept. 1967	1 Sept. 1968	0.48–0.75
2 years	2 May 1967	30 July 1968	0.23–0.44
3 years	26 July 1966	26 Sept. 1968	0.41
	26 May 1967	30 Sept. 1969	0.75

aboveground net production ratio (R/S) obtained in hydroponic cultures may be applied to natural stands. Table 1 proves that particularly the data on two years' production approach the computed value.

An application of the method based on the photosynthetic model to another more productive limosal reed stand yielded a higher R/S ratio: 0.72 for the final and 0.56 for the seasonal maximum value of W_a. The growth of aboveground biomass and leaf area was determined by the same indirect method as in the previous case. The maximum L and W_a were, respectively, 7.9 and 1920 g m^{-2}. The total translocation from underground reserves to aerial organs was not much higher than in the previous case (447 g) though both the leaf area and aboveground biomass were higher by 25%.

In this case, only the annual production of underground biomass was taken into account. The underground biomass originating from past seasons appears only partly in the inflow during early shoot development. Of course, the metabolically active old underground biomass is probably also supplied with and respires newly assimilated materials (see Sect. 3.1.3) but the proportion of the respired biomass is probably small as follows from the analysis given above. The difference between losses (outflow in the spring) and gains (inflow in summer and fall) in underground biomass estimates its losses due to rhizome and root death if one assumes that the underground biomass is in a steady state for a period of several years. This conclusion is, however, only hypothetical and simplifies the real situation. The answer could be obtained by a system analysis of energy transfer in littoral ecosystems involving a sufficiently long time scale and taking into account the variation in all principal environmental conditions affecting the growth of the plants.

It may be concluded that the spring translocation can be estimated from the growth rate of leaf area and aboveground biomass. The time-lag of leaf growth controls the amount of material translocated from the reserves to the aerial organs; the ratio of the growth rate of leaf area to that of aboveground biomass determines the final ratio of underground to aboveground production. Further verification of some of these assumptions is still needed.

References see pp. 285—291.

3.1.6 Growth Analysis of Fishpond Littoral Communities

J. KVĚT

Growth analysis was used in our IBP studies on the primary production of pond littoral vascular plants and their communities as a tool for evaluating the seasonal dynamics of their net production and vertical productive structure. The methods of growth analysis have been treated comprehensively by Blackman (1968), Květ et al. (1971), Ondok (1971b), and Evans (1972). They were originally elaborated mainly for crops, annual plants and herbaceous perennials whose underground biomass and its changes are assessed relatively easily. Hence, growth analysis is applicable without great modifications to such pond vascular plant communities as those of tenagophytes (see Sect. 1.2) and other plants forming temporary communities on emerged shores and bottoms. It can also be easily applied to duckweeds *(Lemnaceae)* and other lemnids provided their passive movements on the water surface are prevented (see Sects. 1.2 and 3.5). Certain modifications of the methods of growth analysis are inevitable if they are applied to perennial herbaceous plants with large and bulky rhizome and root systems, among them to the helophytes growing in pond littorals. The storage of organic matter in the long-lived rhizomes and its translocation and use for growth of the short-lived aerial shoots represent the principal complication (see Sects. 3.1.3, 3.1.4 and 3.1.5). In view of the difficulties, growth analysis of helophytes and their communities has been mostly confined to the aerial parts. The techniques of sampling for the primary biomass data are reviewed in Section 3.1.2. The methods of measuring leaf area have recently been reviewed by Květ and Marshall (1971). Of these, photoplanimetric measurement and a modified dot-counting method by Negisi et al. (1957) were used most frequently. Stem surface area was estimated only in some instances; in *Phragmites*, the assimilatory surface area of a whole stem with leaf sheaths was assumed to be equal to about 10% of its leaf area (i.e., that of the leaf blades, Ondok, unpublished data). When stem surface area was estimated in other species, "geometrical" methods and calculations were used. But in many instances, the assimilatory surface area of the stems and leaf sheaths was not involved in the growth-analytical calculations at all, which is justifiable in view of Gloser's findings (see Sect. 3.2.2) of a small photosynthetic activity in leaf sheaths of *Phragmites*. The application of growth analysis to reedswamp plant communities has been discussed by Ondok (1969) and Květ (1971). In agreement with the concepts expressed in the latter paper, this section also treats growth analysis as a combination of the originally British "classical" approach with the evaluation of the "productive" vertical stand structure in relation to the interception of incoming radiation as pioneered by the Japanese school, starting with Monsi and Saeki (1953).

3.1.6.1 Helophytes and Their Communities

General Remarks. The seasonal dynamics of aboveground net production have been studied by the methods of growth analysis in almost monospecific stands formed by most of the dominants of the various associations belonging to the foederatio *Phragmition communis* (see Sect. 1.1). Growth analytical studies were also made in stands of pure *Bolboschoenus maritimus* ssp. *maritimus* and in a *Caricetum gracilis*. The studies were undertaken in both South Bohemia (Opatovický and other fishponds in the Třeboň basin) and South Moravia (Nesyt and Hlohovecký fishponds, and marshes of the former Lake of Kobylí—see Fiala and Květ, 1971). Growth analysis of complete young polycormones of *Typha angustifolia, T. latifolia* and *Phragmites communis* was performed by Fiala (1970b, 1971b, 1973a; see also Sect. 3.1.3). Some of the results have already been published by Květ et al. (1969), Ondok (1969), Dykyjová et al. (1970), Fiala (1970a, b, 1971b), Květ and Svoboda (1970), Ondok et al. (1970), Ondok and Přibáň (1970), Husák (1971), Květ (1971), Dykyjová and Hradecká (1976), but quite a few others have not yet been published. Little is known on the chlorophyll content in the pond helophytes, except, perhaps, *Phragmites*, and on the chlorophyll index (g chlorophyll m^{-2} stand area) in their communities (see Medina, 1964; Rejmánková, 1973). The primary data on the biomass (*W*, mostly in g m^{-2}) and leaf area (*A*) or leaf area index (LAI) were obtained either by the direct harvest technique (with the necessary subsampling of leaves for the leaf area measurements) or by various indirect techniques (see Ondok, 1971b, and Sect. 3.1.2). The growth characteristics, namely the rate of aboveground dry matter production (= crop growth rate, CGR, in g m^{-2} d^{-1}), relative growth rate (RGR, in g g^{-1} d^{-1} or %), or that of the leaf area (RGR_A in d^{-1} or %) and net assimilation rate related to leaf area (NAR, in g m^{-2} d^{-1}) were mostly calculated as interval values from the primary data on *W* and LAI which had either been obtained directly or had been read off from fitted smooth growth curves of *W* and LAI, depending on the technique used for assessing the primary values. The integral growth characteristics of leaf area duration (LAD, in m^2 d m^{-2}; see Watson, 1952) and biomass duration (BMD, in g d m^{-2}, see Květ and Ondok, 1971) were also calculated in some instances, and the long-term averages of NAR, RGR, and LAR (see below) were calculated from them according to Ondok (1971c) and Ondok and Květ (1971). The NAR, when calculated solely from increments in aboveground biomass, is of dubious value because of the impossibility of separating its two components, namely the translocation of assimilates to and from the rhizome reserves and the net assimilation of the shoots. An attempt to overcome this difficulty is made in Section 3.1.5. The values of NAR derived merely from increments in shoot biomass may be meaningful only during periods of rapid vegetative growth of shoots with a richly developed assimilatory apparatus: equal shares of the total net production may then be assumed to be used for above- and belowground growth, or even more may be used for shoot growth. Finding on the seasonal dynamics of carbohydrate content in *Phragmites* rhizomes and on the seasonal changes in bulk density and lateral bud formation in *Typha* (Fiala 1973a, b; see also Sect. 1.3) appear to support this assumption.

The instantaneous growth-analytical characteristics such as the leaf area ratio (LAR = *A*/*W*, in dm^2 g^{-1}) and its components, the specific leaf area (SLA = *A*/leaf dry weight, in dm^2 g^{-1}) and leaf weight ratio (LWR = leaf dry weight/*W*, in g g^{-1}) as well as other biomass ratios were calculated in the way described, e.g., by Květ et al. (1971). For evaluating the vertical stand structure, the whole canopy was divided up into 0.4 m thick horizontal layers (Fig. I in Introduction to this Sect.).

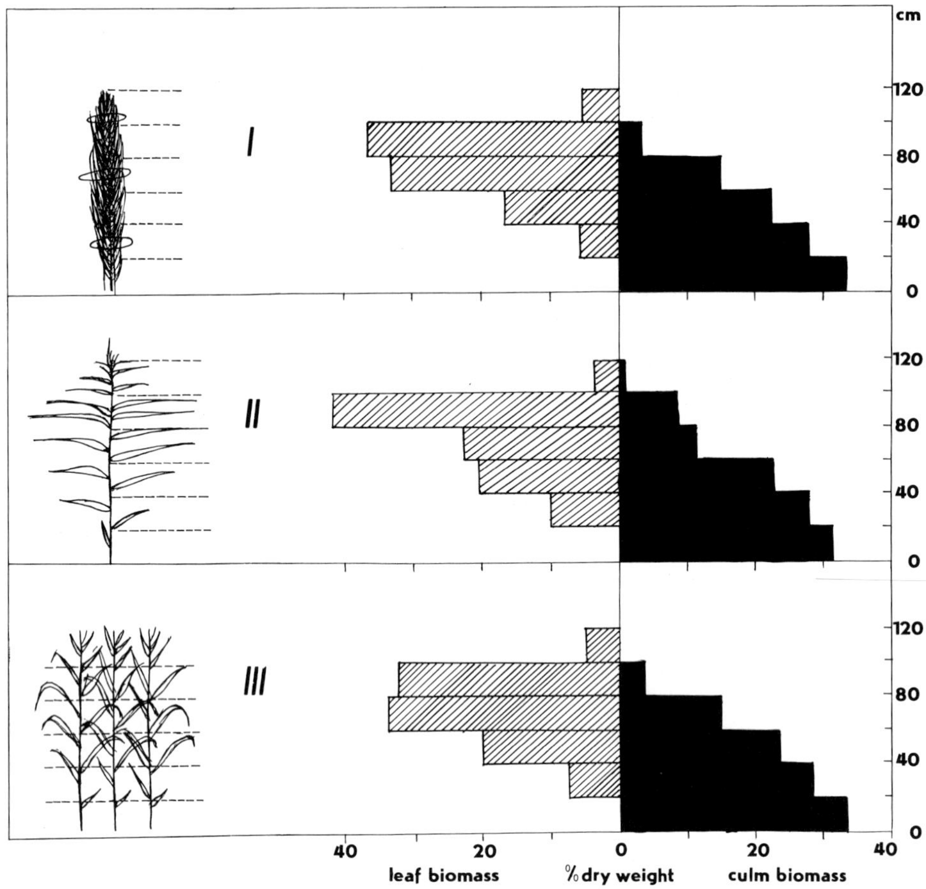

Fig. 1. Effect of the technique of stratified harvesting on resulting vertical distribution of leaf and culm biomass in a young stand of *Phragmites communis*. *(I)* Segments with leaves stretched; *(II)* Segments with whole undamaged leaves; *(III)* Segments with leaves in natural position (Dykyjová, unpublished)

This was done with harvested shoots, with their leaves, if present, stretched upwards; this standardized the sampling procedure but led to a certain overall upward shift of the leaf canopy as compared with its natural position (Fig. 1). The leaf and stem dry weight, mostly also leaf area and, exceptionally, stem surface area were assessed for each layer separately. The LAR and other instantaneous growth-analytical characteristics were also calculated from these values, separately for each layer. If seasonal changes in the vertical stand structure were followed the LAD, BMD and average LAR were also calculated for each layer separately (see Květ et al., 1969). Cumulative LAI (from the top downwards) may be calculated for each level, by 0.4 m, within the canopy as well.

Several examples will now be given of growth-analytical investigations in communities of reed-belt helophytes and of tall sedges.

A growth analysis based on direct destructive harvesting (4 plots 1 m^2 on each date of the aboveground biomass and using interval calculations of the CGR,

Table 1. Analysis of growth of aboveground parts in stand of *Phragmites communis* at Opatovický fishpond in 1966 (S-form, limosal stand in an accumulation biotope). (After Dykyjová and Hradecká, 1976)

Interval	W g m^{-2}	Wl g m^{-2}	LAI m^2 m^{-2}	LAR dm^2 g^{-1}	SLA dm^2 g^{-1}	CGR gm^{-2} d^{-1}	RGR mg $g^{-1}d^{-1}$	NAR g $m^{-2}d^{-1}$	ash % dry wt.
20 April–									
–23 May	1,025	233	3.83	0.37	1.64	31.0	140	(16.12)[a]	8.5
–10 June	1,473	461	7.83	0.53	1.69	24.8	20	4.21	8.0
–7 July	1,851	603	10.09	0.54	1.67	18.0	11	2.01	7.5
–15 July	2,143	757	8.14	0.38	1.06	20.8	9	2.27	8.0
–1 Aug.	2,272	724	7.05	0.31	0.96	7.5	4	0.99	8.0
–8 Sept.	2,731	655	5.47	0.20	0.83	12.0	5	0.93	7.0

Symbols and abbreviations: see text.
[a] Indicates intense mobilization of reserves from rhizomes.

RGR, RGR_A and NAR) is demonstrated in Table 1. The data illustrate the growth dynamics of the "S"-form of a *Phragmites* stand at the Opatovický fishpond as recorded during one growing season (see also Sects. 2.4, 3.1.1 and 3.4.2). The high values of CGR, RGR, RGR_A and NAR occurring at the beginning of the growing season evidently result from assimilate supply from the rhizome reserves to the growing shoots. The growth characteristics recorded between the third and fifth harvests, when the shoots were vigorously growing and the LAI attained its maximum, may be regarded as more or less realistic values that can be used in comparisons between various *Phragmites* stands. The respective long-term and average characteristics of the stands "S" (limosal) and "V", (littoral) at the Opatovický fishpond are as follows: LAD:74547 and 62197 d; BMD:199.7 and 163.3 d kg m^{-2} RGR:0.010 and 0.007 g g^{-1} d^{-1}; LAR:0.373 and 0.381 dm^2 g^{-1} for the interval from 20 April to 1 September.

The higher productivity of the S stand thus seems to be mainly due to its greater LAI. There was surprisingly little difference between the two stands in LAR and SLA, both of which illustrate the plants' investments of net production into the development of the assimilatory apparatus. The coefficient of solar energy conversion η may be estimated from the interval values of CGR transformed into those of energy content per m^2 (using the transformation of 1 g shoot dry weight = 4.3 Kcal = 18 KJ; see Sect 3.1.8 and Dykyjová and Přibil, 1975) which, in turn, are related to the sums of incoming radiation for the interval given. The peak seasonal values of η would then amount to 2.8 and 4.7% of incident PhAR (0.45 × global radiation), between 10 June and 1 July, for the V and S stands respectively, assuming that the whole increment has taken place in the shoots.

Both the primary data and the derived growth characteristics are clearly subject to considerable sampling errors (Table 2). Errors of this kind are presented graphically as confidence limits = ± twice the standard deviation in Figure 2 derived by Pokorný from data by Květ et al. (1969).

Growth analysis based on indirect assessments of biomass and leaf area index from measurements taken on the same shoots, possibly in permanent plots during the growing season, removes the random variation between individual bio-

Table 2. Stand of *Phragmites communis* at Nesyt fishpond in 1966. (Calculated by J. Květ and J. Pokorný from the data by Květ et al., 1969)

Date[a]	Value[b]	Canopy layers (cm above bottom level)[c]										All layers
		−40	−80	−120	−160	−200	−240	−280	−320	−360	−400	
24 May	W_s	32.0	31.3	31.3	30.4	61.4	95.0	194.0	—	—	—	28.9
	W_l	—	77.9	60.8	39.7	39.9	60.9	69.2	—	—	—	29.2
15 July	W_s	19.6	19.2	23.4	21.8	30.5	38.8	59.9	114.5	206.3	—	26.5
	W_l	—	—	85.0	51.7	30.1	28.2	37.6	74.0	93.9	214.3	25.6
26 Sept.	W_s	13.2	11.9	14.7	11.0	19.5	17.4	27.3	70.4	157.1	—	12.8
	W_l	—	—	—	109.0	70.3	9.5	12.7	34.4	69.7	—	14.5

[a] Corresponding to early, peak and final stages of stand development.
[b] Coefficients of variation (%) in stem (W_s) and leaf (W_l) dry weight in individual 40 cm canopy layers during growing season.
[c] Canopy structure assessed in four plots by m^2 on each date.

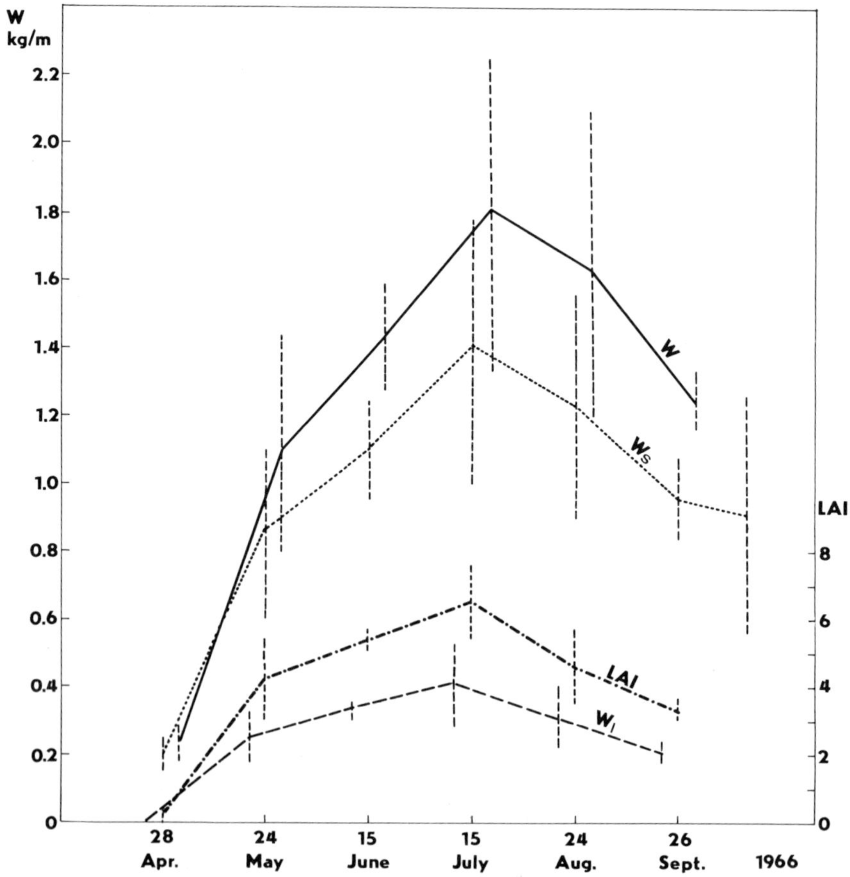

Fig. 2. *Phragmites communis*, Nesyt, 1966. Seasonal course of dry weight of total shoot biomass *W*, stem biomass W_s (with leaf sheaths), leaf biomass W_l and leaf area index LAI. Vertical bars indicate confidence limits (=arithmetic mean ± twice standard error). (Adapted from Květ et al., 1969, by J. Pokorný)

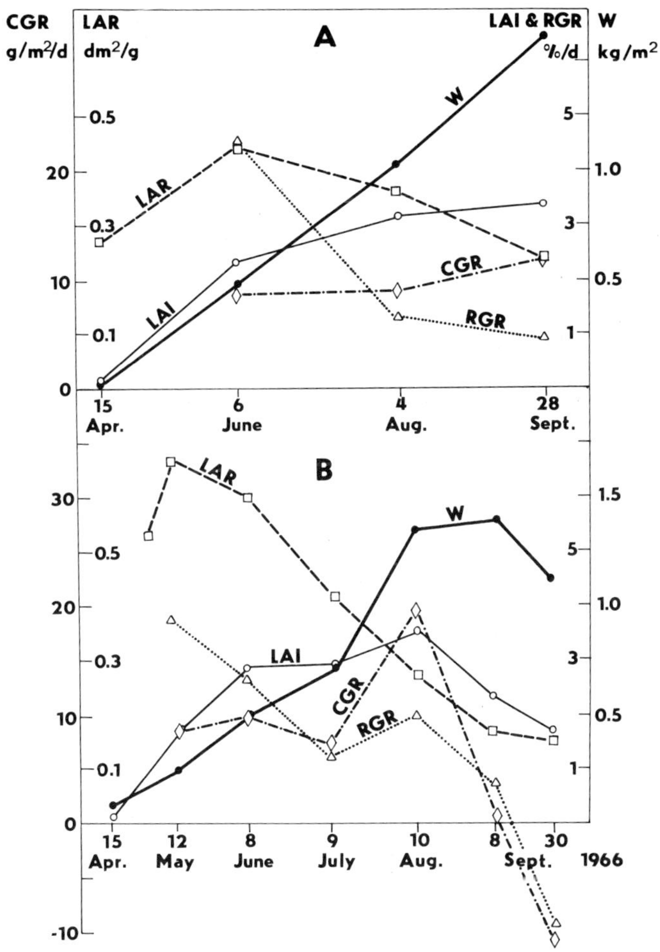

Fig. 3A and B. *Typha latifolia*, Lake of Kobylí, 1966. Growth analysis of aboveground organs performed by destructive harvest method and by an indirect method (B), employing measurements of intact shoots in four permanent quadrats of 0.5 m^2 each. Symbols: See text. (After Květ et al., 1969, and J. P. Ondok et al., unpublished)

mass samples (see Ondok, 1971b). Care, however, must be taken that the shoots measured in the permanent plots are representative of the whole stands investigated. In 1966, Květ et al. (unpublished data) performed a growth analysis of a *Typhetum latifoliae* community in this way in the marshes of the former Lake of Kobylí in South Moravia, some 30 km north of Lednice. The results may be compared with those of a rather rough growth analysis of the same stand based on destructive sampling of 4×1 m^2 plots on four dates during the growing season (Květ et al., 1969; Květ, 1975a). Figure 3 illustrates the similarities and differences between the two sets of results. The main difference is observed towards the end of the growing season: while the shoot biomass increased substantially and the LAI did slightly (or did not change) in the "destructive" samples, both values declined in the permanent plots. The derived growth characteristics behaved accordingly.

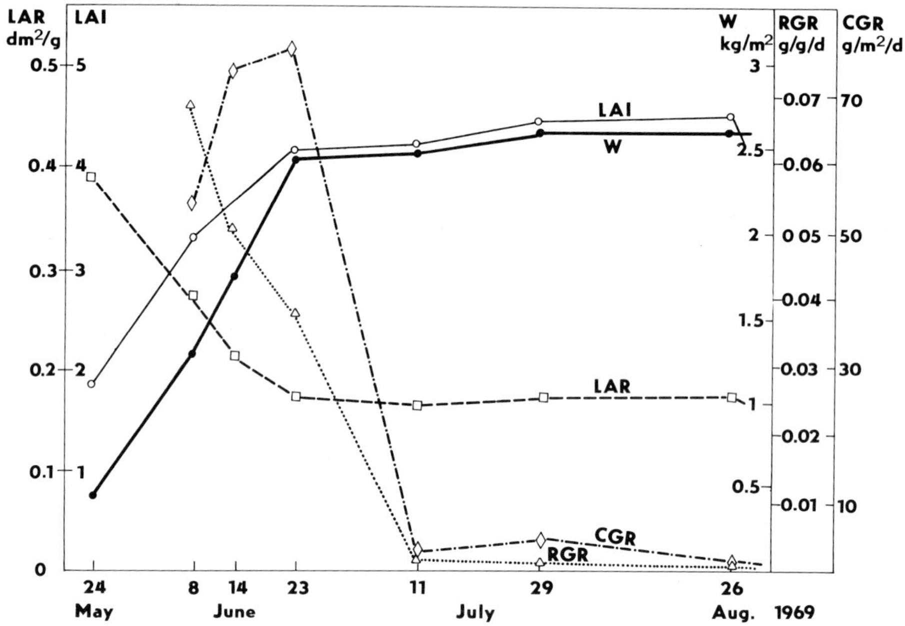

Fig. 4. *Typha angustifolia*, Nesyt, 1969. Growth analysis of aboveground organs performed by an indirect method employing measurements of intact shoots in a permanent plot 4 m^2 in size. (After Husák, 1971)

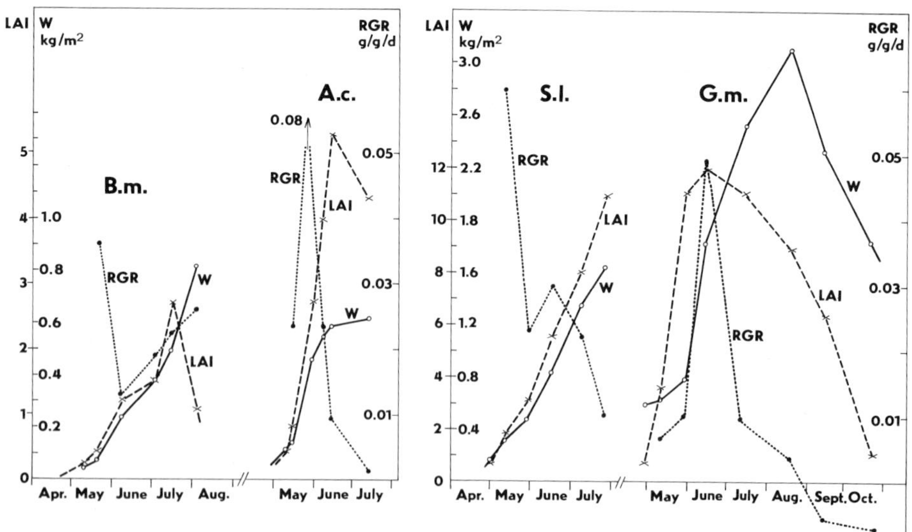

Fig. 5. *Bolboschoenus maritimus (B.m.)* in 1968, *Acorus calamus (A.c.)* in 1970, *Schoenoplectus lacustris (S.l.)* in 1968, and *Glyceria maxima (G.m.)* in 1971, all from Opatovický fishpond. Growth curves of total aboveground biomass *W*, leaf area index *LAI*, and seasonal course of relative growth rate of shoots *RGR*. Note different scales for *W* and *LAI* for (*B.m.* + *A.c.*) and (*S.l.* + *G.m.*) (Dykyjová, unpublished)

The decline in the permanent plots may, however, be an artefact due to frequent trampling, damaging or destruction of the *Typha* rhizomes penetrating the permanent plots from outside, particularly in late summer and early fall (see Fiala, 1973a, and Sect. 3.1.3). The differences in seasonal maximum stand density between the permanent plots and the "destructive" samples seem to support this explanation.

The growth analysis of a monospecific *Typhetum angustifoliae* performed by Husák (1971) in the Nesyt fishpond in 1969 was based on indirect estimates of biomass and leaf area index from shoot-length measurements and assessments of stand density in a relatively large ($4\,m^2$) permanent plot. The results of this analysis are presented in Figure 4. In a more detailed analysis, flowering and non-flowering shoots would have to be treated separately; on the whole, however, the aboveground biomass of the stand seems to have attained a steady state soon after flowering-time.

Growth analysis has also been made in communities of monodominant *Schoenoplectetum lacustris*, *Glycerietum maximae*, *Acoretum calami* and in *Bolboschoenus maritimus* ssp. *maritimus* growing in the littoral of the Opatovický fishpond in 1968–1970. The dense invasion stands had been selected. Figure 5 presents the growth curves of shoot biomass, LAI, and RGR for the four communities. The differences between them are evident. The high LAI values in the *Schoenoplectetum* and *Glycerietum* are worth noting. In *Glyceria*, the growth of the aerial shoots seems to be more synchronized in the relatively winter-cold climate of Central Europe than in the milder climate of South England; see also the data by Hájková and Květ (1970) from South Moravian wet grassland and by Westlake (1966).

The more vigorous growth and greater production of the communities belonging to the foederatio *Phragmition communis* is apparent. But the true reedswamp communities formed by closed *Phragmites*, *Glyceria maxima*, *Typha angustifolia* and, locally in accumulation biotopes, also by *T. latifolia* and *Schoenoplectus lacustris*, appear superior to the communities of *Typha latifolia* or *Schoenoplectus* in erosion biotopes or to those of *Acorus* and of *Sparganium:* for a detailed study of production see Dykyjová and Ondok (1973). The greater production of the true reedswamp dominants seems to be associated with their capacity to form large and closed continuous stands in fishpond littorals. The relatively less productive stands of the other species, frequently only patchy or rather loose, are confined to those biotopes where none of the potentially more vigorous dominants can thrive. The habitat requirements of communities belonging to the different associations of the foederatio *Phragmition* are compared in Section 1.2.

Results of comparative studies on the vertical productive structure and its dynamics in the fishpond reed-belt communities, assessed by the stratified harvest method, have been published by Květ et al. (1969), Dykyjová et al. (1970), Květ and Svoboda (1970), Ondok et al. (1970), Ondok and Přibáň (1970), Dykyjová (1971c, 1973a), and Květ (1971). Figure 1 in the Introduction to this Section illustrates both the method and the data obtained. Little attention, however, has been paid to the variation in these data resulting from sampling errors. The variation in data from individual replicate plots is illustrated in Table 2 for the *Phragmites* stand investigated at Nesyt in 1966 by Květ et al. (1969). This rather wide variation has to be taken into account particularly when estimating the

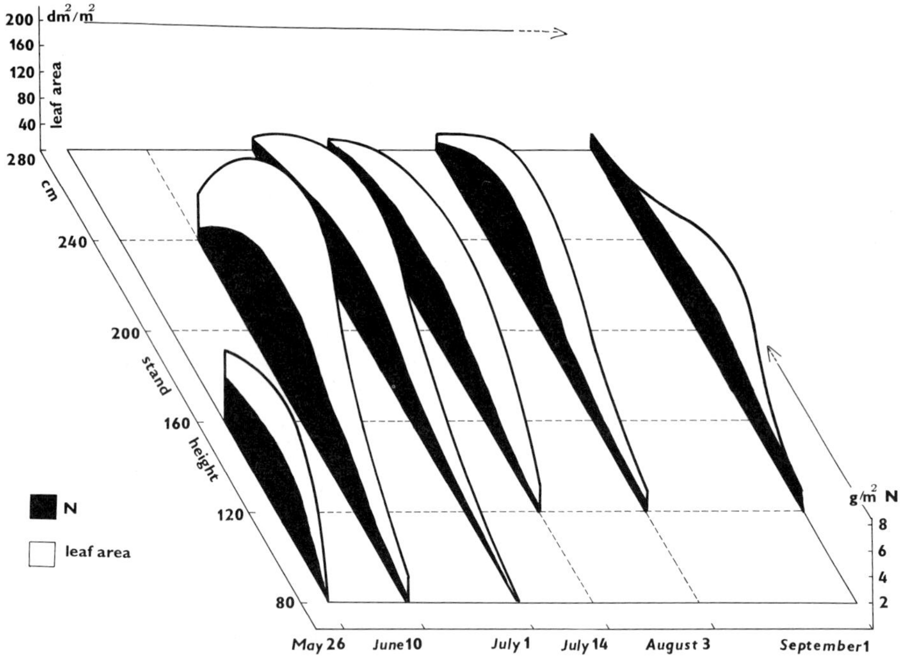

Fig. 6. *Phragmites communis*, littoral biotope in Opatovický fishpond, 1966. Seasonal course of stratified (by 40 cm horizontal layers) leaf area and nitrogen content per unit ground area (g m^{-2} N). (Data by Dykyjová, unpublished)

vertical profiles of cummulative LAI and interpreting this and other characteristics of vertical stand structure in terms of radiation attenuation (see Sect. 3.2.3). The data on the vertical distribution of shoot biomass have also been used for evaluating the vertical distribution of mineral nutrient content in the stands (Květ, 1971, 1973c, 1975a; see also Sect. 3.4.1) and for an assessment of the effects of cutting on the fishpond reed-belt stands (Husák, 1971; see also Sect. 7.1). For the littoral *Phragmites*, "V" form, from the Opatovický fishpond, the seasonal course of the vertical distribution of LAI in relation to the nitrogen content in the leaves is illustrated in Figure 6. Fuller data on the net production as related to nutrient availability during the season are given in Sections 3.4.1 and 3.4.2.

References see pp. 285—291.

3.1.6.2 Communities of Emerged Fishpond Shores and Bottoms

M. REJMÁNEK and J. VELÁSQUEZ

The communities of emerged fishpond shores and of summer-drained fishpond bottoms, described and characterized in Sections 1.2 and 7.1, mostly consist of species populations of rapidly growing annual, biennial or perennial plants with rather shallow root systems. The analysis of the growth and seasonal dynamics is thus relatively easy in these communities and their components. The data acquired provide deeper insight into the reaction of the plants and communities to various microhabitats of the emerged bottom; they may also help to understand the variation in species composition of those communities and their dynamics (ecesis of certain populations, successions, etc.). The depth of the nutrient-rich mud deposited on the bottom and soil moisture content represent the principal factors determining the growth and development of these plant populations and communities. The following text contains two examples of growth analysis, one providing a comparison of growth dynamics in two communities colonizing different microhabitats in a summer-drained pond, the other showing the reaction of one monodominant community to the actual ecophase.

The seasonal development of the vegetation colonizing a summer-drained fishpond bottom was followed by Rejmánek (1974, 1975) in the Nový fishpond (some 15 km south of Třeboň) in the growing season of 1971.

Two homogeneous communities were compared. One, colonizing rather wet habitats, may be classified as *Scirpetum radicantis* Hejný (= *Eleocharito (ovatae)-Caricetum cyperoidis* Klika 1935 fac. *Scirpus radicans);* the other community, colonizing drier sites, belongs to the ass. *Bidenti-Polygonetum hydropiperis* (W. Koch, 1926) Lohmeyer 1950 fac. *Alopecurus aequalis.* The species composition of the communities and approximate proportions of the populations involved are expressed by the Braun-Blanquet scale in the following reléves (25 m^2, 20 July 1971, $E_1 = 100\%$):

1. *Eleocharis ovata* 3, *E. acicularis* 1, *Carex bohemica* 2, *Scirpus radicans* 3, *Alopecurus aequalis* 2, *Glyceria maxima* +, *Rorippa palustris* 1, *Alisma plantago-aquatica* +, *Polygonum lapathifolium* 1, *Elatine triandra* +, *E. hydropiper* r, *Ranunculus sceleratus* +, *Rumex maritimus* +, *Juncus bulbosus* +, *Peplis portula* +, *Ranunculus trichophyllus* r, *Chenopodium polyspermum* +, *Callitriche palustris* r, *Lythrum hyssopifolia* r, *Lemna gibba* +, *Spirodela polyrhiza* r.

2. *Polygonum hydropiper* 5, *P. lapathifolium* 1, *P. minus* +, *Alopecurus aequalis* 2, *Bidens frondosus* 1, *Rorippa palustris* 1, *Carex bohemica* +, *Ranunculus sceleratus* r, *Lycopus europaeus* 1, *Scirpus radicans* +, *Rumex maritimus* r, *Galium palustre* r, *Alnus glutinosa* (juv.) +.

The primary values recorded in either community were both the shoot and the underground dry weight and the assimilatory surface area of all species present in two or three representative quadrats of 30 cm × 30 cm each. The following characteristics were calculated for each interval between harvests from these primary values: total biomass *W*, leaf area index LAI, leaf area ratio LAR, root/shoot ratio R/S, total litter *d*, crop growth rate CGR, relative growth rate RGR, net assimilation rate NAR, and net efficiency of PhAR conversion η assuming an average energy content of 4.3 kcal (= 18 KJ) per 1 g dry biomass (see Sect. 3.1.8).

The results are presented in Figures 1 and 2. The two communities differ little in the seasonal course of their total biomass and in maximum biomass (580 and 620 g m^{-2}, respectively, on 12 August). Pronounced differences were, however, observed in their LAI and LAR. These differences are largely due to the considerable share of stem dry weight in the biomass of *Polygonum hydropiper* in community 2, i.e., the *Bidenti-Polygonetum hydropiperis.* The R/S ratio was somewhat greater in *Scirpetum radicantis* (community 1), containing a substanial share of

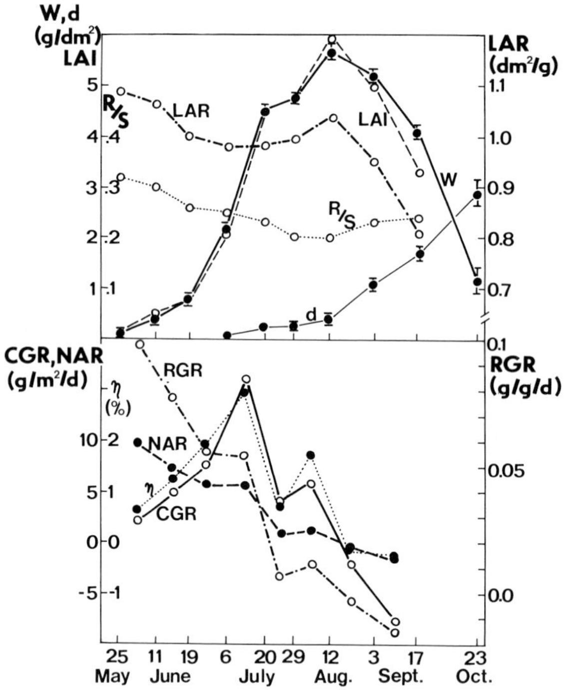

Fig. 1. Course of biomass, *W*, *LAI*, and growth characteristics during seasonal development of vegetation in a summer-drained fishpond: Community 1: *Scirpetum radicantis*. Symbols: See text

Scirpus radicans, a geophyte. The decrease in biomass toward the end of the growing season was first slower in the drier *Bidenti-Polygonetum* community, where litter and detritus accumulated suddenly only in October after the early fall frosts.

The lower parts of Figures 1 and 2 illustrate the different patterns of growth and development in these two communities. Community No. 1, adapted to permanently wet soil conditions, reacted to a dry period (20–29 July) by a marked decrease in all four growth characteristics; their increase followed when the water supply had improved. Community 2, adapted to drier conditions, remained unaffected during the dry period and its biomass continued to grow exponentially; so did the interval values of CGR and NAR. The highest values of CGR amounted to 16.0 and 10.2 g m^{-2} d^{-1}, and those of NAR to 9.7 and 3.0 g m^{-2} d^{-1} in the *Bidenti-Polygonetum* and *Scirpetum* respectively. The relatively frequent positive correlation between RGR and LAR (Eagles, 1971) has been ascertained only in the latter community, the different course of LAR in the two communities being responsible for this. The courses of the coefficient η and of CGR run mostly parallel in both communities; their maximum interval values are 2.88% (community 1) and 1.84% (community 2) of incident PhAR.

The highest value of total biomass ascertained in a single replicate quadrat sampled in the summer-drained Nový fishpond was 1060 g m^{-2} in community 1 where the dominant constituted 92% of the total biomass. With about 40% of the

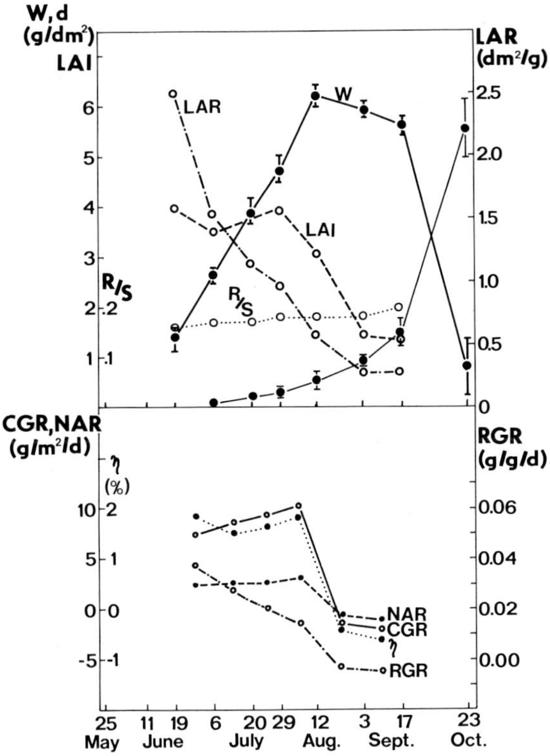

Fig. 2. Course of biomass, *LAI*, *W*, and growth characteristics during seasonal development of vegetation in a summer-drained fishpond: Community 2: *Bidenti-Polygonetum hydropiperis*. Symbols: See text

bottom surface being practically bare, the biomass and LAI values as well as the growth characteristics of the two communities presented here may be regarded as averages representative of the whole summer-drained bottom of the Nový fishpond.

The nearly monospecific community of *Eleocharis acicularis*, belonging to the ass. *Litorello-Eleocharitetum acicularis* Malcuit 1929 was followed by Velásquez (1975) on the emerged shore of the sky-fed Přední Svinětický fishpond (some 10 ha in area) near Bavorov in the South Bohemian České Budějovice and Vodňany basin. In the growing season of 1973, a gradient of ecophases (see Sect. 1.1) developed along the fishpond shores. The reaction of the dominant *Eleocharis acicularis* to the ecophases in which it occurred—limosal (water depth 5 to 10 cm) and terrestrial—was examined by the methods of growth analysis. Sample quadrats of 25 cm × 25 cm were taken to the depth of 0.2 m below bottom level in fivefold replicate in each ecophase on five dates during the growing season. The *Eleocharis acicularis* plants were divided into photosynthetically active and inactive material, the former including the green shoot parts, the latter comprising the shoot parts without chlorophyll, rootstocks and roots. The dry weight of each fraction was assessed separately. In addition, the chlorophyll content and chloro-

Table 1. Growth analysis of *Eleocharis acicularis* at peak of growing season. (A) Monospecific stands of *Eleocharidetum acicularis*: shoot growth in natural conditions; stand in limosal and terrestrial ecophase in the Přední Svinětický fishpond, South Bohemia, 1973. (B) Growth of whole tufts including their underground parts, in experimental cultures; controls and those treated with two doses of complex fertilizer; Třeboň, 1974. See text for explanation of abbreviations; CI, chlorophyll index (= chlorophyll amount per unit ground area); $\overline{\text{LAR}}$, mean LAR calculated as LAD/BMD. (From Velásquez, 1975)

(A)	Interval	W $g\,m^{-2}$	CI $g\,m^{-2}$	LAI $m^2\,m^{-2}$	CGR $g\,m^{-2}\,d^{-1}$	RGR $mg\,g^{-1}\,d^{-1}$	$\overline{\text{LAR}}$ $dm^2\,g^{-1}$
Limosal ecophase	21 June–10 July	130	0.67	8.3	1.9	17	4.7
	10 July–31 July	522	2.48	11.0	18.7	66	3.0
	31 July–24 Aug.	644	4.32	12.0	5.9	10	1.9
Terrestrial ecophase	21 June–10 July	118	0.51	2.5	1.5	11	2.0
	10 July–31 July	224	0.55	4.2	5.0	33	2.0
	31 July–24 Aug.	469	1.06	5.2	10.2	30	1.4

(B) Fertilizer dose $g\,m^{-2}$	Interval	W $g\,m^{-2}$	LAI $m^2\,m^{-2}$	LAR $dm^2\,g^{-1}$	NAR $g\,m^{-2}\,d^{-1}$	CGR $g\,m^{-2}\,d^{-1}$	RGR $mg\,g^{-1}\,d^{-1}$
Control	24 July–14 Aug.	1,093	4.2	0.38	5.4	18.4	20
56	24 July–14 Aug.	1,732	7.5	0.43	5.1	36.4	8
450	24 July–14 Aug.	1,905	13.2	0.69	4.3	46.1	33

		BMD $g\,m^{-2}\,d$	LAD $m^2\,m^{-2}\,d$	$\overline{\text{LAR}}$ $dm^2\,g^{-1}$	NAR $g\,m^{-2}\,d^{-1}$	RGR $mg\,g^{-1}\,d^{-1}$
Control	29 June–14 Aug.	35,705	117	0.33	4.9	16
56	29 June–14 Aug.	57,992	235	0.41	5.2	20
450	29 June–14 Aug.	54,090	333	0.62	4.2	25

phyll index (g chlorophyll m^{-2}) as well as assimilatory surface area of the shoots, expressed as LAI, were assessed in the photosynthetically active fraction; details of the methods are described by Velásquez (1975). Because of the great spatial variation in the biomass of the non-assimilatory parts, growth analysis was only applied to the assimilatory parts (= green shoots). The principal results are given in Table 1. They prove the favorable effect of the limosal ecophase on the growth of *Eleocharis acicularis;* this has also been confirmed experimentally (Velásquez 1975). The seasonal maxima of shoot biomass and LAI recorded in the limosal ecophase were surprisingly high but the amount of ash material contained in the plants or sticking to their surfaces was also substantial: between 14 and 42%, mostly about 22% of dry weight. The net efficiency of solar energy conversion η into shoot biomass was calculated on the basis of the energy content reported by Boyd and Vickers (1971) for *Eleocharis quadrangulata:* 4.0 kcal (= 16.7 KJ) per ash-free dry weight. The highest interval value of $\eta = 2.4\%$ of PhAR (= 0.45 times global radiation) was recorded in the limosal ecophase between 10 and 31 July.

The response of *Eleocharis acicularis* in limosal ecophase to the addition of fertilizer to the substrate (fishpond mud) was followed in cultivation experiments at Třeboň in 1974. Quadrats of young *Eleocharis*, each 625 cm^2 in size, were transferred from an emerged fishpond bottom into cultivation trays. Herbasyn 2, a commercial fertilizer with the following composition: N 10%, P 16.5%; K 23%; Mg 0.8%; B 0.04%; Mn 0.1%; Cu 0.04%, was applied to the plants in doses of 56, 112, 225, and 450 g m^{-2} at monthly intervals from June to September. No fertilizer was applied to the control. The methods of sampling for the primary data and their subsequent treatment were the same as in the field observations reported before, except that only three quadrats were sampled on each harvest date.

The experiment was started in early June, and five harvests were taken between 29 June and 2 September. Table 1 presents some of the results for the 56 and 450 g m^{-2} fertilizer treatments in comparison with the control. The positive effect of the fertilizer on the development of the assimilatory apparatus is evident but the higher fertilizer dose was mostly probably excessive.

The growth analysis of *Eleocharis acicularis*, both in the field and in cultures, has proved that the limosal ecophase and ample nutrient supply provide the best conditions for its growth and development. This, however, holds true not only for many other plants colonizing emerged fishpond shores and bottoms, but also for plants growing frequently in deeper water such as *Sagittaria sagittifolia, Butomus umbellatus, Bolboschoenus maritimus* (hydroochthophyta according to Hejný, 1960), as proved by Hroudová (unpublished). Limosal ecophase is also essential for young vegetative stages of a great many fishpond littoral plants and thus represents one of the factors determining the great diversity in pond littoral vegetation (Hejný, 1960). Growth analysis makes it possible to assess and understand the effects of such complex factors as ecophases in more detail.

References see pp. 285—291.

3.1.7 Primary Data on Biomass and Production Estimates in Typical Stands of Fishpond Littoral Plant Communities

J. Květ and Š. Husák

Numerous estimates of biomass have been made in studies on the productivity in pond littoral plant communities, using one or other of the methods presented in Section 3.1.2. Most of the data originate from the pond reed-belt communities dominated by helophytes, on which the main interest was focused. The productivity of other plant communities occurring in or fringing fishponds has so far received less attention. In view of the difficult estimation of underground biomass, particularly in helophyte communities (see Sect. 3.1.2.5), the majority of data here characterize only aboveground biomass. Table 1 presents examples of data on biomass recorded in a wide range of fishpond plant communities at the peak of their development.

This table as well as other published data (Dykyjová, 1971a, b, 1973a; Dykyjová and Hradecká, 1973, 1976; Dykyjová and Květ, 1970; Fiala, 1970b, 1973a, b; Fiala et al., 1968; Fiala and Květ, 1971; Hradecká and Květ, 1973; Hejný, 1960; Hroudová, unpublished; Husák, 1971; Husák and Hejný, 1973; Husák

Table 1. Biomass in 3 groups of fishpond plant communities assessed at the peak of their development in July to August, in several fishponds of the Č. Budějovice-Vodňany basin (S. Bohemia). (Data by S. Hejný and K. Gregor)

1. Reedswamp communities (foed. *Phragmition communis*)		Biomass [a] (dry wt. g m^{-2})
Phragmitetum communis	Invasion stand	1,115
Glycerietum maximae	Invasion stand	986
Glycerietum maximae	Regenerating stand	690
Schoenoplectetum lacustris	Invasion stand	1,100
Typhetum angustifoliae	4 invasion stands	1,221–1,389
Typhetum angustifoliae	Initial stand	186
Typhetum latifoliae	Invasion stand	1,287
Typhetum latifoliae	Initial stand	60
Sparganietum erecti	Invasion stand	1,036
Sparganietum erecti	Regenerating stand	283
Sparganietum erecti	Initial stand	315
Acoretum calami	2 invasion stands	874–1,013
Acoretum calami	Regenerating stand	263
Equisetetum fluviatilis	Invasion stand	1,244
2. Short-term emergent communities (foed. *Oenanthion aquaticae*). Stands of dominant		
Bolboschoenus maritimus ssp. maritimus	2 stands	535–613
Oenanthe aquatica	Developed stand	285
Sagittaria sagittifolia	2 initial stands	16–34
3. Floating plant communities (foed. *Lemnion minoris, Nymphaeion albae*). Stands of dominant		
Spirodela polyrhiza	Developed stand	134
Nymphoides peltata	Developed stand	176
Trapa natans	Developed stand	107
Potamogeton natans	2 pure stands	196–256
Potamogeton natans with *P. lucens*, *P. pusillus* or *Sagittaria s.*	3 mixed stands	112–201

[a] Groups 1 and 2: aboveground biomass; group 3: total biomass.

and Květ, 1973; Květ, 1973c, 1975a, Květ and Ondok, 1973; Ondok, 1970a, 1971a, 1973a; Rejmánková, 1973a, b, 1975a; Straškraba, 1968; Véber and Kupka, 1973), and those contained in other sections of this book (Sect. 3.1.3, 3.1.6, 3.4, and 3.5) confirm that in fishponds, as in other standing waters, the plant biomass increases, in general, from the hydrophyte and pleustophyte communities of the sublittoral, through the communities formed by rooted macrophytes of the hydroochthophyte and ochthohydrophyte life forms of helophytes which constitute the prevailingly littoral to limosal zones of the reed-belt. Further landward, the plant biomass again decreases with increasing distance from the shoreline and/or the groundwater table—as long as the graminoid uliginosophytes are the predominant life form and shrubs or trees (*Salix* spp., *Alnus*, etc.) are absent. A pronounced gradient of biomass is also observed in the dominant helophyte synusium in any sufficiently broad reed belt. The biomass of the undergrowth synusium then depends, in the first place, on the canopy density of the dominant. Figure 1 illustrates this zonation of the plant biomass in the reed-belt of the Nesyt fishpond at the peak of the growing season. *Phragmites* or *Typha angustifolia*, and

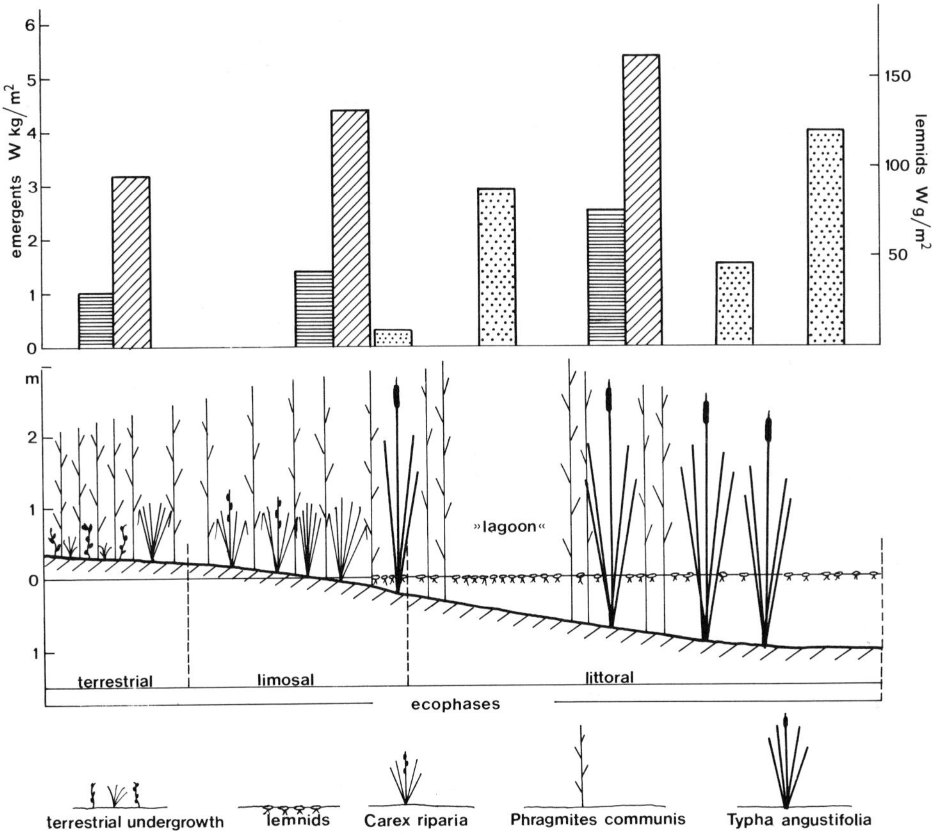

Fig. 1. *Top:* Distribution of macrophyte biomass: [horizontal hatching] total above and [diagonal hatching] total underground biomass of emergent vegetation; [dotted] total biomass of lemnids (mainly *Lemna gibba*) along a transect across reed-belt of Nesyt fishpond; *Bottom:* Sketch of transect at "normal" water level (zero value on the vertical scale) showing principal dominants and undergrowth components and distribution of ecophases. (Compiled from original data by Š. Husák, J. Květ, and E. Rejmánková)

a *Lemna gibba*-dominated community (*Lemnetum gibbae*) form, respectively, the dominant and undergrowth synusia here. The very small algal biomass in the water is not indicated in this scheme (see Sect. 1.3).

In relative terms, the seasonal maximum biomass characterizes the primary production in most communities of rooted emergent macrophytes (see Sect. 3.1.2). However, in some floating and submerged communities, a turnover factor of up to 1.5 to 2 may be assumed (Rich et al., 1971); the annual net production may surpass the seasonal maximum biomass by this factor, whose value is likely to increase with increasing actual length of the growing season. If this is shorter than three months the factor will probably differ little from 1.

Estimation of the absolute values of annual net production from the biomass data is therefore complicated in most instances, not only in perennial helophytes as shown in Section 3.1.2. Yet an attempt has been made to work out the ranges of

Table 2. Comparison of the principal monospecific reedswamp communities (foed. *Phragmition communis*) and other types of emergent, floating, or submerged macrophyte communities. (Compiled from data by several authors for community types I and II, and from Rejmánková, 1973b, c, 1975a; see also Sect. 3.5; for types III and IV)

I Community type Foed. *Phragmition communis*	Seasonal max. biomass kg m^{-2}			R/S biomass ratio	Net production kg m^{-2} $year^{-1}$ [a]			R prod./ S biom. ratio	Average ash content %	Net org. product. kg m^{-2} $year^{-1}$ [b]
	abovegr. S	undergr. R	total		abovegr.	undergr.	total			
Phragmites communis	0.6–3.5	1.6–8.0	2.2–9.5	0.9–2.0	0.6–3.7	0.5–3.0	1.1–6.7	0.5–1.0	10	1.0–6.0
Schoenoplectus lacustris	0.8–3.0	0.8–4.5	1.6–7.5	2.3–3.9	0.8–3.1	1.0–3.0	1.8–6.1	1.0–1.2	10	1.6–5.5
Glyceria maxima	0.6–2.6	0.7–4.7	1.3–5.5	1.1–7.6	0.7–3.4	0.3–1.4	1.0–4.8	0.4	10	0.9–4.3
Typha angustifolia	1.0–3.0	0.9–3.6	1.9–5.1	0.9–1.2	1.1–3.4	0.8–2.7	1.9–6.1	0.8–0.9	10	1.7–5.5
Typha latifolia	0.5–2.0	0.2–1.4	0.8–3.4	0.4–0.6	0.6–2.4	0.2–0.8	0.8–3.2	0.4	10	0.7–2.9
Sparganium erectum	0.9–1.6	0.3–1.4	1.5–2.8	0.5–1.0	1.0–1.9	0.3–0.5	1.3–2.4	0.3	11–13	1.2–2.2
Acorus calamus	0.4–0.6	0.7–1.1	1.0–1.7	1.8	0.4–0.7	0.2–0.5	0.6–1.2	0.5–0.9	11–13	0.5–1.1
II Foed. *Bolboschoenion maritimi* and *Oenanthion aquaticae*; dominant stands of										
Bolboschoenus maritimus ssp. *compactus*	0.2–0.9	0.4–2.0	0.6–2.9	1.6–2.2	0.2–1.0	0.2–1.9	0.4–2.9	1.0–1.9	10	0.4–2.6
B. *maritimus* ssp. *maritimus*	0.5–1.3	0.8–2.8	1.3–4.1	1.6–2.2	0.5–1.4	0.5–2.7	1.0–4.0	1.0–1.9	10	0.9–3.6

				turnover factor					
III Foed. *Lemnion minoris* and *Hydrocharition morsus-ranae*									
Lemnetum gibbae, Nesyt	—	—	0.15	2.0	—	0.3	—	29	0.2
Lemno-Utricularietum, facies with *Riccia rhenana*, Nesyt	—	—	0.15	2.0	—	0.3	—	50	0.15
Riccietum fluitantis, Opatovický	—	—	0.1	2.0	—	0.2	—	25	0.15
Spirodelo-Lemnetum gibbae, Mníšek village pond near Třeboň	—	—	0.5	1.5	—	0.75	—	25	0.6
Lemnid synusium in the reed-belt, Nesyt	—	—	0.01–0.1	1.0–2.0	—	0.01–0.2	—	35	0.05–0.15
IV Hydrophyte communities									
Potamogetoneto (pectimati) Zannichellietum pedicellatae, Nesyt	—	—	0.2	1.3	—	0.3	—	27	0.2
Batrachietum rionii, Nesyt	—	—	0.1	1.5	—	0.15	—	25	0.1
Hippuridetum vulgaris, Nesyt	—	—	0.8	1.2	—	1.0	—	15	0.85

[a] Estimated from ratios between net annual undergrounc production and seasonal maximum aboveground biomass (R prod./S biom.; see Sect. 3.1.2.5) or from probable biomass tumover factors (see Sect. 3.1.7).
[b] Estimated values of annual production of organic matter.

annual net production in those communities for which this seemed feasible. Table 2 presents the resulting data. The communities dominated or formed by *Phragmites communis*, *Glyceria maxima*, *Typha angustifolia*, and *Schoenoplectus lacustris* appear as the most productive and efficient ones.

The long-term (annual) production seems to be related, to some extent, to the range of habitats occupied by each community or dominant species population. Within each set of plant life-forms, the euryplastic dominants seem to form more productive communities. These dominants therefore tend to prevail in intensively managed fishponds, probably taking advantage of the increased supply of mineral nutrients (see Sect. 3.4), and forming dense stands effectively intercepting the incident solar radiation (see Sects. 2.2 and 3.1.8).

But even the less plastic (stenecious) dominants occupy otherwise vacant niches and form communities which may be more productive, at least for certain periods of the growing season, than the communities which the euryplastic species populations would form if they were to occupy the same niches (e.g., *Sparganium erectum*, a relatively stenecious species, as compared with *Phragmites communis* in deep and loose sapropel soils).

The short-term variation in the production rates of the dominant plants and of their communities can only be revealed by an analysis of their growth and production processes as determined by both internal—particularly ontogenetic—and environmental factors (see Sects. 3.1.5, 3.1.6, 3.2, 3.3, 3.4, and 3.5).

References see pp. 285—291.

3.1.8 Determination of Energy Content and Net Efficiency of Solar Energy Conversion by Fishpond Helophytes

D. DYKYJOVÁ

For an evaluation of photosynthetic efficiency in primary production processes in littoral ecosystems, the energy equivalent of the biomass of their primary producers has to be determined. Straškraba (1968) has concluded on the basis of his own and other authors' data that the average energy content in several aquatic macrophytes is equal to 4.48 kcal (= 18.75 kJ) per 1 g ash-free dry biomass. Similarly, Boyd (1970b) concluded that nearly all macrophytes examined by him had much the same energy content. Yet a number of data recently collected as well as newly published data seem to prove that the energy content differs significantly in the ash-free organic matter of various kinds of plant material when sampled successively during a growing season (Dykyjová and Přibil, 1975; Boyd, 1970b; Kaul and Vass, 1972; Imhof and Burian, 1972; Grabowski, 1973). When examining the net efficiency of solar energy conversion, especially for short time intervals, it therefore seems necessary to analyze separately individual plant-material samples taken on different dates during a growing season.

The conversion efficiency for a whole growing season is correlated most closely with the seasonal maximum of the biomass produced during that season.

As the variation in energy content is relatively small in relation to the biomass of a whole stand, numerous authors have calculated the efficiency by means of constant conversion factors such as 4.6 kcal g^{-1} of ash-free dry matter (Westlake, 1965). In first investigations and preliminary communications (Dykyjová, 1971a, b) such average values were also used for calculating the net conversion efficiency in different helophyte communities. A systematic analysis of their energy content was accomplished later for a comparative evaluation of seasonal changes in ecological efficiency.

Analytical Method. The energy content was determined in the samples collected for the analysis of production and of the ash and mineral nutrient content. The sampling techniques were identical with those used for estimating the biomass (see Sect. 3.1.2). The combustion heat was determined with a new type of combustion semimicrocalorimeter, developed and constructed in the laboratories for instrument development of the Czechoslovak Academy of Sciences; for details, see Dykyjová and Přibil (1975). The net efficiency of solar energy conversion by the plant stands was calculated by relating the energy content in the organic matter produced to the sum of incident global radiation as recorded with a Kipp and Zonen solarimeter (see Sect. 2.1). Both the net primary production and radiation input are related to an area of 1 m^2 and to the same time interval, from the beginning of the growing season (date of shoot emergence) to the date of sampling:

$$\eta(\%) = \frac{\text{kcal organic matter m}^{-2} \times 100}{\text{kcal total global radiation m}^{-2}}$$

This coefficient may also be related to incident photosynthetically active radiation (PhAR 380–720 nm), assuming that PhAR represents 45% of the global radiation.

Results. The energy content in different helophytes and in their different organs assessed on different dates during the growing season is evident from Table 1. The energy content is highest in young shoots in spring and decreases during the growing season. Fruits, flowers and inflorescences have a higher energy content than either the vegetative aboveground parts or underground storage organs. The latter contain mainly glycids. The decrease in energy content with the ageing of vegetative plant organs corresponds to other authors' data (Hughes, 1971; Brzoska, 1971, and others). It also reflects the seasonal decrease in the contents of energy-rich substances such as proteins, phosphatids and other lipids, and the increase in the contents of energy-poor cell-wall materials such as cellulose, pectins or lignin.

The net conversion efficiency of solar energy calculated for a whole growing season is correlated positively with the seasonal maximum of biomass and negatively with the number of days of the growing season involved in the calculation. The shorter the growing season and the smaller the sum of incident radiation, the higher the coefficient of energy conversion, and vice versa. Several estimates of the coefficient for shorter time intervals during the growing season are thus of greater value. Data on the energy content in reed-belt helophytes from both South Bo-

Table 1. Examples of seasonal changes in energy contents of macrophytes from the littoral of the Opatovický fishpond. (From Dykyjová and Přibil, 1975)

Phragmites communis

Organs Sampling date	Ash %	Culms kcal g^{-1}		Ash %	Leaves kcal g^{-1}		Ash %	Rhizomes+roots kcal g^{-1}	
		dry weight	org. matter		dry weight	org. matter		dry weight	org. matter
June	9.5	4.37	4.83	8.3	4.70	5.12			
July	5.65	4.32	4.58	8.3	4.72	5.15			
Sept.	4.50	4.46	4.59	8.2	4.25	4.62			
May							11.2	4.09	4.59
Oct.							9.2	4.30	4.72

Glyceria maxima, shoots

Sampling date	Ash %	kcal g^{-1}	
		dry weight	org. matter
May	11.6	4.29	4.86
July	9.85	4.26	4.73
Sept.	8.45	4.25	4.64
Oct.	7.12	4.18	4.51

Typha angustifolia, shoots

Sampling date	Ash %	kcal g^{-1}	
		dry weight	org. matter
May	9.85	4.21	4.67
July	6.25	4.21	4.49
September	6.36	4.35	4.66
October	7.0	4.72	5.07

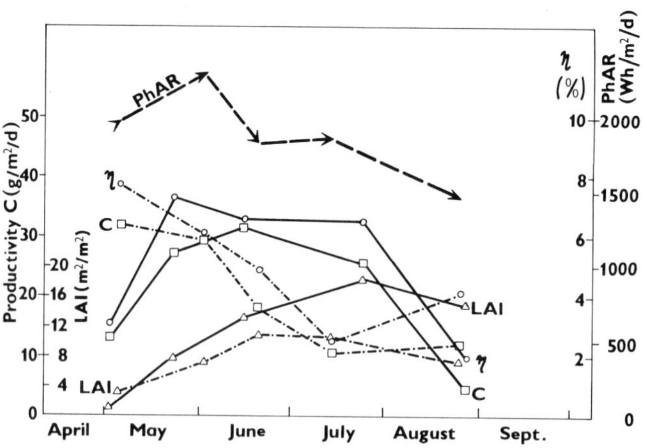

Fig. 1. Production characteristics and net conversion efficiency of PhAR (η) in littoral stands of *Phragmitetum communis* (———) and *Typhetum angustifoliae* (—·—·—·)

hemian and South Moravian fishponds are given in Dykyjová (1971a, b) and in Dykyjová and Přibil (1975); some examples are presented in Table 2 and Figure 1.

Communities of helophytes in the littorals of intensely managed fishponds display a similar net production to artificially intensified crop cultures. Their foliage is mostly adapted to an efficient interception of incident radiation (see

Table 2. Examples of production characteristics and solar energy conversion in helophyte stands in the littoral of Opatovický fishpond

Stands	Harvest date	Days of growing season	Type of biotope	Biomass g d.w./m^2	kcal g dry w.	Σ Global radiation kcal (m^2) season	Σ Global radiation kcal (m^2) day	η Global radiation %	η PhAR radiation %
Phragmites communis	12 July, 1972	82	erosion zone, littoral	1,030	4.42	342.020	4,170	1.33	2.95
Phragmites communis	1 July, 1966	72	limosal, accumulation biotope, eutroph.	1,850	4.53	266.570	3,709	3.16	7.01
Phragmites communis	8 Sept., 1966	140	limosal, accumulation biotope, eutroph.	3,250	4.34	481.200	3,437	2.93	6.50
Typha angustifolia	28 July, 1967	98	accumulation littoral biotope	2,830	4.22	410.120	4,184	2.91	6.47
T. a. maximum value	12 Aug., 1966	113	littoral accumulation biotope	4,040	4.20	397.300	3,516	4.27	9.48
Schoenoplectus lacustris	15 Sept., 1964	137	littoral erosion biotope	2,700	4.23	527.100	3,920	2.16	4.79
S. l. maximum value	26 July, 1971	86	accumulation biotope	4,210	4.21	380.830	4,428	4.65	10.32
Glyceria maxima	12 July, 1972	87	littoral erosion biotope	896	4.20	348.700	4,008	1.07	2.37
G. m. maximum value	12 July, 1972	87	limosal accumulation biotope	2,690	4.27	348.700	4,008	3.29	7.30
Bolboschoenus maritimus	3 July, 1966	63	littoral erosion biotope	378	4.33	253.400	4,022	0.64	1.42
B. m. maximum value	26 July, 1971	86	littoral accumulation biotope	866	4.40	362.420	4,168	1.06	2.35
Acorus calamus	14 June, 1971	54	erosion biotope	729	4.30	178.030	3,297	1.76	3.90
Sparganium erectum	26 July, 1971	86	accumulation biotope	1,884	4.25	362.420	4,168	2.21	4.90

Sect. 2.2). Their high biomass recorded in small plots and the derived high efficiency coefficients of the conversion of incoming PhAR presented in earlier publications have been questioned. Nevertheless, the coefficients of PhAR = 4–7% correspond to those calculated for intensified crops such as maize, sugar beet or sugar cane (Nichiporovich, 1966; Loomis et al., 1971); they have also been confirmed by Velasquez (1975) for *Eleocharis acicularis* in fertilized experimental cultures at Třeboň (see Sect. 3.1.6).

Optimized experimental hydroponic cultures of reed-belt helophytes (see Sect. 3.1.4) have also proved a high potential production in their two- to three-year-old polycormones. But the natural reed-belt communities in fertilized pond littorals also display a high net production and net efficiency of solar energy conversion (Fig. 1), comparable with that recorded in intense outdoor mass cultures of unicellular algae (*Scenedesmus quadricauda*) in the same area and under the same climatic conditions (Dykyjová, 1971a).

References see pp. 285—291.

3.2 Photosynthesis

3.2.1 Estimation of Net Photosynthetic Efficiency from Growth Analytical Data

J. P. Ondok

3.2.1.1 Parameters Used for Estimating Net Photosynthetic Efficiency

The instantaneous net efficiency of assimilation in any plant stand may be expressed by the ratio: dW/Ldt where dW is the increment in total stand biomass (expressed in terms of dry weight) over the time interval dt and L is leaf area index (leaf area per unit ground area) representing the main part of assimilatory plant tissues. Integration of this ratio over a certain growth interval $(t_2 - t_1)$ gives the "net assimilation rate" (NAR) which express the mean net efficiency of photosynthesis within this interval:

$$\mathrm{NAR} = \int_{t_1}^{t_2} \frac{dW}{L\,dt}\,dt. \qquad (1)$$

Various formulae for the calculation of NAR obtained under different assumptions concerning the growth curves of W and L are given in various papers and manuals (see, e.g., Květ et al., 1971; Ondok and Květ, 1971).

In order to facilitate comparisons between results obtained in different sampling intervals and under different conditions (outdoor and greenhouse cultivation or natural stands) another parameter may be introduced; this is defined as the ratio of NAR to PhAR irradiance during the time interval for which NAR has been calculated. This parameter is called "net assimilation efficiency" (NAE) and expresses the efficiency of net photosynthesis in relation to PhAR incident on unit horizontal area:

$$\mathrm{NAE} = \mathrm{NAR}/\bar{E} \qquad (\mathrm{g\,m}_l^{-2}/\mathrm{MJ\,m}_h^{-2}) \qquad (2)$$

where $\bar{E}$ means the interval value of PhAR irradiance and the subscripts l and h distinguish between unit area of leaves and unit horizontal area. NAE depends on both the physiological properties of the plants and habitat conditions.

If the distribution of irradiance on the leaf surfaces is known (this is possible in a more detailed analysis of radiation regime, see Ondok, 1973b) the net efficiency of photosynthesis can be expressed more precisely by the term:

$$\text{NALE} = \text{NAR}/\bar{E}_l \quad \text{(g/MJ)} \tag{2}$$

where NALE is net assimilation leaf efficiency, and $\bar{E}_l$ is the mean interval value of incident PhAR related to the leaf surface area.

If the energy content in the biomass is substituted for its dry weight in NAE, the resulting value will be η, the well-known coefficient of solar energy conversion frequently used in comparisons of plant production efficiency (see Sect. 3.1.8). The same substitution in NALE produces a conversion factor c which indicates the ratio between the energy bound in the biomass and the radiant energy intercepted by leaves. The difference between η and c is obvious.

3.2.1.2 Results Obtained in Hydroponic Cultures

NAR has been determined, for example, in *Phragmites* stands by Dykyjová et al. (1970); Květ et al. (1969); Květ and Svoboda (1970); and in *Typha angustifolia* and *T. latifolia* by Květ et al. (1969). Aboveground biomass was used in the calculations of NAR instead of total biomass as it is hardly possible to assess the successive increments of underground biomass. NAR calculated in this way has therefore only a restricted value, since the share of biomass translocated from rhizomes into aerial plant organs in spring and in the opposite direction in summer and fall was not involved. The result was an overestimated NAR in spring and an underestimate in summer and fall. This problem is encountered in most littoral helophyte communities.

It is possible to estimate NAR in cultivated stands, e.g., in hydroponic cultures. This method was used in the years 1966–1969 for studying the growth of both above- and underground biomass in several littoral helophyte species and NAR was also calculated on the basis of the data obtained. It is still uncertain to what extent the NAR values obtained in cultures can be extrapolated to natural stands. The main difference between cultivated and natural stands may occur in the ratio between aboveground and underground biomass, while the total annual net production by cultures of helophyte species (e.g., *Typha angustifolia*, *T. latifolia*, *Glyceria maxima*, *Schoenoplectus lacustris*) can attain or even exceed that of natural stands. Since only increments in total biomass are involved in the

Table 1. NAR and NAE in *Typha angustifolia* and *T. latifolia* cultivated in field conditions

Typha angustifolia			*Typha latifolia*		
Sampling interval	NAR $g\,m^{-2}\,d^{-1}$	NAE $g\,m_l^{-2}/MJ^{-1}\,m_h^{-2}$	Sampling interval	NAR $g\,m^{-2}\,d^{-1}$	NAE $g\,m_l^{-2}/MJ^{-1}\,m_h^{-2}$
4 June–15 July	12.7	1.39	4 June–16 July	14.4	1.58
15 July–16 Aug.	20.9	2.49	16 July–20 Aug.	11.5	1.42
16 Aug.–15 Sept.	10.8	1.64	20 Aug.–12 Sept.	17.6	2.59
15 Sept.–30 Oct.	7.2	1.48	12 Sept.–14 Nov.	2.9	—

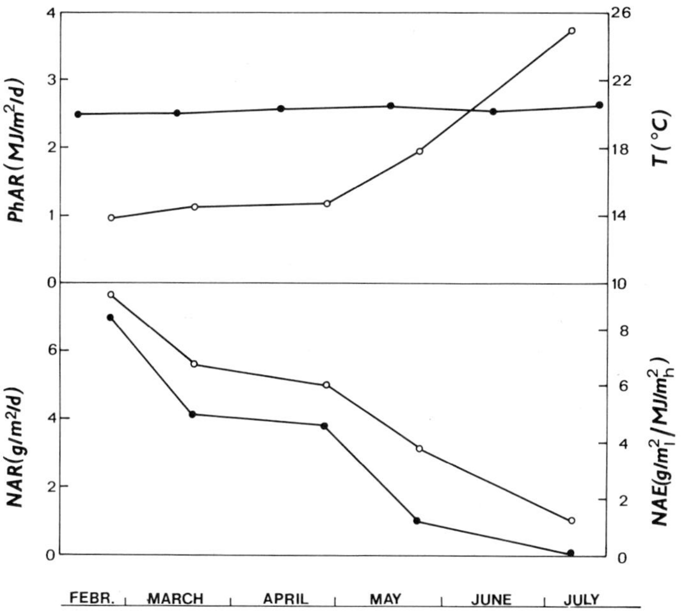

Fig. 1. *Top*: Course of air temperature (●——●) and daily mean sums of PhAR for individual sampling intervals (○——○); *Bottom*: Course of *NAR* (○——○) and *NAE* (●——●) for hydroponic cultures of *Typha latifolia*

calculation of NAR it may be estimated in natural stands by means of values obtained from cultures. The following sections provide several examples of this approach.

Typha latifolia. The species is suitable for cultivation both in hydroponic and in field experiments. It was cultivated in 1968 from one-year-old seedlings in large shallow trays (1 × 10 m). The solid substrate of diatomite earth was periodically irrigated with a nutrient solution. The details on the cultivation conditions and method are given in Section 3.1.4. During the cultivation season from January to June the air and substrate temperatures and the incoming solar radiation (PhAR) were recorded. The greenhouse was only partially air-conditioned; the air temperature fluctuated between 20 and 30° C. On several successive dates five replicate polycormones were harvested and the average dry weight of total biomass and leaf area per polycormone determined. NAR was calculated using formula (10.15) given by Květ et al. (1971). Figure 1 shows the resulting course of NAR in correlation with the daily mean values of PhAR computed for individual harvest intervals from daily records. The NAR values are high for the small amount of PhAR intercepted by the plants cultivated in a greenhouse, which transmit only about 50% of PhAR or even less, in January to July (for details see Dykyjová et al., 1972).

Schoenoplectus lacustris. This species was cultivated from rhizome cuttings in a greenhouse in the same year and under the same conditions as was *Typha latifolia*. Figure 2 shows the resulting course of NAR and NAE obtained from the

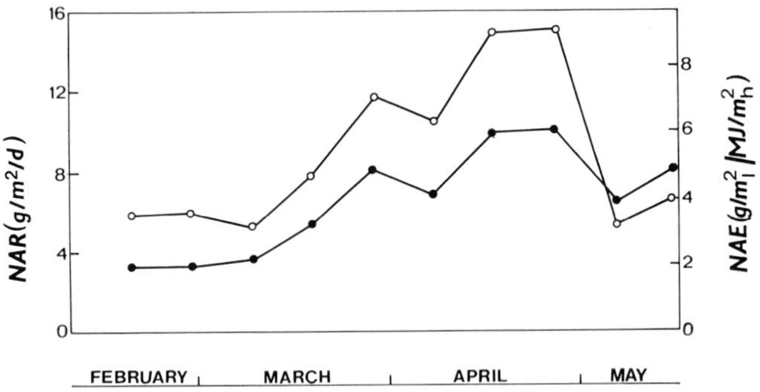

Fig. 2. Course of *NAR* (○——○) and *NAE* (●——●) calculated for *Schoenoplectus lacustris* cultures

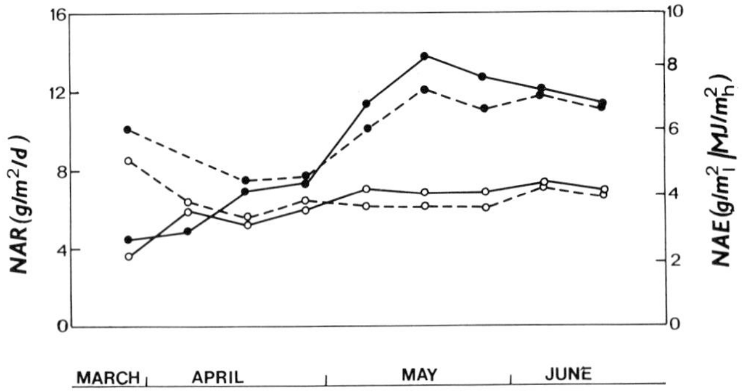

Fig. 3. Course of *NAR* (———) and *NAE* (-----) calculated for cultures of *Typha angustifolia* (—●—) and *Acorus calamus* (—○—)

sample data. The slower decrease of NAR and NAE as compared with that in *Typha latifolia* is due to less self-shading in *Schoenoplectus* cultures. This species practically does not form leaves and the assimilatory area is equal to the stem surface area. The mean value of NAE lies in the range of 2–8 g m_l^{-2}/MJm_h^{-2}.

Typha angustifolia and *Acorus calamus*. These two species were cultivated under similar conditions to the previous two in the year 1969. On each sampling date, six polycormones were harvested, the total biomass (dry weight) and leaf area per 1 polycormone were determined to obtain data for the calculation of NAR and NAE. Their course during the cultivation period is given in Figure 3. Higher values of both NAR and NAE were obtained in *Typha angustifolia* than in *Acorus calamus*. The mean NAE is 6 g m_l^{-2} MJm_h^{-2} in *Typha angustifolia* and near to 4 in *Acorus*. The absence of a marked decrease in NAE may be explained by little self-shading in both species, which did not attain the shoot height and biomass of the natural stands. With regard to the definition of NAE we should expect this to decrease with increasing PhAR during the period of experimentation. Such course has been found in *Typha latifolia* (see Fig. 1). The fluctuations in the course of NAE in the other three species are probably due to interactions with other factors (e.g., temperature) or to sampling error involved mainly in the leaf area measurements.

3.2.1.3 Results Obtained in Field Cultures

Reports on the estimation of net photosynthetic efficiency in natural helophyte communities using the methods of growth analysis are found only exceptionally in the literature. This is due to the small amount of data on the seasonal changes in underground biomass. Highly valuable data on the seasonal increments in underground organs were obtained by Fiala (1973d) for *Typha angustifolia* and *Typha latifolia* (see Sect. 3.1.3).

These data can be used for calculating seasonal changes in NAR both in *Typha angustifolia* and *T. latifolia*. Such changes are recorded in Table 1, together with estimates of NAE related to monthly means of the daily sums of PhAR incident at Lednice during the growing seasons of 1969 and 1970 (see Fiala, 1973d). The NAR is quite high, with somewhat higher values in *Typha angustifolia* than in *T. latifolia*. The high net photosynthetic efficiency of these species contributes to the explanation of their high productivity. As NALE could not been evaluated it cannot be decided whether this high productivity is to be attributed to a highly favorable radiation regime in the *Typha* polycormones or to a high photosynthetic activity of their leaf tissues. In view of the relatively low extinction of radiation within a *Typha* canopy (see Ondok, in press, 1977), the former explanation seems more probable. Another reason may be the rapid allocation of assimilates to actively growing parts of the *Typha* polycormones (Fiala, 1973d). The high productivity of *Typha angustifolia* has also been assesed by Ondok and Dykyjová (1973) in its fully developed stands in South Bohemian ponds.

3.2.1.4 Estimation of NALE in a Phragmites Stand

Net assimilation leaf efficiency (NALE) was calculated for a *Phragmites* stand using data from measurements made on June 17. The average PhAR irradiance on the leaf surface was computed on the basis of results given by Ondok (1973a, and 1977) where the theoretical part of this computation is treated. The distribution of both the direct sun and diffuse sky radiation is given in Figure 4. The distribution of PhAR is expressed by fractions of leaf area index (LAI) for the diffuse and by fractions of sunlit foliage area (SLI) for the direct radiation. The average irradiance on leaf surface $(\bar{E})_l$ was computed from the distributions in Figure 4 as the mean weighted by SLI and LAI fractions. For a noon sun, elevation $\bar{E}_l$ amounts to 5.19 kJ m_l^{-2}, i.e., 10.4% of PhAR above the stand. As $\bar{E}_l$ is related to one day, the changes of $\bar{E}_l$ with sun elevation have to be established. This is done in Figure 5 where relative values of $\bar{E}_l$ are compared with E above the stand. This ratio of $\bar{E}_l/E_o$ varies only little (within a range of 4%) during the day; its daily average is also affected by leaf inclination and orientation. On the date of measuring, the *Phragmites* leaves were mainly vertically inclined. $\bar{E}_l$ was also computed for other types of leaf arrangement (uniform and mainly horizontal distribution of leaf inclination). The differences were surprisingly small, not exceeding 2%. Similarly, only small differences have been found in E_l for various conditions of cloudiness. Thus, for the case considered, the average value of 12% may be used for estimating NALE from NAE.

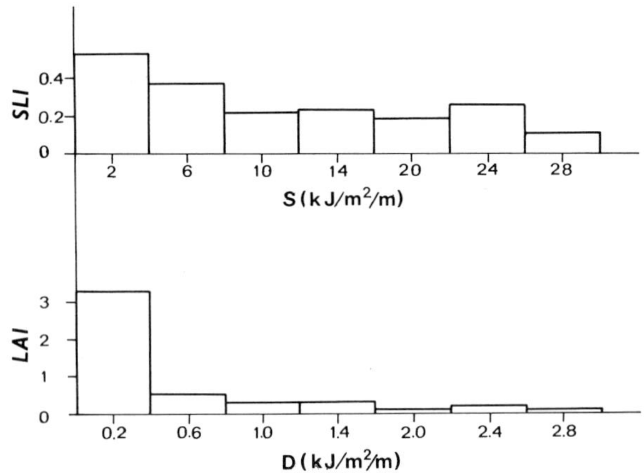

Fig. 4. Distribution of direct sun radiation (S) on sunlit foliage area and of diffuse sky radiation (D) on leaf area (SLI, sunlit foliage area index; LAI, leaf area index). Middle values of S and D given on horizontal scale

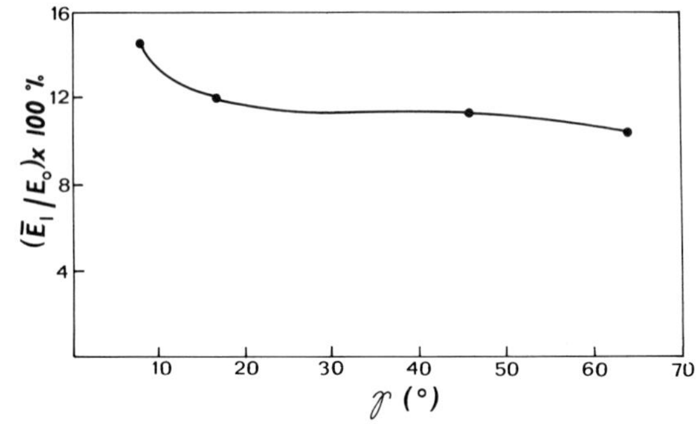

Fig. 5. Dependence of $\bar{E}_l/E_o$ on sun elevation (γ). $\bar{E}_l$, mean PhAR irradiance on leaf surface; E_o, PhAR irradiance above the stand

Table 2. Values of NALE and c as dependent on NAR for a reed stand on 17 June (LAI=4.7)

NAR($g\ m^{-2}\ d^{-1}$)	NALE ($g\ m_l^{-2}/MJ^{-1}\ m_h^{-2}$)	c(%)
1	0.60	1.2
2	1.20	2.5
3	1.80	3.8
4	2.40	5.0
5	3.00	6.3
6	3.60	7.5
7	4.20	9.2
8	4.80	10.1
9	5.40	11.3
10	6.00	12.6

The daily sum of incident PhAR on 17 June is 13.9 MJ/m_l^2 under clear sky conditions. $\bar{E}_l$ then amounts to 1.67 MJ/m_l^2. In this case, e.g., a NAR equal to $3.0\,g\,m^{-2}\,d^{-1}$ would result in a NALE of 1.8 g/MJ. Table 2 shows the corresponding values of NAR and NALE, valid for the stand and date considered. The values of NAR that were not calculated as the increments in underground biomass were not determined; they will probably not exceed the $5\,g\,m^{-2}\,d^{-1}$, when taking into account the LAI of 4.76. NALE will then not exceed the value of $3.0\,g\,MJ^{-1}$.

When calculating the conversion factor *c* the energy content in the *Phragmites* biomass is taken from the data by Dykyjová et al. (1970) as equal to $18.4\,kJ\,g^{-1}$. Higher values ranging between 20.0 and $21.7\,kJ\,g^{-1}$ have been obtained in further investigations on seasonal changes in energy content in selected helophyte species (Dykyjová and Přibil, 1975; see also Sect. 3.1.8). The conversion factor *c* listed in the third column of Table 2 was calculated on the basis of an energy content of $21.0\,kJ\,g^{-1}$.

References see pp. 285—291.

3.2.2 Net Photosynthesis and Dark Respiration of Reed Estimated by Gas-Exchange Measurements

J. GLOSER

Gasometric measurements of net photosynthesis and dark respiration were performed with the aim of assessing some basic data necessary for mathematical modeling of canopy photosynthesis and primary production in *Phragmites communis* stands. The intention was not only to find absolute values of these processes in different organs, but especially to define the dependence of both photosynthesis and respiration on the principal abiotic variables.

Separate measurements of gas exchange in individual organs (leaves, internodes with sheaths, and inflorescences) were preferred to enclosing whole shoots in a big chamber (as done, e.g., by Burian, 1969). The separate measurements ensured precisely defined and homogeneous environmental conditions for each structure in the course of measurement.

Plants and Methods. The measurements were made at the Opatovický fishpond near Třeboň in a dense invasion monospecific stand of ass. *Phragmitetum communis* at littoral ecophase. On the same site, intensive studies of stand structure, microclimate, plant growth, and mineral uptake were performed (see Sects. 2.1 and 3.4). The apparatus for gas-exchange measurements was situated within the stand about 20 m off-shore, in close vicinity to the plants studied (Fig. 1).

The period of measurement was July and August, 1973. Some special experiments were repeated the next year (July and August, 1974) at the Hlohovecký fishpond (near the Nesyt fishpond) in South Moravia. In each case, the vegetative

Fig. 1. Leaf chamber with an infrared gas analyzer situated within the reed stand. (Photo: J.P. Ondok)

growth of the aboveground shoots had been nearly completed: no new leaves were formed, but *Phragmites* flowered during August.

The net photosynthetic and dark respiration rates were calculated from the values of CO_2 uptake and output. The design of the gasometric apparatus and leaf chamber has been described by Gloser (1977); only a brief outline of the system will, therefore, be given.

The apparatus was of an open type with an infrared gas analyzer. Air was taken from the height of about 3 m above the reed canopy, and it was passed through a mixing vessel (50 l) to the leaf chamber. A parallel air stream was allowed to flow directly to the infrared gas analyzer (URAS 2), where its CO_2 concentration was compared with that of the air leaving the leaf chamber. Absolute values of CO_2 concentration in the system-entering air were measured periodically using a calibration gas mixture in steel cylinders as reference.

The leaf chamber was equipped with a heat exchanger (water as transfer medium) and with an efficient cylindrical fan. The walls of the chamber were made of acrylic plastic. The enclosed leaf was irradiated from both sides, so that the natural pattern of leaf irradiance was not changed. Leaf and air temperatures were measured with thermocouples and recorded with a multipoint recorder, as were the CO_2 analysis data and irradiance.

Irradiance was measured with two instruments: a Kipp solarimeter for a continuous record of global radiation, and a selenium photocell with selective

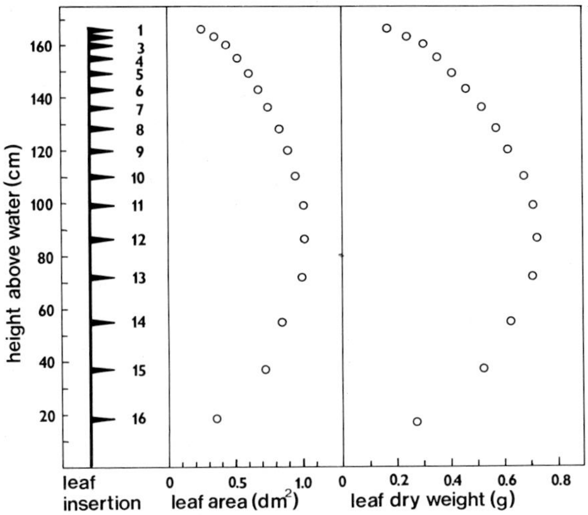

Fig. 2. Schematic diagram of vertical distribution of leaf insertion (leaves numbered from apex) and area and weight of individual leaves. Opatovický fishpond, July 1973, depth of water 55 cm. Average values of five plants

filters for a precise determination of radiation in the photosynthetically active part of the spectrum (400–700 nm). The latter sensor was exposed under the same conditions as the leaf measured. It is only the values of the photosynthetically active radiation (PhAR) that are employed in presenting the results of the photosynthesis measurements.

Vertical Gradients. The average number of leaves was 16 per plant. Most leaves were situated in the uppermost 0.5 m layer (Fig. 2), although the maximum area and weight per individual leaf (not an integral per layer!) were found in older leaves situated below this layer. There were no marked differences in the specific leaf area (= leaf area / leaf weight ratio) between the leaves of different insertion level: most values fell within the range of 1.35–1.50 dm^2/g.

The CO_2 exchange characteristics were related to both leaf area and leaf weight as well as to individual leaf laminae (Fig. 3). Surprisingly, the photosynthetic activity was very little affected by leaf ageing. With the exception of the two oldest leaves, no significant differences in CO_2 uptake occurred under standardized conditions, regardless of the reference unit employed (leaf area or leaf weight). Walker and Waygood (1968) found a similar vertical distribution of photosynthetic activity in a *Phragmites communis* plant.

Photosynthesis per individual leaf laminae (or photosynthetic capacity, according to Walker and Waygood, 1968) is distinctly correlated with both leaf area and leaf weight.

Dark respiration was measured during the night, starting from two hours after sunset, when the rate of CO_2 output was already stabilized. The dark respiration rate per unit leaf area decreases from the youngest to the basal leaves (Fig. 3), but in very old leaves, exceptionally high rates were detected in some instances. Calculations of dark respiration per unit leaf weight show the same trend.

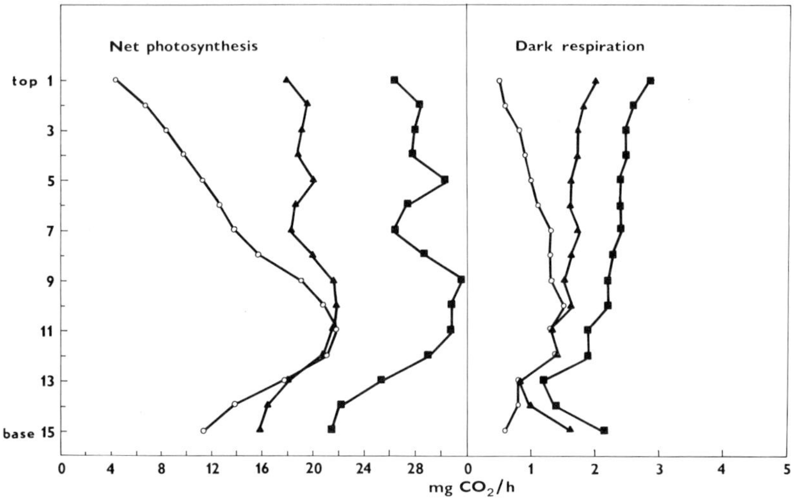

Fig. 3. Gas exchange characteristics of leaves of *Phragmites communis* of different insertion (numbered from top). Rates of net photosynthesis interpolated for standard conditions (irradiance 300 Wm^{-2}, leaf temperature 20° C and CO_2 concentration in the leaf chamber 300 vpm); mean values from 6 plants. Net photosynthesis and dark respiration per total leaf (-○-○-), dm^{-2} leaf area (-▲-▲-), and g^{-1} dry weight (-■-■-). Opatovický fishpond July 1973

Besides the leaves (= leaf laminae), CO_2 exchange characteristics were measured in other aboveground parts (internodes, inflorescences). The aim of these measurements was to find the relative importance of each structure in the CO_2 exchange of the whole plant and to obtain data for modeling canopy photosynthesis.

The internodes were, of course, totally covered by the leaf sheaths, and CO_2 exchange was measured in this natural complex. The projected area (not surface area) of the internodes was used in calculations of gas exchange rates: the same procedure was applied to the inflorescences.

The net photosynthetic rate of the internodes with leaf sheaths was lower than that of the leaf laminae even at high levels of irradiance (Table 1). A decrease in irradiance brings about a steep decrease in the internode net photosynthesis (Gloser, 1977). Their compensation irradiance is therefore high (about 40 Wm^2, in leaves only 4–10 Wm^{-2} at the same temperature). Their dark respiration rate per unit area is similar to that in the leaves. When their gas exchange rates are calculated per unit dry weight, extremely low values will result as the area/weight ratio is much smaller in these structures than in the leaf laminae.

The inflorescences are characterized by low rates of net photosynthesis and by high rates of dark respiration. This unfavorable balance between assimilatory and respiratory processes is well reflected in the extraordinarily high value of compensation irradiance (see Table 1).

Dependence on Irradiance and Temperature. The relationship between net photosynthetic rate and irradiance was measured at various leaf temperatures within the interval of 6–33° C. The results of these measurements represent principal

Table 1. Gas exchange characteristics of internodes and inflorescences in *Phragmites communis*

	Compensation irradiance Wm^{-2}	Net photosynthesis[a]		Dark respiration[b]	
		mg CO_2 dm^{-2} h^{-1}	mg CO_2 g^{-1} h^{-1}	mg CO_2 dm^{-2} h^{-1}	mg CO_2 g^{-1} h^{-1}
Internodes + leaf sheaths:					
apical parts of shoot	38	10.4	1.3	3.7	0.5
central parts of shoot	36	7.0	0.7	2.3	0.2
basal parts of shoot	42	5.0	0.4	2.5	0.2
Inflorescences	145	8.1	4.3	10.2	5.4

[a] Measured at irradiance of 380 W/m^2, at plant surface temperature of 18–20° C.
[b] Measured at 18° C.
Measurements taken at Opatovický fishpond, July 1973.

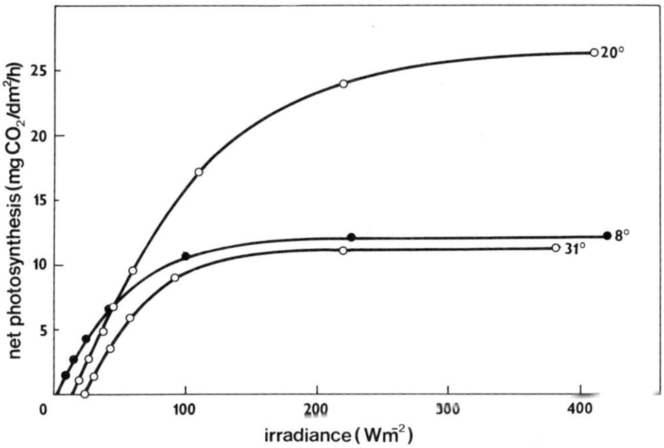

Fig. 4. Relationship between net photosynthetic rate and irradiance at three different temperatures (° C). Fifth leaf from top. Hlohovecký fishpond, July 1974

input data for mathematical modeling of leaf and canopy photosynthesis. They are treated in detail by Gloser (1977), and only a brief summary will be given here.

At optimum temperatures for net photosynthesis (20–25° C), the compensation irradiance was about 4–12 Wm^{-2}. At these temperatures, photosynthesis was not saturated by radiant energy even during the midday hours when irradiance was above 400 Wm^{-2}. On the other hand, saturation was usually attained at temperatures above and below the optimum (Fig. 4). The compensation irradiance increased with leaf temperature from 2 Wm^{-2} at 6° C to about 20 Wm^{-2} at 33° C.

The dependence of net photosynthetic rate on leaf temperature is better demonstrated in Figure 5, but the depicted relationship is only valid for high irradiance. At lower irradiance levels, especially below 100 Wm^{-2}, the shape of the temperature curve changes substantially: the temperature optimum is transposed to lower temperatures (Gloser, 1977).

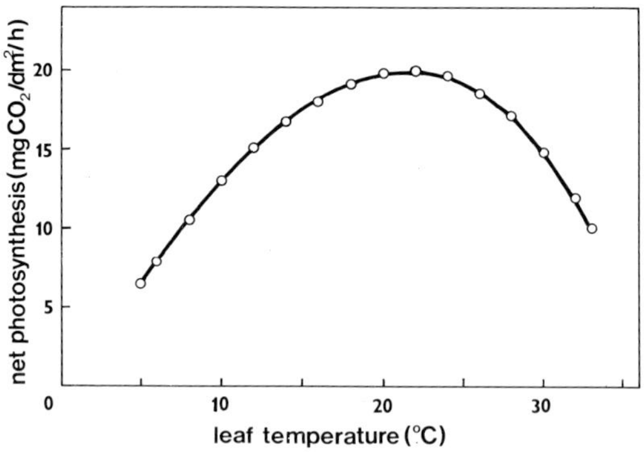

Fig. 5. Relationship between net photosynthetic rate and leaf temperature. Mean values from five leaves (fifth leav from the top). Irradiance 320 to 410 Wm^{-2}. Hlohovecký fishpond, July 1974

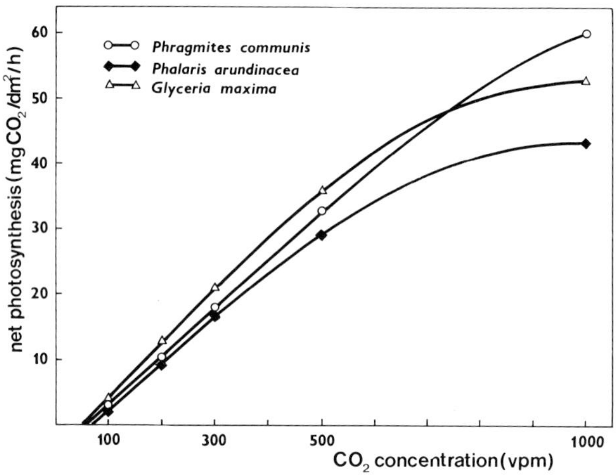

Fig. 6. Relationship between net photosynthetic rate and CO_2 concentration in the leaf chamber. Irradiance 315 Wm^{-2}, leaf temperature $21 \pm 1°$ C. Hlohovecký fishpond, July 1968

CO_2 Dependence and Diffusion Resistance of Leaves. The photosynthetic response to changes in CO_2 concentration in the ambient air and the leaf-diffusion characteristics were studied in the laboratory using an apparatus similar to that described by Nátr and Gloser (1967).

The net photosynthetic rate increases linearly with increasing CO_2 concentration in the leaf chamber nearly to the highest concentration applied (1000 vpm). At 1000 vpm CO_2, the photosynthetic rate was higher in all cases in *Phragmites* than in any other grass species tested by Gloser (1973).

The compensation concentration of CO_2 was about 50–60 vpm, which is an usual range for plants possessing the C_3 biochemical mechanism. The slope of the linear part of the CO_2 curve is determined by the sum of resistances to CO_2 transfer from the bulk air to the reaction centers in the chloroplasts. By an analysis of the diffusion process, it was possible to split the sum of resistances to

Table 2. Net photosynthetic rate, CO_2 compensation concentration, and leaf diffusion characteristics at Hlohovecký fishpond, July, 1968

Plant	P_N	C_C	r_s	r_M	$\frac{r_M}{r_s}$	$\frac{r_{H_2O}}{r_{CO_2}}$
Phragmites communis	23.1	58	1.8	3.5	1.9	0.28
Glyceria maxima	22.3	57	1.5	4.2	2.7	0.26
Phalaris arundinacea	18.1	45	1.9	6.1	3.3	0.19

P_N, net photosynthetic rate in mg CO_2 dm^{-2} h^{-1}; C_C, CO_2 compensation concentration in vpm; r_s, stomatal resistance to CO_2 transfer in s cm^{-1}; r_M, mesophyll resistance to CO_2 transfer in s cm^{-1}; r_M/r_s ratio; and ratio between total resistances to vapor (r_{H_2O}) and CO_2 transfers (r_{CO_2}). Leaf temperature $21 \pm 1°$ C; irradiance 315 $W\,m^{-2}$; mean values of five measurements.

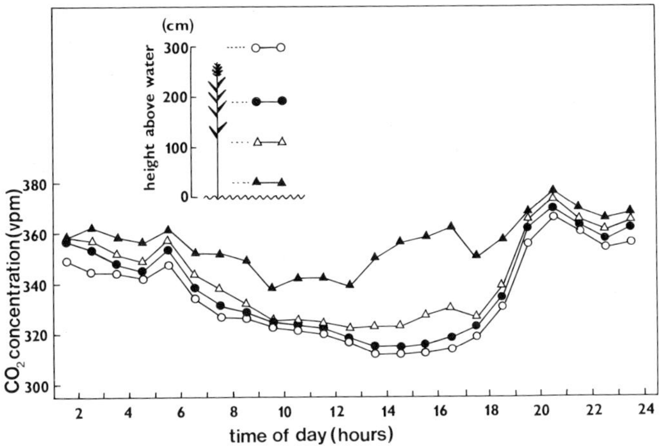

Fig. 7. Diurnal changes of CO_2 concentration in air above and in a *Phragmites* stand. Location of air sampling tubes in the stand is schematically depicted

CO_2 transfer into the following components: the boundary-layer resistance of air (= about 1.2 *s* cm^{-1}), stomatal resistance (r_s) and mesophyll resistance (r_M).

The mesophyll resistance was assumed equal and calculated as residual substomatal resistance. In the leaves of *Phragmites*, it was always higher than r_s, although the r_M/r_s ratio was not as high as in some other grass species (Table 2).

The laboratory measurements of the dependence of net photosynthesis on CO_2 concentration have revealed that a 30 vpm increase or decrease in CO_2 concentration in the air may induce a change in net photosynthesis of about 10%. It seems, therefore, quite important to know the magnitude of diurnal fluctuations of CO_2 concentration in the air within the reed canopy.

The measurements of CO_2 profiles were taken in the *Phragmites* stand at the Hlohovecký fishpond in the first half of August, 1974. The air was sampled continuously at four different levels both within and above a dense undisturbed reed canopy. After passing through polyamide tubing (connecting lines, about 15 m in length) and damping bottles (2 l each), the air samples were analyzed with an infrared gas analyzer at one-minute intervals.

A typical example of diurnal changes in CO_2 concentrations is presented in Figure 7. Permanently highest was the concentration in the lowest air layer (30 cm above water level). Within the leaf canopy, the air composition was highly dependent on the diurnal changes of CO_2 concentration in the above-canopy layer of air.

The CO_2 concentration was never markedly less (i.e., by at least 10 vpm) within the leaf canopy than was the above-canopy concentration. The CO_2 depletion in the *Phragmites* stand caused by its photosynthetic activity is probably very small and unimportant in limiting the canopy photosynthesis.

References see pp. 285—291.

3.2.3 Modeling of Photosynthetic Production in Littoral Helophyte Stands

J. P. ONDOK and J. GLOSER

3.2.3.1 Basic Assumptions

A model of CO_2 uptake and dark respiration will be derived for a pure stand of *Phragmites communis* using the data on radiation regime given in Section 2.1, and those on CO_2 uptake and respiration contained in Section 3.2.2. The model has empirical character in some points and mimics the principal functions of a photosynthetic system and their dependence on various factors. Stress is laid upon a detailed elaboration of the regime of radiation interception which determines the input into the system. Several simplifications have been made in this first version of the model with respect to the importance of individual factors influencing photosynthesis.

Changes in CO_2 concentration have not been involved, as measurements in the field have revealed only slight fluctuations in CO_2 concentration in the littoral reed stand (see Sect. 3.2.2). The effect of water saturation deficit on photosynthesis has not been included in view of its slight importance in emergent helophytes (Rychnovská, 1967, 1973). The differences, both positive and negative, between leaf surface temperature and air temperature within the stand ranged between 0 and 3° C. In the model, leaf surface temperature has therefore been replaced by air temperature recorded within the stand. Pearcy et al. (1974) recorded differences between leaf surface temperature and air temperature inside a reed stand of 2–3° C in the morning and 5–8° C at midday, the leaf surface always being cooler.

In the radiation regime, the component of scattered radiation is not considered in order to simplify the computation procedure. This component plays an unimportant role in the radiation regime, as is evident from the results reported in Section 2.1.

The respiration of underground organs is not included in the model as the relevant data are lacking. The net photosynthesis is thus diminished to an

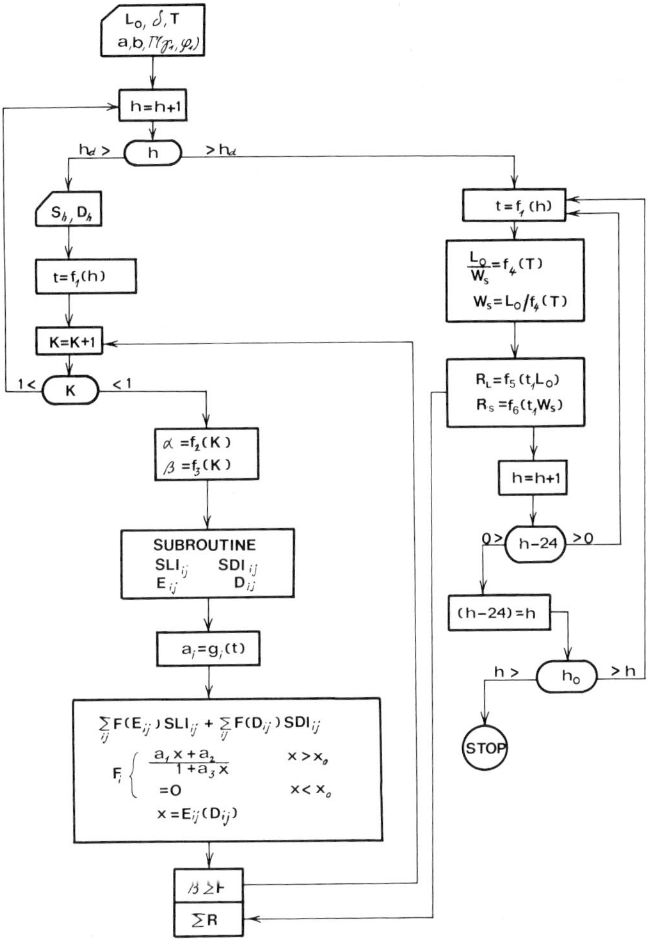

Fig. 1. Block diagram of model of the daily course of CO_2 uptake and dark respiration in *Phragmites* leaves and stems. For explanation see text

unknown extent, under conditions favorable for respiration and unfavorable for photosynthesis. Neither have the assimilation of the stem surfaces and the assimilation and respiration of inflorescences been included, for the sake of simplicity, although the relevant data were available. The share of the stem surfaces, covered mainly by leaf sheaths, in the total assimilatory surface area is small except during a short period of early stand development in spring. In a mature stand, this share does not exceed 10% and rapidly decreases when the leaves start to become dry. Even the irradiance of the green surfaces is relatively poor, except for the uppermost canopy layer. Finally, the rate of CO_2 uptake by leaf sheaths is much lower than that of leaf blades (see Sect. 3.2.2). It can be reasonably admitted that this rate does not exceed 3% of the total CO_2 uptake in a developed stand. Nevertheless, the assimilation of the stem surfaces will be introduced in a refined version of the model.

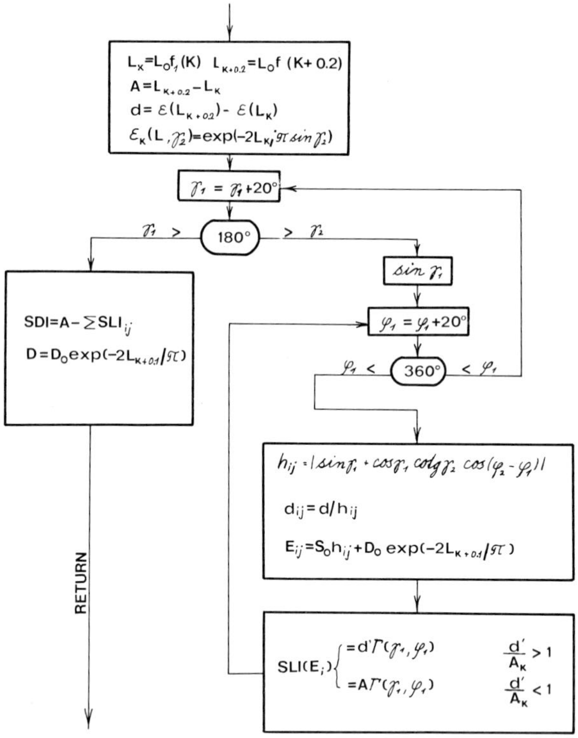

Fig. 2. Block diagram of subroutine in photosynthetic model. For explanation see text and Section 2.1

The inflorescences photosynthesize only from about mid-July, but they may then play an important role as in some other grasses (Thorne, 1959). The exclusion of their CO_2 uptake from the present model is to some extent compensated by disregarding their dark respiration (for the measured data see Sect. 3.2.2).

3.2.3.2 Derivation of the Model

The radiation inputs are given either as hourly sums of direct sun and diffuse sky radiation during a day, or they may be simulated by the method described in Section 2.1. The choice of this time unit involves certain simplifications because the short-term changes in irradiance due to changing cloudiness are not evident from the hourly sums of radiant energy. The model structure itself is of course independent of the time scale, and any other shorter time unit can be incorporated in the model if appropriate data and parameters are at hand.

A block scheme of the model of the daily course of CO_2 uptake and dark respiration is presented in Figure 1 and 2. At the beginning, the following parameters are put in: T—time expressed in 10-day intervals with T_0—21 March, which serves for calculation of the seasonal course of stem biomas W_s (expressed as dry weight); L_o—leaf area index, which may also be calculated as a function

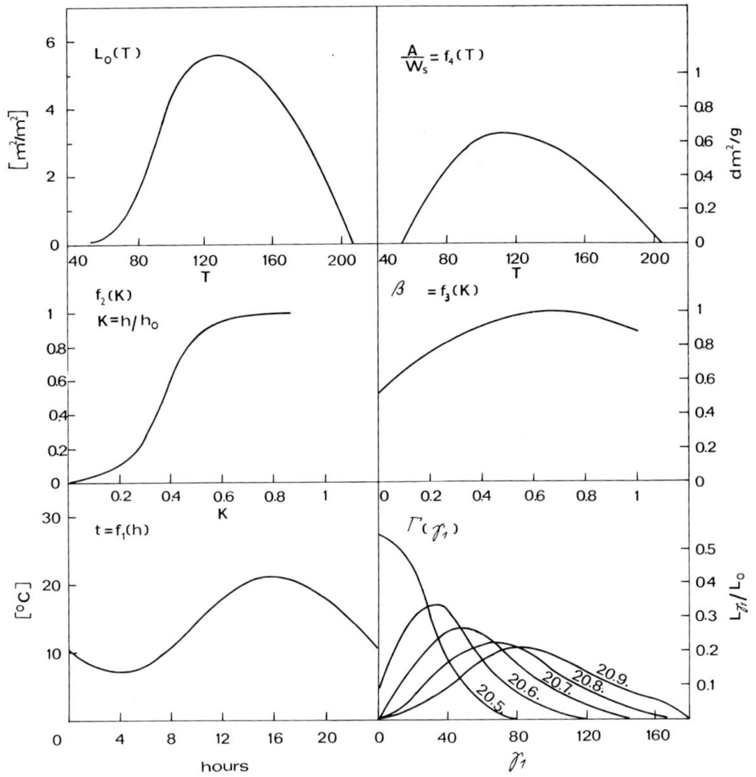

Fig. 3. Functions used in the model obtained from the fit to empiric data

of T (see Fig. 3); δ—sun declination; and a, b parameters of a function describing the daily course of air temperature inside the stand:

$$f_1(\mathrm{h}) = a \sin(\mathrm{h} - 150^\circ) + b \tag{1}$$

where $f_1(\mathrm{h})$ is the function of the day hour (h), $a = (t_{\max} - t_{\min})/2$ and $b = (t_{\min} + t_{\max})/2$, $t_{\max}$, $t_{\min}$ being the daily maximum and minimum air temperature inside the stand. The parameters a and b correspond with the daily temperature range as well as with the value of daily mean air temperature. The last input value of $\Gamma(\gamma_1, \varphi_1)$ is a frequency function of leaf inclination (γ_1) and orientation (φ_1). A uniform orientation is assumed and the seasonal course of the frequency function for leaf inclination is given in Figure 3.

The first day-hours' cycle begins with h_o, which is the hour including the time of sunrise. A decision command serves for a program branching towards computation of the dark respiration after the cycle has finished ($\mathrm{h} = \mathrm{h}_d$). For each day hour, the data on global radiation are converted into PhAR and air temperature is calculated using $f_1(\mathrm{h})$. The present model employs the average temperature course during a day, a and b being determined for each month from monthly averages of $t_{\max}$ *and* $t_{\min}$. The daily fluctuations of t and its vertical distribution inside the stand will be introduced into a more refined model.

The next cycle is that for individual canopy layers. After the computation has been finished for all layers, the program returns to the next day hour. In the present model, a stand is divided into five layers, but it may be divided into more layers if the speed capacity of the computer is sufficient. Each layer is defined by k, which is a relative fraction from the total stand height, measured from top downward. The vertical gradient of cumulative leaf area index (L) is expressed by the function:

$$L = L_o f_2(k) = L_o \frac{1}{1+138.6 \exp(-13.56\,k)} \quad (2)$$

which has been derived empirically from seasonal values of the vertical gradient of L; L_o is the total cumulative leaf area index. The function $f_2(k)$ is shown in Figure 3. By means of L_o and $f_2(k)$, both the cumulative leaf area index and the leaf area index of a canopy layer may be calculated at each stand level. The extinction of both the direct sun and diffuse sky radiation is computed from the cumulative leaf area index using simplified extinction formulae (Ondok, 1975):

$$S/S_o = \exp(-2L_o/\pi \sin\gamma_2) \quad (3)$$

$$D/D_o = \exp(-2L_o/\pi) \quad (4)$$

where S, D represent, respectively, direct sun and diffuse sky radiation inside the stand, S_o, D_o direct sun and diffuse sky radiation above the stand; γ_2 is sun elevation. This computation of irradiance for each of (m × n) leaf inclination and orientation classes is carried out in the subroutine I (see Fig. 2). Both direct and diffuse sky radiation (E) are computed for sunlit foliage and only diffuse sky radiation is for shaded foliage; corresponding fractions of sunlit leaf area index (SLI) and shaded leaf area index (SDI) are computed as well. The details of the computation are described in Section 2.1.

The CO_2 uptake is calculated by means of a generalized hyperbola:

$$F = \frac{a_1 E + a_2}{1 + a_3 E} \quad (5)$$

where F is the rate of CO_2 uptake in (mg dm^{-2} h^{-1}), E stands for either the direct sun or diffuse sky radiation in Wm^{-2} and a_i, a_2, a_3 are parameters depending on leaf-surface temperature (see Figs. 4 and 5). This dependence is expressed by polynomes of the 5th degree in t, where t is air temperature inside the stand. Since the empirical shape of this dependence is asymetrical the polynomial formulation of the dependence of CO_2 uptake on leaf-surface temperature has been preferred to a normal distribution function sometimes used in photosynthetic models (see Idso, 1969).

The total CO_2 uptake calculated for an individual stand layer is multiplied by a coefficient β which expresses the vertical gradient of photosynthesis and is given by the function:

$$f_3(k) = 0.88 + 0.73k - 1.1k^2 \quad (6)$$

(see Fig. 3). This empirical function has been derived from measured data. After CO_2 uptake has been computed for the whole layer, the program computes it for the next layer.

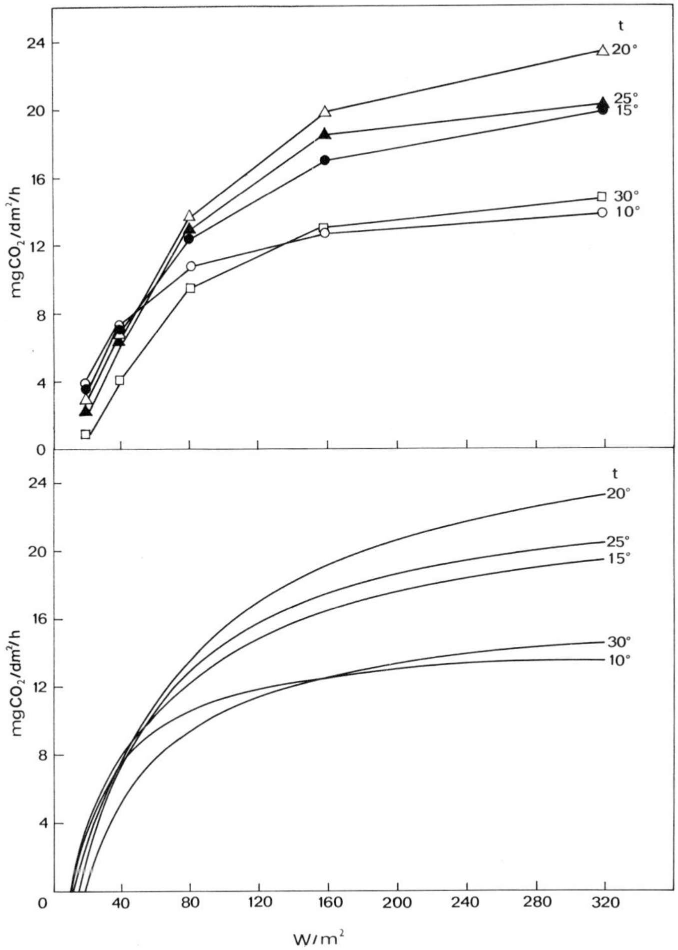

Fig. 4. Radiation dependence of CO_2 uptake in a *Phragmites* stand. *Top:* Measured data; *Bottom:* Fitted data

The computation of dark respiration begins with the calculation of t for each hour. Both the leaf and stem respiration are calculated by means of the empirical functions f_5 and f_6:

$$R_l = f_5(t) = 0.295\,\exp(0.069\,t) \quad (\text{mg } CO_2 \text{ per dm}^2/\text{h}) \tag{7}$$

$$R_s = f_6(t) = 0.055\,\exp(0.069\,t) \quad (\text{mg } CO_2 \text{ per g/h}) \tag{8}$$

where the leaf respiration (R_l) is related to the leaf-surface area whereas the stem respiration (R_s) is related to the stem dry weight, which is determined more easily. The stem dry weight is calculated from L_o and T using a formula fitted to the empirical data:

$$W_s = \frac{L_o}{f_4(T)} \tag{9}$$

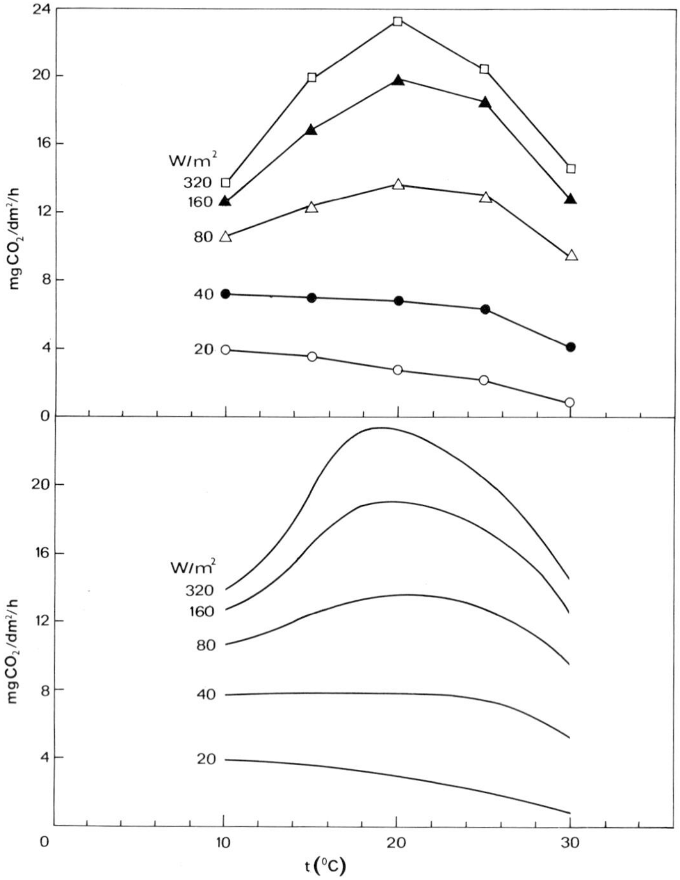

Fig. 5. Temperature dependence of CO_2 uptake in a *Phragmites* stand. *Top:* Measured data; Bottom: Fitted data Measured data from Gloser (unpublished)

where

$$f_4(T) = \frac{L_o}{W_s} = 0.00053\,T^3 - 0.0317\,T^2 + 0.527\,T - 2.01 \qquad (10)$$

Finally, after the dark respiration has been computed the daily totals of both CO_2 uptake and dark respiration are obtained. The program was written for a programable calculator Wang 600-TP with which all the computations were originally carried out.

3.2.3.3 Stand Photosynthesis Investigated by the Model

The model can be used to estimate the seasonal course of CO_2 assimilation in the leaves and dark respiration in both leaves and stems and their seasonal totals. Another application of the model is the study of stand photosynthetic activity under various conditions simulated by enhancing parameters of the model. A few examples of both kinds of application are shown here.

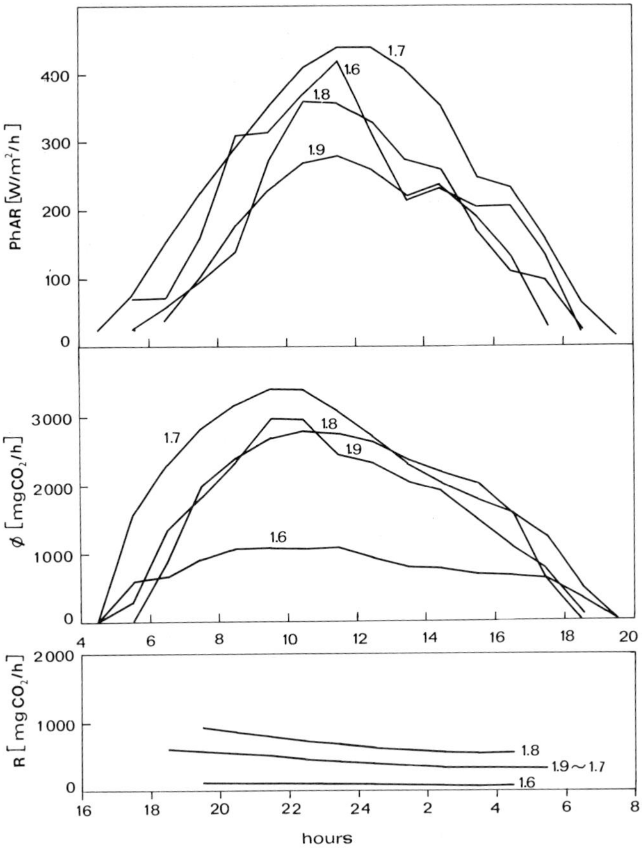

Fig. 6. Seasonal changes in the daily course of PhAR (*top*), CO_2 uptake by *Phragmites* leaves (*center*), and dark respiration in leaves and stems (*bottom*)

The seasonal changes in CO_2 uptake and dark respiration in *Phragmites* leaves and stems are illustrated in Figure 6. An interesting feature of the daily course of CO_2 uptake is the daily maximum commonly attained before noon. A similar course has also been reported by Pearcy et al. (1974) for a *Phragmites* stand. This feature is interpreted by dependence of the photosynthetic light curves on leaf temperature (see Fig. 5). For small irradiance values the photosynthetic rate is relatively higher at low than at high temperatures. Because of the self-shading effect a high proportion of the leaves is only poorly irradiated: hence the resulting CO_2 uptake is higher in the forenoon hours at rather low temperatures than in corresponding afternoon hours when temperatures are higher.

The daily sums of CO_2 uptake obtained from Figure 7 are listed in Table 1. The slight increase in $\bar{F}$ (average daily rate of CO_2 uptake) in September brought about by decreasing L_o (the leaves gradually become dry) is due to simplifications in the model. In expressing the growth of L_o by a curve, the circumstance is neglected that the decrease in L_o is not uniformly distributed between all canopy layers, as the lower leaves begin to dry first. It follows from the extinction curves of radiation in a reed stand that the decreasing leaf area in the lower canopy

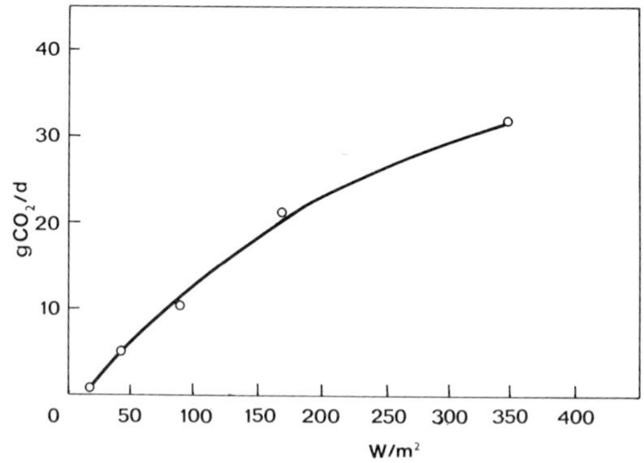

Fig. 7. Dependence of stand photosynthesis on radiation inputs calculated by means of the model. Data relate to 1 July

Table 1. Seasonal changes in daily totals of stand CO_2 assimilation (f_d) leaf and stem dark respiration (R_d), average daily CO_2 assimilation rate ($\bar{F}$) and daily PhAR sums (E_d)

Date	F_d $g\,CO_2\,d^{-1}$	R_d $g\,CO_2\,d^{-1}$	$\bar{F}$ $mg\,CO_2\,dm^{-2}\,h^{-1}$	E_d $kWh\,m^{-2}$
1.6.	11.23	0.72	12.53	2.51
15.6.	28.16	2.16	9.1	2.94
1.7.	31.91	4.34	5.3	3.46
1.8.	23.84	6.77	3.0	2.51
1.9.	24.91	4.89	4.1	1.91

layers influences only slightly the extinction gradient, which is determined mainly by the interception of radiation in higher stand layers. This circumstance will be respected in the refined model although the error in $\bar{F}$ is quite small.

The radiation-saturation levels of the photosynthetic curves for single leaves (see Fig. 3) were not attained in our field measurements. The dependence of CO_2 uptake by the whole stand on incoming radiation is nearer to linearity than the by a single leaf; there is practically no sign of a tendency toward radiation saturation in the canopy as a whole: see Figure 7. The input data of the daily course of irradiance under clear sky conditions were successively diminished by a constant ratio in order to obtain the daily photosynthesis for several points of the radiation scale. Similar results have been obtained for other days. The radiation-saturation levels of CO_2 uptake in individual leaves reported by Burian (1973) are very low (not higher than 220 $W\,m^{-2}$ of PhAR) compared with these results.

The dependence of CO_2 assimilation on leaf temperature is given in Figure 5. The optimum values vary for different values of irradiance and shift from 20° C for the highest irradiance toward lower values for lower irradiance levels. For the irradiance of 20 $W\,m^{-2}$ PhAR, the optimum leaf temperature is equal to or less than 10° C. Pearcy et al. (1974) found the optimum leaf temperature at 30° C. But they do not indicate the irradiance values at which the optimum was evaluated, which might have caused a bias in their estimates. If, however, the optimum of 30° C is correct, it must be explained by specific conditions in the stand investigated.

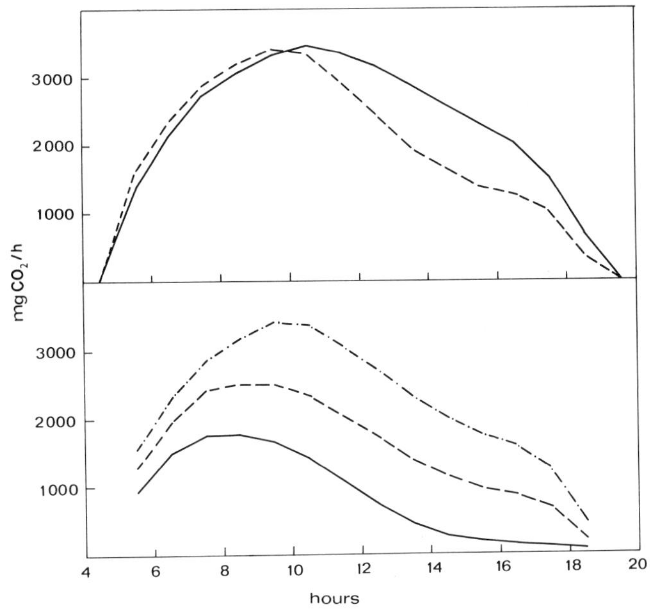

Fig. 8. *Top:* Daily courses of CO_2 uptake for 2 temperature amplitudes: 3° (——), and 8° (– – –); b = 15.7 in both cases. *Bottom:* Daily couses of CO_2 uptake for 3 values of mean daily temperature: 16° (– · – ·), 20° (– – –), and 24° (——); a = 5.9° in all cases

A change in the parameters of the temperature curve *a*, *b* shows that *a* (amplitude) influences strongly the position of the maximum rate and only slightly the daily total of CO_2 uptake, whereas *b* (daily average) affects both the position of the maximum uptake rate and the daily total CO_2 uptake. This is evident from Figure 8.

The leaf area irradiated below the compensation point was estimated by means of the model. The fraction of non-assimilating leaves depends mainly on L_o and on the sun elevation. Two examples giving a general picture of this dependence are shown in Figure 9. A more detailed analysis (e.g., the vertical distribution of non-assimilating leaf area for individual day hours) is possible using the present model.

The influence of foliage geometry (leaf inclination and orientation) is documented only by one example in Figure 10, where the daily course of CO_2 uptake is compared for the erectophile and planophile *Phragmites* foliage. The respective distribution of leaf inclination and orientation are listed in Table 2. The erectophile foliage has a higher CO_2 uptake mainly at higher sun elevations. At low sun elevations, no marked differences are found. In this way, the changing input parameters make it possible to seek the optimum leaf geometry in a stand, the criterium being the highest CO_2 uptake.

The comparison of these results with those reported in the literature is complicated by the variety of methods used and different morphological characteristics and environmental conditions of the stands investigated by various authors. Pearcy et al. (1974) measured the CO_2 uptake under field conditions. Their values of the photosynthetic rate lie within the range of values

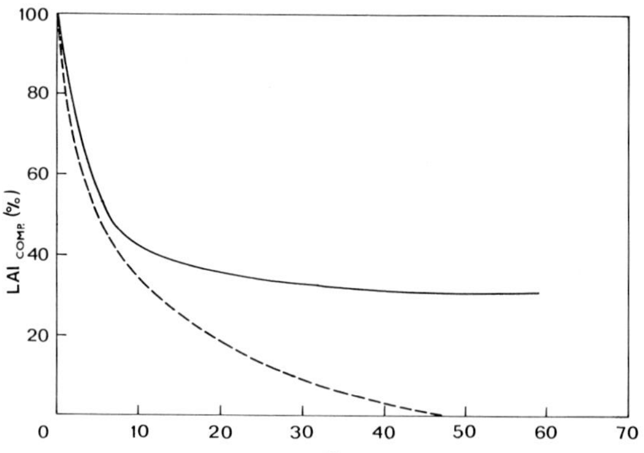

Fig. 9. Dependence on sun elevation of the fraction of non-assimilating leaf area out of total leaf area on 2 dates: 1 July (– – –) and 1 August (——)

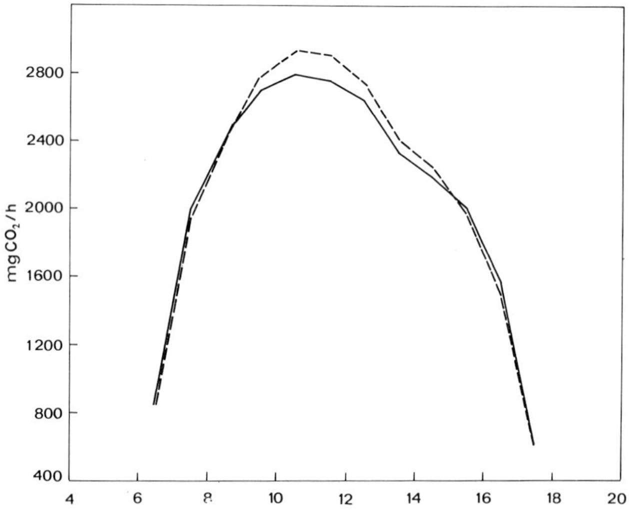

Fig. 10. Comparison of the daily courses of CO_2 uptake by an erectophile (– – –) and planophile (——) foliage, date 1 Sept.

assessed here. The high irradiance occurring in their habitats (see Fig. 1 of Pearcy et al., 1974) admits an explanation by adaptation of the plants. The highest value of CO_2 uptake reported amounts to 12.4 mol $CO_2/dm^2/min$ which corresponds with 32.7 mg $CO_2/dm^2/h$.

The results of measurement of CO_2 uptake in *Phragmites* leaves by the radioisotope method reported by Walker and Waygood (1968) are mainly concerned with the differences in assimilation between individual leaves on a shoot. The different reference bases for expressing CO_2 uptake rate are compared: leaf area, leaf fresh weight and chlorophyll content. The rate of CO_2 uptake reported by them, nearly 6 mg CO_2 dm^{-2} h^{-1}, seems low for the described cloudless conditions, but precise comparison with the results reported here is not possible as the

Table 2. Density function of leaf inclination for erectophile and planophile *Pragmites* foliage used in calculation of daily course of CO_2 uptake in Figure 10

Class interval of γ_1 (°)[a]	Erectophile foliage[b]	Planophile foliage[b]
0– 20	0.40	0.02
20– 40	0.37	0.10
40– 60	0.16	0,19
60– 80	0.06	0.22
80–100	0.01	0.19
100–120	0	0.12
120–140	0	0.08
140–160	0	0,05
160–180	0	0.03

[a] Measured from the vertical.
[b] Leaf orientation assumed to be distributed uniformly.

irradiance on leaf surface is not given. Walker and Waygood failed to find significant differences in photosynthetic rate between individual leaves. Such gradients have been assessed here (see Sect. 3.2.2), and Rychnovská (1967) and Burian (1973) also report similar results. The existence of such gradients is to be expected, as they reflect the influence of leaf development and ageing on the assimilation rate (see, e.g., Šesták et al., in press, 1977).

Burian (1969, 1971, 1973) measured CO_2 uptake by a reed stand in a macrocuvette and that by single leaves in a leaf cuvette. Several objections to his macrocuvette may be raised. The PVC foil cover used has a high permeability for CO_2 and the insufficient ventilation (0.3 m^3 of air exchanged in one hour) could hardly prevent overheating of the macrocuvette. The radiation regime inside the macrocuvette enclosing 20 shoots must have been different from that outside the cuvette. The method of calculation of stand photosynthesis from data on leaf photosynthesis is not described. Burian's conclusion of a stronger dependence of CO_2 uptake on air temperature than on irradiance contradicts his finding (Burian, 1973) of a slight variation in CO_2 uptake with temperature variation in June and July. His comparison of production measured by the harvest method with that calculated from gas exchange measurements is based on excessive simplifications concerning the growth and respiration of underground organs.

A feedback exists between the model activities and empirical work. The "experimentation" with a model makes the scientist aware of interesting features of the system dynamics which are not obvious from the empirical data. It helps him to formulate new problems and to organize his investigations. On the other hand, the adequacy of each model has to be verified by confrontation with reality. The model proposed here is to be viewed in this way. Although its first version is based on several simplifications and needs further refinement, some of the conclusions are of importance for further investigations.

References see pp. 285—291.

3.3 Water Relations. Water Balance, Transpiration, and Water Turnover in Selected Reedswamp Communities

M. Rychnovská

3.3.1 Physiological Features of Detached Shoots and Leaves of Phragmites communis

The water relations were mainly studied in pure stands of *Phragmites communis* and in plant communities dominated by this highly adaptable plant, as well as in communities of *Glyceria maxima (Glycerietum maximae)*. For understanding the ecology of *Phragmites* communities, it is essential to know the physiological processes which determine the productivity, viability and adaptability of this species to the given biotope.

Wetland biota are adapted to surplus water in their environment, but even are sometimes confronted with water stress. The aim of the study has been to find out the amplitude of water relations in plant communities and biotopes of *Phragmites* and ecologically related plant species and to contribute to the understanding of adaptability to the occasional water stress in *Phragmites communis*.

3.3.1.1 Water Content in Whole Shoots, Leaf Blades and Leaf Discs at Full Turgidity

Mature shoots of *Phragmites communis*—fully turgid—show considerable fluctuations in their water content expressed in terms of percentage of dry weight. The variation in water content of *Phragmites* in whole shoots of *Phragmites* populations in different biotopes is considerable. Two facts are surprising: (1) plants from a stand at littoral ecophase and hence well-supplied with water mostly show a slightly lower water content than those from a stand at terrestrial ecophase; (2) the water content sometimes changes quite rapidly most probably according to water supply. A higher water content was observed after a period of rain, with no visible drops of water left on the leaf surfaces; the structure of the shoots obviously enables them to absorb surface water. The water content in whole shoots and leaf blades (expressed as percentage of dry weight) changes in range of values of 110–213% and 194–213% respectively.

The water content is less variable in discs cut out from the leaf blades. Unfortunately, only few data are available on the water content in fully turgid leaf blades. But existing data suggest that their water content is highly stable. The water content in leaf discs is mostly less than in whole leaf blades (ranges of values 158–196%). In comparison with *Glyceria maxima* and other grasses of wet biotopes, *Phragmites* seems "mesophytic".

Measurements in a *Glyceria maxima* stand have shown a water content in the shoots of about 324%. The following data on water content (% of dry weight) in South Moravian populations of *Glyceria maxima* originate from studies on alluvial meadows: 511% (in May), 386% (in July, 1971), and 372% (in June, 1968) (Rychnovská et al., 1972). Other grasses growing in adjacent biotopes show the following values: *Phalaris arundinacea* 272%, *Alopecurus pratensis* 169% (Rychnovská et al., 1972).

3.3.1.2 Gradients of Water Content in Leaves According to Their Level of Insertion and Along the Leaf Blade

In studies on water relations in *Phragmites communis* under field conditions, it is necessary to sample its shoots or their parts and to test them by different field and laboratory methods. In pure or nearly pure *Phragmites* populations there is no problem of obtaining homogeneous material for dealing with whole shoots of a certain size (standard, substandard, etc.). In many instances, however, it is only parts of shoots, leaf blades or discs cut out from the leaves that are suitable for laboratory studies. Detailed analyses of ecophysiological gradients along the shoots and leaf blades have therefore been made (Rychnovská, 1967, 1973; Tuschl, 1970). They will now be summarized.

1. Gradient along the shoot: consecutive leaves, according to the level of their insertion, show a characteristic gradient in their size, dry weight, specific leaf area (A_s = leaf area/leaf weight), fresh weight and water content. A similar gradient,

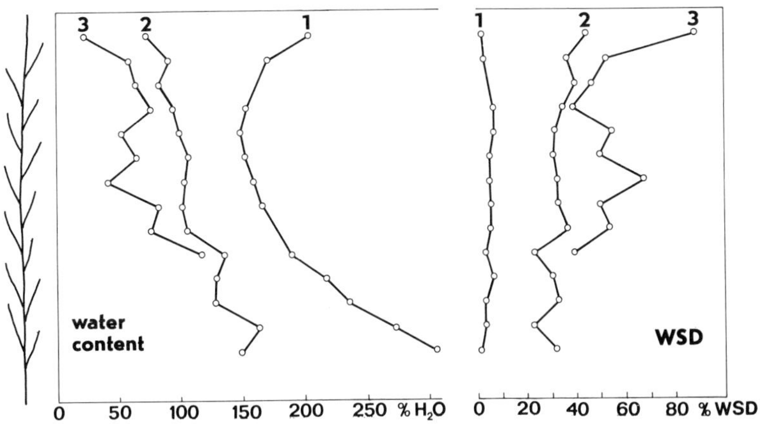

Fig. 1. Physiological heterogeneity of consecutive leaves in *Phragmites communis*. Water content in % of dry weight *(left)* and water saturation deficit (WSD) in percentage of full water content *(right)*. Shoot 1, littoral stand, intact plant at noon; shoots 2 and 3, littoral stand, shoots detached from root and rhizome by a cut below water surface and left in stand for several hours (2) and one day and night (3)

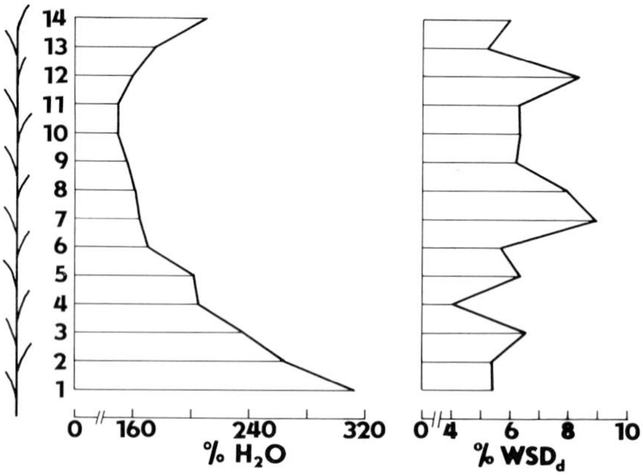

Fig. 2. Heterogeneity of water content (in % dry weight) and WSD (%) in consecutive leaves of the littoral type of *Phragmites communis* according to their level of insertion (Opatovický fishpond, 20 July 1972, at 16 h)

Table 1. Gradient in water content and WSD along leaf blade in *Phragmites communis*. Mean value for five medium leaves; determined in leaf discs 8 mm in diameter. Plants at stage of incipient drying (Hýskovský fishpond, 8 July 1965)

Leaf parts	Water content in % dry wt.	WSD
Apical part	141–144	23.0–20.9
1st folded zone	145	22.0
Transition	147	20.5
Medial wrinkled zone	149–158	19.9–18.1
2nd folded zone	161	18.3
Transition	160	18.4
Smooth zone	162–165	18.2– 5.8
Basal part	171	4.5

from base to apex, is observed when water saturation deficit develops. The gradients are never linear: the basal and apical leaves differ more sharply from the other leaves. The gradient under normal physiological leaf conditions, however, is regular and smooth. On the other hand, the leaves seem to act independently under water stress: the curve becomes zig-zag (Fig. 1). These differences in the behavior of the leaves can be registered in either intact or detached shoots, in a natural canopy or under standard conditions in the laboratory. Figure 2 demonstrates the water content and water saturation deficit (WSD) in consecutive leaves of flooded littoral *Phragmites* at the Opatovický fishpond. The shoots were sampled with those used for the determination of transpiration.

2. Gradient along the leaf blade: marked physiological differences are found even within one leaf blade. The water content in its basal parts is nearly twice that in the apical parts; on the other hand, the specific leaf area (A_s) of the basal part of a leaf blade is only half that in its upper part. When incipient drying occurs, the

WSD develops gradually from the leaf apex to its base (Rychnovská, 1967). Table 1 gives an example of the physiological heterogeneity in one leaf blade of *Phragmites*.

3.3.1.3 Desiccation Rates of Shoots and Detached Leaf Blades

Table 2 provides data on experiments performed in order to assess the rates of shoot and leaf desiccation in *Phragmites* populations occurring in different habitats. The aim was to clarify the influence of the water factor and the response and adaptability of the plants to it. As the experiments were performed in field laboratories without any control of the air conditions, it was only possible to compare two phenotypes examined together. Each time, the two phenotypes examined belonged most probably to one genetically identical population. The greatest difference between two populations was observed at the Hlohovecký fishpond near Lednice (Series A) in South Moravia. The littoral stands of *Phragmites* occur here in the reed-belt of the Hlohovecký fishpond in littoral ecophase, while the terrestrial stands of *Phragmites* occur in a small patch amidst dry grassland communities (Foed. *Mesobromion*) on loess (Květ, 1973b, Rychnovská, 1973). Underground water carried by an impermeable layer was obviously available to the *Phragmites* patch situated about 500 m from the fishpond. The vegetation of

Table 2. Phenotypic differences and desiccation of detached shoots of *Phragmites communis* under laboratory conditions

	Series A Hlohovecký pond July 7, 1969		Series D Opatovický pond July 12, 1972	
	Terrestrial (T)	Littoral (L)	Terrestrial (T)	Littoral (L)
(a) Fresh weight of 1 shoot FW (g)	44.81	43.28	73.96	76.32
(b) Dry weight of 1 shoot Ws (g)	18.40	15.64	22.47	26.71
(c) Dry weight of leaves Wl (g)	8.65	7.53	6.80	7.77
(d) Wl:Ws (%)	47.10	48.20	30.54	29.60
(e) Number of leaves (average)	11.6	12.0	10.8	13.6
(f) Dry weight of 1 leaf (average) (g)	0.74	0.62	0.63	0.57
(g) Specific leaf area (cm^2/g)	114.7	145.0	146.8	148.8
(h) Water content (% of Ws) in shoots	144.06	179.50	212.38	187.44
Water loss after 180 min (A) and 300 min (D) of desiccation, expressed as % of				
(i) Water content of shoot	23.25	19.58	16.71	19.50
(j) Dry weight Ws	33.57	35.45	39.08	36.34
(k) Fresh weight FW	13.74	12.61	11.60	12.66
(l) Dry weight of leaves Wl	71.11	73.60	129.00	123.17
Desiccation rate 0–30 min expressed in				
(m) mg/g (Ws)/min	2.57	2.40	2.66	1.93
(n) mg/g (H_2O)/min	1.77	1.32	1.11	1.04

Comparison between series can be made in the structural characteristics (a–h). Rates of water loss and other functional characteristics (i–n) may be compared only between phenotypes of the same series.

the Opatovický fishpond (Series D) has been studied and described in various ways. Its phenotypes and ecotypes of *Phragmites* have been described by Dykyjová (1971a; see also Sect. 3.1.1). The same two ecologically different biotopes as those studied by Dykyjová and Hradecká (1973, 1976; see also Sect. 3.1.1), were chosen for further analyses of water relations in *Phragmites*. In the littoral *Phragmites* stand, the water was 0.8 m deep at the time of measurement. The terrestrial to limosal biotope was situated at the outer edge of a reed-belt and was colonized by patches of pure stands of *Phragmites* and *Glyceria maxima* respectively. Water level was only slightly below the soil surface.

At the beginning of these experiments some xerophytic features were presumed in the terrestrial *Phragmites* types such as a lower water content, smaller A_s and a slower desiccation rate. The results contradict this hypothesis in several respects.

The water content in the shoots tends to be higher in the terrestrial types in Series D, A_s is almost equal, and the proportion of leaves in the shoot biomass is much the same in each series. The only conspicuous "xeromorphic" features are the lower water content in the shoots and a smaller A_s in the terrestrial *Phragmites* type in the Series A (Table 2).

The functional characteristics of the *Phragmites* shoots contradict the above-mentioned hypothesis as well. After several hours of desiccation, the water loss is generally slightly greater in the terrestrial than in the littoral *Phragmites* types. But this is by no means an unambiguously xerophytic feature. A similar tendency can be traced at full turgidity during the first minutes of desiccation exposure: its rate was greater in all terrestrial phenotypes when related back to their shoot and leaf dry weight or water content.

These experiments prove an extremely poor phenotypical adaptation of the *Phragmites* shoots to the presumed water stress. No other explanation can be offered than an absence of water stress under natural conditions. A highly balanced water input through the underground organs exists obviously in all the *Phragmites* types studied, hence the same performance of the shoots regardless of their apparent distance from the water table.

Genetically fixed differences certainly exist between the water relations in various ecotypes of *Phragmites communis*. Our results imply this tendency; for instance, the water content in the shoots is lower in the ecotype originating from drier climatic conditions (Hlohovecký fishpond). Further observations are needed in this direction.

3.3.2 Water Relations in Intact Shoots of Phragmites communis and Their Response to Environmental Factors

Daily variation in the transpiration of the *Phragmites* shoots has been analyzed in order to obtain further knowledge on the adaptability of *Phragmites communis* and for an evaluation of transpiration of reedswamp communities. The transpiration rate was determined gravimetrically. The method description and some preliminary results have been published by Květ (1973b) and Rychnovská

Table 3. Transpiration of *Phragmites communis* under natural conditions in South Moravian biotopes (Lednice fishponds)

Series:	A		B
Locality (fishpond):	Hlohovecký July 7, 1969		Nesyt July 23, 1975
Type of biotope:	Terrestrial	Littoral	Littoral
(a) Aboveground living biomass ($g\ m^{-2}$)	1,170[a]	2,068	1,465
(b) Stand density (living shoots m^{-2})	60[a]	180	59
(c) Dry weight (W) per shoot analyzed (g)	19.5	14.4	24.7
(d) Shoot water content in % dry weight	180.4	193.5	166.6
Transpiration rate of shoots (daily mean values) expressed in			
(e) $mg\ g^{-1}$ shoot dry weight min^{-1}	4.77	6.56	8.19
(f) $mg\ g^{-1}$ water content in shoots min^{-1}	2.68	3.49	5.09
(g) $mg\ g^{-1}$ water content in leaves min^{-1}	6.42	7.64	—
(h) Maximum transpiration rate $mg\ g^{-1}$ shoot dry weight min^{-1}	11.63	12.50	14.76
(i) Daily transpiration ($g\ g^{-1}$ shoot dry weight d^{-1})	5.15	5.63	5.16
(j) Water turnover rate ($g\ g^{-1}$ water content in shoots d^{-1})	2.89	2.99	3.18
(k) Water turnover rate ($g\ g^{-1}$ water content in leaves d^{-1})	6.93	6.56	—
(l) Daily transpiration of the stand[a] ($kg\ m^{-2}\ d^{-1}$)	6.03	5.01	7.57

[a] Estimates only.

and Šmíd (1973). The data given in Tables 3 and 4 present examples of the measurements, giving only the daily mean values. Detailed analyses of the transpiration curves and relevant climatological data will be published later.

3.3.2.1 Water Saturation Deficit (WSD): Its Daily Variation in the Shoots, Leaves, and Leaf Discs

Several methods can be used to determine the WSD. The classical method after Stocker (1929) is not suitable for *Phragmites*. A *Phragmites* shoot separated from its root system never attains full turgidity by suction of water through its stem only. The method according to Vassiljev (1931) is therefore more acceptable. This method assumes the morning water content to be 100% and relates the WSD to it. The great variation in shoot water content is a serious drawback of this method as only parallel "paired" individuals can be evaluated. The leaf-disc method (Čatský, 1960) is not suitable for leaves with aerenchyma or loose parenchyma, as water infiltrates into the mesophyll through the cut edges of the discs; the resulting WSD values are higher than in intact leaves. But *Phragmites* represents an outstandingly suitable material for the leaf disc method. On the other hand, *Phragmites* leaf blades show a pronounced gradient in their water content (see Table 1). The discs can be cut out only from their wider basal parts. Conse-

Table 4. Transpiration of *Phragmites communis* and *Glyceria maxima* under natural conditions at the Opatovický fishpond (Series D) in South Bohemia on July 20, 1972

Ecophase:	Limoso-terrestrial		Littoral	
Shoots of: (their size):	*Phragmites*	*Glyceria*	*Phragmites* (standard)	*Phragmites* (substandard)
(a) Aboveground living biomass ($g\ m^{-2}$)	1,210	1,122	780	
(b) Stand density (living shoots m^{-2})	74	—	61	
(c) Dry weight (W) per shoot analyzed (g)	24.8	—	28.6	10.4
(d) Shoot water content in % dry wt.	119.7	279.1	104.6	169.3
Transpiration rate of shoot (daily mean values) expressed in				
(e) $mg\ g^{-1}$ shoot dry weight min^{-1}	10.72	14.46	10.75	10.47
(f) $mg\ g^{-1}$ water content in shoots min^{-1}	8.92	6.05	10.03	7.12
(h) Maximum transpiration rate ($mg\ g^{-1}$ shoot dry wt. min^{-1})	16.46	25.25	17.98	15.78
(i) Daily transpiration ($g\ g^{-1}$ shoot dry wt. d^{-1})	8.20	11.52	8.84	7.63
(j) Water turnover rate ($g\ g^{-1}$ water content in shoots d^{-1})	7.00	4.82	8.25	5.05
(l) Daily transpiration of the stand ($kg\ m^{-2}\ d^{-1}$)	9.92	12.93	6.42	

quently, they always show a somewhat higher water content and smaller WSD than whole leaf blades.

A considerable WSD can be determined in all *Phragmites* populations (Table 5), regardless of whether they grow in water or on dry land. The entire shoots show a WSD of about 16% in the littoral and of some 22% in the dry biotope at the Hlohovecký pond at noon and 20–21% WSD in the Opatovický pond, in both the biotopes compared. A somewhat smaller WSD (15%) was found only in the substandard shoots. Highly interesting is a comparison of the data on the WSD obtained from the water content in the leaves. In both *Phragmites* types from the Hlohovecký fishpond, the WSD in the leaves amounts to about 0.8–0.75 of that determined in the whole shoots. The culms thus appear to play a considerable role in stabilizing the water content in *Phragmites* shoots.

The WSD values obtained by the leaf-disc method are rather low, never exceeding 7%; in the *Phragmites* stand at the Nesyt fishpond, this method did not reveal any WSD at all. The most probable reason, inherent in the method itself, has been explained above (see also Rychnovská, 1967).

Another conclusion can be derived from the knowledge of the daily variation of WSD in the shoots. The rich development of the leaf apparatus seems closely matched with the absorption and transport of water through the stem. Thus it

Table 5. Daily course of water content and water saturation deficit in terrestrial and littoral shoots of *Phragmites communis* (Hlohovecký fishpond, July 7, 1969)

Interval (h)	Terrestrial					Littoral				
	WC_s	WC_l	WSD_s	WSD_l	WSD_d	WC_s	WC_l	WSD_s	WSD_l	WSD_d
6– 9	202	194	0	0	0	213	200	0	0	0.3
9–12	192	173	5.0	10.9	3.0	192	190	9.9	5.0	2.7
12–15	157	160	22.3	17.6	5.3	178	176	16.5	12.0	1.4
15–18	173	160	14.4	17.6	2.9	189	176	11.3	12.0	0
18–21	169	160	16.4	17.6	2.0	193	181	9.4	9.5	0

WC_s Water content of shoots (% dry wt.).
WC_l Water content of leaves (% dry wt.).
WSD_s Water saturation deficit in shoots.
WSD_l Water saturation deficit in leaves.
WSD_d Water saturation deficit in leaf discs.

seems that water transport may limit the other physiological functions in a *Phragmites* stand or when its transpiration is extremely intense.

3.3.2.2 Transpiration Rate

The pertinent process of transpiration is not easily ascertained quantitatively. Direct methods employing intact plants are difficult to apply to the big shoots of *Phragmites*. Indirect aerodynamic methods are not sufficiently sensitive and cannot be applied to the mostly relatively small patches or narrow strips of *Phragmites* stands in pond littorals. The old direct gravimetrical method was therefore used. The cut-off entire shoots, or their whole transpiring portions, were weighed, exposed in the stand for 3–4 min and weighed again. The transpiration was expressed in terms of water loss per unit both of dry weight and total water content, or water content in the leaves. The former calculation was used to estimate the daily water vapour output of the whole stand, the latter was to estimate the water turnover rate (WTR) per day. The curves characterizing the daily course of transpiration were plotted from measurements taken at approximately 15 min to hourly intervals. An example is presented in Figure 3.

From the results presented in Table 3 and 4, the following conclusions can be drawn: The xerophytic behavior of the terrestrial *Phragmites* populations is evident at the Hlohovecký pond (A). The transpiration rates (water loss per minute) are lower in all cases here than those found in the littoral shoots. Because of rapid evaporation of the morning dew, the period of transpiration during the day is, however, longer in the terrestrial *Phragmites* patch. The daily water output is thus much the same in both the terrestrial and the littoral stand. Similarly, the daily transpiration per 1 m^2 of the littoral reed has been estimated at 5–6 kg water in the Hlohovecký fishpond and 7.5 kg in the Nesyt fishpond. The South Bohemian reedswamp stands have been found to transpire more intensely. For standard shoots of average height, the difference between the littoral and terrestrial ecophases is small. The substandard (short) shoots differ much more because of the suppressed water output, obviously caused by the canopy microclimate (see

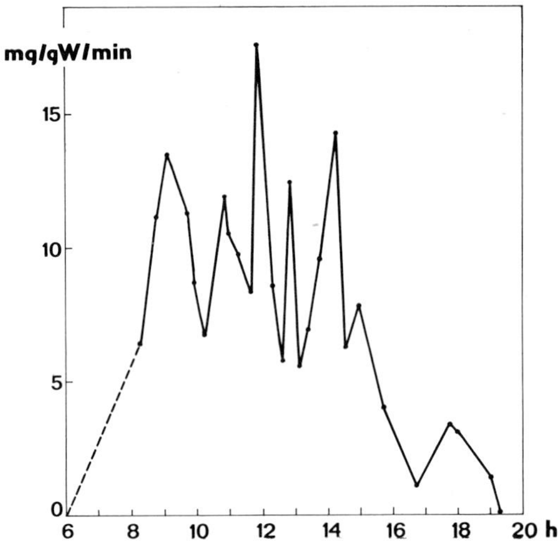

Fig. 3. Daily course of transpiration in *Phragmites communis;* Nesyt fishpond (B) 23 July 1975, littoral stand. Transpiration rate expressed as water loss in mg g^{-1} dry weight min^{-1}

Sect. 2.1.2). On the other hand, the limosal population of *Glyceria maxima* shows the highest transpiration recorded in these measurements. At the Opatovický fishpond, the daily transpiration of the littoral *Phragmites* stand ("V" form), amounted to about 6.4 kg m^{-2}; the limoso-terrestrial Phragmites stand ("S" form) transpired nearly 10 kg m^{-2}, and the limosal stand of *Glyceria* nearly 13 kg m^{-2}.

The WTR is another characteristic enabling a deeper understanding of plant water relations. This value tells us how many times the water content is changed in a plant or plant part per day. The most important results are presented in Table 3 and 4, showing that the WTR fluctuated around the value of 3 in whole shoots: the water pool is thus renewed in a shoot three times a day. When the daily water loss is related to the water content in the leaves, considerably higher values result: 6.5–6.9 times. Such a high WTR demands a continuous water supply. In view of the much lower WTR in the whole shoot, the stem with leaf sheaths appears to play the role of a water pool buffering the great transpiration losses. For this reason, the high values of WSD in the entire shoot do not manifest themselves in the leaves. The negative water balance on a hot midday is buffered by the water pool in stems and does not affect the leaf blades. *Glyceria maxima* shows a similar adaptation, containing even more water in its whole shoots than *Phragmites*.

3.3.2.3 Sources of Homeostatic Water Relations in Shoots

The examples presented above point to the importance of the entirety of the *Phragmites* shoot. The importance of the root system for the water balance in the shoots has been demonstrated by Rychnovská (1973). Shoots separated from their roots by a cut made below the water surface and with their bases left submerged, wilt and dry up irreversibly within several hours. After 20–30 h, the WSD reaches

Table 6. Desiccation rate of leaf blades attached to or detached from the stem of *Phragmites communis* under standard laboratory conditions (Hlohovecký fishpond, July 7, 1969)

	Leaf blades attached	Leaf blades detached
Water loss in mg g^{-1} dry wt. min^{-1}		
during first 60 min	1.06	1.84
during 6 h	0.75	1.51
Time needed to attain a water saturation deficit of		
10%	94.3 min	54.3 min
25%	333 min	165 min

40–50% in the leaves, as estimated by the leaf-disc method. This performance is not common among mesophytic and hydrophytic grasses or broad-leaved plants.

Another example demonstrating the importance of the entire plant organism for the water relations is shown in Table 6. The experiment was performed with highly uniform *Phragmites* shoots. Their leaves were separated from the stem and left in the laboratory for desiccation. A parallel series was exposed with the leaves left attached to the stem. The desiccation rate was probably much the same during the first minutes of exposure; after one hour, however, the water loss was much higher from the detached leaves; within 6 h their water loss became twice as high as that from the attached leaves. The desiccation curves of the detached leaves correspond with cuticular transpiration; presumably, its value is the same. The water content was twice as high in the leaves attached to the stems after the same period of desiccation: the probable explanation lies in the water-storage function of the internodes.

3.3.3 Water Relations in Different Fishpond Littoral Communities in Comparison with Grassland Communities

The daily transpiration of the *Phragmites* and *Glyceria* stands per m^2 is considerable. Here it is compared with the estimated transpiration in mesophytic or grassland communities in South Moravia.

A mesophytic alluvial meadow community dominated by *Alopecurus pratensis* transpired 10.5 kg m^{-2} day^{-1} (6 June 1968, at Dolní Věstonice). A similar meadow type later showed the following values: 4.9 and 5.35 kg m^{-2} day^{-1} (25 May and 29 July 1971, at Lanžhot). A moist meadow with dominant *Phalaris arundinacea* transpired 9.6 kg m^{-2} day^{-1} at Dolní Věstonice, while the highest value was 6.7 kg m^{-2} day^{-1} in the same community type at Lanžhot. A wet meadow dominated by *Glyceria maxima* transpired 8.3 at Dolní Věstonice; its transpiration was considerably lower at Lanžhot, reaching only 3.8 (25 May 1971) and 4.3 kg m^{-2} day^{-1} (29 July 1971).

All these measurements indicate that the transpiration of reedswamp communities roughly corresponds with that of wet and mesophytic meadows in the same region (Rychnovská, 1972; Rychnovská et al., 1972). The data presented here are also in agreement with those calculated by Šmíd (1975). His estimates of daily evapotranspiration based on the heat-budget approach and made for the flooded littoral stand of *Phragmites communis* at the Nesyt fishpond in June to August, 1973, are: 5.6, 6.9, and 5.5 kg m^{-2} day^{-1}. The transpiration of these dense stands usually exceeds the evaporation from the open water. Their effect on the local climate therefore should not be neglected, particularly in the rather dry and prevailingly agricultural region of South Moravia.

The water relations have been studied only to a limited degree in other pond littoral plant species than *Phragmites* and *Glyceria maxima*. Rychnovská (in Hejný, 1969), and Květ (1975b) measured the transpiration rate and other characteristics of water relations in altogether 13 species colonizing emerged fishpond shores. A rather rapid water turnover in their transpiring parts was characteristic of most of them and in some species growing in relatively dry soil (e.g., *Trifolium repens*, *Potentilla anserina*), adverse effects of water saturation deficit on transpiration were observed.

Methodological Remarks. Although transpiration measurements are commonly made with cut-off shoots, some remarks should be added on their practical application to *Phragmites*. Its shoots are several meters long and it is hardly possible to place them on one balance. Therefore, two scale balances (sensitivity 0.01 g, maximum weight 200 g) were mostly used and both weights added. The *Phragmites* shoot was inserted into a polyethylene sleeve immediately after cutting and stretched between the two balances. The time of exposure was measured between its drawing-out and reinsertion into the sleeve. The second weighing proceeded in the same manner. The weight of the sleeve was checked after each measurement. In this way, it was possible to handle even tall shoots relatively easily.

References see pp. 285—291.

3.4 Nutrient Uptake by Littoral Communities of Helophytes

D. Dykyjová

3.4.1 Absorption, Accumulation and Translocation of Nutrients in Plant Tissues

In order to examine the relations betveen the nutrient availability and net production in helophyte communities in fishpond littorals, chemical analyses were made in their water, sapropels, bottom soils, and plant biomass. To compare the chemistry of different habitats, only monocenotic communities (their dominating species) were analyzed, as in main production analyses. The importance of the water chemistry for mineral nutrition of rooted aquatic plants cannot be denied. The accessory aquatic roots of emergent plants rooted in the bottom, such as *Phragmites* or *Typha*, play an important part in the additional water and nutrient supplies to the plant; for details and references see Dykyjová and Hradecká (1976). The finely branched aquatic roots develop in the submerged basal parts of the stems. They are formed especially in erosion biotopes on poor sandy bottoms. In fertilized fishponds, these roots can acquire a significant portion of mineral nutrients which cannot be absorbed via the roots growing in nutrient-poorer deeper layers of the bottom. This is why even the communities of *Phragmitetum communis* growing in erosion habitats of the Opatovický fishpond such as the *Phragmites* V biotope have a high biomass and production. These morphological and physiological adaptations of helophytes to the absorption of nutrients from both the bottom soils and the aquatic environment allow a more intense nutrient uptake than is the case in dry land plants.

The chemical composition of water, sapropels, and bottom soils is discussed in Sections 2.3 and 2.4. Any comparisons of the uptake and accumulation of mineral elements in the plant tissues can only be made on the assumption that the nutrient content as ascertained in the abiotic environment corresponds with their available pool, which can be transported into the living plant tissues. The methodological problems of nutrient extraction from the bottom soils were discussed in Section 2.4.

3.4.1.1 Analysis of Plant Biomass

The plant biomass analyses were performed according to Koppová et al. (1955) from an extract obtained by wet combustion of dried plant material in concentrated H_2SO_4 + 35% H_2O_2; nitrogen by the micro-Kjeldahl technique; phosporus and magnesium colorimetrically, with ammonium vanadate and molybdate, and with titanium yellow, respectively; calcium, potassium and sodium by flame photometry. The analyses were made at the chemical laboratory of the Forestry Research Institute, Station Opočno under the supervision of Mr. J. Vacek.

3.4.1.2 Seasonal Changes in the Contents of Mineral Nutrients

Another source of difficulties in comparisons between different kinds of plant material is the great variation in nutrient contents in the plant organs according to their ontogenetic age and metabolic activity. The developmental rate of certain species, such as in the stands of *Bolboschoenus maritimus* or *Acorus calamus*, is quite fast and they pass through their reproductive phenophase in July. In the stands of *Sparganium erectum* on the other hand, sprouting and growth are retarded, with the shoots emerging from the water only in early June. Populations of *Glyceria maxima* start to sprout in early spring and the formation and growth of the tillers continue during the whole growing season, sometimes even during mild winter periods. The different rates of development in individual species may cause errors in comparisons between their biomass and nutrient contents based only on single samples taken on the same date. In many cases, it was possible to detect samples taken too early or too late according to the percentage nitrogen content or the K/Ca ratio. Both these assays are more sensitive to metabolic age than a mere morphological aspect of a community. During early growth, uptake of the essential nutrients is more rapid than the rate of net organic production, but later the uptake rates decline. The percentage contents of ash (not considering the so-called residual ash, see Květ, 1973a) and of the essential nutrients such as N, P, and K, decrease during seasonal development: see Figure 1 and Table 1. Calcium content, on the other hand, increases during the growing season, especially in the leaves. The nutrient dynamics in emergent macrophytes are discussed by Boyd (1970a), Bayly and O'Neil (1972a, b), Dykyjová (1973), Dykyjová and Hradecká (1976), Květ 1973a, b), Mochnacka-Lawacz (1974), and others.

The young and photosynthetically highly active leaves or whole shoots contain higher percentages of mineral nutrients than the stems. Some nutrients, such as P and Mg, become accumulated in panicles or fruits. Underground storage organs accumulate energy-rich organic matter, but relatively less of mineral elements. The same gradient of mineral nutrient contents, corresponding to the average age of the organs analyzed, is also assessed by the method of stratified harvesting (see Sect. 3.1.6). A shoot of *Phragmites* displays the youngest tissues at its apex. The uppermost portions of the culms with corresponding leaves contain therefore higher amounts of N, P and non-residual ash as well as Ca. Potassium content, on the other hand, remains higher in the basal portions of the culms, in ontogenetically youngest tissues (see Fig. 2). The K/Ca ratio is also highest in the young apical tissues. The assimilatory organs of *Schoenoplectus lacustris* are, in fact, stems with the young growing tissues located at their bases. This situation is illustrated in Table 2 by the highest K/Ca ratio, contrary to *Phragmites*, at the base. Figure 3 present the seasonal courses of ash content in the culms of *Phragmites* growing in the limosal S biotope in the Opatovický fishpond. The seasonal

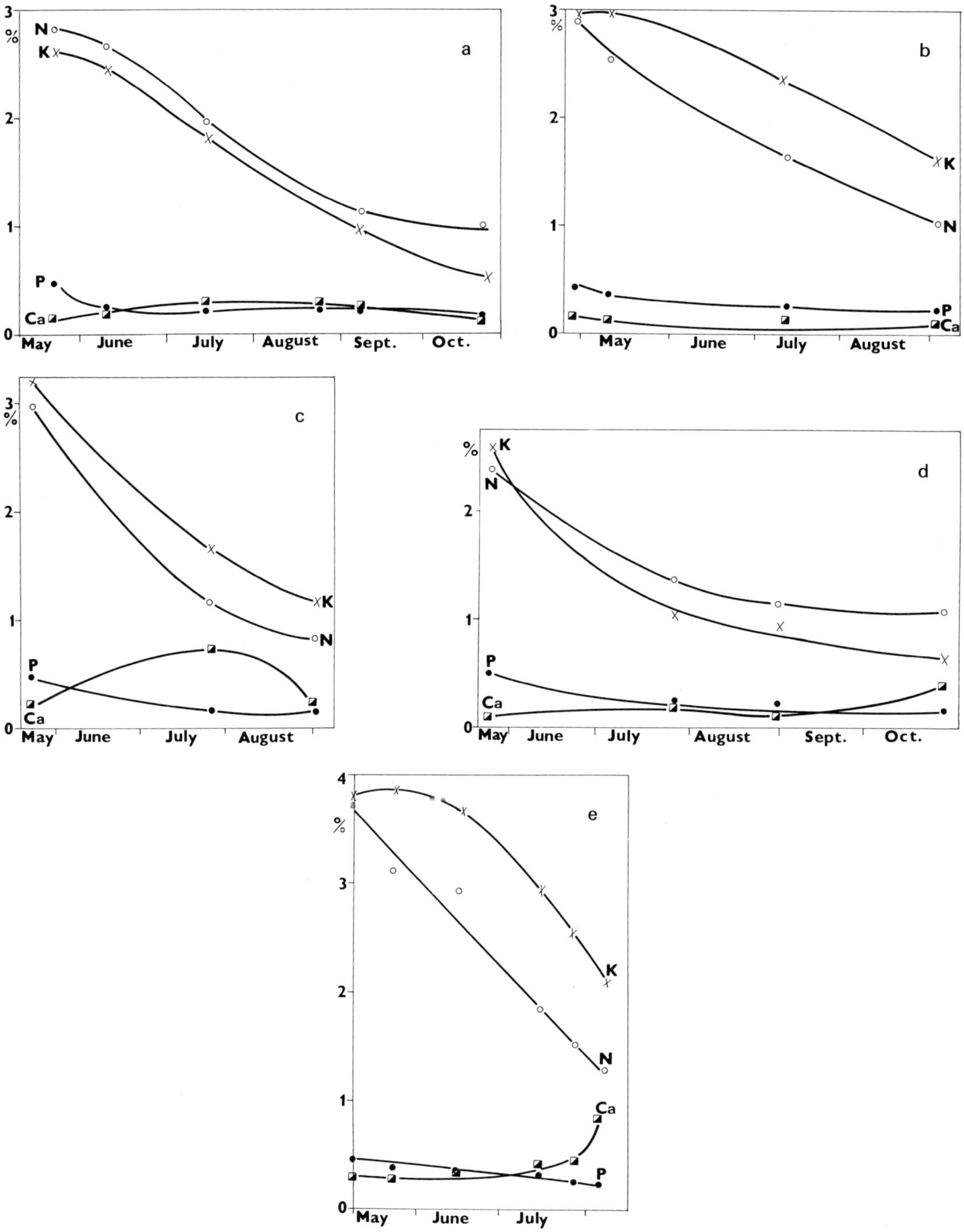

Fig. 1a–e. Seasonal course of the percentage content of mineral nutrients in shoots of: (a) *Phragmites communis;* (b) *Glyceria maxima;* (c) *Typha angustifolia;* (d) *Schoenoplectus lacustris;* (e) *Acorus calamus*

Table 1. Seasonal changes in macronutrient contents of pure stands of helophytes in the fishpond littoral (in %) dry wt. and as total uptake in g m^{-2})

Date	N		P		K		Ca		Mg		Ash
	%	g m^{-2}	%	g m^{-2}	%	g m^{-2}	%	g m^{-2}	%	g m^{-2}	%
Phragmites shoots, Opatovický fishpond											
23 May	2.77	28.4	0.48	4.9	2.76	27.4	0.14	1.5	0.14	1.5	8.0
10 June	2.65	38.1	0.24	3.7	2.46	36.3	0.17	2.5	0.12	1.9	7.5
1 July	2.13	40.9	0.28	4.1	1.65	30.6	0.28	5.1	0.09	1.7	8.3
15 July	1.91	40.8	0.22	4.7	1.74	37.6	0.29	6.3	0.09	2.7	8.0
25 Aug.	1.63	28.6	0.23	5.1	1.26	22.0	0.28	5.0	0.08	1.5	7.5
8 Sept.	1.13	30.9	0.18	5.3	0.93	25.5	0.27	7.4	0.08	2.2	6.0
23 Oct.	1.00	18.1	0.17	3.2	0.55	10.0	0.18	3.3	0.08	1.5	6.6
Phragmites rhizomes and roots, Opatovický fishpond											
17 May	1.60	59.5	0.19	7.3	1.45	54.2	0.06	0.8	0.07	2.7	11.2
14 Aug.	1.05	35.4	0.11	3.8	1.11	37.3	0.03	0.4	0.06	2.0	—
8 Oct.	1.07	64.0	0.12	7.4	1.13	67.1	0.02	1.2	0.06	4.0	9.2
Typha angustifolia, Opatovický fishpond											
23 May	2.90	9.7	0.49	1.6	3.20	10.7	0.20	0.6	0.16	0.5	9.85
30 June	1.78	30.16	0.38	6.4	1.97	33.3	0.69	11.66	0.12	2.0	6.43
26 July	1.15	32.7	0.16	4.5	1.65	46.8	0.73	20.6	0.18	5.2	6.25
1 Sept.	0.80	24.5	0.15	4.6	1.16	35.3	0.18	5.5	0.13	3.9	6.36
Glyceria maxima, Opatovický fishpond											
29 April	2.94	17.2	0.42	2.4	2.96	17.3	0.14	0.8	0.12	0.7	—
10 May	2.76	17.4	0.40	2.5	3.20	20.1	0.16	1.0	0.15	0.9	11.6
12 July	1.64	44.2	0.26	7.1	2.30	61.8	0.12	3.50	0.11	3.4	9.8
4 Sept.	1.05	21.0	0.23	5.4	1.60	32.1	0.08	1.7	0.13	2.6	8.4
20 Oct.	1.12	19.4	0.21	3.63	1.49	25.8	0.37	6.4	0.18	3.1	9.8

Table 2. Vertical gradient of the K/Ca ratio in the shoots of *Schoenoplectus lacustris*

Date	19 July		
Stand layer (cm above bottom)	K	Ca	K/Ca
	mg/100 g dry wt.		
0– 40	1,096	95	11.5
40– 80	1,396	274	5.0
80–120	1,530	405	3.7
120–160	1,640	458	3.5
160–200	1,605	446	3.5
200–240	1,525	431	3.5
240–280	—	—	—
Spikelets	854	106	8.0
Leaves	1,467	120	12.2

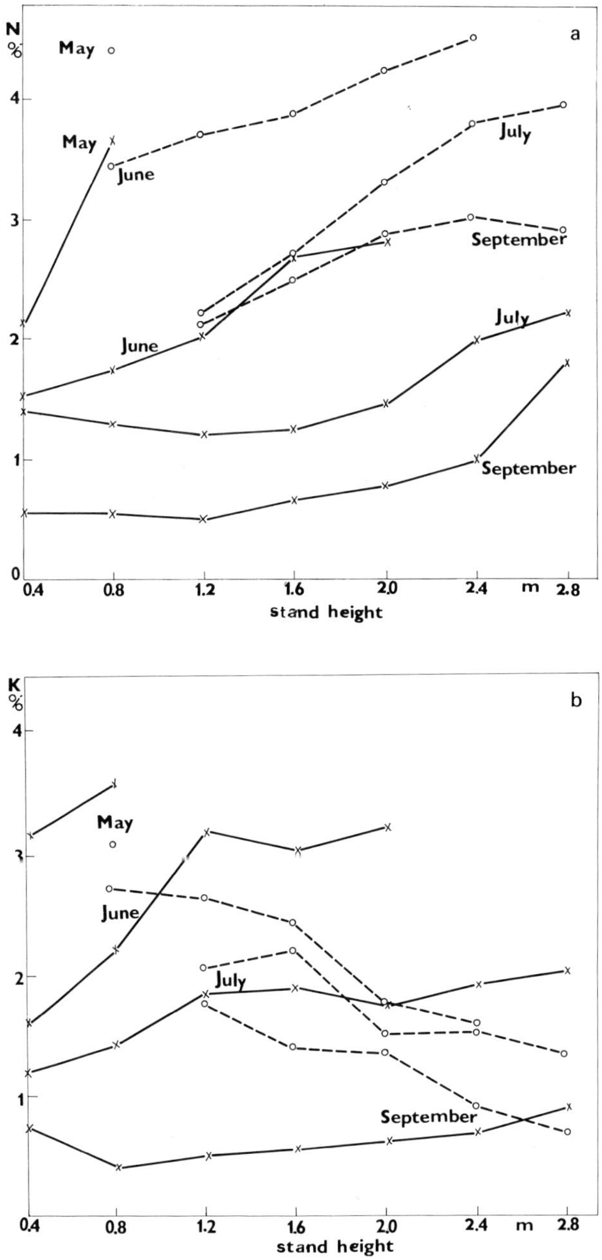

Fig. 2. Two gradients of the N (a) and K (b) contents in the shoots of *Phragmites communis* in successive horizontal stand layers from base to apex during growing season (from May to September). *Full lines:* Culms with leaf sheaths; *Dashed lines:* Leaves (= leaf blades)

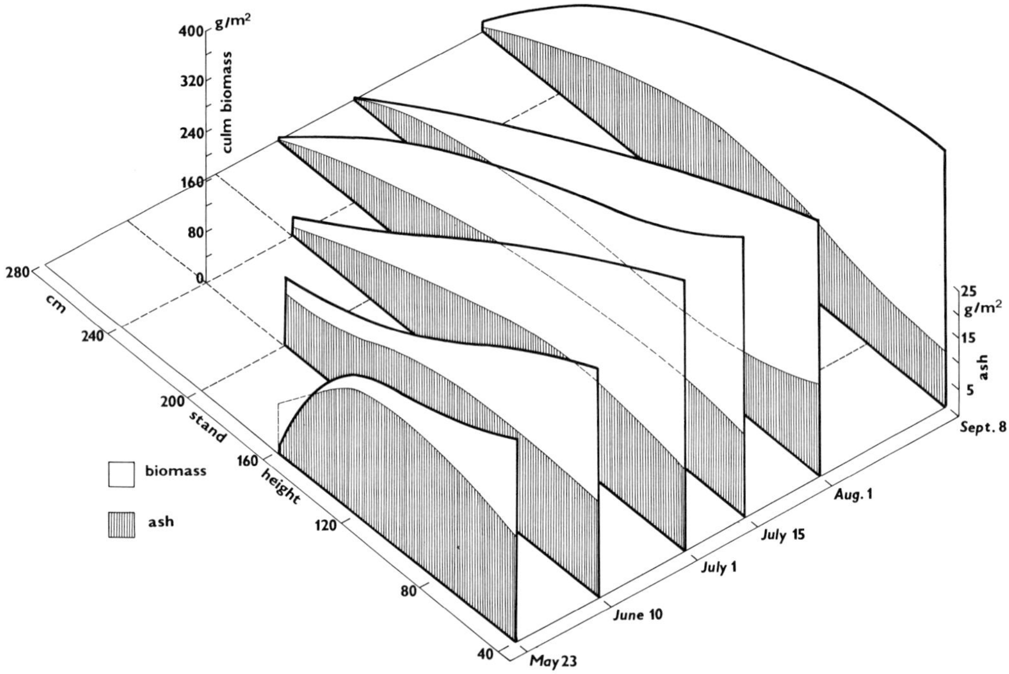

Fig. 3. Culm biomass in successive horizontal stand layers in the limosal (S) *Phragmites* ecotype and corresponding seasonal changes in ash content *(shaded areas)*

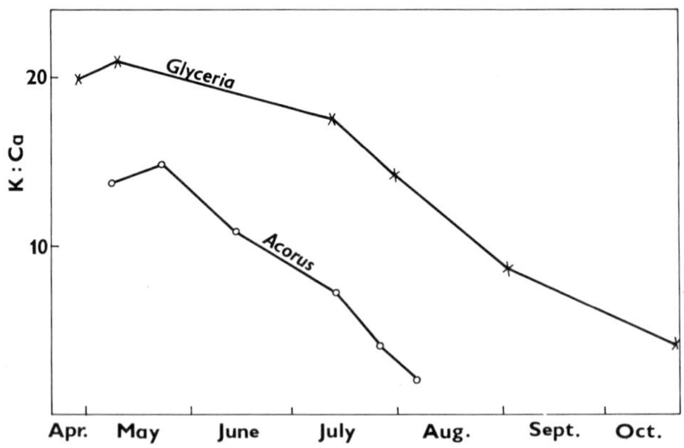

Fig. 4. Seasonal changes in the K/Ca ratio in the shoots of *Glyceria maxima* and *Acorus calamus*

development of leaf area and the changes in biomass are presented in Table 2 in Section 3.1.6.1. The ash content in the culms with leaf sheaths increases acropetally but declines during the growing season. The same vertical gradients of N, P and K contents and their seasonal decline occur in the leaves per 1 m^2 of stand area. The poorer V form is distinguished by a much smaller leaf area duration and by a different vertical and seasonal gradient of N content (see Fig. 6 in Sect. 3.1.6.1).

Table 3. Examples of seasonal changes in the K/Ca ratio of shoots in three different helophytes

Date	K	Ca	K/Ca	K	Ca	K/Ca	K	Ca	K/Ca
	mg/100 g dry wt.								
Organs	Culms			Leaves			Total shoots		
Phragmites communis, Opatovický fishpond, accumulation zone									
5 May	2,850	147	19.4	2,100	148	14.2	2,768	148	18.7
10 June	2,295	108	21.2	2.211	269	8.2	2,263	171	13.2
1 July	1,650	134	12.3	1,670	570	2.9	1.656	132	5.8
15 July	1,578	120	13.1	1,547	620	2.5	1,560	296	5.2
25 Aug.	1,195	115	10.4	1,420	708	2.0	1,259	284	4.4
8 Sept.	850	89	9.5	1,200	791	1.5	934	270	3.4
21 Sept.	740	96	7.7	1,170	841	1.4	881	261	3.3
21 Oct.	420	74	5.6	688	430	1.6	552	200	2.7
							Total underground		
17 May							1,456	62	23.4
14 Aug.							1,111	29	38.3
8 Oct.							1,131	22	51.4

Typha angustifolia, Nesyt fishpond				*Schoenoplectus lacustris*, Opatovický fishpond, erosion zone			
Leaves				Total shoots			
24 May	1,501	342	4.4	23 May	2,380	110	21.6
23 June	1,192	353	3.3	30 June	2,068	232	8.8
29 July	1,192	409	2.9	26 July	1,360	201	6.7
26 Aug.	682	485	1.4	1 Sept.	1,180	192	6.1
29 Sept.	577	554	1.0	20 Oct.	804	340	1.5

Calcium, entering the plants mainly passively with the transpiration stream, accumulates in the leaves in the form of insoluble compounds, and its content progressively increases with the ageing of the leaf tissues (Table 3, Fig. 1, c, d, e). The high capacity for the accumulation of potassium, especially in ontogenetically young organs, results in a high K/Ca ratio in helophytes. This "potassium plant type" is known to occur in grasses *(Poaceae)* and graminoid plants (families *Cyperaceae* and *Juncaceae*; see Iljin, 1932, 1936; Denayer de Smet, 1964; Horak and Kinzel, 1971). The K/Ca ratio found in emergent aquatic plants is high, but it declines with developmental age of the plant tissues during the growing season (Table 3 and Fig. 4). Duckweeds *(Lemnaceae)*, on the other hand, absorb more Ca in relation to K, like many other floating and submerged plants (Straškraba, 1968; Boyd, 1968; see Table 2 in Sect. 3.5).

It is therefore necessary in each case, when comparing the nutrient uptake by plants growing in different habitats, to consider their developmental stage. The nutrient contents in different helophytes, as presented in Table 1, are related to the stage of their maximum stand development, before the onset of shoot senescence.

Table 4. Chemical composition of plant tissues of *Phragmites communis* growing in halotrophic (Nesyt fishpond) and eutrophic (Opatovický Fishpond) habitats. Summarized according to Husák (1971) and Dykyjová and Hradecká (1976)

Content of nutrients mg/100 g dry wt.	Shoots (young plants)				Shoots (ripe plants)				Underground organs			
	Nesyt		Opatovický		Nesyt		Opatovický		Nesyt		Opatovický	
	Stems	Leaves	Stems	Leaves	Stems	Leaves	Stems	Leaves	Rhizomes	Roots	Rhizomes	Roots
N	992	4,410	2,030	3,835	808	3,160	875	2,940	919	992	840	1,295
P	293	353	455	491	172	309	159	273	406	309	82	152
K	1,123	1,060	2,920	2,400	509	840	933	1,524	709	459	828	642
Na	44	31	64	39	41	31	60	37	184	206	—	—
Ca	76	190	102	108	44	319	180	560	386	104	11	29
Mg	77	216	99	129	38	270	61	181	401	340	62	61

Within the ranges of this intraspecific variation, it can be proved that all helophytes growing in nutrient-rich habitats, whether littoral or limosal, accumulate more mineral nutrients than their populations growing in nutrient-poor habitats. Numerous field estimates (Sect. 3.4.2) as well as experiments in hydroponic cultures (Sect. 3.4.3) are in good agreement with this presumption. Greater nutrient availability enhances not only the accumulation of mineral elements, but also the net production.

It therefore seems paradoxical, in a comparison between the littorals of the Opatovický fishpond and the Nesyt fishpond, that the higher nutrient content in the water, sapropels and bottom soils of Nesyt is not much reflected in a higher accumulation of nutrients in the plant biomass (see Table 4). A probable explanation is that the more vigorous growth of individual shoots in the populations of Phragmites communis and Typha angustifolia in South Moravian biotopes, caused not only by edaphic but also by climatic factors enhancing the organic production, brings about a "dilution" in the percentage contents of the nutrients absorbed.

3.4.2 Control of Helophyte Production by Nutrient Availability

During investigations both the Opatovický and Nesyt fishponds were extensively fertilized and manured (see Sect. 1.1). As mineral nutrients codetermine the efficiency of solar energy conversion by plant communities (see Sect. 3.1.8) an exceedingly dense growth of helophytes results from excessive nutrient supply. The dense growth is multiplied by the regeneration of densely growing new shoots in occasionally cut littoral reed-belts. The growth of emergent macrophytes was especially luxuriant in the accumulation biotopes additionally fertilized by inflowing drainage water from fields or by farm waste-water. The luxuriant colonization of such areas results in further eutrophication of the fishpond through decay of the helophyte litter.

Several accumulation and erosion biotopes of fishpond littorals (see Sect. 2.4 and Figs. 5, 6) are occupied by the same types of monospecific reedswamp communities. It was therefore possible to compare the production and amounts of nutrient taken up by the helophytes in biotopes differing in the supply of available nutrients.

Shoot samples taken on all sites where the seasonal maximum biomass had been estimated (see Sects. 3.1.2 and 3.1.7) were oven-dried, finely ground and analyzed for the contents of the principal macronutrients. Corresponding underground organs, as long as their biomass had been estimated, were analyzed in the same way. For the analytical methods see Section 3.4.1. All the analytical data on the plants relate to the stage of maximum stand development before the onset of senescence of the assimilatory organs. The specific accumulation capacity of each helophyte species growing in different habitats is reflected in the percentage contents of individual nutrients in the dry biomass. The total amount of nutrients accumulated in the stand per unit ground area (so-called mineralomass) is, of course, related to the amount of biomass on the same area. Both the nutrient

Fig. 5a and b. Spring physiognomy of *Phragmites* stands in the limosal (a) and littoral (b) biotopes. Waste-water channel across the limosal stand shows richly accumulated organic material in the soil with remnants of thick rhizomes. Initial reed polycormones in the littoral biotope are without previous year's culms. (Photo: D. Dykyjová)

Fig. 6a and b. Littoral (a) and limosal (b) stands of *Glyceria maxima*. Littoral stand with flowering shoots in July (Photo: D. Dykyjová). Densely tillering limosal stand in early June. (Photo: J.P. Ondok)

concentrations and their amounts taken per 1 m^2 are presented in Table 5. They are related to both the average and highest values of seasonal maximum biomass. The uptake capacity of a whole stand of each monospecific or mixed plant community is estimated as an average value of samples taken along a transversal or longitudinal transect across the stand. The values for the whole littoral of the Opatovický fishpond have been calculated in the same way.

Table 5. Minimum and maximum values of nutrient content in aboveground biomass of reedswamp helophytes of Opatovický fishpond (Dykyjová, 1973c)

	N	P	K	Ca	Mg	Na
	% of dry weight at the stage of maximum developments of the stands					
Phragmites communis	1.77–2.13	0.19–0.28	1.08–1.74	0.03–0.29	0.08–0.17	0.03–0.05
T. angustifolia	1.5	0.16	1.16–1.65	0.18–0.73	0.13–0.18	0.25
T. latifolia	1.43	0.22	1.47	0.21	0.18	
Sparganium erectum	1.42–2.55	0.32–0.48	3.60–4.10	0.18–1.23	0.19–0.29	0.39–0.44
Acorus calamus	1.26–2.92	0.20–0.35	1.85–3.67	0.34–0.85	0.14–0.21	0.09–0.24
Schoenoplectus lacustris	1.03–1.77	0.23–0.34	1.36–1.69	0.07–0.25	0.09–0.15	0.34–0.40
Glyceria maxima	1.29–1.82	0.18–0.31	1.70–2.30	0.12–0.19	0.10–0.13	
Bolboschoenus maritimus	1.36–1.79	0.28–0.35	1.40–2.69	0.08–0.63	0.10–0.17	0.16–0.66
	Amounts of nutrients in g per m^2 stand area					
Phragmites communis	13.7–40.9	1.4– 5.3	7.8–37.4	0.2– 6.3	0.7–2.8	0.6–0.9
T. angustifolia	24.5–46.7	4.5– 6.5	20.0–66.9	3.1–29.5	3.5–8.4	
T. latifolia	50.9	7.7	52.0	7.45	6.32	
Sparganium erectum	23.9–28.7	4.0– 7.4	43.0–67.9	1.9–17.5	2.0–4.5	
Acorus calamus	14.9–21.3	2.1– 3.3	22.4–26.7	2.1–10.5	1.3–1.8	
Scirpus lacustris	17.6–52.9	3.8–11.1	19.6–57.3	0.8– 9.5	1.1–6.3	
Bolboschoenus maritimus	8.4–11.8	1.4– 2.4	7.1–16.6	0.4– 5.5	0.6–1.4	
Glyceria maxima	6.6–49.1	0.9– 8.5	10.2–61.8	0.7– 5.3	0.5–3.5	

Examples from the Opatovický fishpond:

In its highly eutrophic accumulation biotopes influenced by waste-water inflow, the nutrient uptake by the helophytes is always higher than in the poorer biotopes; both the biomass and the percentage nutrient contents in it are higher as well. Figures 7–10 show these characteristics in relation to the nutrient contents in the rhizosphere.

The helophyte species vary in their selective absorption capacity. (Table 6; see also Table 1 in Sect. 3.4.1.) Some species such as *Sparganium erectum*, *Acorus calamus* and *Glyceria maxima* are highly plastic in their nutrient uptake and absorb high amounts of nutrients from the fertile sapropels, like many submerged or floating plants. *Glyceria maxima* (Fig. 8) displays the highest plasticity in nutrient uptake as related to availability. The high absorption capacity for potassium in South Bohemian soils is impressively presented in Figures 8 and 9 for *Glyceria maxima* and *Typha angustifolia*. Other helophytes such as *Phragmites* (Fig. 7) and *Schoenoplectus lacustris* possess a smaller absorption plasticity even in highly eutrophicated biotopes. The South Bohemian acidophilous subspecies *B. maritimus* (L) Palla ssp. *maritimus* Hejný, originally occurring in oligotrophic habitats on sandy soils, prefers either mineral or organomineral soils where sapropel accumulation on the bottom is poor. Its mineral uptake is limited even in eutrophicated habitats (see Fig. 10). Concentration factors, as calculated for submerged aquatic plants and algae by Forsberg (1960), Caines (1965), Boyd (1968, 1969) and others provide a better quantitative understanding of the absorption

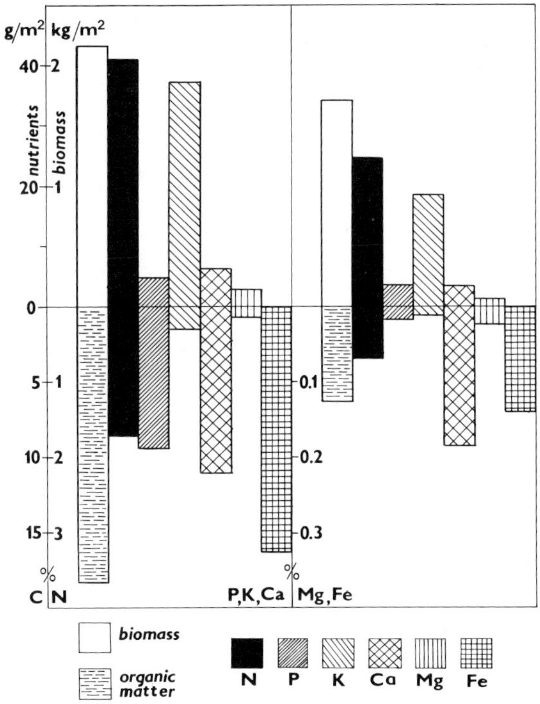

Fig.7

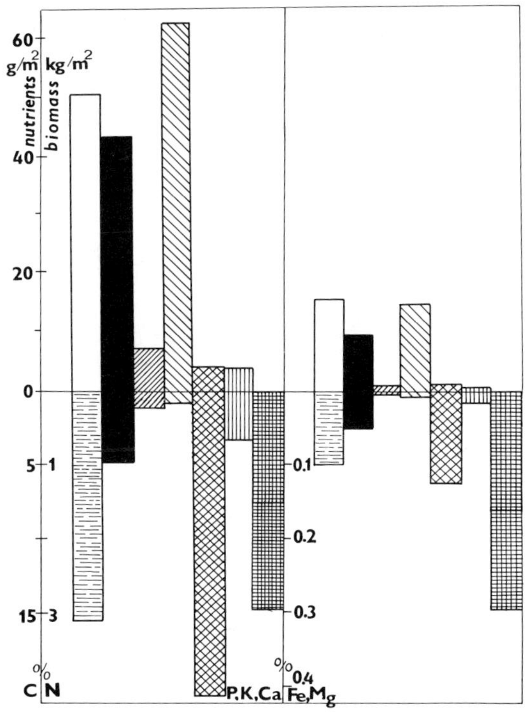

Fig.8

Fig. 7. Aboveground and underground biomass of *Phragmites communis* (in kg per m^2), and nutrient contents in aerial parts as related to percentage organic carbon (*C*) and nutrient contents in soil in accumulation *(left)* and erosion *(right)* biotopes

Fig. 8. *Glyceria maxima*. Details as in Figure 7

capacity. A concentration factor relates the concentration of a nutrient in the plant tissues (in their fresh or dry matter) to the concentration of the nutrient in the water. In aquatic plants, the concentration factors are particularly high for certain heavy metals (Boyd, 1968, 1969). The ash content is also usually higher in submerged plants than in terrestrial plants or emergents (Straškraba, 1968).

The concentration factors are meaningful only for submerged aquatic plants absorbing their nutrients mainly from the surrounding water. For helophytes, rooted and absorbing nutrients mostly from the soil or bottom sediments, it is difficult to define the actually available nutrient pool. The difficulties of the commonly used extracting procedures are discussed in Section 2.4. The concentration factors presented in Table 6 and calculated on the basis of available soil nutrients extracted with 1% citric acid, are therefore only of relative value. But it is clear that specific differences exist in the concentration factors not only between different species, but also between individual stands of the same species according to the size and availability of the nutrient pool in the rhizosphere and interstitial water. In extremely nutrient-poor habitats, the concentration factors are much higher for all nutrients such as those of *Bolboschoenus* rooted in an infertile sandy bottom (see Table 6 and Fig. 10) than in fertile sapropels. The relatively small

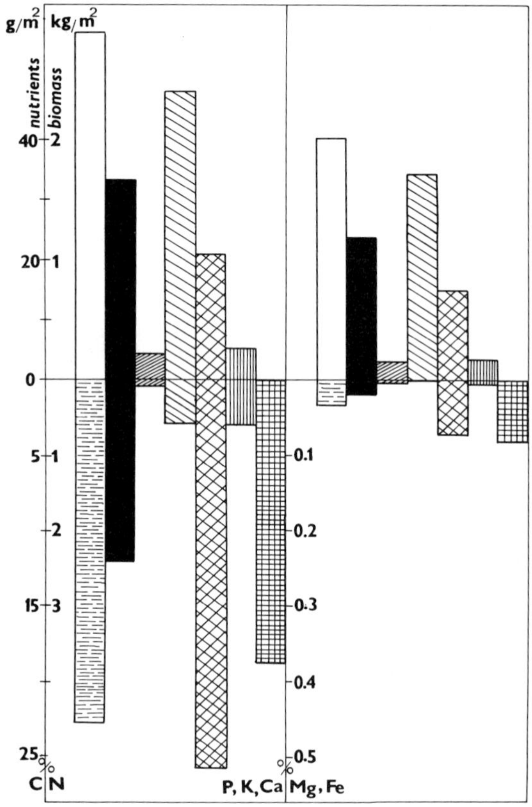

Fig. 9

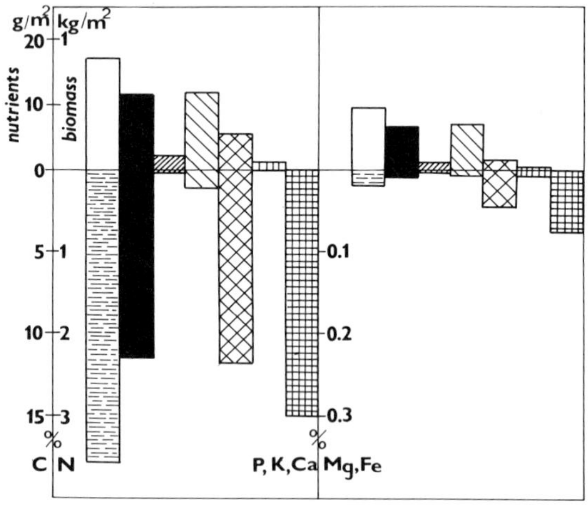

Fig. 10

Fig. 9. *Typha angustifolia*. Details as in Figure 7

Fig. 10. *Bolboschoenus maritimus*. Details as in Figure 7

concentration factors for Ca in perhaps all habitats investigated may be explained only partly by a low absorption capacity for Ca^{2+}. Another probable reason is that the "available Ca", as extracted with 1% citric acid, does not correspond with the real concentration.

The littoral helophytes occupy large areas and their production increases significantly the input of organic matter into the pond. The plant nutrients which are not translocated from the aerial parts to the underground organs at the end of the growing season are transferred to the detritus and later to the bottom sediments or are leached from the litter into the water. Most of the nutrients supplied to the pond in fertilizers appear to be deposited rapidly on the bottom and absorbed by the mud. The manuring and fertilizing of ponds thus prepares favorable conditions for "aquatic culture" of littoral helophytes, which the pond managers regard as undesirable weeds.

The originally oligo- or dystrophic but nowadays fertilized and manured water of the South Bohemian ponds, in comparison with different lakes of Europe (Bernatowicz, 1969; Björk, 1967), contains strikingly higher concentrations of the

Table 6. Nutrient concentrations in shoot biomass of dominant helophytes in Opatovický fishpond (% in dry weight) and concentration factors, C (plant: soil)

Stands	Date	N	C	P	C	K	C	Ca	C	Mg	C	Na	C
Phragmites communis accumulation biotope	15 July	1.91	1.12	0.22	1.1	1.74	49.7	0.29	1.30	0.11	6.47	0.033	3.30
Phragmites communis erosion biotope	15 July	1.53	2.18	0.22	11.5	1.20	109.1	0.20	1.08	0.098	4.08	0.053	6.62
Typha angustifolia accumulation biotope	26 July	1.15	0.54	0.16	17.7	1.65	26.6	0.73	1.70	0.184	4.08	—	—
Typha latifolia accumulation biotope	20 July	2.20	1.0	0.35	3.5	2.53	120.4	0.28	0.91	0.143	4.08	—	—
Schoenoplectus lacustris erosion biotope	26 July	1.05	1.33	0.25	83.3	1.36	33.1	0.20	0.88	0.132	0.80	—	—
Bolboschoenus maritimus erosion biotope	4 July	1.36	19.43	0.28	93.3	1.40	350.0	0.63	14.3	0.161	40.2	0.450	150.0
Glyceria maxima accumulation biotope	12 July	1.04	1.34	0.26	10.8	2.30	135.3	0.13	0.33	0.129	1.95	—	—
Glyceria maxima erosion biotope	12 July	1.32	2.64	0.20	33.3	1.90	172.7	0.15	1.22	0.112	6.58	—	—
Sparganium erectum accumulation biotope	22 July	1.69	0.65	0.32	13.3	3.60	120.0	0.61	1.62	0.228	7.12	0.409	22.7

principal nutrients (see Sect. 2.3). Björk (1967), demonstrating the role of perennial rooted helophytes in a limnic ecosystem, has proved that an occasional addition of nutrients, such as a short-time supply of sewage to the surface soil layer, may have a prolonged effect on helophyte production. The pool of nutrients maintained in the perennial plant system can be enriched for a long time from the rhizosphere. In this way, the reed belts of fertilized fishponds amplify the eutrophication of the water body. The nutrients accumulated annually in the plant biomass in the Opatovický fishpond, represent hundreds of kilograms of pure elements.

The heavily fertilized pond water and bottom can be regarded as the principal reason for the high production of the reedswamp communities and for the progressive eutrophication of both the pelagial and littoral biotopes. An illustrative example of two clones of *Phragmites communis* growing in an erosion and an accumulation biotope (V and S types) is described by Dykyjová and Hradecká (1973, 1976; see also Sects. 3.1.1 and 2.4).

In special comparative measurements of production in *Phragmites* communities (see Sect. 3.1.7) the biomass of reed stand occurring in the littorals of a number of South Bohemian and South Moravian ponds was related to the biotope fertility. Relatively nutrient-poorest is the erosion zone of the largest South Bohemian Rožmberk fishpond.

3.4.3 Verification of Field Data by Experiments in Hydroponic Cultures

Numerous estimates of biomass, production, and nutrient content in helophyte communities have proved their manifold connections with the availability of nutrients. It was therefore desirable to examine these connections in experimental hydroponic cultures in a defined nutrient medium.

In the experiments by Véber and Dykyjová (1973, and unpublished results of 1971–1973), the same hydroponics as described in Section 3.1.4 were used for preliminary assays in which the influence of graduated concentrations of the nutrient solution was followed. In the season of 1970, genetically uniform one-year-old *Phragmites* seedlings and thick rhizome cuttings of *Typha latifolia*, each bearing a single lateral bud, were planted out in six cultivation units. One cascade unit corresponded with one experimental treatment in threefold replicate for either species. The original nutrient solution (see Sect. 3.1.4) was applied in increasing concentrations of 50, 100, 150, and 200%. Tap water was used as control. The level of the nutrient solution, 20 cm above the substrate, simulated the flooding in a shallow littoral biotope. The solution was circulated forcedly by centrifugal pumps twice a week through each cascade unit, the original nutrient solution level was maintained by adding water in order to compensate for water losses due to evapotranspiration, and the nutrient solution was exchanged completely every two to three weeks.

With regard to the long vegetation cycle of the perennial helophytes investigated, the experiments had been planned for a period of three years; it was therefore possible to estimate the net production of underground organs by har-

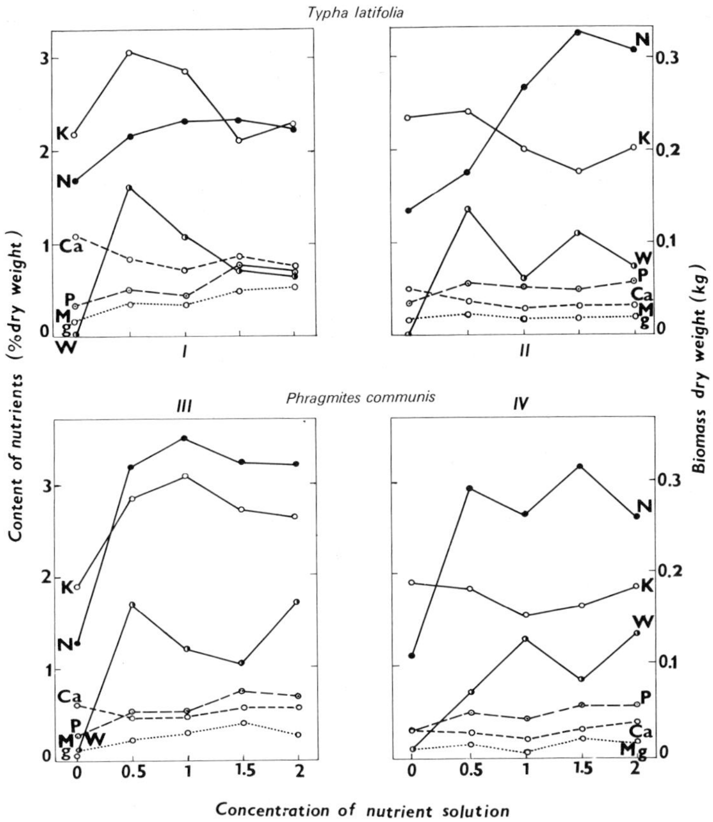

Fig. 11. Biomass (dry weight), *W*, and percentage contents of nutrients in shoots (I, III), and in the underground organs (II, IV) of *Typha latifolia* (*above*) and *Phragmites communis* (*below*), cultivated in the first season (1970) in increasing concentrations of nutrient solution. O, tap water; 0.5 to 2 = 50% to 200% concentration of initial nutrient solution (Véber, 1973)

vesting only a single complete polycormone in each treatment per season. Only the shoot biomass could be harvested from all three replicates in the first season and from two replicates in the next season.

The seedlings and cuttings were planted out in mid-April, one in each cultivation tank in order to enable estimates of the total seasonal net production to be made in a single polycormone per tank area (1 m^2). The results of the two-years experiments were instructive:

Typha latifolia and *Phragmites* growing in tap water, supplied only with the nutrient amounts contained in the water and in the reserve organs, were markedly retarded in their growth. Their offshooting or tillering ceased quite early, their stunted shoots were short and biomass was very small. The premature flowering of many shoots in both *Phragmites* and *Typha* had evidently been provoked by the extremely nutrient-poor environment. Surprisingly, both *Typha latifolia* and *Phragmites* were most productive and developed the largest leaf area in the 50% nutrient solution. *Typha latifolia* also produced the greatest number of offshoots here. However, none of the shoots flowered in *Phragmites* and only one to three

Table 7. Maximum values of nutrients accumulated in the shoots, underground organs, and total biomass (in g per m^2 of cultivation area) in helophytes growing in hydroponic cultures under forced mineral nutrition (optimum concentration of nutrient solution for maximum net production)

Species and concentration of nutrient solution	Organs	Biomass dry weight g m^{-2}	N g m^{-2}	P g m^{-2}	K g m^{-2}	Ca g m^{-2}	Mg g m^{-2}
Typha latifolia,	shoots	4,371	53.5	11.2	59.2	25.7	28.5
3 yr old	underg.	3,860	102.7	26.3	59.2	11.6	11.1
100% solution	total	8,231	156.2	37.5	118.4	37.3	39.6
Bolboschoenus	shoots	985	19.2	3.9	36.8	3.8	4.0
maritimus, 1 yr old	underg.	1,475	38.2	9.7	21.3	1.47	2.1
50% solution	total	2,630	57.4	13.6	58.1	5.30	6.1
Acorus calamus,	shoots	908	24.8	0.5	23.0	4.8	4.6
1 yr old	underg.	466	10.9	3.2	10.5	0.7	0.7
50% solution	total	1,374	35.7	3.7	38.5	5.5	5.3

per replicate did in *Typha*. Completely sterile shoots of both species also developed in all the higher concentrations. The biomass production, leaf area and offshooting declined with increasing concentration of nutrient solution. *Typha latifolia* was affected particularly severely. It seems, therefore, that the full (100%) concentration of the nutrient solution used in the preceding cultivation experiments (described in Sect. 3.1.4) was slightly higher than the optimum. The reaction of *Phragmites* was not quite so straightforward. It attained a secondary peak of biomass production and leaf area in the highest concentration of nutrient solution. This anomalous reaction is difficult to explain, the more so since in the following experimental season all the nutrient-solution concentrations higher than 100% were beyond the optimum for both species.

The effects of the nutrient-solution concentrations on growth are also reflected in the nutrient uptake by the shoots and underground organs (Fig. 11). In the shoots, all nutrient concentrations except that of nitrogen decline with the increasing nutrient-solution concentrations higher than 100%. In the rhizomes, the accumulation of potassium appears suppressed by an enhanced uptake of nitrogen.

With regard to the anomalies in *Phragmites*, Véber (unpublished) extended the scale of concentrations of the original nutrient solution to 0.25% and 300% in the subsequent growing season (1971). Empty cultivation tanks, from which whole polycormones had been harvested in the preceding season, were planted either with young cuttings of *Acorus calamus* or with tubers of *Bolboschoenus maritimus*. *Typha latifolia*, in the second year of its polycormone development, was again most productive in the 50% nutrient solution. The same result was obtained with the young *Bolboschoenus maritimus* and *Acorus calamus* plants. In all higher nutrient concentrations and in all species tested, the biomass of both shoots and underground organs was reduced and the highest (300%) concentration appeared highly toxic, with typical symptoms such as stunted shoots and chlorotic leaves in *Phragmites*, dying off of the leaves and whole shoots in *Acorus*, *Typha*, and *Bol-*

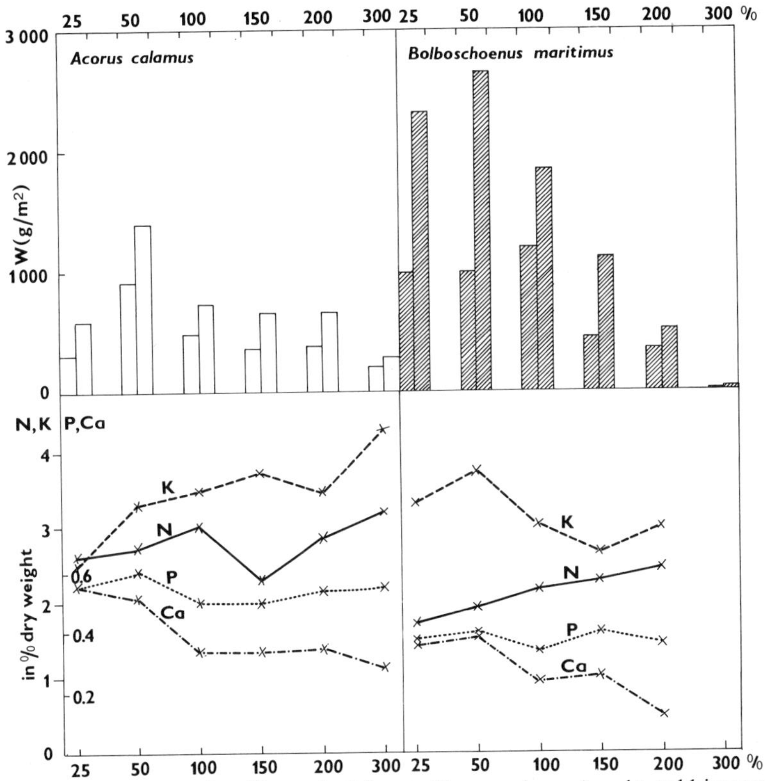

Fig. 12a. *Above*: Net production (dry wt.) of shoots *(lower columns)* and total biomass *(higher columns)* of *Acorus calamus* and *Bolboschoenus maritimus* grown in hydroponic cultures with increasing concentrations of nutrient solutions (abscissa). *Below*: Content of macronutrients (in % dry wt.) in shoots of the same species

Fig. 12b. Hydroponic cultures of three-year-old polycormones of *Typha latifolia* and of one-year-old *Phragmites communis* in outdoor cascade units (season 1972) with increasing concentrations of the nutrient solution. From left to right: 10, 25, 50, 100, 150, and 200% nutrient solution. (Photo: D. Dykyjová)

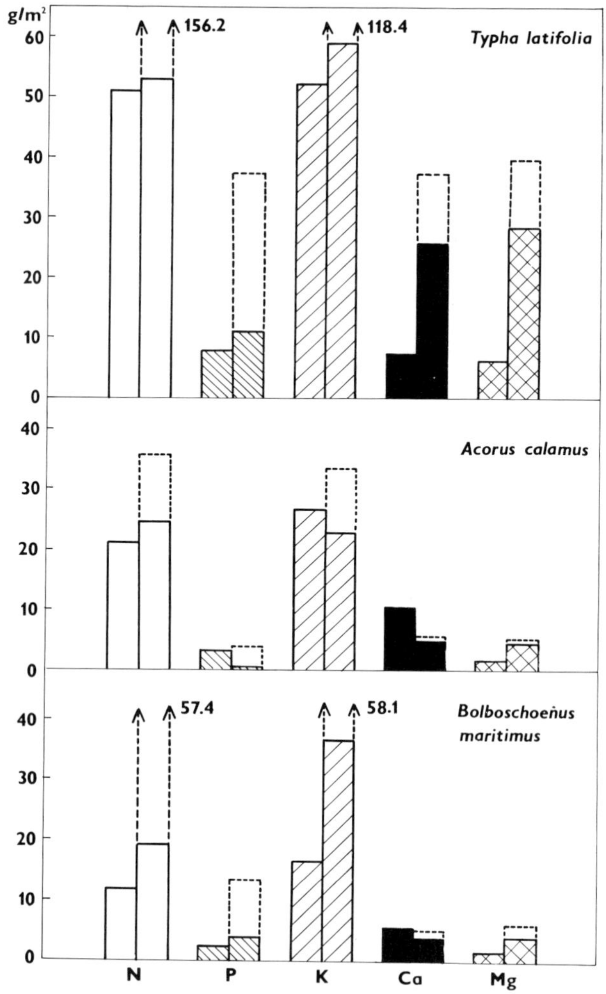

Fig. 13. Uptake of nutrients (in g per stand biomass from 1 m^2) by shoots of *Typha latifolia*, *Acorus calamus*, and *Bolboschoenus maritimus* cultivated with an experimentally forced mineral nutrition (*right columns*), as compared with maximum nutrient uptake in the same species growing in Opatovický fishpond *(left columns)*. *Dashed lines:* total nutrient uptake in stand per 1 m^2, including underground organs (only in hydroponic cultures, *right columns*)

boschoenus. In the damaged plants, the nutrient uptake increased in many cases out of proportion with the declining production and dying of the plant. The biomass analyses of the preceding season demonstrated convincingly that vigorous and more productive plants accumulate more nutrients but resist an excess accumulation of nutrients in higher concentrations. But this resistance was lost when the plants were grown in toxic concentrations manifested by growth anomalies in the season of 1971 (Fig. 12b). The concentrations of the 200% and 300% nutrient solutions correspond, respectively, with a 4% and 6% total nutrient content in the medium, which never occurs in natural habitats. It is particularly the extremely high content of micronutrients (compare the original nutrient solution in Sect. 3.1.4) that may be responsible for the symptoms of toxicity.

In the two subsequent seasons of 1972 and 1973, the 300% nutrient solution was left out and an additional low concentration of 10% was tested (Fig. 12b). Only the two- to three-year-old polycormones, evidently better adapted to higher nutrient concentrations, attained the highest biomass in the original 100% concentration. Table 7 presents the net production in *Bolboschoenus maritimus* and *Acorus calamus:* both species achieved the highest biomass of both above- and underground organs in the 50% nutrient solution.

Another experiment was made by Rejmánková in 1974 with duckweeds (see Sect. 3.5). Velásquez (1975) experimented with *Eleocharis acicularis* grown in trays filled with a mixture of sand and fishpond mud. The nutrient uptake by the fertilized populations was markedly increased in comparison with the control populations (see Sect. 3.1.6.2).

The experiments with duckweeds or *Eleocharis acicularis* have also proved the natural capacity of these plants to accumulate nutrients from a nutrient-rich habitat.

The high values of net production attained by some helophytes at optimum concentrations of the nutrient solution are related to a high accumulation of nutrients both in the shoots and total biomass. In Figure 13 these values surpass the maximum uptake under natural conditions (as compared with the same species in the Opatovický fishpond).

References see pp. 285—291.

3.5 Growth, Production and Nutrient Uptake of Duckweeds in Fishponds and in Experimental Cultures

E. REJMÁNKOVÁ

In his review of the literature about the *Lemnaceae*, Hillman (1961) characterizes duckweeds as "inconspicuous plants of no economic importance", being suitable only as "experimental organisms for morphogenetic, physiological and biochemical research". The situation has now changed markedly; these aquatic plants have been studied thoroughly as: (1) potential means of reducing inorganic nutrients and organic pollution in waste waters (Culley and Epps, 1973); (2) source of animal feed (Schulz, 1962; Taubaev et al., 1971; Culley and Epps, 1973); (3) biological indicators of water quality (Lange and Zon, 1973); (4) model material in plant-population studies (Clatworthy and Harper, 1962). Duckweeds have proved their high production capacity as well as rate of nutrient absorption.

Five species of duckweeds occur in Czechoslovakia; three of them, *Spirodela polyrrhiza*, *Lemna minor*, and mainly *Lemna gibba* have been investigated. *Lemna minor* is restricted to nutrient-poor and cool waters, while *Lemna gibba* predominates (alone or together with *Spirodela polyrrhiza*) in rather eutrophic and warm habitats (Rejmánková, 1975 b). The communities of duckweeds, their habitats and distribution are characterized in Section 1.1. In littorals of large fishponds, duck-

Table 1. Values of seasonal maximum biomass and relative growth rate (RGR) of a *Lemnetum gibbae* community recorded at Opatovický and Nesyt fishponds during 1971–1973

Site		$W_{max.}$	Interval	RGR $g\,g^{-1}\,d^{-1}$		
				min.	mean	max.
Opatovický fishpond	lagoon	20	3 May–28 Aug.	0.0105	0.015	0.183
	Phragmitetum communis	7	3 May–1 Aug.	0.005	0.078	0.143
Nesyt fishpond	lagoon	50–150	15 May–26 June		0.295	
	Typhetum angustifoliae loose stand	50– 90	15 May–30 July	0.113	0.195	0.279
	Typhetum angustifoliae dense stand	9– 10	24 May–26 July	0.006	0.126	0.241
	Bolboschoenetum maritimi	80	12 May–26 July	0.155	0.215	0.288
	Phragmitetum communis		15 May–30 July	0.067	0.100	0.176
		12– 80	24 May–26 June	0.035	0.081	0.118

$W_{max.}$, seasonal maximum biomass, dry weight, g m^{-2}.

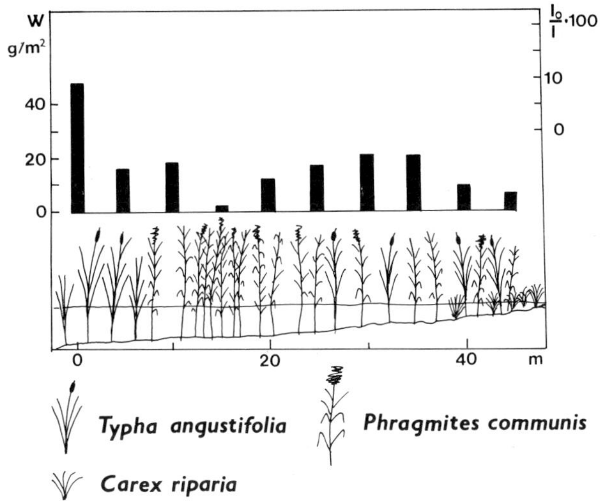

Fig. 1. Biomass of the duckweed synusium (*columns*) along a transect from open water to shore on 28 June 1971. Sketches: structure of the reed stands. (After Rejmánková, 1975a)

weeds form synusia in helophyte communities dominated by *Phragmites communis*, *Typha* spp., *Bolboschoenus maritimus* and other species, or fill the open spaces, "lagoons", within these stands. The spatial structure of the duckweed communities and hence their assimilatory surface are incomparably simpler than in emergent macrophyte communities. Yet duckweeds are highly productive owing to their rapid growth within relatively short life cycles, resembling in this respect unicellular algae. One duckweed frond (leaflet), which is usually regarded as one operating individual, lives for about four weeks, produces a certain amount of daughter fronds, dies and undergoes decay within approximately two weeks; thus the life span of one generation of fronds is about six weeks.

The methods of growth analysis were employed in the studies on production. The following characteristics were used: biomass (dry weight) and relative growth rate (RGR), as described by Květ et al. (1971). For detailed descriptions and discussions of the methods see Rejmánková (1971, 1973a,c). The production of duckweeds was assessed: (1) in the littoral of the large Nesyt and Opatovický fishponds; (2) in four small heavily eutrophicated village ponds in South Bohemia; and (3) in outdoor hydroponic cultures at Třeboň; for details of the hydroponic equipment see Adamcová-Bínová (1968) and Rejmánková (1971).

1. Growth and production of duckweeds in the littorals of the Nesyt and Opatovický fishponds. Table 1 presents the biomass and RGR values of the *Lemna gibba* synusium occurring in different helophyte communities: *Phragmitetum communis*, loose and dense *Typhetum angustifoliae*, in stands of *Bolboschoenus maritimus*, and in lagoons within the reed-belts of the two ponds. The duckweed biomass was smallest in dense stands of both *Phragmites communis* and *Typha angustifolia*; duckweeds covered about a quarter of water surface in these stands. The growth of duckweeds is limited here by very low irradiance at water surface. At Nesyt the duckweed biomass was much higher in loose

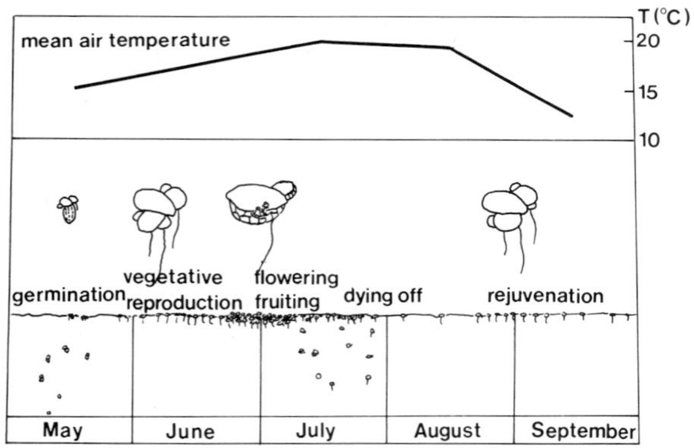

Fig. 2. Scheme of phenological development of duckweeds (*Lemma gibba*) during main growing season (May to September). *Upper part*; Monthly mean air temperatures at Lednice in 1971 and 1972. (After Rejmánková, 1975a)

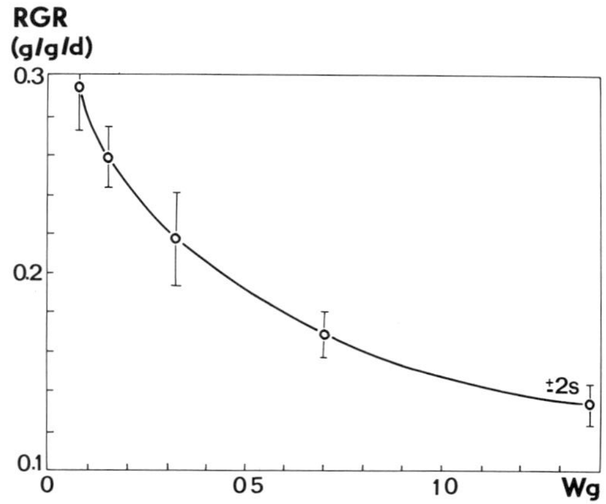

Fig. 3. RGR values plotted against initial frond density expressed in terms of dry weight per 625 cm^2 water-surface area. Třeboň, hydroponic cultivation, 1971. (After Rejmánková, 1973c)

Table 2. Contents of mineral nutrients and K/Ca ratio in natural mixed populations of a *Spirodelo-Lemnetum gibbae* community in a small village pond

Date	N		P		K		Ca		Mg		Na		Ash	K/Ca
	%	$g\,m^{-2}$	%	$g\,m^{-2}$	%	$g\,m^{-2}$	%	$g\,m^{-2}$	%	$g\,m^{-2}$	%	$g\,m^{-2}$	% (dry wt.)	
22 July 1974	4.16	4.94	1.14	1.35	2.80	3.32	1.75	2.08	0.49	0.59	0.44	0.52	14.5	1.6
19 Aug.	3.29	4.94	0.96	1.44	2.57	3.87	2.70	4.06	0.53	0.80	0.40	0.61	14.5	0.9
17 Sept.	2.97	4.88	0.82	1.34	2.55	4.18	2.35	3.85	0.41	0.67	0.31	0.51	15.1	1.1
1 Oct.	3.01	5.18	0.62	1.07	2.46	4.24	2.35	4.04	0.34	0.58	0.25	0.44	14.5	1.0

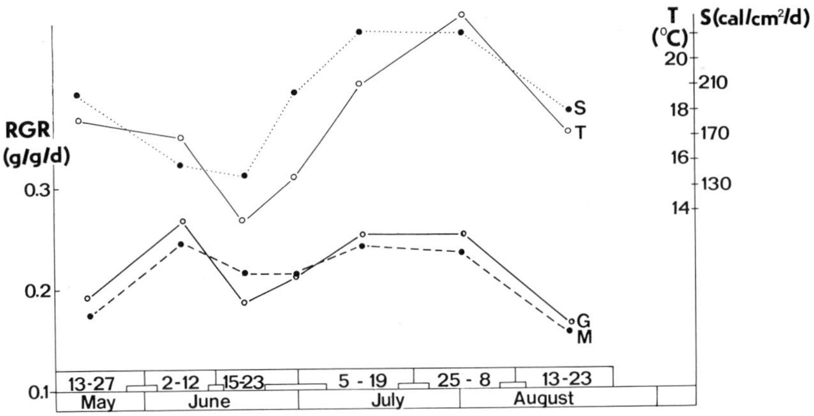

Fig. 4. Comparison of changes in RGR of *Lemna gibba* (*G*) und *Lemna minor* (*M*) with changes of mean water temperature (*T*) and daily PhAR irradiance (*S*)

Typha, namely in 1972, when the weather conditions were very suitable for rapid vegetative reproduction of duckweeds; for the sequence of growth phases in *Lemma gibba* see Fig. 1. A similar situation was observed within the *Bolboschoenus maritimus* stands. During 1972, the duckweeds growing in the lagoons were competitively excluded by filamentous algae, whereas in 1971 and 1973 a thick mass of *Lemna gibba* covered the lagoons and its biomass was very high. The remarkable periodicity in the seasonal development of duckweeds is typical of the Nesyt fishpond (Fig. 2); a similar periodicity was found neither in small village ponds nor in fry or fingerling ponds. The differences in the RGR of duckweeds occurring in different helophyte stands, although less pronounced, are in agreement with the data on biomass. The RGR values estimated in the experimental enclosures within the reed-belt correspond with those occurring outside the enclosures only as long as the surrounding duckweeds are not overcrowded (Rejmánková, 1973c). The growth of duckweed populations is negatively density-dependent after they have attained a 100% cover (Fig. 3); the RGR is therefore somewhat reduced in overcrowded duckweed stands. The data on RGR presented here represent only a measure of the potential growth ability of duckweeds in different parts of the pond littoral. In Table 1, only the positive RGR values are given; later on during the season, RGR fell to zero, with the growth of new fronds being equal to the rate of decay of old fronds, or was even negative. The most probable principal reason for the poor growth of duckweeds at the peak of the vegetation season was their overheating in the lagoons and the drastically reduced irradiance within dense stands of helophytes.

2. Growth of duckweeds from four South Bohemian village ponds located southeast of Třeboň. All four ponds are nutrient-rich (see Table 3). In all ponds, the duckweed populations were formed by *Lemna gibba* and *Spirodela polyrrhiza* (association *Lemno-Spirodeletum* Koch 1954, see Sect. 1.2). The growth of the duckweeds started at the beginning of April by germination of *Spirodela* turions and by vegetative reproduction of hibernating fronds of *Lemna*; as a result of their growth, one to two layers of the plants covered the water surface at the

Fig. 5. (a) Mixed populations of *Spirodela polyrhiza* and *Lemna gibba* from a village pond, cultivated in nutrient solution, scale: 10 cm; (b) Boxes for experimental cultivation of duckweeds. (Photo: E. Rejmánková)

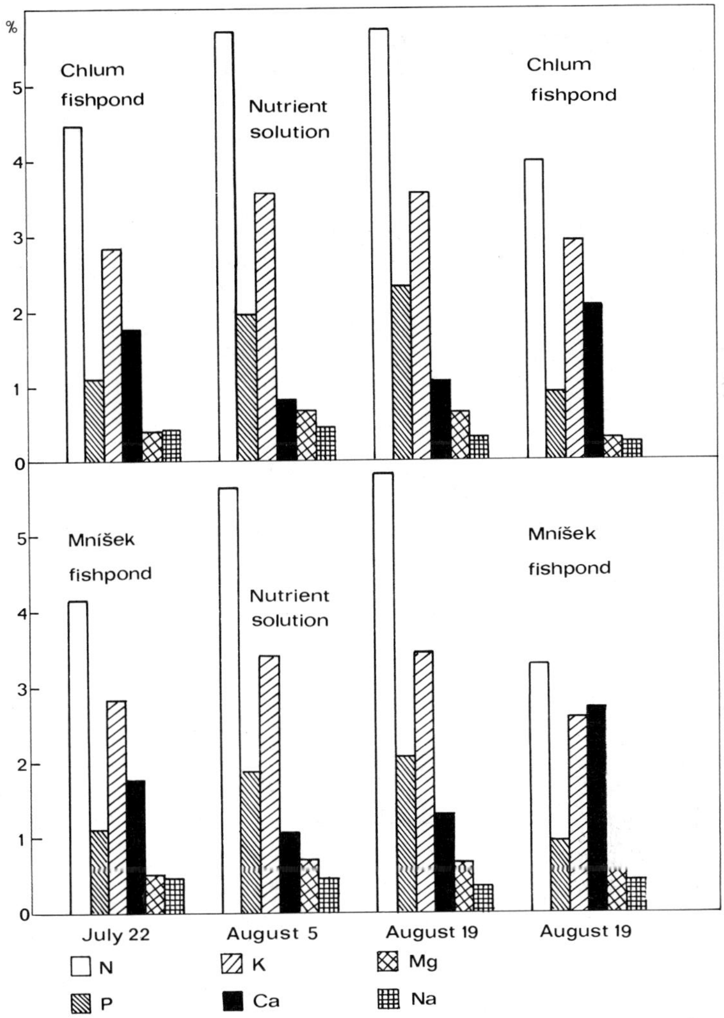

Fig. 6. Nutrient contents (in % dry weight) in biomass of a mixed community of *Lemna gibba* and *Spirodela polyrhiza* sampled in two village fishponds *(left)* as compared with enhanced nutrient uptake after transplantation in nutrient solution *(center)*. *Right*, nutrient content in population in original fishpond biotopes at end of experiment

beginning of June, and the ratio of *Spirodela: Lemna* was about 1:3 at that time. High biomass values were recorded in the community from June to mid-October; dead fronds began to prevail at the end of September.

Mineral analyses of the dry plant material (see Table 2) showed its high nitrogen contents. Our data are in close agreement with those by Culley and Epps (1973) for *Spirodela oligorhiza* growing in an animal waste lagoon.

3. Growth of duckweeds in outdoor hydroponic cultures. Figure 4 shows the RGR values for *Lemna gibba* and *Lemna minor* grown in hydroponic cultivation equipment at Třeboň during the growing season of 1971. The seasonal course of

Table 3. Contents of mineral nutrients (% dry weight) in a natural mixed *Spirodelo-Lemnetum gibbae* community in a village fishpond in comparison with contents in samples of the same populations after transfer to a nutrient solution

Date	Biotope	N	P	K	Ca	Mg	Na
22 July	Mníšek fishpond	4.16	1.14	2.80	1.75	0.49	0.44
5 Aug.	Nutrient solution	5.60	1.85	3.40	1.05	0.69	0.44
19 Aug.	Nutrient solution	5.77	2.05	3.40	1.28	0.67	0.31
19 Aug.	Mníšek fishpond	3.29	0.96	2.57	2.70	0.53	0.40

the RGR roughly follows the course of both the water temperature and the incoming solar radiation. A difference was observed at the beginning of the growing season when the RGR increased in both *Lemna* species, despite the decrease in temperature and radiation. This is probably due to the inherent seasonal periodicity of duckweed growth. During the 1974 season, the RGR values were estimated in mixtures of *Lemna gibba* and *Spirodela polyrrhiza* (the stock cultures originated from the village ponds mentioned above) at about seven-day intervals from July to October (Fig. 5). The observed changes in the *Lemna: Spirodela* ratio are given elsewhere (Rejmánková, 1976). The RGR values were about half those of 1971, and varied far less, except at the end of the season. The lower RGR values of 1974 (range from 0.061—0.128) may have been caused by lower temperatures and smaller radiation input (see Section 2.1.1); also, the plants used for the experiments may have possessed a different production ability. The presence of two species, each with a different temperature optimum for growth, seems to be responsible for the relatively less variable RGR during most of the growing season. Figure 6 and Table 3 show the rapid reaction of the nutrient uptake by the plants after their transplantation into nutrient solution. Their nutrient content, except that of Ca and Na, became much higher than that in the duckweeds from their original habitats both at the beginning (22 July) and end (19 Aug.) of the experiment.

Duckweeds, the morphologically simplest green vascular plants, represent a highly suitable material for experimental study of quite a few important ecosystem processes. The available data on the ecophysiology of duckweeds may serve for the first orientation in the study of their role in both natural and artificial aquatic and wetland ecosystems.

References see pp. 285—291.

References

Adamcová-Bínová, J.: PP/IBP initial level experiments in South Bohemia. Ann. Rep. Algol. Lab. Třeboň **1967**, 161–170 (1968)

Bayly, L. I., O'Neil, T. A.: Seasonal ionic fluctuations in a *Typha glauca* community. Ecology **53**, 714–719 (1972a)

Bayly, L. I., O'Neil, T. A.: Seasonal ionic fluctuations in a *Phragmites communis* community. Can. J. Bot. **50**, 2103–2109 (1972b)

Bernatowicz, S.: Macrophytes in the lake Warniak and their chemical composition. Ekol. Pol. **17**, 447–467 (1969)

Björk, D.: Ecological investigations of Phragmites communis. Studies in theoretical and applied limnology. Folia Limnol. Scand. **14** (1967)

Björk, S.: Chromosome geography and ecology of *Phragmites communis*. Södra Sveriges Fiskeriförening 1961–1962, 1–11, 1963

Blackman, G. E.: The application of the concepts of growth analysis to the assessment of productivity. In: Functioning of Terrestrial Ecosystems at the Primary Production Level. Eckardt, F. E. (ed.). Paris: UNESCO, 1968, pp. 243–259

Bornkamm, M.: Ein Jahresrhythmus des Wachstums bei *Lemna minor* L. Planta **69**, 178–186 (1966)

Boyd, C. E.: Some aspects of aquatic plant ecology. In: Reservoir Fishery Resources Symposium, Athens, Georgia, April 5–7 1967, 1968

Boyd, C. E.: Production, mineral nutrient absorption and biochemical assimilation by *Justitia americana* and *Alternanthera philoxeroides*. Arch. Hydrobiol. **66**, 139–160 (1969)

Boyd, C. E.: Production, mineral accumulation and pigment concentrations in *Typha latifolia* and *Scirpus americanus*. Ecology **51**, 285–290 (1970a)

Boyd, C. E.: Amino acid protein and caloric content of vascular aquatic macrophytes. Ecology **51**, 902–906 (1970b)

Boyd, C. E., Vickers, D.: Relationships between production, nutrient accumulation and chlorophyll synthesis in an *Eleocharis quadrangulata* population. Can. J. Bot. **49**, 833–838 (1971)

Bray, J. R., Lawrence, D. B., Pearson, L. C.: Primary production in some Minnesota terrestrial communities for 1957. Oikos **10**, 38–49 (1959)

Brzoska, W.: Energiegehalte verschiedener Organe von nivalen Sproßpflanzen im Laufe einer Vegetationsperiode. Photosynthetica **5**, 183–189 (1971)

Burian, K.: Die photosynthetische Aktivität eines *Phragmites communis*-Bestandes am Neusiedler See. Sitzungsberichte d. Österr. Akad. Wiss. Math. Naturwiss. Kl., Abt. I, **178**, 43–62 (1969)

Burian, K.: Primary production, carbon dioxide exchange and transpiration in *Phragmites communis* Trin. on the lake Neusiedler See, Austria. Hidrobiologia **12**, 203–218 (1971)

Burian, K.: Das Schilfgürtel-Ökosystem eines Steppensees. In: Ökosystemforschung. Ellenberg, H. (ed.). Berlin-Heidelberg-New York: Springer-Verlag 1973, pp. 61–78

Caines, L. A.: The phosphorus content of some aquatic macrophytes with special reference to seasonal fluctuations and application of phosphate fertilizers. Hydrobiologia **25**, 289–301 (1965)

Čatský, J.: Determination of water deficit in disks cut out from leaf blades. Biol. Plant. **2**, 76–78 (1960)

Clark, J. P., Evans, F. C.: Distance to nearest neighbour as a measure of spatial relationship in populations. Ecology **35**, 445–453 (1954)
Clatworthy, J. N., Harper, J. L.: The comparative biology of closely related species living in the same area. V. Inter- and intraspecific interference within cultures of *Lemna* spp. and *Salvinia natans*. J. Exp. Bot. **10**, 33–54 (1962)
Cottam, G., Curtis, J. T.: The use of distance measures in phytosociological sampling. Ecology **37**, 451–460 (1956)
Culley, D. D., Epps, E. A.: Potential usefulness of duckweeds in waste water treatment and animal feeds. J. Wat. Pollut. Control Fed. **45**, 337–347 (1973)
Denayer de Smet, S.: Note sur la composition minérale des graminées et plantes graminoides des tapis végétaux naturels de Belgique. Bull. Soc. R. Bot. Belg. **97**, 19–25 (1964)
Dykyjová, D.: Ecotypes and ecomorphoses of common reed, *Phragmites communis* Trin. (In Czech). Preslia **43**, 120–138 (1971 a)
Dykyjová, D.: Productivity and solar energy conversion in reedswamp stands in comparison with outdoor mass cultures of algae in the temperate zone of Central Europe. Photosynthetica **5**, 329–340 (1971 b)
Dykyjová, D.: Production, vertical structure and light profiles of incident radiation in littoral stands of reed-bed species. Hidrobiologia **12**, 361–376 (1971 c)
Dykyjová, D.: Specific differences in vertical structures and radiation profiles in the helophyte stands. In: Ecosystem Study on Wetland Biome in Czechoslovakia. Hejný, S. (ed.). Czechosl. IBP-PT-PP Rep. No. **3**, Třeboň, 1973 a, pp. 121–131
Dykyjová, D.: Content of mineral macronutrients in emergent macrophytes during their seasonal growth and decomposition. In: Ecosystem Study on Wetland Biome in Czechoslovakia. Hejný, S. (ed.). Czechosl. IBP/PT-PP Rep. No. **3**, Treboň, 1973 b; pp. 163–172
Dykyjová, D.: Accumulation of mineral elements in the biomass of reedswamp species. In: Czechosl. IBP/PT-PP Report No. 3, Třeboň, 1973 c, pp. 151–161
Dykyjová, D., Hradecká, D.: Comparative investigation on the microclimate in two reed-bed biotopes and its relation to the ecotype, productivity and trophic conditions of habitat. Pol. Arch. Hydrobiol. **20**, 111–119 (1973)
Dykyjová, D., Hradecká, D.: Production ecology of *Phragmites communis* Trin. I. Relations of two ecotypes to microclimate and nutrient conditions of habitat. Folia Geobot. phytotaxon. Bohemoslov. **11**, 23–61 (1976)
Dykyjová, D., Květ, J.: Comparison of biomass production in reedswamp communities growing in South Bohemia and South Moravia. In: Productivity of Terrestrial Ecosystems and Production Processes. Dykyjová, D. (ed.). Czechosl. IBP/PT-PP Rep. No. 1, Praha, 1970, pp. 71–79
Dykyjová, D., Ondok, J. P.: Biometry and the productive stand structure of coenoses of *Sparganium erectum* L. Preslia **45**, 19–30 (1973)
Dykyjová, D., Ondok, J. P., Hradecká, D.: Growth rate and development of the root/shoot ratio in reedswamp macrophytes in winter hydroponic cultures. Folia Geobot. Phytotaxon. Bohemoslov. **7**, 259–268 (1972)
Dykyjová, D., Ondok, J. P., Přibáň, K.: Seasonal changes in productivity and vertical structure of reed-stands (*Phragmites communis* Trin.). Photosynthetica **4**, 280–287 (1970)
Dykyjová, D., Přibil, S.: Energy content in the biomass of emergent macrophytes and their ecological efficiency. Arch. Hydrobiol. **75**, 90–108 (1975)
Dykyjová, D., Véber, K., Přibáň, K.: Productivity and root/shoot ratio of reedswamp species growing in outdoor hydroponic cultures. Folia Geobot. Phytotaxon. Bohemoslov. **6**, 233–254 (1971)
Eagles, C. F.: Changes in net assimilation rate and leaf-area ratio with time in *Dactylis glomerata* L. Ann. Bot. N.S. **35**, 63–74 (1971)
Evans, G. C.: The Quantitative Analysis of Plant Growth. Oxford: Blackwell 1972
Fiala, K.: Seasonal changes in growth of the underground organs in *Typha latifolia*. In: Productivity of Terrestrial Ecosystems. Production Processes. Dykyjová, D. (ed.). Czechosl. IBP/PT-PP Rep. No. **1**, Praha, 1970 a, pp. 99–100
Fiala, K.: Rhizome biomass and its relation to shoot biomass and stand pattern in eight clones of *Phragmites communis* Trin. In: Productivity of Terrestrial Ecosystems. Production Processes. Dykyjová, D. (ed.). Czechosl. IBP/PT-PP Rep. No. **1**, Praha, 1970 b, pp. 95–98

Fiala, K.: Comparison of seasonal changes in the growth of underground organs of *Typha latifolia* L. and *T. angustifolia* L. Hidrobiologia **12**, 235–240 (1971 a)
Fiala, K.: Seasonal changes in the growth of clones of *Typha latifolia* L. in natural conditions. Folia Geobot. Phytotaxon. Bohemoslov. **6**, 255–270 (1971 b)
Fiala, K.: Growth and production of underground organs in *Phragmites* (Ms, in Czech). Thesis, Inst. of Botany, Czechosl. Ac. Sci., Brno, 1973c
Fiala, K.: Underground biomass and estimation of annual rhizome increments in two polycormones of *Phragmites communis* Trin. In: Littoral of the Nesyt Fishpond. Květ, J. (ed.). Studie ČSAV **15**, Praha, 1973d, pp. 135–137
Fiala, K.: Seasonal changes in the growth and total carbohydrate content in the underground organs of *Phragmites communis* Trin. In: Ecosystem Study on Wetland Biome in Czechoslovakia. Hejný, S. (ed.). Czechosl. IBP/PT-PP Rep. No. **3**, Třeboň, 1973b, pp. 107–110
Fiala, K.: Growth and production of underground organs of *Typha angustifolia* L., *Typha latifolia* L., and *Phragmites communis* Trin. Pol. Arch. Hydrobiol. **20**, 59–66 (1973a)
Fiala, K.: Underground organs of *Phragmites communis* (Cav.) Trin. ex Steudel, their growth, biomass, and net production. Folia Geobot. Phytotaxon. Bohemoslov. **11**, 225–259 (1976)
Fiala, K.: Underground organs of *Typha angustifolia* L. and *Typha latifolia* L., their growth, propagation and production. Acta Sci. Nat. (Brno) (In press 1978)
Fiala, K., Dykyjová, D., Květ, J., Svoboda, J.: Methods of assessing rhizome and root production in reed-bed stands. In: Methods of Productivity Studies in Root Systems and Rhizosphere Organisms. Leningrad: Nauka 1968, pp. 36–47
Fiala, K., Květ, J.: Dynamic balance between plant species in South Moravian reedswamps. In: The Scientific Management of Animal and Plant Communities for Conservation. Duffey, E., Watt, A. S. (eds.). Oxford: Blackwell Sci. Publ. 1971, pp. 241–269
Forsberg, G.: Subaquatic macrovegetation in Öslysjön, Djursholm. Oikos **11**, 183–199 (1960)
Gloser, J.: Ecological study of photosynthesis in dominant grass species of alluvial meadows near Lanžhot. (In Czech. Ms.) Thesis, Inst. of Botany, Czechosl. Acad. Sci., Brno, 1973
Gloser, J.: CO_2-exchange characteristics in *Phragmites communis* derived from measurements in situ. Photosynthetica **11**, 139—147 (1977)
Grabowski, A.: The biomass, organic matter contents and calorific values of macrophytes in the lakes of the Szeszuga drainage area. Pol. Arch. Hydrobiol. **20**, 269–282 (1973)
Greig-Smith, P.: Quantitative Plant Ecology. London: Butterworth 1964
Hájková, A., Květ, J.: Analysis of primary productivity in two types of inundated meadows. In: Productivity of Terrestrial Ecosystems. Production Processes. Dykyjová, D. (ed.). Czechosl. IBP/PT-PP Rep. No. **1**, Praha, 1970, pp. 47–50
Hejný, S.: Ökologische Charakteristik der Wasser- und Sumpfpflanzen in den Slowakischen Tiefebenen (Donau- und Theißgebiet). Vyd. SAV, Bratislava, 1960, pp. 492
Hejný, S.: *Coleanthus subtilis* (Tratt.) Seidl in der Tschechoslovakei. Folia Geobot. Phytotaxon. Bohemoslov. **4**, 345–399 (1969)
Hillman, W. S.: The *Lemnaceae*, or duckweeds. Bot. Rev. **27**, 221–287 (1961)
Horak, O., Kinzel, H.: Typen des Mineralstoffwechsels bei den höheren Pflanzen. Österr. Bot. Z. **119**, 475–495 (1971)
Hradecká, D.: Common reed (*Phragmites communis* Trin.) in South Bohemia, South Moravia and South Slovakia. Morphology of the inflorescences and flower wraps. In: Ecosystem study on Wetland Biome in Czechoslovakia. Hejný, S. (ed.) Czechosl. IBP/PT-PP Rep. No. **3**, Třeboň, 1973, pp. 47–53
Hradecká, D., Květ, J.: Morphological and production characteristics of three clones of *Phragmites communis* Trin. from the Nesyt fishpond. In: Květ, J. (ed.) Littoral of the Nesyt fishpond. Studie ČSAV **15**, Praha, 1973, pp. 97—101
Hughes, M. K.: Seasonsal calorific values from a deciduous woodland in England. Ecology **52**, 923–926 (1971)
Husák, Š.: Productivity and Structure of Intact and Cut Invasion Stands of *Phragmites communis* Trin. and *Typha angustifolia* L. at the Nesyt Fishpond (In Czech). Thesis, Bot. Dept. Fac. Sci. Purkyně Univ., Brno, 1971
Husák, Š., Hejný, S.: Marginal plant communities of the Nesyt fishpond (South Moravia). Pol. Arch. Hydrobiol. **20**, 461–467 (1973)
Husák, Š., Květ, J.: Zonation of higher-plant shoot biomass in the reed-belt of the Nesyt fishpond. Studie ČSAV **15**, Praha, 1973, pp. 73–81

Idso, S. B.: A theoretical framework for the photosynthetic modelling of plant communities. Advan. Frontiers Plant Sci. **23**, 91–118 (1969)

Iljin, W. S.: Zusammensetzung der Salze in den Pflanzen auf verschiedenen Standorten. Kalkpflanzen. Beih. Bot. Zentralbl. **50**, 95–137 (1932)

Iljin, W. S.: Zur Physiologie der kalkfeindlichen Pflanzen. Beih. Bot. Zentralbl. **54**, 569–568 (1936)

Imhof, G., Burian, K.: Energy flow studies in a wetland ecosystem. Special Publication, Austrian Acad. Sci. for International Biological Program (IBP). Vienna, 1972

Jakrlová, J.: Primary production and plant chemical composition in flood-plain meadows. Acta Sci. Nat. (Brno) **9** (9), 1–52 (1975)

Kaul, V., Vass, K. K.: Production studies of some macrophytes of Srinagar lakes. In: Productivity Problems in Freshwaters. Proc. IBP-UNESCO Symp. PWN, Warsaw-Kraków, 1972, pp. 725–731

Kershaw, K. A.: Quantitative and Dynamic Ecology. London: Edward Arnold Publ. 1964

Koppová, A., Pirkl, J., Kalina, J.: The determination of mineral elements in plant material by quick accurate methods. (In Czech with German summary). Věd. Pr. Výzk, Úst. Rostl. Výroby v Praze-Ruzyni, Praha, 1955, pp. 119–127

Kožená-Toušková, I.: Composition of nests of birds breeding in the *Phragmition* community. Acta. Sci. Nat. (Brno) **7** (7), 1–36 (1973)

Květ, J.: Growth analysis approach to production ecological investigations in reedswamp plant communities. Hidrobiologia **12**, 14–40 (1971)

Květ, J.: Mineral nutrients in shoots of reed (*Phragmites communis* Trin.). Pol. Arch. Hydrobiol. **20**, 137–147 (1973a)

Květ, J.: Transpiration of South Moravian *Phragmites communis*. In: Littoral of the Nesyt Fishpond. Květ, J. (ed.). Studie ČSAV **15**, Praha, 1973b, pp. 143–146

Květ, J.: Shoot biomass, leaf area index and mineral nutrient content in selected South Bohemian and South Moravian stands of common reed, (*Phragmites communis* Trin.). Results of 1968. In: Ecosystem Study on Wetland Biome in Czechoslovakia. Hejný, S. (ed.). Czechosl. IBP/PT-PP Rep. No. **3**, Třeboň, 1973c, pp. 93–95

Květ, J.: Growth and mineral nutrients in shoots of *Typha latifolia* L. Symp. Biol. Hung. **15**, 113–123 (1975a)

Květ, J.: Transpiration in seven plant species colonizing a fishpond shore. Biol. Plant. **17**, 434–442 (1975b)

Květ, J., Hudec, K.: Effects of grazing by gray-lag geese on reedswamp plant communities. Hidrobiologia **12**, 351–359 (1971)

Květ, J., Marshall, J. K.: Assessment of leaf area and other assimilating plant surfaces. In: Plant Photosynthetic Production. Manual of Methods: Šesták, Z., Čatský, J., Jarvis, P. G. (eds.). The Hague: Dr. W. Junk, N. V., 1971, pp. 517–555

Květ, J., Ondok, J. P.: The significance of biomass duration. Photosynthetica **5**, 417–420 (1971)

Květ, J., Ondok, J. P.: Zonation of higher-plant shoot biomass in the littoral of the Opatovický fishpond. In: Ecosystem Study on Wetland Biome in Czechoslovakia. Hejný, S. (ed.). Czechosl. IBP/PT-PP Rep. No. **3**, Třeboň, 1973, pp. 87–92

Květ, J., Ondok, J. P., Nečas, J., Jarvis, P. G.: Methods of growth analysis. In: Plant Photosynthetic Production. Manual of Methods. Šesták, Z., Člatský, J., Jarvis, P. G. (eds.). The Hague: Dr. W. Junk N. V., 1971, pp. 343–391

Květ, J., Svoboda, J.: Development of vertical structure and growth analysis in a stand of *Phragmites communis* Trin. In: Productivity of Terrestrial Ecosystems. Production Processes. Dykyjová, D. (ed.). Czechosl. IBP/PT-PP Rep. No. 1, Praha, 1970, pp. 95–98

Květ, J., Svoboda, J., Fiala, K.: Canopy development in stands of *Typha latifolia* L. and *Phragmites communis* Trin. in South Moravia. Hidrobiologia **10**, 63–75 (1969)

Lange, L., Zon, J. C. J.: Proposal for a numerical description of the development of aquatic macrophytic vegetation as an aid for the assessment of water quality. Wasser- und Abwasser-Forsch. **4**, 1–4 (1973)

Loomis, R. S., Williams, W. A., Hall, A. E.: Agricultural productivity. Ann. Rev. Plant Physiol. **22**, 431–468 (1971)

McNaughton, S. J.: Ecotype function in *Typha* community type. Ecol. Monogr. **36**, 297–325 (1966)

Medina, E.: Über die Beziehungen zwischen Chlorophyllgehalt, assimilierender Fläche und Trockensubstanzproduktion einiger Pflanzengemeinschaften. Ph. D. Thesis, University of Stuttgart-Hohenheim, 1964

Mochnacka-Ławacz, H.: Seasonal changes of *Phragmites communis* Trin. Part II. Mineral contents. Pol. Arch. Hydrobiol. **21**, 369–380 (1974)

Monsi, M., Saeki, T.: Über den Lichtfaktor in den Pflanzengesellschaften und seine Bedeutung für die Stoffproduktion. Jpn. J. Bot. **14**, 22–52 (1953)

Nátr, L., Gloser, J.: Carbon dioxide absorption and dry weight increase in barley leaf segments. Photosynthetica **1**, 19–27 (1967)

Negisi, K., Satoo, T., Yagi, K.: A method for the rapid measuring of leaf areas. (In Japanese). Jpn. Forest Soc. (Tokyo) **39**, 380–384 (1957)

Nelson, N.: A photometric adaptation of the Samogyi method for the determination of glucose. J. Biol. Chem. **153**, 375–380 (1944)

Nichiporovich, A. A.: Photosynthetic systems of high productivity (In Russian). Nauka, Moscow, 1966

Ondok, J. P.: Die Probleme der Anwendung der Wachstumsanalyse auf Forschungen von *Phragmites communis* Trin. Hidrobiologia **10**, 87–95 (1969)

Ondok, J. P.: Growth analysis applied to the estimation of gross assimilation and respiration rate. Photosynthetica **4**, 214–222 (1970a)

Ondok, J. P.: The horizontal structure of reed stands *(Phragmites communis)* and its relation to productivity. Preslia **42**, 256–261 (1970b)

Ondok, J. P.: Horizontal structure of some macrophyte stands and its production aspects. Hidrobiologia **12**, 47–55 (1971a)

Ondok, J. P.: Indirect estimation of primary values used in growth analysis. In: Plant Photosynthetic Production. Manual of Methods. Šesták, Z., Čatský, J., Jarvis, P. G. (eds.). The Hague: Dr. W. Junk, N. V., Publishers, 1971 b, pp. 392—141

Ondok, J. P.: Calculation of mean leaf area ratio in growth analysis. Photosynthetica **5**, 269–271 (1971c)

Ondok, J. P.: Vegetative propagation in *Scirpus lacustris* L. Biol. Plant. **14**, 213–218 (1972)

Ondok, J. P.: Average shoot biomass in monospecific helophyte stands of the Opatovický fishpond. In: Ecosystem Study on Wetland Biome in Czechoslovakia. Hejný, S. (ed.). Czechosl. IBP/PT-PP Rep. No. **3**, Třeboň, 1973a, pp. 83–85

Ondok, J. P.: Photosynthetically active radiation in a stand of *Phragmites communis* Trin. III. Distribution of irradiance on sunlit foliage area. Photosynthetica **7**, 311–319 (1973b)

Ondok, J. P.: Photosynthetically active radiation in a stand of *Phragmites communis* Trin. IV. Stochastic model. Photosynthetica **9**, 201–210 (1975)

Ondok, J. P.: Regime of global and photosynthetically active radiation in helophyte stands. Studie ČSAV, Praha, **10**, 1—112 (1977)

Ondok, J. P., Dykyjová, D.: Assessment of shoot biomass of dominant reed-beds in Třeboň basin. Methodical aspects. In: Ecosystem Study on Wetland Biome in Czechoslovakia. Hejný, S. (ed.). Czechosl. IBP/PT-PP Rep. No. **3**, Třeboň, 1973, pp. 79–82

Ondok, J. P., Dykyjová, D., Přibáň, K.: Vertical structure and light penetration into the reed stand during seasonal ontogenesis. In: Productivity of Terrestrial Ecosystems. Production Processes. Dykyjová, D. (ed.). Czechosl. IBP/PT-PP Rep. No. **1**, Praha, 1970, pp. 89–90

Ondok, J. P., Květ, J.: Integral and differential formulae in growth analysis. Photosynthetica **5**, 358–363 (1971)

Ondok, J. P., Přibáň, K.: Seasonal dynamics of the production in reed stand (*Phragmites communis* Trin.) determined by nondestructive method. In: Productivity of Terrestrial Ecosystems. Production Processes. Dykyjová, D. (ed.). Czechosl. IBP/PT-PP Rep. No. **1**, Prague, 1970, pp. 91–94

Pazourková, Z.: Caryology of some forms of *Phragmites communis* Trin. In: Ecosystem Study on Wetland Biome in Czechoslovakia. Hejný, S. (ed.). Czechosl. IBP/PT-PP, Rep. No. **3**, Třeboň, 1973, pp. 147–149

Pearcy, R. W., Berry, J. A., Bartolomew, B.: Field photosynthetic performance and leaf temperatures of *Phragmites communis* under summer conditions in Death Valley, California. Photosynthetica **8**, 104–108 (1974)

Pelikán,J., Svoboda,J., Květ,J.: On some relations between the production of *Typha latifolia* and a muskrat population. Zool. Listy **19**, 293–302 (1970)

Pénzes,A.: Über die Morphologie, Dynamik und zönologische Rolle der sproßkolonienbildenden Pflanzen (Polycormone). Fragm. Florist. Geobot. (Kraków) **6**, 505–515 (1960)

Raicu,P., Staicu,S., Stoian,V., Roman,T.: The *Phragmites communis* Trin. chromosome complement in the Danube Delta. Hydrobiologia **39**, 83–89 (1972)

Rejmánek,M.: Species diversity and productivity of plant communities. (Ms.) 1st Int. Congr. Ecol., The Hague, Sept., 1974

Rejmánek,M.: On the Ecological Theory of the Population Structure of Communities. (In Czech). Thesis, Dept. Bot. Charles Univ., Prague, 1975

Rejmánková,E.: The influence of temperature and irradiance on the growth and production of duckweeds (*Lemna gibba* L., *Lemna minor* L., and *Spirodela polyrrhiza* L., Schleiden). (In Czech). Thesis. Dept. Bot., Charles Univ., Prague, 1971

Rejmánková,E.: Seasonal changes in the growth rate of duckweeds (*Lemna gibba* L.) in the littoral of the Nesyt fishpond. In: Littoral of the Nesyt Fishpond. Květ,J. (ed.). Studie ČSAV **15**, Prague, 1973a, pp.103–106

Rejmánková,E.: Biomass of submerged macrophytes growing in the Nesyt fishpond. In: Littoral of the Nesyt Fishpond. Květ,J. (ed.). Studie ČSAV **15**. Praha, 1973b, pp.107–110

Rejmánková,E.: Biomass, production and growth rate of duckweeds (*Lemna gibba* and *L.minor*). In: Ecosystem Study on Wetland Biome in Czechoslovakia. Hejný,S. (ed.). Czechosl. IBP/PT-PP Rep. No. **3**, Třeboň, 1973c, pp. 101–106

Rejmánková,E.: Biology of duckweeds in a Pannonian fishpond. Symp. Biol. Hung. **15**, 125–131 (1975a)

Rejmánková,E.: Comparison of *Lemna gibba* and *Lemna minor* from the production ecological viewpoint. Aquat. Bot. (Amsterdam) **1**, 423–427 (1975b)

Rejmánková,E.: Germination of seeds of *Lemna gibba*. Folia Geobot. Phytotax. Bohemoslov. **11**, 261–267 (1976)

Rich,P.H., Wetzel,R.G., Thuy,N. van: Distribution, production and role of aquatic macrophytes in a southern Michigan marl lake. Freshwater Biol. **1**, 3–21 (1971)

Rychnovská,M.: A contribution to the autecology of *Phragmites communis* Trin. 1. Physiological heterogeneity of leaves. Folia Geobot. Phytotaxon. Bohemoslov. **2**, 179–188 (1967)

Rychnovská,M.: Transpiration of meadow communities. In: Ecosystem Study on Grassland Biome in Czechoslovakia. Rychnovská,M. (ed.). Czechosl. IBP/PT-PP Rep. No. **2**, Brno, 1972, pp.37–43

Rychnovská,M.: Some physiological features of water balance in littoral and terrestrial *Phragmites communis* Trin. In: Littoral of the Nesyt fishpond. Květ,J. (ed.). Studie ČSAV **15**, Praha, 1973, pp.147–151

Rychnovská,M., Květ,J., Gloser,J., Jakrlová,J.: Water relations in three zones of grassland. Acta Sci. Nat. (Brno) **6** (5), 1–38 (1972)

Rychnovská,M., Šmíd,P.: Preliminary evaluation of transpiration in two *Phragmites stands*. In: Ecosystem Study on Wetland Biome in Czechoslovakia. Hejný,S. (ed.). Czechosl. IBP/PT-PP, Rep. No. **3**, Třeboň, 1973, pp.111–119

Schierup,H.H.: Preliminary investigations on variation in biomass of *Phragmites communis* Trin. Ms, Univ. of Aarhus, 1970

Schulz,B.: Wasserlinsen *(Lemnaceae)*. Wittenberg: Die Neue Brehm Bücherei, Vol. 304, 1962

Šesták,Z., Čatský,J., Solarová,J., Tichá,I., Zikmundová,H.: Carbon dioxide transfer and photochemical activities as factors of photosynthesis during ontogenesis of bean leaves. In: Genetic Aspects of Photosynthesis. Nasyrov,Y.S., Šesták,Z. (eds.). The Hague: Dr. W.Junk N.V., In press, 1977

Šesták,Z., Jarvis,P.G., Čatský,J.: Criteria for the selection of suitable methods. In: Plant Photosynthetic Production. Manual of Methods. Šesták,Z., Čatský,J., Jarvis,P.G. (eds.). The Hague: Dr. W.Junk N.V., 1971, pp. 1–48

Šetlík,J., Šust,V., Málek,I.: Dual purpose open circulated units for large-scale culture of algae in temperate zones. I. Basic design considerations and scheme of a pilot plant. Algol. Stud. (Třeboň) **1**, 111–166 (1970)

Šmíd,P.: Microclimatological characteristics of reedswamps at the Nesyt fishpond. In: Littoral of the Nesyt Fishpond. Květ,J. (ed.). Studie ČSAV **15**, Praha, 1973, pp.29–38

Šmíd, P.: Evaporation from a reedswamp. J. Ecol. **63**, 299–309 (1975)

Stocker, O.: Das Wasserdefizit von Gefäßpflanzen in verschiedenen Klimazonen. Planta **7**, 382–387 (1929)

Straškraba, M.: Der Anteil der höheren Pflanzen an der Produktion der stehenden Gewässer. Mitt. Int. Ver. Theor. Angew. Limnol. **14**, 212–230 (1968)

Taubaev, T.T., Neskubo, P.M., Abdiev, M., Normuchamedov, Ch.: Use of *Lemna minor* as fodder for small cattle. In: Cultivation of algae and water macrophytes in Uzbekistan. Tkačenko, I.P. (ed.). Taškent: 1971, pp. 136–138 (In Russian)

Thorne, G.N.: Photosynthesis of laminae and sheaths of barley leaves. Ann. Bot. N.S. **23**, 365–370 (1959)

Tuschl, P.: Die Transpiration von *Phragmites communis* Trin. im geschlossenen Bestand des Neusiedler Sees. Wiss. Arbeiten, Burgenland (Eisenstadt) **44**, 126–186 (1970)

Vassiljev, J.M.: Über den Wasserhaushalt von Pflanzen der Sandwüste im südöstlichen Kara-Kum. Planta **14**, 225–309 (1931)

Véber, K., Dykyjová, D.: Analysis of the production of selected species of littoral macrophytes under controlled conditions of mineral nutrition. Ann. Rep. Algol. Lab., Treboň, 1970, 169–176 (1973)

Véber, K., Kupka, Z.: Stands and crops of reed (*Phragmites communis*) in Czechoslovakia. In: Ecosystem Study on Wetland Biome in Czechoslovakia. Hejný, S. (ed.). Czechosl. IBP/PT-PP Report no. **3**, 1973, Třeboň, pp. 255—256

Velásquez, J.: Communities and production ecology of *Eleocharis acicularis* (L.) R. et Sch. Thesis, Inst. of Botany, Czechosl. Acad. Sci., Průhonice, 1975

Walker, J.M., Waygood, E.R.: Ecology of *Phragmites communis*. I. Photosynthesis of a single shoot in situ. Can. J. Bot. **46**, 549–555 (1968)

Watson, D.J.: The physiological basis of variation in yield. Adv. Agron. **4**, 101–145 (1952)

Westlake, D.F.: Some basic data for investigations of the productivity of aquatic macrophytes. Mem. Ist. Ital. Idrobiol., 18 suppl., 229–248 (1965)

Westlake, D.F.: The biomass and productivity of *Glyceria maxima* I. Seasonal changes in biomass. J. Ecol. **54**, 745–753 (1966)

Westlake, D.F.: Primary production of freshwater macrophytes. In: Photosynthesis and Productivity in Different Environments. Cooper, J.P. (ed.). (International Biological Programme 3). Cambridge: Cambridge Univ. Press, 1975, pp. 189–206

Fig. I. Covers of *Cladophora fracta* deposited on the shore after a fall of water level in Nesyt fishpond. (Photo: Š. Husák)

Section 4

Structure and Functioning of Algal Communities in Fishponds

In fishpond littoral ecosystems microphytes are generally inferior in importance to the macrophytes. The spatial pattern, community structure, and developmental dynamics of the microphytes are determined largely by the structure and developmental dynamics of the macrophytes. However, the microphytes may participate appreciably in the total material and energy budget of the littoral ecosystem. Algae comprise numerous types (growth forms, species) whose ways of life differ from those of rooted higher plants. During certain seasonal phases, algae are capable of intercepting the radiant energy which, otherwise, would be lost to the ecosystem. Although they are at a disadvantage in competition with higher plants, their development may also influence the development of the macrophytes.

In order to understand the role which microphytes play in pond littoral ecosystems it is necessary to pursue their seasonal development not only in the littoral, but also in the pelagial, i.e., in the open-water area of a fishpond. Life cycles of many dominant algae may proceed alternately in the littoral and pelagial: therefore exact data on the changes in their biomass are often difficult to obtain. The same applies to algal production budgets in individual subsystems within the littoral as defined by various macrophyte vegetational units. Algae tend to blur boundaries between adjoining ecosystems, which may appear distinct if defined only according to the dominant macrophytes (Fig. I).

4.1 Structural Elements. Principal Populations of Algae. Spatial Distribution

P. Marvan, J. Komárek, H. Ettl, and J. Komárková

The algal flora of a fishpond is generally quite rich, including elements from ecologically different microbiotopes. This holds true especially for the pond littoral ecosystems. In managed fishponds, two distinct zones of the littoral, an outer landward and an inner waterward zone, may be distinguished, both usually differing in the structure of macrophyte communities, in water chemistry as well as in the composition of the algal microvegetation (See Fig. I in Introduction to this Sect. and Fig. 11 in Sect. 2.3). These differences arise and/or are supported by intense long-term management of the fishponds. In the central part of a pond, this management generally creates a biotope differing considerably from the environment in which the fishpond had been set up. Hence it is in the littoral that the ecological pressures of the original and changed environments meet; frequently, a distinct boundary arises between the outer littoral, where the original conditions tend to prevail, and the far more modified inner littoral.

The algal microvegetation of a fishpond littoral markedly reflects this zonation. Its species composition is primarily determined by the character of the geological substrate, macrophytic vegetation cover, climatic conditions, etc., and may be quite different in different fishponds. Human activities (mainly fertilizer application) tend to wipe out original differences in species composition between individual fishponds. However, they also affect most strongly the central fishpond area and the inner littoral, while its outer parts are affected far less (Bílý, 1964; Komárková and Přibil, 1973). Substantial differences may thus arise in the microvegetation of the two littoral zones. On the other hand, we can meet quite similar algal communities in fishponds of originally very different character.

A comparison of the algal flora in the two specimen fishponds, Opatovický in South Bohemia and Nesyt in South Moravia, has been made in order to disclose differences in the participation of the fundamental structural elements of the microvegetation in these two ponds (Fig. 1). They are both similarly managed and highly eutrophic and yet distinguished by different geological and environmental features.

The collective term "littoral microvegetation" denotes several types of algal assemblages differing in both their species composition and specific features of their development dynamics. The basic algal structural elements in fishponds are the following (see Ettl et al., 1973b; Marvan et al., 1973):

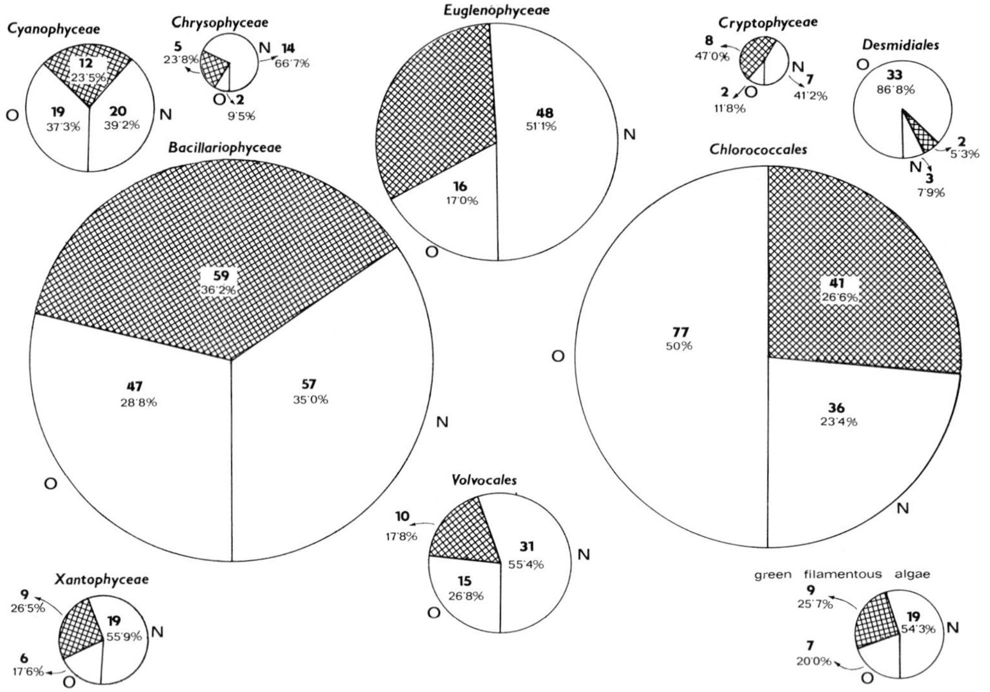

Fig. 1. Representation of different algal groups in Opatovický (O) and Nesyt (N) fishponds. Diameter of each circle directly proportional to sum of species of respective group. Shaded sectors: species common to both ponds; upper figures: species numbers; lower figures: percentages of species. Of all algal species identified, 26.4% occur in both fishponds

1. Phytoplankton.
2. Epipelon (strata in the zone of contact between sediments and water).
3. Epilithon.
4. Epiphyton (epiphytes on aquatic higher plants or their dead parts).
5. Free-floating (or loosely attached to higher plants) clusters of filamentous algae.

The algal microvegetation consists of a complicated mosaic of these elements, whose spatial distribution and species composition is subject to permanent change. Marked boundaries between the outer and inner littoral also occur in the two specimen fishponds; the different bottom substrates and climatic conditions determine the differences between the algal communities of their littorals in general.

Several indexes have been used in the subsequent characteristics and comparisons between the microphyte communities. These indexes are defined on p. 65–66.

1. Phytoplankton: Plankton communities show several characteristic features. They alternate in time in dependence on changes in environmental factors and seasons, often in regularly repeated seasonal successions (G. M. Smith, 1924). On individual sites, they develop more or less homogeneously in space as their individual components are motile, actively or passively. Their dominants are species populations or groups of them, among which any ecologically suited element may prevail in accord with the set of species and local microhabitats available. But the latter are difficult to identify. The animal component of the ecosystem often exercises a decisive effect on the community. The planktonic communities tend to be different in fishponds with a different geological substrate. The differences may sometimes become obscured in intensely managed fishponds with a controlled inflow (source of other algal species), which are drained completely every one to three years and where the intense management varies each year as to fertilizer application, wash-out from surrounding areas, fish-feeding, character and quantity of fish stock, etc. Yet it has been possible to identify a certain series of plankton communities in managed fishponds of the South Bohemian basins. The different development of these communities in individual years can be related to the fishpond management. A survey of the most frequent plankton communities occurring in the "main" production fishponds is given in Table 1 (Komárek, 1973b).

There is generally little horizontal differentiation in the species composition of the plankton in the pelagial of fishponds smaller than some 200 hectares. In bays of large fishponds, the microvegetation may be influenced by local inflows. An example is the eutrophicated western bay of Nesyt. The species composition of the phytoplankton in the inner littoral is usually similar to that in the pelagial (Straškraba, 1963; Komárek, 1973a; Komárková and Komárek, 1975). A different plankton sometimes develops in the outer littoral, especially if it is isolated from the main water body by a wide reed-belt. This situation is found in the Opatovický fishpond (Fig. 11 in Sect. 2.3).

A comparison of the phytoplankton structure between the erosion and accumulation types of littoral biotopes and the pelagial of the Opatovický fishpond is shown in Table 2. Differences between the two types of littoral biotopes are negligible, but the pelagial community differs from the two littoral communities, as indicated by the values of I_g based on the presence of typically planktonic elements. Values of the homotoneity index H for the planktonic communities of the pelagial and of three types of helophyte stands in the littoral vary between 60 and 70% in summer (Komárková and Komárek, 1975).

The effect of vertical zonation is almost absent in shallow fishponds except for a temporary mass occurrence of blooms of blue-greens near the water surface during periods of calm summer weather.

2. *Epipelic Mats in the Contact Zone of Bottom Sediments with Water:* In eutrophic fishponds, the development of epipelic mats of autotrophic algae is limited. They are absent in summer partly because of great water turbidity due to plankton development, and partly because the fishpond bottom is constantly disturbed, mainly by the feeding activities of carp. Distinct epipelic mats are thus sometimes present in the littorals and, to a greater extent, in shallow bays (0.1–0.6 m deep) of such fishponds where the bottom is covered with a layer of black organic mud. As a rule, they occur only for a short period in spring (March to May) when the water is still clear and transparent and the fish activity is low (Table 3).

Species of these benthic communities belong predominantly to three taxonomic groups: (1) *Cyanophyceae:* mainly several species of *Oscillatoria* which

Table 1. Main summer communities of phytoplankton in eutrophic ponds (South Bohemia)

Dominants	*Ceratium hirundinella*	*Aphanizomenon fl.-aq.*	*Aphanizomenon fl.-aq.*
Subdominants	*Volvox aureus* *Melosira granulata*	*Ceratium hirundinella* *Volvox aureus*	
Preferential species of groups of species	*Ceratium hirundinella* ——— alternative *Volvox aureus* *Eudorina elegans* *Melosira granulata*	*Aphanizomenon fl.-aq.* ——— alternative *Ceratium hirundinella* *Volvox aureus*	*Aphanizomenon fl.-aq.* *f. flos-aquae*
Groups of phytoplankton (alternative)	*Bacillariophyceae* (altern. subdom.) *Volvocales (Volvox, Eudorina)* (altern. subdom.) *Chrysophyceae* + *Cryptophyceae* + *Chlorococcales* + *Cyanophyceae* (water-blooms) (+)	*Ceratium hirundinella* (<20%) *Volvox aureus* (<20%) *Eudorina elegans* (+) *Fragilaria* (+) *Anabaena* or *Microcystis* (+) nannoplankton <1000 cells ml^{-1}	*Bacillariophyceae* <5% *Cryptophyceae* <5% *Chlorococcales* (<5%) nannoplankton (+)
Fertilization	natural eutrophication organic –	natural eutrophication organic –	mineral + + (partic. PO_4) organic (+)
Physical and chemical characteristics (N and P in mg/l)	transpar. = (0.6) 1.4–>3 m pH = 6.9–8.3 (>9?) alkal. = 0.9–2.7 mval N-aos = 0.47–1.61 NO_3-N = <0.1 NH_3-N = <0.1 (0.12)	transpar. = 0.8–3.0 m pH = 7.0–9.7 alkal. = 0.9–2.0 mval N-aos = 0.69–1.6 (3.25 ?) NO_3-N = <0.05–(0.31) NH_3-N = 0.1–0.7	transpar. = 0.5–0.8 m pH = (7.0 ?) 8.7–9.6 alkal. = 1.1–2.0 mval N-tot. = (0.8) 1.0–2.0 (3.0) N-aos = 0.92–3.25 NO_3-N = ±0.19 NH_3-N = <0.1–(0.51) P-tot. = (0.05) 0.1–0.3
Zooplankton	*Daphnia longispina* *Daphnia pulicaria*	*Daphnia longispina* *Daphnia pulicaria* *(Daphnia hyalina)* *(Cyclopidae)* *(Cyclops vicinus)*	*(Daphnia longispina)* *Daphnia pulicaria* *(Daphnia hyalina)*
Fish	low fish stock <350 (700)/ha	low fish stock (small fish and fry lacking)	low or high fish stock (only large fish)
Reservoir characteristics	backwaters medium-sized ponds	backwaters medium-sized ponds (to 60 ha)	medium-sized to large fisponds (to 500 ha)
Season	June–Sept.	May–Oct. (climax association)	May–Oct. (climax association)

mesotrophic and
oligotrophic stagnant waters
←

Cyanophyceae (water-blooms) *Chlorococcales*	*Chlorococcales*	*Chlorococcales*	phytoflagellates
			Chlorococcales
Chlorococcales g. et sp. div. ——— alternative *Microcystis* sp. div. *Gomphosphaeria naeg.* *Anabaena* sp. div. *Aphanizomenon fl.-aq.* f. *klebahnii*	*Chlorococcales* g. et sp. div.	*Chlorococcales* g. et sp. div. *Volvocales* g. et sp. div. ——— alternative *Euglenophyceae* g. et sp. div. *Merismopedia tenuiss.*	*Volvocales*, part *Chlamydomonas* sp. div. *Euglenophyceae* (*Euglena, Phacus, Trachelomonas;* sp. div.) ——— alternative *Chlorococcales* g. et sp. div.
Cyanophyceae (water-blooms) part. $>50\%$ *Chlorococcales* 10–50% *Bacillariophyceae* $<20\%$ phytoflagellates $<20\%$ *Cyanophyceae* (plankt., no water-blooms) (+)	*Chlorococcales* $>75\%$ *Bacillariophyceae* + ($<20\%$) phytoflagellates + *Cryptophyceae* (+) *Ceratium hirundinella* (+) *Cyanophyceae* (plankt., no water-blooms) (+)	*Chlorococcales* $>75\%$ *Volvocales* $<20\%$ *Euglenophyceae* $<20\%$ *Bacillariophyceae* + *Cyanophyceae* (plankt., no water-blooms) (+)	*Volvocales* and *Euglenophyceae* together $>80\%$ *Chlorococcales* + *Bacillariophyceae* (+) *Cyanophyceae* (plankt., no water-blooms) (+)
mineral + + organic +	mineral + + organic + +	mineral + + organic + + (part. duck-farming)	mineral (+) organic + + + (part. duck-farming)
temp. (inic.) = $>14°$ C transpar. = 0.4–1.9 m pH = 7.1–8.2 (9.5) alkal. = (1.3) 1.8–6.0 mval N–tot. = 1.0–2.0 (3.7) N–aos = 0.88–1.9 (2.7) NO_3–N = <0.1 = 1.5 (5.1) NH_3–N = <0.1–0.3 (0.4) P–tot. = 0–0.3 (3.2) PO_4–P = 0–0.34	temp. (inic.) = $>16°$ C pH = 7.3–8.0 alkal. = 2.0–4.0 mval N–aos = 0.42–0.81 NO_3–N = <0.1 NH_3–N = <0.1 P–tot. = 0.5–1.0 PO_4–P = 0.08–0.9	temp. (inic.) = $>12°$ C pH = 7.5–9.2 alkal. = 1.23–2.30 mval N–aos = ±1.32–2.18 NH_3–N = ±0.13–0.61	transpar. = max. 0.5 m NH_3–N = higher content!
Bosmina longirostris *Ceriodaphnia pulchella* *Diaphanosoma brachyur.* *Daphnia hyalina* *Daphnia cucullata* *Cyclopidae* *Rotatoria*	*Bosmina longirostris* *Ceriodaphnia affinis* *Diaphanosoma brachyur.* (*Daphnia pulicaria*) *Daphnia hyalina* *Cyclopidae* *Rotatoria*	*Bosmina longirostris* *Ceriodaphnia affinis* *Diaphanosoma brachyur.* *Daphnia hyalina* *Cyclopidae* *Rotatoria*	*Daphnia magna* *Daphnia pulex* *Moina micrura* *Cyclopidae* *Infusoria*
high fish stock $>(350)$ 1000 ha^{-1} (small fish or fry present)	high fish stock? $>(300)$ 20,000 ha^{-1}? (small fish or fry present)	high fish stock? (small fish or fry present)	(small fish present?)
small to large fishponds	small to large reservoirs	small to medium-sized reservoirs	small ponds
May–Oct. (climax association)	May–Sept. (climax association)	(climax association)	(Feb.) March–June (Aug.) July–Nov.

polytrophic waters ⟶

Table 2. Indexes I_s (see Sect. 1.3) of algal communities in the pelagial, and the erosion and accumulation littorals in Opatovický fishpond

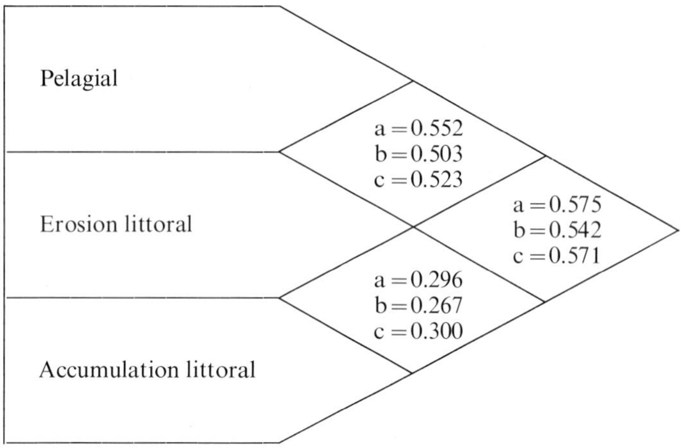

a, planktonic species; b, facultative planctonic species; c, all species.

dominate the mats at the peak of their development (hence the name "Oscillarieta" for these communities, Prát and Cyrus, 1948); (2) *Bacillariophyceae:* generally richer in species, but prevailing in the biomass only at the beginning of the vegetation of benthic mats; (3) *Euglenophyceae:* free-living in microbiotopes of the mats represented by both colored and apochloric species (they are dominant mainly on sites with a strong allochthonous pollution).

In the fishponds of both South Bohemia and South Moravia, two types of Oscillarieta occur, both developing from March to June:

(a) Benthic mats dominated by *Oscillatoria limosa* and *O. tenuis*, sometimes also by *O. brevis*, *O. chalybea*, *O. angusta* and *O. splendida*. These mats become detached from the bottom at the end of the growing season (in May), float on the surface and are driven by wind to the windward shore of the fishpond. The flora of subdominant diatoms differs in dependence on local conditions. Typical of Oscillarieta are *Navicula radiosa*, *N. cryptocephala*, *Cymbella ventricosa*, *Synedra ulna*, *Nitzschia sigmoidea*, *N. acicularis*, *Fragilaria capucina*, and others. Of the sulfur bacteria, the incidence has been recorded of *Beggiatoa alba*, of *Euglenophyceae* several species of *Euglena* and *Trachelomonas*. In Nesyt, this community occurs in the shallow and highly eutrophicated western bay. In Opatovický, it forms continuous carpets in the shallow and eutrophicated accumulation biotopes in the bays along the eastern shore.

(b) Mats developing in the littoral at a depth of approximately 0.5 m on sites rich in decomposing macrophyte litter. The content of H_2S is high, and sulfur bacteria (*Lamprocystis roseo-persicina*, *Chromatium okenii*, *Beggiatoa alba*, *Lampropedia hyalina*, *Thiothrix nivea*, and others) participate considerably in the community. The bacteria form white or pink strata and, sometimes, a continuous contact layer between the surface mats of blue-green algae and the black organic mud of the substrate. The dominant blue-greens are mainly *Oscillatoria tenuis*, less frequently *O. limosa*, *O. chalybea* and *Spirulina major*. Most frequent of the

Euglenophyceae are *Euglena viridis, E. pisciformis, E. intermedia, E. spirogyra, E. hemichromata, E. subehrenbergii, Distigma tortuosum, D. proteus, Cyclidiopsis acus, Menoidium pellucidum, Astasia hypolimnica, Anisonema dubium, A. acinus* and *Petalomonas involuta.* The diatom flora is considerably less varied than in the Oscillarieta of the eutrophic bays. The accompanying diatom flora (*Navicula cincta, N. cryptocephala* v. *veneta, N. simplex, Nitzschia frustulum, Synedra minuscula,* etc.) indicates slightly halophilous character of this community. It was observed only in Nesyt. It is of interest that similar communities of filamentous blue-green algae with sulfur bacteria and *Euglenophyta* are common in saline habitats on the sea-shore where, however, the dominant species are different.

3. Epilithon. An epilithic vegetation is generally of little importance in a typical fishpond littoral. Marvan et al. (1973) give a brief description of the structure of the epilithon in Nesyt. In spring, blue-greens (*Pleurocapsa minor* and *Phormidium autumnale*), the ulotrichal *Stigeoclonium* sp., and numerous diatom species (*Gomphonema olivaceum, Navicula gracilis, Rhoicosphenia curvata,* etc.) colonize the surfaces of limestone boulders. Later, they are replaced by filamentous algae (*Cladophora, Oedogonium*). The vertical zonation of this algal vegetation is marked. The flat and muddy or sandy littorals of South Bohemian fishponds or the granite pavements of their dams are not suitable for the development of an epilithic algal vegetation. Periphytic species occurring on macrophytes (*Rivularia aquatica, Stigeoclonium* spp., diatoms) are feebly developed in these biotopes.

4. Epiphyton on Macrophytes. The term epiphyton refers here to the littoral microvegetation attached to submerged parts of aquatic macrophytes. Its dominants are sessile, unicellular and filamentous algae (in fishponds: *Homoeothrix stagnalis, Characium* spp., *Stigeoclonium farctum* and several diatom species). Their accessory components are free-living species without organelles of attachment (chlorococcal algae, diatoms, etc.), which may sometimes outweigh the typically epiphytic algae both in number of species and biomass.

The species composition of the epiphyton is different in individual fishponds, but the structure of epiphytic communities colonizing submerged stems of various species of the littoral macrovegetation may be uniform within one fishpond. In Opatovický, the differences in the species composition of the epiphytic microvegetation growing on submerged parts of various helophytes were not substantial. This is in accord with data from lakes (e.g., Masjuk, 1957; Bohr, 1962). But there may be local quantitative differences, some depending on substrate (firm stems of *Typha* or smooth and fragile stems of *Schoenoplectus*; young or dead leaves of *Phragmites*, etc.), others on the prevailing wind direction (erosion and accumulation littoral), and yet others on density of the helophyte canopy (see Sect. 2.2). In general, epiphytes are much better developed on dead than on live parts of the macrophytes.

In Nesyt, four types of epiphytic communities occur both on young and old shoots of *Phragmites communis* and *Typha angustifolia* (Marvan et al., 1973). Their brief floristic characteristics are given in Table 4. The diatomaceous type and the type dominated by *Stigeoclonium*, both occurring on plant parts formed during the current year, are floristically similar to each other and poor in species (about 30, see Table 5). The diatomaceous type is frequently represented by a single

Table 3. Seasonal development of bottom strata and free-floating clusters of *Oscillarieta* in Opatovický fishpond

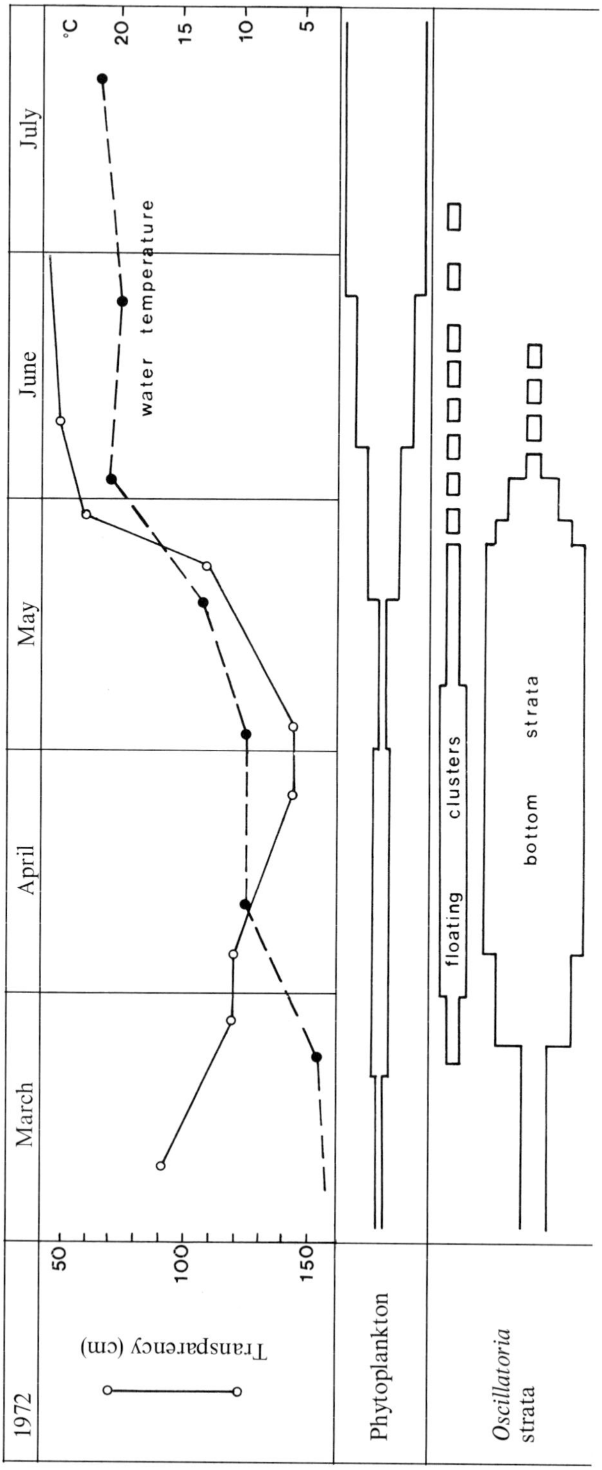

Dominant *Oscillatoria* species	*Oscillatoria limosa* 5 *Oscillatoria tenuis* 1 *Oscillatoria angusta* 1	*Oscillatoria limosa* 4 *Oscillatoria tenuis* 1 *Oscillatoria brevis* +	*Oscillatoria limosa* 2–4 *Oscillatoria tenuis* + *Oscillatoria splendida* 1		
Number of species from other groups of algae	*Desmidiales* 2 *Chlorococcales* 6 *Euglenophyceae* 3 (*Trachelomonas*) *Chrysophyceae* 2 (*Chrysococcus*) *Bacillariophyceae* 21	*Desmidiales* 1 *Chlorococcales* 6 *Bacillariophyceae* 19	*Desmidiales* 7 *Chlorococcales* 11 *Euglenophyceae* 5 (*Euglena, Phacus*) *Bacillariophyceae* 15		
Subdominant diatoms	*Gomphonema acuminatum* 4 *Gomphonema truncatum* 4 *Navicula cryptoceph.* 3 *Nitzschia sigmoidea* 4 *Synedra ulna* 4		*Bacillariophyceae* (together) 4 (without subdominant species)		

Table 4. Dominant species in the periphyton on *Phragmites communis* and *Typha angustifolia* in Nesyt fishpond

1. Epiphyton with dominant diatoms	*Cocconeis placentula* *Gomphonema parvulum* *Navicula gracilis* *Rhoicosphenia curvata* *Homoeothrix stagnalis*	Spring–fall on young leaves
2. *Stigeoclonium* type	*Gomphonema parvulum* *Navicula cryptocephala* *Navicula gracilis* *Synedra ulna*	June–September mainly on *Typha angustifolia* (on young and old leaves)
3. *Oedogonium* type (*O. varians, O. capillare*)	*Fragilaria vaucheriae* *Melosira varians* *Rhoicosphenia curvata* *Synedra fasciculata* *Synedra ulna* *Spirogyra* sp.	Spring–fall on old leaves and stems of *Phragmites* and *Typha*
4. *Cladophora* type (*C. fracta, C. glomerata*)	*Cocconeis pediculus* *Rhoicosphenia curvata* *Synedra fasciculata* *Oedogonium* spp.	May–October on litter

Table 5. Indexes I_s (see Sect. 1.3) of different types of epiphytic communities on *Typha angustifolia* in Nesyt fishpond

	Diatom type	*Stigeoclonium* type	*Oedogonium* type
Stigeoclonium type	0.978		
Oedogonium type	1.161	1.056	
Cladophora type	1.262	1.216	1.110

dominant accompanied by a few sporadically represented other species. This type occurs even in rather heavily shaded microbiotopes. In contrast, the *Stigeoclonium* type favors well irradiated biotopes. Characteristic species of the two remaining types of epiphyton are green filamentous algae of the genera *Oedogonium* or *Cladophora*. These communities comprise considerably more species and colonize previous year's litter. The representation of accompanying species is rather different in both types of epiphyton (Table 5). Characteristic of both types is their easy transition to a free-floating stage. As in Opatovický, no distinct differences were found in the species composition between the epiphytes on *Phragmites communis* and *Typha angustifolia* in Nesyt. This applies especially to the *Cladophora* type (I_S=0.79). The diatomaceous type (*Rhoicosphenia curvata, Cocconeis pediculus, Synedra fasciculata, S. pulchella, Gomphonema parvulum, G.*

truncatum, etc.) displays appreciably more marked differences ($I_S = 1.09$) due to its poorly developed species composition. The stage with dominant *Stigeoclonium* is much scarcer on *Phragmites*.

In Opatovický, the types of epiphyton characterized by a massive development of *Oedogonium* and *Cladophora* are absent, and *Oedogonium* occurs only sporadically. Several diatoms important in Nesyt (*Amphora veneta, Cocconeis pediculus, Gomphonema olivaceum, Navicula gregaria, N. gracilis, Synedra pulchella, S. fasciculata*, etc.) are either absent or occur only occasionally in Opatovický. In contrast, epiphyton samples from Opatovický contain, in addition, numerous species of chlorococcal algae originating from the open water.

All autochthonous communities of epiphyton developing on submerged stems of the littoral helophyte vegetation display a characteristic zonation caused mainly by the vertical distribution of irradiance in the vegetatively coloured water (Komárek, 1973a).

5. *Free-Floating Clusters of Filamentous Algae*. Communities of filamentous algae, either floating on the water surface or attached loosely to higher plants, may form an important component of the littoral microvegetation in stagnant waters (Bohr, 1962; Pieczyńska, 1971b; Kostikova, 1972; Marvan et al., 1973, etc.). Their incidence is scarce in the pelagials of managed fishponds, and their biomass and production usually do not outweigh those of the phytoplankton. This holds true particularly for the intensely managed fishponds of the Třeboň basin. Filamentous green algae *(Hydrodictyon, Spirogyra, Cladophora)* develop in the fishponds only if these are understocked or after some undesirable interference with conditions in the fishpond (e.g., after stocking large production ponds with fry). Normally, filamentous algae are confined to overgrown parts of the littoral where they locally form populations floating on the open water surface of the "lagoons" within continuous helophyte stands *(Tribonema, Microspora, Spirogyra, Zygnema)*. This is also the situation in Opatovický, but in Nesyt and other South Moravian fishponds the production of filamentous algae may, under certain conditions, surpass that of phytoplankton not only in the littoral, but in the whole fishpond. The species composition of these clusters depends on the water chemistry changing during the year (Fig. 2) and also being different in various parts of the fishpond. In Table 6 the indexes I_C represent the percentage of those samples in which both species occur together, from the total number of species in which at least one of the species in question occurs. The indexes thus characterize the affinity of different dominants to each grouping, influenced by environmental factors and by their seasonal changes.

The principal communities of floating filamentous algae regularly present in the two fishponds compared may by classified as follows:

A. Clusters of filamentous algae dominated by *Tribonema* species:

In spring, soon after the ice disappears, these clusters form a well-defined community in shallow water (less than 0.3 m deep) in the zone of tall sedges and in the outer reed-belt on sites where sufficient radiation reaches the water surface (Fig. 2).

In Opatovický, this community is acidophilous (pH = 4.6–6.7); it occurs here at water temperatures from 6–14° C and when average daily sums of PhAR irradi-

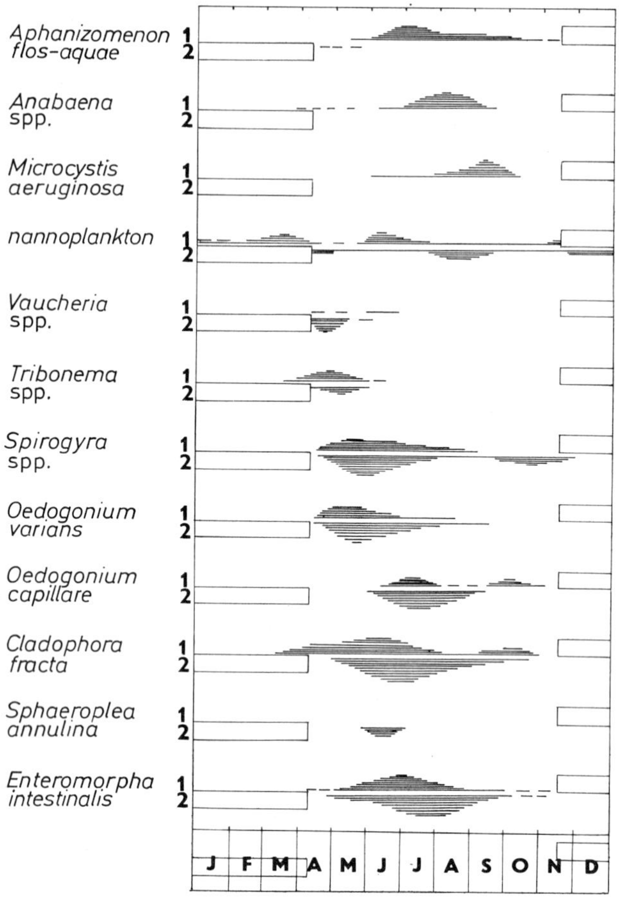

Fig. 2. Seasonal dynamics of development of principal dominant algal species in Nesyt fishpond; 1, 1971; 2, 1972; white strips, period of drawdown of fishpond

ance are higher than 3 MJ m^{-2}. As soon as the temperature rises and the developing higher plant canopy casts more shade on the water surface, the community disappears. Its characteristic species are *Tribonema vulgare*, *T. ambiguum*, *T. minus*, *T. viride*. The acidophilous character of the community is confirmed by the presence of numerous species of desmids, *Ophiocytium ilkae*, and of the diatoms *Eunotia curvata*, *E. glacialis*, *E. pectinalis* v. *minor*, *Pinnularia subcapitata*, and others.

The outer littoral of the Nesyt fishpond is colonized by fewer acidophilous species: *Tribonema vulgare*, *Microspora quadrata*, *Microthamnion kuetzingianum*, *Eunotia curvata*. In addition, clusters of *Tribonema* are accompanied by numerous halophilous and mesohalobic diatoms such as *Navicula salinarum*, *N. simplex*, *Nitzschia commutata*, *N. frustulum*, *Hantzschia vitrea*. The coexistence of acido- and halophilous elements is typical of this community in Nesyt (see also p. 61) as well as in other South Moravian fishponds. The dominant components of the community (*Tribonema viride*, *T. vulgare*, *T. angustissimum*) are found relatively frequently together with *Diatoma tenue* and with species of *Fragilaria* and *Vaucheria;* far less frequent is their common incidence with algal dominants of the inner littoral such as *Cladophora, Oedogonium* or *Enteromorpha* (see Table 6).

B. Stands and floating clusters of *Vaucheria:*

The incidence of submerged stages of several species of this genus is rather infrequent in the fishpond littorals. However, after a sudden flooding of their stands on emerged bottoms, this genus may temporarily turn into a prominent dominant and start to form free-floating stages. In Nesyt, a mass development of *Vaucheria* community with dominating *Vaucheria debaryana, V. geminata, V. arrhyncha* was observed in May, 1972 (Fig. 2). In its species composition, it is similar to the previous community and to that with dominant *Fragilaria;* significant is a higher representation of diatoms of the epipelic strata (*Navicula cryptocephala* and several species of the genus *Nitzschia*), also comprising numerous halophilous to mesohalobic elements *(Navicula cincta, N. salinarum, N. simplex, Nitzschia frustulum, Hantzschia spectabilis, H. vitrea, Amphora veneta, Diatoma tenue).* Other filamentous algae (*Tribonema*, filamentous diatoms *Melosira varians* and *Fragilaria vaucheriae*) only seldom participate in this community. *Vaucheria* itself accompanies clusters of other filamentous algae only exceptionally.

C. Communities with dominant *Fragilaria:*

Mass development of populations of *Fragilaria* species takes place in the littorals of numerous fishponds and lakes in spring and fall. In the Mazurian Lakes (Pieczyńska, 1976), these populations take a great part in the periphyton in stands of *Phragmites*. In the Opatovický fishpond, *Fragilaria construens, F. vaucheriae*, and *F. capucina* are abundant in spring (from April to May) and in late summer. These species are found in the reed-belt and form fine, cottonwool-like tufts loosely attached to submerged parts of the higher plants. In fall, they become an important component of the algal community associated with *Utricularia* (Fig. 3). In Nesyt, *Fragilaria construens* is absent, and ephemerous communities with dominant *Fragilaria vaucheriae* and/or *F. capucina*, attached loosely to leaves of *Phragmites* or *Potamogeton pectinatus*, were observed only in spring (May to June). Both species also accompany clusters of other filamentous algae, particularly of *Cladophora* and *Enteromorpha*. They occur frequently together with other colonial diatoms, mainly *Melosira varians* and *Diatoma tenue*.

D. Free-floating clusters with dominant *Zygnematales:*

Populations of *Zygnematales* constitute important elements of the microvegetation in various types of stagnant waters. They frequently form species-poor communities with a minimum of accompanying species, both epiphytic or free-living, which usually comprise several species of the dominant genus. In Nesyt, this type of community is present mainly in the highly eutrophic western bay, where its typical biotopes are shallows with a muddy bottom and warm water. Of the accompanying species, most important are epipelic diatoms, numerous members of *Euglenophyceae* and other flagellates. Pure stands of *Spirogyra* were found less frequently near a fertilizer depot. In other parts of the fishpond the incidence of *Spirogyra* is mainly confined to clusters in which it is accompanied by other filamentous algae, particularly by *Oedogonium varians* in spring, and later by *Cladophora fracta*. The species composition of these mixed communities is gener-

Table 6. Indexes of common occurrence I_C between dominants in floating clusters of filamentous algae in Nesyt fishpond

	Oedogonium varians	Spirogyra spp.	Cladophora spp.	Enteromorpha intestinalis	Oedogonium capillare	Sphaeroplea annulina	Fragilaria spp.	Melosira varians	Diatoma tenue	Vaucheria spp.
Oedogonium varians										
Spirogyra spp.	0.524									
Cladophora spp.	0.341	0.364								
Enteromorpha intestinalis	0.162	0.167	0.241							
Oedogonium capillare	0.148	0.115	0.224	0.041						
Sphaeroplea annulina	0.200	0.160	0.135	0.025	0.185					
Fragilaria spp.	0.138	0.180	0.256	0.122	0.022	0.000				
Melosira varians	0.161	0.207	0.129	0.060	0.128	0.030	0.231			
Diatoma tenue	0.096	0.078	0.080	0.000	0.036	0.000	0.107	0.148		
Vaucheria spp.	0.037	0.038	0.013	0.028	0.000	0.000	0.103	0.154	0.154	
Tribonema spp.	0.127	0.154	0.119	0.049	0.029	0.000	0.310	0.161	0.312	0.250

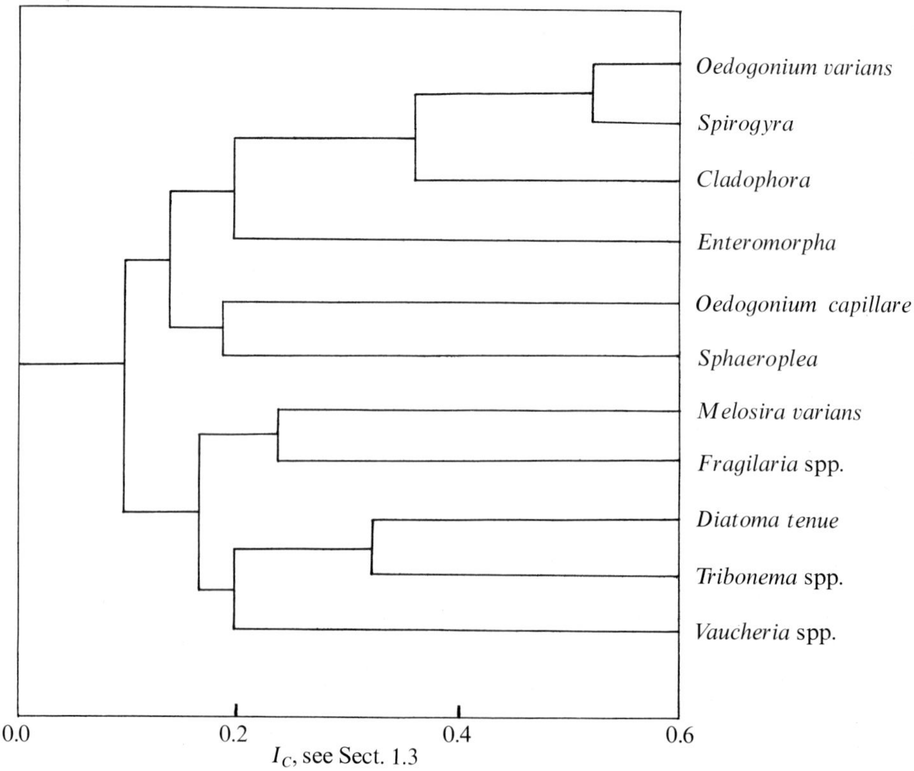

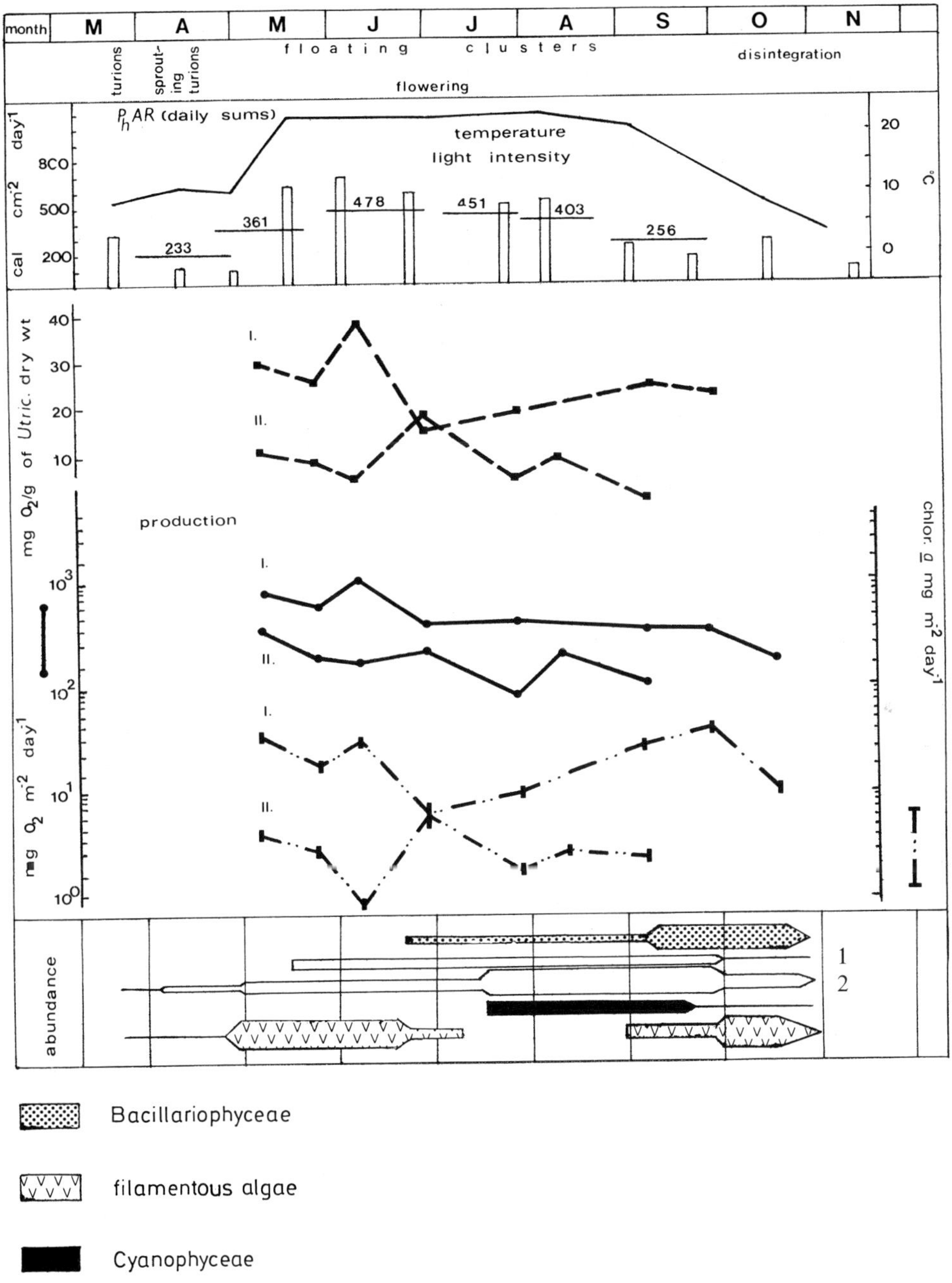

Fig. 3. Course of primary production of periphyton on floating clusters of *Utricularia* at 2 localities (I and II) in Opatovický fishpond in 1972. Data of irradiance: monthly means of daily sums of global radiation and daily sums of PhAR on days of measurement. (After Komárková and Komárek, 1975)

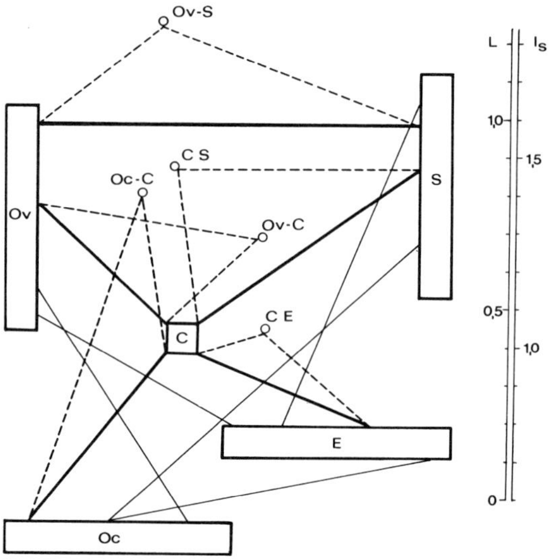

Fig. 4. Comparison of similarity in species composition of free-floating communities of filamentous algae with dominating *Cladophora* (*C*), *Enteromorpha* (*E*), *Oedogonium capillare* (*Oc*), *O. varians* (*Ov*) and *Spirogyra* (*S*). The length of connecting lines $L = I_s - 0.6$; o = intermediate types (mixed clusters of *Oedogonium varians* and *Spirogyra*, etc.). Only most common combinations plotted

ally nearer to that of clusters with dominant *Cladophora fracta* or *Oedogonium varians* (Table 7, Fig. 4).

In the Opatovický fishpond, mass development of *Spirogyra* is less frequent than in Nesyt. It was recorded in stands of *Glyceria maxima* near the open water, which provide microbiotopes similar to those in Nesyt. In addition, *Spirogyra* spp. occur as accessory species in other algal communities, especially in clusters of *Zygnema*. The latter were observed only sporadically in Nesyt.

E. Floating communities with dominant *Oedogonium:*

Several species of the genus *Oedogonium* build up important communities of floating algae in the Nesyt fishpond. With regard to the dominant species it was possible to distinguish two types differing both in species composition and time of occurrence:

i. Communities with dominant *Oedogonium varians* form the spring wave of the mass development of filamentous algae. As indicated in Table 7 and Fig. 4, they are often associated with *Spirogyra*. The most frequent epiphytes of *O. varians* are *Fragilaria vaucheriae* and *Synedra fasciculata;* other accessory species are *Melosira varians* and *Navicula gracilis*.

ii. Floating clusters with dominant *Oedogonium capillare* occur later and usually persist for longer periods. They are most frequently associated with *Sphaeroplea annulina* followed by *Cladophora fracta*. But clusters with dominant *Oedogonium capillare*, and those with dominant *Cladophora fracta* differ appreciably ($I_S = 1.44$; Table 7). During the period of its maximum development, the most frequent epiphyte on *Oedogonium* is *Amphora veneta*.

Table 7. Indexes I_S (see Sect. 3.1) of communities of filamentous algae in the Nesyt fishpond

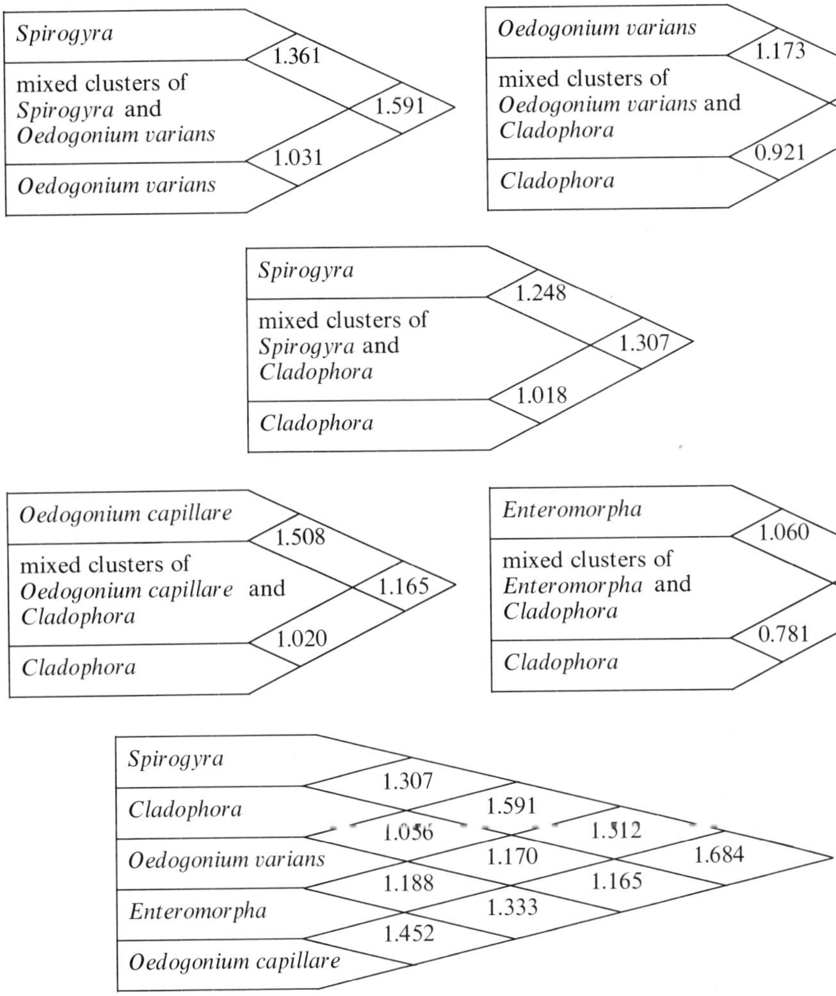

F. Communities with dominant *Cladophora:*

The dominant species of this community, *Cladophora fracta*, and possibly also *C. glomerata*, represent the most productive component of the algal vegetation in Nesyt in summer. They occur in the reed-belt and along its edges; sometimes, they form large mats on the open water surface and on the bottom below this surface, or they are loosely attached to submerged macrophytes such as *Potamogeton pectinatus* and *Zannichellia pedicellata*. The *Cladophora* species either form pure stands or constitute groups with other condominants, particularly with *Enteromorpha intestinalis* and *Oedogonium capillare*. The most important epiphytes are *Synedra ulna*, *S. pulchella*, *S. fasciculata*, *Cocconeis pediculus*, *Rhoicosphenia curvata*, *Amphora veneta*, *Gomphonema parvulum*, *Aphanochaete repens*, and *Homoeothrix stagnalis*. During the peak development of *Cladophora*, the epiphytes are unable to keep up with its growth and their number on the thallus is small. The gradual disappearance of *Cladophora* is accompanied by a mass development of filaments of *Thiothrix*.

G. Communities with dominant *Enteromorpha intestinalis*:

In Nesyt, *Enteromorpha intestinalis* represents an important element of this algal vegetation. Its occurrence in South Moravian waters is evidently due to their slightly increased salinity. Lagoons within the reed-belt provide the best conditions for the development of *Enteromorpha*, but clusters of their thalli may also cover large areas of the open water surface at their peak development in summer. Young developmental stages of *Enteromorpha intestinalis* frequently occur together with *Cladophora fracta*. The number of ephiphytes on *E. intestinalis* is smaller than on *C. fracta*. The most frequent accessory species are *Amphora veneta*, *Synedra* sp. div., and epiphytic species of *Nitzschia* and *Characiopsis*.

It is thus evident that the effects of a similar management of the Nesyt and Opatovický fishponds bring about a preferential incidence of several communities of the same kind: water blooms of blue-green algae, epipelic Oscillarieta, certain types of epiphyton, e.g., with dominant *Stigeoclonium*, etc. The remarkable differences in the original character of the two fishponds persist between their assemblages of numerous less frequent accompanying species.

Another difference pertains to the shares of various communities among the basic structural elements of the microvegetation. In Opatovický, the most important elements are planktonic algae present in both the pelagial and the reed-belt for most of the year, with dominant chlorococcal and/or blue-green algae. In Nesyt, the number of planktonic algal species is much the same, but many species occur solely in the eutrophicated western bay enriched with phytoplankton from the principal inflow (Mikulovský brook; Marvan and Sládeček, 1973, 1974). This is rich in chlorococcal algae, centric diatoms and in flagellates belonging to different taxonomic groups. Filamentous green algae, however, take up a prominent position in Nesyt, being the most productive component of the littoral algal vegetation. Their communities comprise relatively few dominants capable of rapid growth and reproduction under favorable conditions. If the fishpond is not filled up completely, floating filamentous algae may spread over its whole area.

The relative shares of the phytoplankton and filamentous algae may be influenced by fishpond management (see Hrbáček, 1962, 1964; Hrbáček and Novotná-Dvořáková, 1965, etc.). If the fishpond is understocked with fish, its phytoplankton is often highly diverse but its production is poor. A rich crustacean zooplankton reduces the number of small algal species, while large species may persist. In such a fishpond poor in phytoplankton, the bottom is irradiated sufficiently to create favorable conditions for the development of filamentous algae. On the other hand, in a densely stocked fishpond the secondary producers are rapidly consumed. In addition, the water is enriched both by mineral fertilizers and organic substances from the bottom, released by predation of large fish at the bottom and by an increased bacterial activity. A strong vegetative coloration of the water is brought about by chlorococcal and volvocal algae in addition to *Euglenophyceae* and water blooms of *Microcystis* and *Anabaena* while filamentous algae are suppressed. Mass development of filamentous algae (*Cladophora*,

Spirogyra, Hydrodictyon), caused evidently by a small fish stock, has repeatedly been observed in numerous fishponds of the Třeboň basin. In Nesyt, a rich zooplankton (Losos and Heteša, 1971; Sládeček, 1973) is also responsible for a reduced standing crop of phytoplankton. But Nesyt is not understocked, and the great importance of filamentous algae in this fishpond is also due to the chemical properties of its water (see Sect. 2.3). In view of the pond's increased total hardness and alkalinity, its pH remains permanently alkaline, increasing locally (particularly in the vicinity of clusters of filamentous algae) up to 10 or more. Under these conditions, extremely low concentration of free CO_2 may be responsible for the low phytoplankton production and, on the other hand, for a strong enhancement of the alkalitrophic *Cladophora fracta, Enteromorpha intestinalis* and other filamentous algae. It seems that even the development of water bloom in Nesyt has to be preceded by a decrease in water hardness through a partial biogenic decalcification frequently taking place in summer. In years when the Nesyt fishpond is filled up, floating algae tend to develop less abundantly. The same applies to the alkalitrophic macrophyte communities of the ass. *Potamogetoneto pectinati–Zannichellietum pedicellatae.* This favors the survival of clusters of filamentous algae near the water surface. This phenomenon has never been observed in the far less alkaline Opatovický fishpond.

References see pp. 335—337.

4.2 Dynamics of Algal Communities

P. Marvan, J. Komárek, H. Ettl, and J. Komárková

4.2.1 Phytoplankton

Plankton elements take up an important position in many types of littoral macrophyte communities in most fishponds (Fig. 1). The dynamics of their development appear to be influenced not only by changes in the physical, chemical and biotic factors in the littoral community, but also by those in the pelagial and by the intensity of water exchange between the two zones (Straškraba, 1963). The course of changes in the composition of phytoplankton (Fig. 2 and Fig. 2, Sect. 4.1) may be altered by fishpond management.

In the Opatovický fishpond, after an inconspicuous spring peak in the development of diatoms (*Asterionella formosa*, *Melosira granulata* v. *angustissima*, *Synedra* spp., *Stephanodiscus* spp.) there is a temporary decrease in the standing crop of phytoplankton, associated with a greater development of its zooplankton (April to May). The summer community with a rich representation of chlorococcal algae and with several species of blue-greens forming water bloom, develops from late May or early June and persists until September. It is followed by a renewed increase in diatoms (*Stephanodiscus* spp.) and flagellates (*Trachelomonas*, *Chrysophyceae* g. sp. div.). The plankton vegetation is discontinued either by the autumnal discharge of water from the fishpond for cropping the fish stock or it passes through an inconspicuous maximum of diatoms to a winter community with chrysomonads, diatoms (*Stephanodiscus*) and *Koliella* spp.

Water-blooms are formed by species common in eutrophic fishponds, such as *Aphanizomenon flos-aquae* f. *klebahnii*, *Anabaena flos-aquae*, *A. lemmermannii*, *A. circinalis*, and *Microcystis aeruginosa*. These develop from the middle of May and attain their maximum between the first half of June and September, when they start to disappear. Changes in their biomass are inconspicuous during the summer months. The water-bloom is often carried to the littoral, where it forms a continuous cover over the water surface. The accumulated mass of water-bloom then starts to decompose.

The phytoplankton of the Nesyt fishpond (Fig. 2, Sect. 4.1) shows similar seasonal changes in relative abundance of the principal groups of algae and in the total numbers of algae, but the latter are smaller than in Opatovický on the average (see Sect. 4).

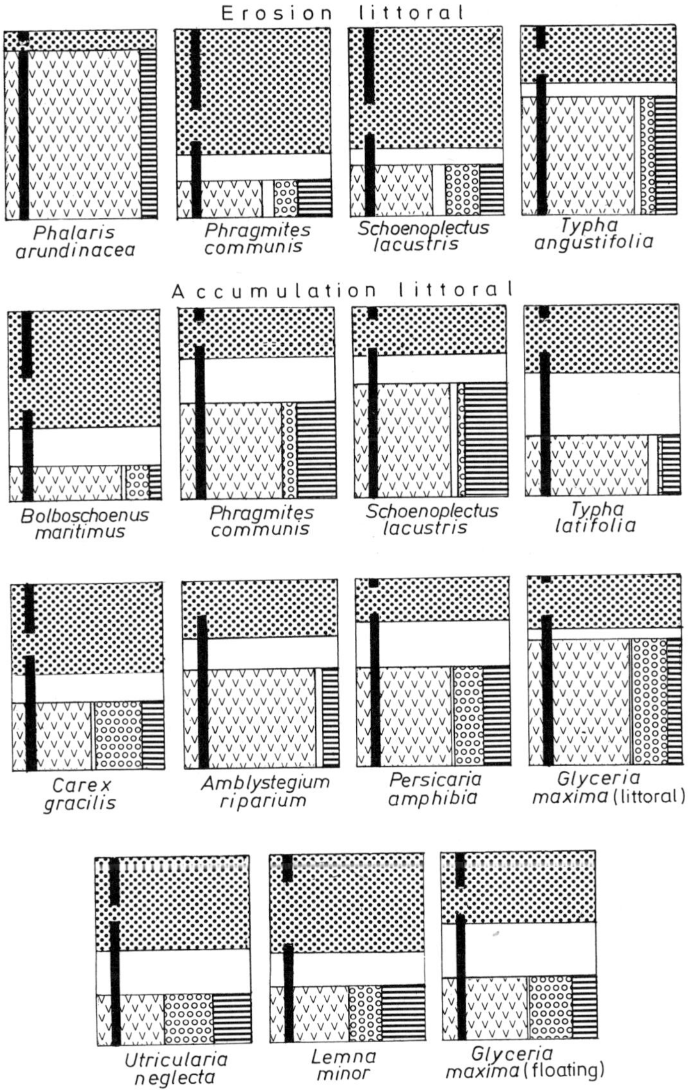

Fig. 1. Relative frequency of planktonic and benthic algae accompanying various types of macrophyte communities in Opatovický fishpond. Whole area of rectangles = 100%. Length of black columns on left gives frequency of species predominant in the plankton *(above)* and in the periphyton *(below)*

▒ typical planktonic species. ▤ free-living unicellular benthic algae. ▒ free-living filamentous algae. ▒ sessile algae. ☐ intermediate types

The spring peak in the number of phytoplankton is followed by a sudden fall usually resulting from the predatory activity of the zooplankton and from competition with non-planktonic algae and submerged macrophytes. Marked differences exist between the developmental dynamics of the principal phytoplankton components in the first and second year of the two-year rotation in fishpond management. As indicated by Losos and Heteša (1971) and by results from 1970–1972 (Komárek et al., 1973), water-bloom starts later but persists longer in the first year of the rotation, and *Aphanizomenon flos-aquae* becomes its principal component. Other species, particularly *Anabaena* spp., are absent or greatly suppressed in these years. In some years, an enormous development of the submerged macrovegetation as well as of littoral floating algae suppresses completely the onset of water-bloom. In the second year of the rotation, water-bloom starts earlier. *Aphanizomenon flos-aquae* is again the first to start its mass development, but later it is usually replaced by other dominants: first by *Anabaena* and then by *Microcystis*.

The incidence of plankton in the helophyte stands is frequently due solely to a horizontal mixing of pelagial and littoral water. In Opatovický, however, the share of numerous typically planktonic algae was repeatedly observed to be higher in periphytic samples than in the surrounding water. This applies, for example, to stands of *Utricularia*. A similar situation was observed by Brandl et al. (1970), who found in other fishponds a considerably higher representation of planktonic algae amidst different submerged plants than outside these stands. Although shading reduces the phytoplankton production here, the conditions for the development of planktonic algae may be more favorable in the littoral than in the pelagial. For example, water-bloom starts earlier in Nesyt in the sheltered and shallow lagoons inside the reed-belt than in open water, and a temporarily increased abundance of other planktonic algae was recorded here as well. However, the conditions for phytoplankton development deteriorate here later because of a mass development of littoral filamentous algae.

4.2.2 Epiphyton

The regime of irradiance and the nature of the higher-plant substrate represent the most important factors determining the numbers of epiphyton and its spatial distribution as well as the developmental dynamics of the individual epiphytic types (see p. 301; Tab. 4, Sect. 4.1). In both Opatovický and Nesyt, more favorable conditions for development of macroscopically visible epiphytes are found only in loose helophyte stands.

In the Opatovický fishpond, the seasonal course of the vegetation of sessile communities was much the same on all sites investigated (Fig. 2). As a rule, the usual diatom maxima occur in spring, but they are less pronounced in the highly eutrophicated fishponds than in rather oligotrophic lakes (see, e.g., Meschkat, 1934). These maxima are followed by a dominance of blue-green, chlorococcal and green filamentous algae in summer and, again, by inconspicuous maxima of diatoms in fall. In winter, the littoral is often either not flooded or frozen down to the bottom and, therefore, specific winter species do not develop here. Local modifications of this scheme occur mainly in association with the growth and development of the macrophyte vegetation determining the irradiance of the biotope.

In both Opatovický and, particularly, Nesyt, the epiphyton provides an important source of free-living developmental stages of several algae. This applies

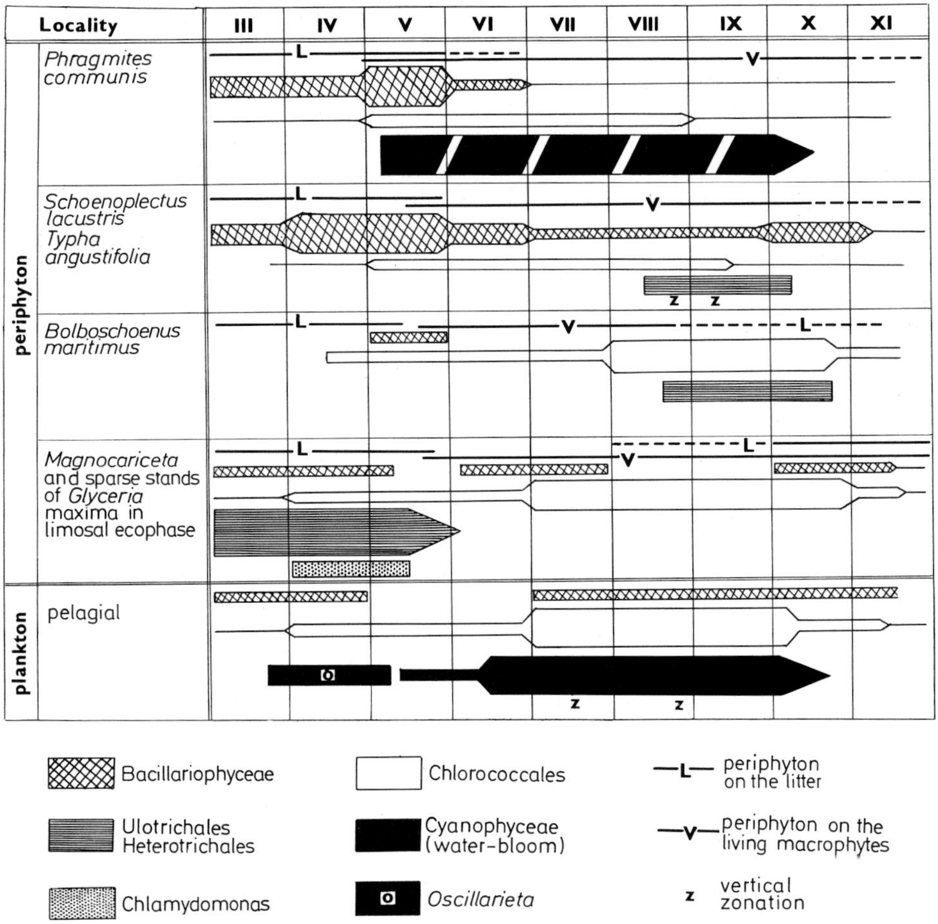

Fig. 2. Seasonal changes in abundance of main algal groups in different localities of Opatovický fishpond during vegetation seasons of 1971 and 1972. (After Komárek, 1973a)

mainly to green filamentous species of the genera *Oedogonium* and *Cladophora*, and to young developmental stages of *Tribonema*, prior to the phase of their mass occurrence in free-living clusters. The same applies to numerous diatoms. Many species of *Synedra*, *Nitzschia*, and *Gomphonema*, after detachment from the epiphyton, become important components of the littoral bioseston, and adapt themselves morphologically to this way of live. Several diatoms forming filamentous or chain-like colonies also start life cycles as epiphytes (*Fragilaria*, *Diatoma tenue*; Marvan, 1972). The capability to adapt to a free-floating way of life appears to be of great importance for the microphytes occurring in the macrophyte communities, facilitating the changes in their position in the ecosystem structure and a better use of the ecological niches available. Epiphytes living on the helophyte litter are a far more important source of natant algal clusters than epiphytes on the live shoots.

4.2.3 Floating Clusters of Filamentous Algae

The production of these components of the algal vegetation is determined by the dynamics of a relatively small number of dominants. As in the phytoplankton, the migration of algal clusters on the water surface is responsible for a continuous exchange of algal matter between the littoral and pelagial. Certain dominants originate in the epiphyton (or epilithon, see Sect. 4.1), but the complete life cycle of other dominants takes place in free-floating stages (*Spirogyra*, *Enteromorpha*); others may even form a component of the communities colonizing emerged bottoms (*Vaucheria*). Nevertheless, the developmental dynamics of these algae have numerous features in common.

In both Opatovický and Nesyt, *Tribonema* spp. are the most important filamentous algae of the outer littoral. At the peak of their development, *Tribonema* clusters attain a biomass of 5–80 g m^{-2} (dry weight). However, the output of this organic matter to the open water is only small. Further development of filamentous algae is limited in the other littoral by shade below the helophyte canopies and by lack of water in some years. A seasonal sequence of several algal dominants is usual in the inner fishpond littoral. Species of *Fragilaria* and *Spirogyra* dominate in Opatovický, both being restricted only to the reed-belt. *Fragilaria* is responsible for most of the periphyton production in spring, and its other peak occurs in fall. *Spirogyra* attains the maximum biomass in early summer, mostly in the lagoons within the reed-belt. Penetration of the *Spirogyra* clusters into the pelagial is thus prevented mechanically. In Nesyt, the development of individual dominants is greatly influenced by competition for nutrients and radiation. During the growing season of 1972, after a partial winter drainage of the fishpond, the following sequence of the dominants was recorded:

Vaucheria sp.: Mass development of its submerged stands occurred soon after refilling of the pond with water in spring, especially in the flooded upper parts of the littoral. In early May, *Vaucheria* formed an almost continuous carpet on the bottom of the lagoons. These algae partly passed over to the floating stage, but only few of them entered the pelagial. Their vegetation season was short, and they were soon replaced by another algal assemblage.

Oedogonium varians and *Spirogyra*: The initial stages of *O. varians* develop mainly as epiphytes on macrophyte litter. An outburst starts as soon as the water temperature reaches 10° C. In 1972, *O. varians* and *Spirogyra* formed the second wave of mass incidence of littoral algae, which culminated some three to four weeks after refilling of the pond. During the phase of intense growth of *O. varians*, zoospores are formed, but these colonize a different substrate. Older epiphytes composed of filaments measuring several decimeters are readily torn off and carried far out into the open water. By contrast to *Oedogonium varians*, *Spirogyra* originates in shallow and unshaded parts of the littoral, mainly in the western bay. The period of its mass development coincides with that of *O. varians*. Both algae become distributed throughout almost the entire fishpond. Responsible for this widespread distribution was an enormous development of submerged macrovegetation. At the end of May, the estimated maximum total biomass of both algal species was roughly 100 t dry matter in the whole fishpond, corresponding to an average of about 40 g m^{-2} (Lhotský and Marvan, 1975). Their mass development was terminating at this time, as documented by their starting decay. Locally, the amount of concentrated algal matter attained more than 800 g m^{-2} of dry *biomass*.

Cladophora fracta: Its thalli appear on the water surface in the Nesyt fishpond soon after the ice-thaw, but the species attains its maximum development when the wave of *Oedogonium* and *Spirogyra* is receding. The source of the floating stages are epiphytes on dead shoots, mainly of *Typha* and *Phragmites;* the alga frequently frees itself together with pieces of crumbled substrate. During the floating stage, the structure of the thallus undergoes changes, caused by prevailing intercalary cell division. Released clusters migrate on the water surface and, under favorable conditions, form extensive continuous carpets on sites with a well-developed submerged vegetation. In mid-June 1972, a rough estimate of total biomass of *Cladophora* in Nesyt indicated approximately 170 t, i.e., 60 g m^{-2} of dry matter. At the beginning of July, the fishpond is colonized by young and photosynthetically highly active thalli, and by older and less active clusters. The thalli are heavily encrusted with calcium carbonate, whose concentration in the water thus becomes somewhat reduced. Direct insolation is responsible for a local complete decay of chlorophyll and the death of a certain number of clusters on the water surface. At the beginning of August, *Cladophora* starts to recede and is replaced by another dominant. Measurements of oxygen production disclosed its low photosynthetic activity at this time. The number of bleached and dead clusters increases, and the surviving thalli form characteristic pyriform akinetes toward the end of the growing season (in 1972 this occurred as early as July).

Oedogonium capillare: Another dominant of floating algal communities started later than *Oedogonium varians* and persisted for a longer period. In 1972, its maximum development coincided with that of *Cladophora*, while in 1971 it was still abundant in October. Its numbers among epiphytes are lower than those of *Oedogonium varians*.

Enteromorpha intestinalis: In summer, this species replaces *Cladophora* and forms the wave of mass development of littoral algae. It hibernates in the form of old tube-shaped thalli which produce thin and at first uniseriate lateral branches in spring. Young developmental stages formed by isolated branches resemble the thalli of *Cladophora* and are often found together with them in one cluster. Gradually, during summer, the lateral branches widen and break off. At this stage, the thalli are almost void of epiphytes. Clusters of young thalli mostly float on the water surface or concentrate in shallows and mix with other filamentous algae. They increase gradually in volume and form long tubes which are about 2 cm wide. At this stage, they may prevail among the algae, and epiphytes (particularly *Amphora veneta*) become more abundant on the thalli. Although *Enteromorpha* is best adapted to mid-summer conditions, there is evidence of an unfavorable effect of strong insolation on its thalli floating on the water surface, as in *Cladophora*.

Not unlike planktonic algae, floating clusters of filamentous algae may also be carried passively to poorly irradiated sites where the community respiration exceeds its photosynthesis. In the fishpond littorals, there is an uninterrupted transition from this extreme case to fully irradiated microbiotopes. Some measurements performed at Nesyt in 1972 indicate that slight shading of algal clusters may favorably influence their photosynthetic activity. This is the case in loose helophyte stands such as those of *Bolboschoenus*. On the other hand, on the windward site, these stands favor the accumulation of drifted and gradually decomposing algal matter. A shoal of zooplankton, particularly large *Cladocera*, has frequently been observed in the vicinity of these clusters. An evaluation of the mass and energy budgets appears to be more complicated in ecosystems with dominant filamentous algae in the microvegetation than it is in ecosystems with dominant phytoplankton. The synthesis and decomposition of algal matter may be spatially separated and an exchange of matter between the individual sub-systems becomes an important feature of the nutrient cycling in the whole fishpond.

References see pp. 335—337.

4.3 Primary Production and Functioning of Algae in the Fishpond Littoral

J. KOMÁRKOVÁ and P. MARVAN

Three principal types of producers can be distinguished in a pond littoral: (1) macrophytes; (2) littoral plankton; (3) periphyton, including epiphyton as well as filamentous algae floating amidst higher water plants (see Tamas, 1964; Marvan et al., 1973, and others). The shares of individual structural elements of the vegetation in the total littoral primary production vary considerably (Straškraba, 1963; Pieczyńska and Szczepańska, 1966; Komárková, 1973; Komárková and Komárek, 1975; see Table 1) and may change markedly during the year (Figs. 1, 2). An investigation of the ecological function of algae in fishponds is impeded by a number of methodological problems. It is inconvenient to use without modification the radiometric ^{14}C method for estimating the algal primary production in heavily fertilized fishponds and in their polluted littorals. For these reasons, the oxygen method of light and dark bottles was employed, by which the gross primary production was estimated with sufficient accuracy. In 1972, a method suggested by Pieczyńska (1965) was used in a study on primary production by the periphyton in various kinds of helophyte communities at the Opatovický fishpond (for details see Komárková and Komárek, 1975). In the subsequent years, Assman's (1951) method was employed when working in a *Phragmites* stand where the submerged stem portions are not green except for a short period in spring. Oxygen concentrations were measured by the Winkler method and, in summer, by a modification of this method after Bruhns (Standard Methods, 1965). Chlorophyll was estimated by a method recommended by Vollenweider (1969).

Table 1. Share of 3 components of littoral producers in total primary production in littoral zones

	Eutrophic Lake Mikołajskie, NE Poland (Pieczyńska and Szczepańska, 1966)	Fishponds near Blatná, SW Bohemia (Straškraba, 1963)	Opatovický fishpond near Třeboň, S Bohemia (Komárková and Komárek, 1975)
Macrophytes	57%	70%	53 –83.5%
Phytoplankton	20%	7%	9 –36%
Periphyton	23%	21%	5.5–11%

The assessed values of oxygen production were related either to water surface area or to its unit volume.

In Nesyt, production was measured mainly in floating algae. The amount of algae used for exposure was determined as wet weight after a short-time (1 min) centrifugation of the cluster in a battery-propelled field centrifuge (Kubíček, 1969), and later converted to dry weight. Daily production was estimated from the measured oxygen production and the result related to unit dry weight. Only small amounts of algae were used in order to avoid the influence of mutual shading of the filaments on the final result. In addition, several other methods described by Ettl et al. (1973a) were used.

Algal production was calculated from the data on oxygen production expressed in mg l^{-1} under the assumption that 1 g O_2 produced corresponds to 3.12 kcal (= 3.64 Wh), for the pelagial.

4.3.1 Phytoplankton

In the open water of the Opatovický fishpond, according to measurements in 1972, the primary production by the phytoplankton was 3600 kcal m^{-2} (net production, calculated as 75% of gross production).

For Nesyt, no systematic measurements are available for an estimation of the annual primary production of phytoplankton. According to an estimate based on earlier measurements by Heteša (in Pelikán et al., 1973), the annual production by the phytoplankton is about 2800 kcal m^{-2}; this is considerably less than in the Opatovický fishpond. A rough calculation indicates that in 1972 the total phytoplankton production in Nesyt hardly surpassed 500 kcal m^{-2}. The reasons for this difference in the role of phytoplankton in the two ponds are discussed in Section 4.2.

In order to understand the role played by algae in the littoral of the Opatovický fishpond, five sites within the littoral helophyte communities which remained flooded for the whole year, were selected for study. On these sites, the number of the helophyte shoots per m^2 corresponded with their average density and biomass in each respective community. Two sites each, in two stands of *Glycerietum maximae*, one site each in stands of *Bolboschoenus maritimus*, *Schoenoplectetum lacustris*, and *Phragmitetum communis*, and one site in the open-water area of the fishpond were investigated. In 1973 to 1975, for similar studies, two transects across two helophyte communities most typical of South Bohemian fishponds were selected, *Phragmitetum communis* and *Glycerietum maximae* (Figs. 11 and 12 in Sect. 2.3).

The annual phytoplankton production and its comparison with macrophyte and periphyton production are shown in Tables 2 and 3. Algae grow for the whole year, which helophytes do not. The highest production was found in spring. According to measurements in the Opatovický fishpond, the lowest estimate of net production derived from the difference between the initial oxygen concentration and that in the light bottle (each value calculated separately for each sample) proved to be 70% of the gross primary production in the pelagial (an annual

Table 2. Annual net and gross production in the pelagial and in stands of various primary producers (macrophytes) in the littoral of Opatovický fishpond in 1972

	Macrophytes			Phytoplankton			Periphyton		
	Density of stand Plants m^{-2}	NPP (DW) g m^{-2}	NPP kcal m^{-2}	GPP (O_2) g m^{-2}	NPP (lowest estim.) kcal m^{-2}	NPP/GPP (lowest ratio) %	GPP (O_2) g m^{-2}	NPP (lowest estim.) kcal m^{-2}	NPP/GPP (lowest ratio) %
Pelagial				1,651	*3,606*	70.0			
Phragmites	61	872	*3,913*	179	*171*	30.7	136	*250*	52.7
Schoenoplectus	157	650	*2,604*	543	*939*	55.4	316	*349*	58.0
Bolboschoenus	72	334	*1,321*	379	*779*	65.8	262	*238*	63.8
Glyceria[a]	202	970	*4,682*	289	*471*	52.2	148	*236*	55.3
Utricularia[a]		6.9	*232*				64.4	*143*	71.2
Glyceria[b]	189	659	*3,109*	277	*458*	53.1	170	*253*	46.2
Utricularia[b]		2.5	*85*				23.7	*72*	67.8

[a] Eutrophic bay.
[b] Heron bay; see Fig. 5 in Sect. 1.1.

Table 3. Primary production in *Phragmites* transect [gross primary production (GPP) of algae and net primary production (NPP) of macrophytes] in 1974 (1. April 1974–31. March 1975). (Cf. Fig. 11 in Sect. 2.3)

Locality		1	2	3	4	5	6
GPP of algae							
O_2	water	158	119	256	575	464	1,322
g m^{-2} yr^{-1}	periph.	38	109	31	—	82	—
NPP of *Phragmites*							
DW g m^{-2} yr^{-1}		1,400	2,001	2,640	—	1,762	—
Depth (m)		0.14	0.16	0.32	0.52	0.71	1.12
Density of stems m^{-2}		125	233	104	80		
Max. height of stand (m)		2.50	2.70	2.95	—	2.45	—

mean; Straškraba, 1963; Fott, 1972). The data on the annual net primary production (NPP) by the littoral phytoplankton growing under helophyte canopies were calculated in the same way. Ratios between the lowest net and gross production estimates, expressed in percentage terms, are given in Table 2. Negative net production values have also been found below a fully developed *Phragmites* canopy. On the other hand, the phytoplankton net primary production reached 90% on shallow sites in spring. Data on net production by the helophytes refer to their seasonal maximum biomass converted into energy values using the equivalents given by Dykyjová and Přibil (1975). In this case, the difference between seasonal maximum biomass and annual net aboveground production was regarded as negligible in all helophyte communities involved except *Glyceria*, where a 10% correction was added (see Sect. 3.1.7).

In all instances, the helophyte production in the littoral surpasses considerably that of the phytoplankton (Table 2). As the phytoplankton represents the most important source supporting the trophic steps leading to fish production, and the phytoplankton development is controlled by the density of the helophyte communities, an excessively wide shallow littoral zone is regarded as undesirable in intensely managed fishponds (Straškraba, 1963; Komárková and Komárek, 1975). If the total primary production by algae in a whole managed fishpond is considered from this aspect, the participation of the littoral in aquatic primary production is relatively small in comparison with the production in the open water (Straškraba, 1963; Komárková and Komárek, 1975). These data contradict those from several lakes inferring that the littoral zone is more productive (Assman, 1951; Wetzel, 1964; Pieczyńska, 1971a) and its phytoplankton or periphyton play an important part in the primary production of the whole lake. The oligotrophy of the lakes appears to explain this discrepancy.

No helophyte community is ever fully homogeneous and the algal production in it varies with its varying canopy density. In order to prevent this source of error, algal production was studied along transects. In Table 3 and Figure 1, a comparison is given of the production by phytoplankton and other productive components on five sites within a *Phragmitetum communis* stand. A marked division of this transect into two parts is evident (see also p. 138). The sites 1 and 2, situated in the outer littoral, beyond a bench of drifted detritus, are characterized by a small phytoplankton production but by high production by periphyton occur-

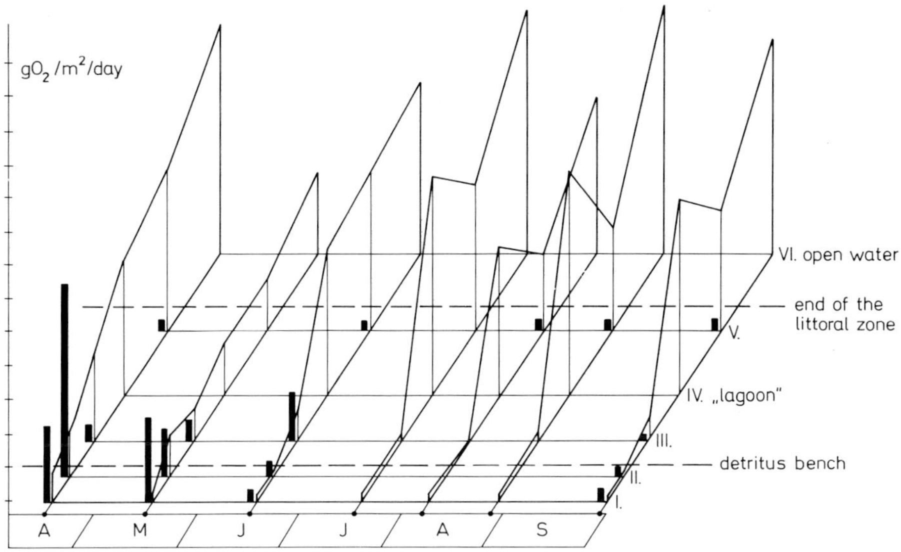

Fig. 1. Gross primary production of microphytes (seasonal course) in transect across *Phragmites* stand in Opatovický fishpond (1973). *Black columns:* periphyton production; cf. Fig. 11. Sect. 2.3.

ring mainly in spring. In Figure 1, the production values are related to unit area of water surface; with regard to the shallowness of the outer littoral (sites 1 and 2) the data may seem distorted, but they have been confirmed by data related to unit volume of water. The inner littoral is far more important for the total primary production in the pond. It markedly interacts with the pond pelagial and the phytoplankton is relatively more important here. A similar situation was also found in other helophyte communities situated in the inner littoral (Komárková and Komárek, 1975).

The limitation of algal production in the pond littoral by insufficient irradiance below the helophyte canopies has been confirmed by a depression in phytoplankton production on site 5 in summer (Fig. 1). According to chemical analyses (see Sect. 2.4) the phytoplankton production was hardly limited by a lack of nutrients in the water. An exception may have been the period following the spring peak of the diatoms and filamentous algae. Straškraba and Pieczyńska (1970) have also proved that poor irradiance controls algal production in the littoral in summer. They recorded 5–55% of relative illuminance at the water surface in a developed *Phragmites* stand. In the Opatovický fishpond, such a high transmission through helophyte canopies during the peak season was never encountered. Data on irradiance by Ondok (1977; see also Sect. 2.2) together with parameters characterizing the development of the helophyte stands (see Sect. 3.1.5) and the corresponding data on total primary production by algae, illustrate an inversely proportional relationship between the developmental stage of the stand and algal production (Fig. 2). Other data, given in Section 2.2, confirm that the PhAR irradiance at the water surface did not exceed 5% during the peak of the growing season (July–Aug.). Nonetheless it was possible to detect some oxygen production by the phytoplankton even in this poorly irradiated water.

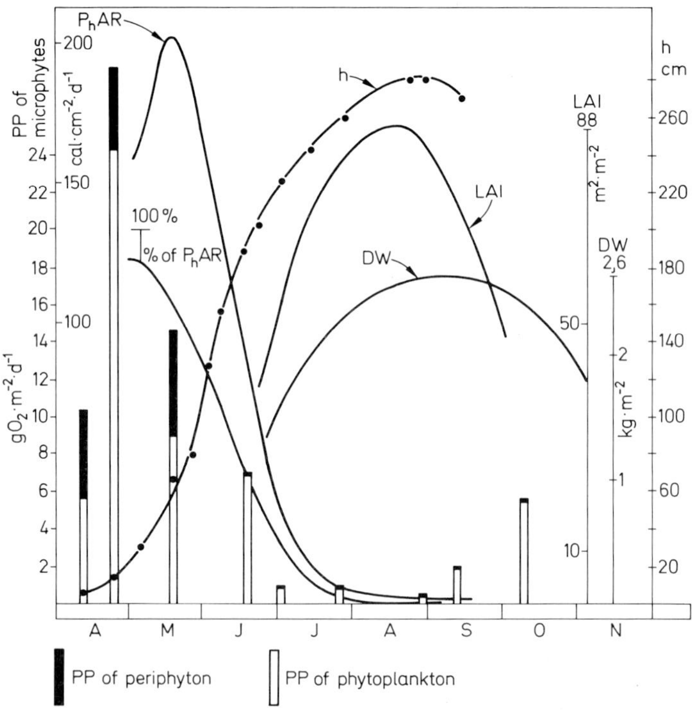

Fig. 2. Dependence of gross production of algae *(columns)*, PhAR falling on water surface and development of *Phragmites* stand in Opatovický fishpond in 1973; *LAI*, leaf area index; *DW* shoot biomass (in dry weight); *h*, height of *Phragmites* stand above bottom; PhAR measured at water surface and above stand

Table 4. Percentage of daily PhAR falling on water surface in stands of littoral helophytes in Opatovický fishpond, in comparison with irradiance above stand

Irradiance (PhAR) above the stand Season	Σ PhAR 588–3,230 kcal m^{-2} d^{-1} June–Aug. 1966	Σ PhAR (297)–1,865–(3,270)[a] kcal m^{-2} d^{-1} July–Aug. 1972
Glyceria maxima (terrestr.)	0.69– 0.77%	
Glyceria maxima (floating)		7.1–30.9%
Phragmites communis	0.55– 0.80%	1.2–23.1%
Schoenoplectus lacustris		72.6%
Typha angustifolia	15.50–19.00%	15.2–38.6%
References	Dykyjová, 1971	Ondok, 1977[a]

Measured in high vegetation season, on sunny and cloudy days; sparse and dense stands measured.
[a] Data depend on helophyte shoot density.

The data in Table 4 may be compared with the annual oxygen production per liter by the phytoplankton living in these stands, and in unshaded open water, the latter producing nearly 60 g O_2 l^{-1} $year^{-1}$ (calculated as O_2 dm^{-2} $year^{-1}$ divided by depth). Within the *Glycerietum maximae* stands the phytoplankton production attains 87.6% of this value, and 38.4, 82.4, and 58.2% within *Phragmite-*

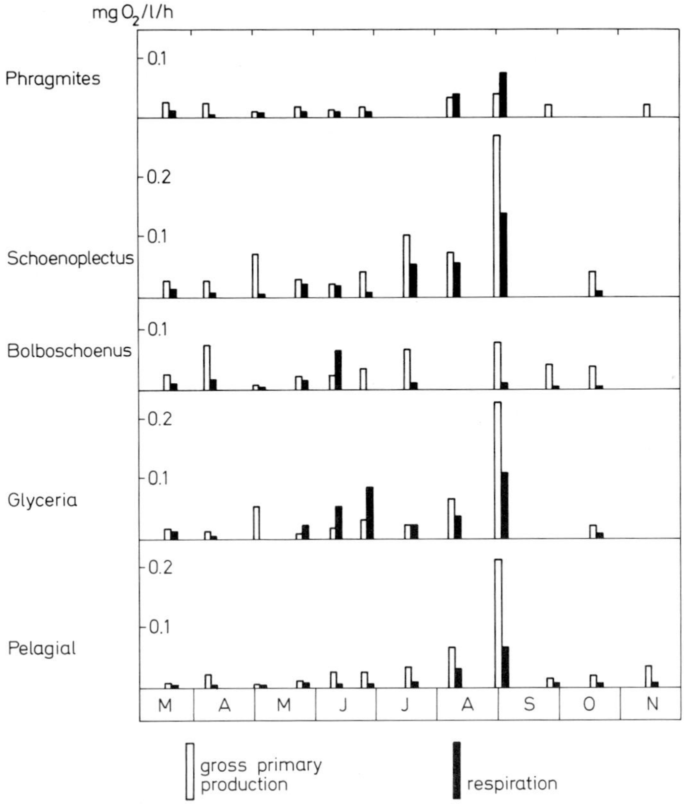

Fig. 3. Comparison of gross primary production and respiration of nannoplankton in different types of littoral macrophyte stands in Opatovický fishpond in 1972

tum communis, *Schoenoplectetum lacustris*, and *Bolboschoenus maritimus* stand respectively. The high phytoplankton production amidst the shoots of *Glyceria* is to be ascribed to a high, though short-term production of chlamydomonads and small fragments of free-floating filamentous algae in spring; in the remaining stands, the production is evidently reduced in proportion to shading.

In Figure 3 and Table 5 several values of gross primary production (GPP) are compared with the respiration rate of the nannoplankton (algae, zoonannoplankton and bacteria) in various types of littoral helophyte communities. Oxygen seems to be produced practically throughout the year, although anerobic conditions occur temporarily on certain sites. But in numerous localities, this oxygen frequently appears to be used up immediately for respiration by the algae, bacteria and zooplankton as well as in various processes of decomposition and chemical oxidation. During summer, the decomposition processes are intense over large areas in the littoral (see Sect. 5). The total gross primary production and oxygen consumption were also compared on various sites along the transects in the *Phragmitetum* and *Glycerietum* communities after exposure of the water samples in dark bottles for 24 h. The results shown in Table 6 confirm that in several instances, oxygen produced by algae in dense stands of *Phragmites* cannot satisfy

Table 5. Ratios of lowest net and gross production estimate. Nannoplankton in littoral and pelagial of the Opatovický fishpond during 1972

Date	Pelagial	Stands of:			
		Phragmites	*Bolboschoenus*	*Schoenoplectus*	*Glyceria*
23 March	56	57	58	55	33
11 Apr.	85	85	78	73	66
4 May	61	64	83	76	64
22 May	92	96	28	26	−58
6 June	72	41	−10	16	−55
27 June	82	53	78	78	−69
27 July	72	31	86	48	1
7 Aug.	54	−17	70	29	50
5 Sept.	67	−87	77	48	52
26 Sept.	50	27	76	26	52
19 Oct.	83	70	71	77	58
15 Nov.	71	68	90	77	63

Table 6. Comparison of phytoplankton and total gross primary production (GPP_p and GPP_{tot}) with respiration of nannoplankton (R_n) in the *Phragmites* transect in spring and summer of 1973 in Opatovický fishpond

Locality	1		2		3		4 (lagoon)		5		6 (open water)	
	R_n	GPP_p GPP_{tot}	R_n	GPP_p GPP_{tot}	R_n	GPP_p GPP_{tot}	R_n	GPP_p	R_n	GPP_p GPP_{tot}	R_n	GPP_p
25 Apr.	7.0	6.6	3.5	7.7	1.6	5.1	0.9	5.6	2.9	5.9	3.3	5.6
	>	25.0		39.6		6.1				6.4		
14 May	5.9	2.0	2.0	8.7	1.6	2.5	2.2	2.7	—	2.0	1.6	2.4
	>	28.3		18.9		4.0				2.0		
7 June	0.5	0.9	1.7	0.8	1.3	1.6	2.2	5.3	7.5	4.9	2.5	3.4
		2.4	>	2.6		4.0			≫			
5 July	4.2	0.6	3.0	0.0	3.0	0.3	4.1	13.0	2.5	7.4	1.8	8.3
	≫	0.6	≫	0.6	≫	0.4				7.4		
2 Aug.	4.1	0.3	1.3	0.2	—	0.0	9.3	6.8	—	3.2	1.3	4.1
	≫	0.3	≫	0.2	≫	0.0	>			3.5		
29 Aug.	3.9	1.1	1.6	0.0	2.8	0.9	4.7	11.5	5.3	4.7	4.1	7.6
	≫	1.1	>	3.4	≫	0.9			≫	5.2		

>, ≫, nannoplankton respiration outweighs oxygen production of nannoplankton or total production; data expressed in mg $O_2\,l^{-1}\,d^{-1}$; for site characteristics see Figure 11 in Sect. 2.3.

even the total needs of the nannoplankton, whose daily oxygen budget becomes negative. Although the light and dark bottle method does not inform on the share of algal respiration in the total respiration it is evident that, in extreme instances, even the oxygen budget of the phytoplankton itself may become negative (see also Straškraba and Pieczyńska, 1970).

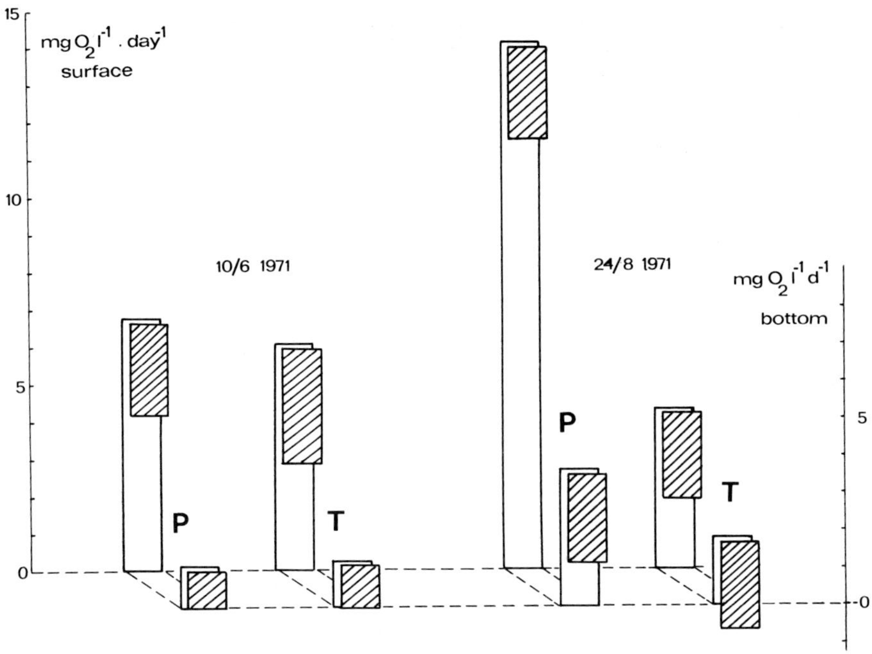

Fig. 4. Oxygen balance of nannoplankton in the pelagial (*P*) and in a loose littoral stand of *Typha* (*T*) in Nesyt fishpond, from surface *(left scale)* and bottom *(right scale)* measurements (after Heteša, unpublished); *white columns:* daily net oxygen production; *lined columns:* daily respiration

Figure 4 shows the results of two similar measurements in Nesyt. These were made in the open water and in a loose *Typhetum angustifoliae* in which the relative irradiance was about 70% at water surface. At a depth of approximately 0.8 m, the respiration of the whole aquatic community (including its heterotrophic components) surpassed its daily gross production. In the whole water column, the respiratory losses amounted to almost 75% of total gross production, as compared with only 27% in the open water. In dense stands of *Typha* and *Phragmites*, which are typical of Nesyt and produce some 2.5 to 3 kg m^{-2} of total biomass per year (see Sect. 3.1.7), the contribution of phytoplankton to the total primary production is evidently negligible. In *Phragmites* stands not cut in winter, this statement applies even to the early spring season. The littoral phytoplankton of Nesyt does not participate significantly in the total primary production even in loose stands of *Typha angustifolia* or *Bolboschoenus maritimus*, in which synusia of floating filamentous algae or duckweeds frequently become dominant (see Sect. 3.5).

In Opatovický, the production capacity of the algae decreases considerably during the summer months, but their biomass is quite high. Table 7 gives a survey of data (in terms of chlorophyll concentration in water) obtained from the four sites investigated. During the development of the cyanophycean water-bloom, the water contained more than 800 mg chlorophyll l^{-1}, corresponding roughly to a

Table 7. Chlorophyll *a* content in phytoplankton in three littoral stands and in pelagial in Opatovický fishpond in 1972 (μg l^{-1} of chlorophyll *a*)

Date	Pelagial	*Phragmites*	*Schoenoplectus*	*Glyceria*
23 March	30.6	56.6	84.4	100.1
11 Apr.	45.6	39.3	60.8	24.4
4 May	1.7	23.4	22.4	—
22 May	1.6	76.9	49.5	50.4
6 June	23.3	19.4	26.1	70.4
27 June	57.1	37.3	49.4	72.1
27 July	45.4	45.4	42.7	53.8
8 Aug.	58.3	73.4	70.2	117.9
5 Sept.	407.3	822.7	619.9	508.3
26 Sept.	84.5	595.0	418.5	—
19 Oct.	53.3	220.2	158.1	102.0
15 Nov.	51.3	90.2	—	—

biomass of 40 mg l^{-1} of fresh weight (=approximately 8 mg l^{-1} of dry weight). The low photosynthetic activity of the water-blooms in the littoral has already been pointed out (Komárková and Komárek, 1975). Straškraba (1963) also has emphasized that there may be a striking disproportion between plankton biomass and production in the littoral zone. In summer phytoplankton brought into the helophyte communities from the open water or from other better irradiated parts of the littoral becomes the only source of phototrophic planktons in dense stands of the fishpond reed-belt. The participation of phytoplankton in the oxygen production within the reed-belt thus also depends on conditions for its development in the adjoining areas. This dependence is most pronounced with water-blooms.

Water-bloom originating in the open water is carried by the waves deep into the reed-belt, where it starts to decompose. This can be observed particularly in stands with leafless stem bases (*Schoenoplectus*, *Phragmites*, partly also *Bolboschoenus*) situated along windward fishpond shores. The decomposition is less rapid in this shaded water than it is in the warmer and strongly irradiated open water. The helophyte communities growing along the shores of the Opatovický fishpond appear to collect part of the excessive water-bloom and to act as a sanitary zone in the fishpond. After mineralization, the nutrients dissolved in the water are returned from the littoral to the open water (Schröder, 1975).

A division of the littoral transect into its outer and inner zones has been found correct even with respect to the oxygen regime. In the outer zone (cf. Pieczyńska, 1971b—eulittoral), on sites 1 and 2, algal production represents the principal source of oxygen in the water, followed by exchange of oxygen with the air. But even here, anerobic conditions sometimes occur in summer. In the inner zone (sites 3—6), large amounts of oxygen are obtained by exchange of water between the littoral and pelagial. Abundant algae brought in from the open water continue to produce oxygen in the less shaded parts of the reed-belt and in lagoons within the helophyte stands. The actual oxygen concentration increases from site 1 to site 3, while it remains much the same further along the inner part of the transect, corresponding to that in the open water.

4.3.2 Periphyton

In Opatovický, the measurements of primary production by the periphyton were made simultaneously with those of phytoplankton production. Production by the epiphyton was measured as well as that by free-floating or loosely attached filaments of diatoms and other algae. The importance of the epiphyton itself, growing on live stems and leaves of helophytes, varies: it is negligible in some types of communities *(Phragmitetum communis, Glycerietum maximae)*, while it acquires significance in others (*Schoenoplectetum, Bolboschoenetum* stands). In the annual budget, however, the epiphyton production is always inferior to that of loosely attached algal forms which participate mainly in the spring outburst of algae (see Sect. 4.2). For 1972, data are available on gross production by periphyton on the same six sites where phytoplankton production was followed (Table 1). Net production was calculated by employing the same conversion factors as with the phytoplankton (Sect. 4.3.1). Periphyton was also studied in detail along the transect in the *Phragmitetum* and *Glycerietum maximae* stands; for the results see Table 2.

In dense stands of *Phragmitetum communis* periphyton becomes a significant producer only along the relatively better irradiated stand edges, cut sites, etc. Except for a short period in spring, the share of the periphyton in total algal primary production in the reed-belt is practically negligible (Fig. 1). This also applies to the periphyton in other helophyte communities along the fishpond shores, although their radiation régime is more favorable (Komárková, 1973; Komárková and Komárek, 1975).

Measurements made in Opatovický in 1972 illustrate this situation. In the summer, the autochthonous periphyton participated negligibly in the primary production by littoral algae. The spring (second half of April and May), however, was marked by an outburst of diatoms (epiphytes, particularly filamentous species) occurring during a temporary decline in the standing crop of phytoplankton in the open water. The decomposing plant remnants also release nutrients, CO_2 and heat. The development of epiphytes and clusters of filamentous and later of free-floating algae *(Tribonema, Microspora)* is promoted by the presence of a rich crustaceoplankton grazing on the phytoplankton and bacteria (see also Sect. 4.2).

The finding that allochthonous phytoplankton may participate in the gross production to a greater degree than autochthonous epiphyton applies to the situation in the Opatovický fishpond but cannot be generalized. Pieczyńska (1965) gives slightly lower values of primary production by the littoral phytoplankton than by the periphyton in the epilittoral of Lake Mikołajskie. In Nesyt, the production of the floating littoral algae may surpass that of the phytoplankton even on sites outside the reed-belt (see also Sect. 4.2).

As already emphasized, the production of littoral macrophytes clearly outweighs that of the microphytes, including the periphyton (see also Table 1). For example, in a *Phragmitetum communis* stand, the share of periphyton in the total primary production attains only 6.4%, in a stand of *Glycerietum* only 5.0%, etc. However, there is a periphyton community whose production is rather high in comparison with that of its substrate plant. This is the periphyton on floating clusters of *Utricularia*, it also has a particular species composition. In Opatovický

Table 8. Potential production of filamentous algae in Nesyt fishpond, measured by method of light and dark bottles

Date (1972)	Dominants	Exposed	$mg\ O_2\ g^{-1}\ h^{-1}$			$mg\ O_2\ g^{-1}\ d^{-1}$		
			NPP	R	GPP	NPP	R	GPP
5 May	*Vaucheria* spp.	in pelagial	1.20	1.22	2.42	3.2	29.3	32.5
		in *Typha* stands	1.00	1.22	2.22	0.5	29.3	29.8
26 May	*Oedogonium*: *Spirogyra* (3:1)—young	in pelagial	6.72	2.19	8.91	74.9	52.6	127.5
	Spirogyra: *Cladophora*: *Oedogonium* (3:2:1)—old	in pelagial	1.70	4.08	5.78	−15.2	97.9	82.7
14 June	*Cladophora* collected near SE shore	near SE shore	12.14	6.08	18.62	126.1	145.9	272.0
		near N shore	13.28	4.90	18.18	148.0	117.6	265.6
	Oedogonium collected near N shore	near SE shore	9.41	3.81	14.03	113.5	91.4	205.0
		near N shore	6.27	3.94	10.21	54.6	94.6	149.2
6 July	*Cladophora* collected in S bay	in S bay	34.10	1.73	35.83	480.9	41.5	522.4
		near SE shore	35.30	1.64	36.94	499.2	39.2	538.6
	Cladophora collected near SE shore	in S bay	3.10	2.02	5.12	26.2	48.5	74.6
		near SE shore	5.75	2.20	7.95	63.1	52.8	115.9
8 Aug.	*Cladophora* (encrusted)	in pelagial	2.08	4.10	6.18	−15.5	98.4	82.9
		in *Bolboschoenus* stand	3.58	4.10	7.68	4.6	98.4	103.0
	Enteromorpha	in pelagial	5.13	2.40	7.53	43.4	57.6	101.0
		in *Bolboschoenus* stand	10.65	2.40	13.05	117.4	57.6	175.0
6 Sept.	*Enteromorpha*	in pelagial	7.66	1.22	8.88	81.8	29.3	111.1
	different species (mixed)	in pelagial	3.48	1.46	4.94	26.8	35.0	61.8

R, respiration; GPP, gross primary production; NPP, net production (lowest estimate); both related to unit dry weight.

in 1972, the production of periphytic algae on clusters of *Utricularia* floating in two different *Glycerietum maximae* stands amounted to 64.4 and 23.7 g $m^{-2} O_2$ respectively, i.e., 143 and 72 kcal m^{-2} of net primary production per year, while the production of the *Utricularia* itself was, respectively, 6.9 and 2.5 g m^{-2} of dry weight, i.e., 232 and 85 kcal m^{-2} per year (Fig. 3 in Sect. 4.1).

In 1972, the conditions were exceptionally favorable for the development of floating algae in Nesyt (see Sect. 4.2). Such outbursts of filamentous algae may also occur in South Bohemian fishponds where, however, wrong management is the factor responsible. The possibility of migration of algal clusters greatly complicates the elaboration of an overall production budget. The results of direct measurements of oxygen production do not provide sufficiently representative values in view of a considerable heterogeneity in the spatial distribution of the algal biomass. Apart from a number of other factors, the production estimates may also be distorted by disturbance of the spatial structure of the cluster during measurement. The data in Table 8 correspond with the potential photosynthetic activity when mutual shading does not play any important role. Such data may therefore serve only in the construction of hypotheses on the possible share of floating algae in the material and energy budget of a littoral ecosystem.

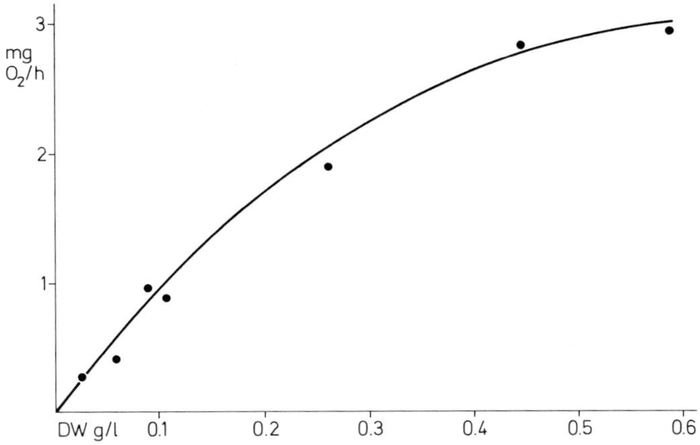

Fig. 5. Dependence of O_2 production on biomass concentration (dry weight per unit volume) of *Enteromorpha intestinalis* (Nesyt fishpond in 1973)

The photosynthetic activity of the algal clusters appears to change considerably during their development. The phase of explosive development is evidently followed by a prolonged period of retarded rates of both cell division and net production. As shown by the study on cell growth and division in individual filaments of *Cladophora* (Ettl, 1973; Ettl et al., 1973a), one generation period may be much shorter than 24 h during the phase of rapid growth. Equivalent data are not available for filaments of *Oedogonium* and *Spirogyra*, but a comparison of changes in their standing crop during their spring outbursts indicates that their rates of cell division are also high. By contrast, numerous measurements have confirmed a remarkably low oxygen production despite the small mutwal shading of the filaments (Table 8).

The low photosynthetic activity of older filaments may be ascribed to their decreased nutrient supply or to the transition of the thallus to a less active state. A limitation, caused mainly by lack of nutrients, has been indicated by measurements of oxygen production by the same algal material placed in the water originating from two different parts of the fishpond, but exposed to much the same conditions of irradiance and temperature. In addition, significant differences were also disclosed between algae of the same species but collected from different parts of the fishpond and exposed in the same water and to the same physical conditions. Apparently, these internal factors play a significant role, mainly in filaments of *Cladophora*, the thalli of which are more differentiated than in the other algae concerned.

The effect of reduction in the specific growth rate of the thalli is increased by mutual shading of the filaments in a dense cluster under certain ecological conditions. Figure 5 illustrates the dependence of oxygen production on the percentage dry weight content in *Enteromorpha intestinalis*. Wind is probably responsible for the formation of extensive algal carpets in which the density of biomass may attain up to 900 g dry weight m^{-2}. The net efficiency of solar energy conversion by such densely concentrated algal biomass is close to zero or even negative. The same applies to floating algae occurring along edges of a dense reed-belt especially when the algal clusters have become old and their photosynthetic activity poor.

In submerged macrophyte communities of the foederatio *Potamogetonion pusilli* (see Sect. 1.2), filamentous algae become concentrated in large masses mainly because their clusters are supported by the bodies of the vascular plants which prevent their sinking into deeper water with an unfavorable radiation regime. Contrary to emergent communities, the floating algae frequently become important producers in submerged communities and their development may limit that of the vascular plants. Difficulties in obtaining representative data on their primary production are similar to those encountered in helophyte communities. A rough estimate, based on maximum values of specific oxygen production and on changes in biomass indicates, for the filamentous algae within a community of *Potamogetoneto pectinati—Zannichellietum pedicellatae* in Nesyt, that their total annual production does not surpass some 0.2–0.3 g m^{-2} of dry weight.

A stabilized location of the algal clusters on the water surface also has its drawbacks in that it brings about chlorophyll decay and bleaching of the algal thalli. In Nesyt, this effect is most marked when the water surface remains calm. Rather than open water, loose stands of emergent macrophytes such as *Typha* or *Bolboschoenus* provide more favorable conditions for the development of filamentous algae. The favorable effect of shading has been confirmed for *Enteromorpha* (see Table 8).

The filamentous algae take up large amounts of mineral nutrients (Lhotský and Marvan, 1975). They also shade out temporarily and mostly locally other components of the aquatic community in the fishpond and their massive incidence is undesirable even from the viewpoint of fishpond management. In carefully managed fishponds, however, filamentous algae usually do not develop so as to seriously impede the function of the fishpond.

References see pp. 335—337.

References

Assman, A. V.: Role of periphyton in the organic production in water reservoirs. Dokl. AN SSSR **76**, 906–908 (1951)

Bílý, J.: Die Algen der staatlichen Naturreservation Řežabinec bei Ražice. Bacillariophyceae. (In Czech.) Sb. Jihočes. Mus., Č. Budějovice, přír. **4**, 77–106 (1964)

Bohr, R.: Phytosociological studies on the periphyton in Lake Mamry. Stud. Soc. Sci. Torun. Sect. D (Bot.) **6** (1): 1–44 (1962)

Brandl, Z., Brandlová, J., Poštolková, M.: The influence of submerged vegetation on the photosynthesis of phytoplankton in ponds. Rozpr. Česk. Akad. Věd, Řada Mat. Přír. Věd **80**, 33–62 (1970)

Dykyjová, D.: Production, vertical structure and light profiles in littoral stands of reed-bed species. Hidrobiologia **12**, 361–376 (1971)

Dykyjová, D., Přibil, S.: Energy content in the biomass of emergent macrophytes and their ecological efficiency. Arch. Hydrobiol. **75**, 80–108 (1975)

Ettl, H.: Cytological investigation methods for the growth of *Cladophora glomerata*. Pol. Arch. Hydrobiol. **20**, 181–184 (1973)

Ettl, H., Březina, V., Marvan, P.: Methodical notes on assessment of productivity in littoral algae. In: Littoral of the Nesyt fishpond. Květ, J. (ed.). Studie ČSAV **15**, Praha, 111–115 (1973a)

Ettl, H., Marvan, P., Komárek, J.: Algal biocenoses of the Nesyt fishpond in Southern Moravia. Ann. Rep. Algol. Lab., Třeboň, 1970, 143–147 (1973b)

Fott, J.: Observation on primary production of phytoplankton in two fishponds. In: Proc. IBP-UNESCO Symp. Productivity Problems of freshwaters, 1972. Warsaw, Krakow: PWN, 1972

Hrbáček, J.: Species composition and the amount of the zooplankton in relation to the fish stock. Rozpr. Česk. Akad. Věd, Řada Mat. Přír. Věd **72** (10): 1–116 (1962)

Hrbáček, J.: Contribution to the ecology of waterbloom forming blue-green algae. Verh. Int. Verein. Theor. Angew. Limnol. **15**, 837–846 (1964)

Hrbáček, J., Novotná-Dvořáková, M.: Plankton of four backwaters related to their size and fish stock. Rozpr. Akad. Věd, Řada Mat. Přír. Věd **75** (13): 1–65 (1965)

Komárek, J.: Seasonal changes in the algal microflora of Opatovický fishpond (South Bohemia). In: Ecosystem Study on Wetland Biome in Czechoslovakia. Hejný, S. (ed.). Czechosl. IBP/PT-PP Rep. No. **3**, Třeboň, 185–196 (1973a)

Komárek, J.: The communities of algae of Opatovický fishpond (South Bohemia). In: Ecosystem Study on Wetland Biome in Czechoslovakia. Hejný, S. (ed.). Czechosl. IBP/PT-PP Rep. No. **3**, Třeboň, 179–184 (1973b)

Komárek, J., Ettl, H., Marvan, P.: A review of algae in Opatovický fishpond (South Bohemia) in 1971–1972. In: Ecosystem Study on Wetland Biome in Czechoslovakia. Hejný, S. (ed.). Czechosl. IBP/PT-PP Rep. No. **3**, Třeboň, 175–178 (1973)

Komárková, J.: Primary production of phytoplankton and periphyton in Opatovický fishpond (South Bohemia) in 1972. In: Ecosystem Study on Wetland Biome in Czechoslovakia. Hejný, S. (ed.). Czechosl. IBP/PT-PP Rep. No. **3**, Třeboň, 197–212 (1973)

Komárková, J., Komárek, J.: Comparison of pelagial and littoral primary production in a South Bohemian Fishpond (Czechoslovakia). Symp. Biol. Hung. **15**, 77–95 (1975)

Komárková, J., Přibil, S.: Chemical and physical properties of pelagial and littoral water in Opatovický fishpond. In: Ecosystem Study on Wetland Biome in Czechoslovakia. Hejný, S. (ed.). Czechosl. IBP/PT-PP Rep. No. **3**, Třeboň, 15–27 (1973)

Kostikova, L. E.: Filamentous algae of the Kiev reservoir. In: The Kiev Reservoir, Naukova Dumka, Kiev (In Russian), 234–248 (1972)

Kubíček, F.: Anwendung einer Batterienzentrifuge für die Bestimmung der Biomasse. Biológia (Bratislava) **24**, 245–249 (1969)

Lhotský, O., Marvan, P.: Biomasseproduktion und Bindung der Nährstoffe bei einigen Massenvorkommen von Algen in der Natur. In: Izučenie Intensivnych Kultur Vodoroslei. Nečas, J. (ed.). Třeboň, (1975)

Losos, B., Heteša, J.: Hydrobiological studies on the Lednické rybníky ponds. Acta Sci. Nat. (Brno) **5**, 1–54 (1971)

Marvan, P.: Zur Frage der Kettenbildung bei benthisch lebenden Fragilariaceen. Arch. Hydrobiol., Suppl. **43**, Algolog. Stud. **8**, 289–316 (1972)

Marvan, P., Ettl, H., Komárek, J.: Littoral algal vegetation of the Nesyt fishpond. In: Littoral of the Nesyt Fishpond. Květ, J. (ed.). Studie ČSAV **15**, Praha, 63–66 (1973)

Marvan, P., Sládeček, V.: Main source of pollution of the Nesyt fishpond. In: Littoral of the Nesyt fispond. Květ, J. (ed.). Studie ČSAV **15**, Praha, 159–160 (1973)

Marvan, P., Sládeček, V.: Sources of pollution of the Nesyt fishpond (In Czech). Bull. Met. Střed. Vodohospod. Labe **25**, 100–110 (1974)

Masjuk, N. P.: *Chlorococcales* of Lake Tur in the Volyň-area. (In Ukrainian). Ukr. Bot. Zh. (Kijev) **14**, 72–86 (1957)

Meschkat, A.: Der Bewuchs in den Röhrichten des Plattensees. Arch. Hydrobiol. **27**, 436–517 (1934)

Ondok, J. P.: Régime of global and photosynthetically active radiation in helophyte stands. Studie ČSAV, Praha, **10**, 1—112 (1977)

Pelikán, J., Květ, J., Úlehlová, B.: Principal constituents and relationships in the reed-belt ecosystem at the Nesyt fishpond. In: Littoral of the Nesyt Fishponds. Květ, J. (ed.). Studie ČSAV **15**, Praha, 17–23 (1973)

Pieczyńska, E.: Variation in the spring production of plankton and periphyton in the littoral zone of lakes. Bull. Acad. Pol. Sci. Cl. II, Sér. Sci. biol. **13**, 219–225 (1965)

Pieczyńska, E.: Mass appearance of algae in the littoral of several Mazurian lakes. Mitt. Int. Ver. Theor. Angew. Limnol. **19**, 59–69 (1971 a)

Pieczyńska, E.: Ecology of the lake shore (In Polish). Univ. Warsz., Inst. Zool., Zakł. Hydrobiol., (1971 b)

Pieczyńska, E., Szczepańska, W.: Primary production in the littoral of several Mazurian Lakes. Verh. Int. Ver. Theor. Angew. Limnol. **16**, 372–379 (1966)

Prát, S., Cyrus, Z.: Oscillatorietum. Věstn. Král. Česk. Spol. Nauk, Mat.-Přír. **7**, (1948)

Schröder, R.: Release of plant nutrients from reed borders and their transport into the open waters of the Bodensee-Untersee. Symp. Biol. Hung. **15**, 21–27 (1975)

Sládeček, V.: A swarm of the zooplankton in the Nesyt fishpond. In: Littoral of the Nesyt fishpond. Květ, J. (ed.). Studie ČSAV **15**, Praha, 71–72 (1973)

Smith, G. M.: Ecology of the plankton-algae in the Palisades Interstate Park, including the relation of control methods to fish culture. Roosev. Wild Life Bull. **2** (2), 95–195 (1924)

Straškraba, M.: Share of the littoral region in the productivity of two fishponds in Southern Bohemia. Rozpr. Česk. Akad. Věd, Řada Mat. Přír. Věd **73** (13), 1–64 (1963)

Straškraba, M., Pieczyńska, E.: Field experiments on shading effect by emergents on littoral phytoplankton and periphyton production. Rozpr. Česk. Akad. Věd, Řada Mat. Přír. Věd **80** (6), 1–32 (1970)

Tamás, G.: Beiträge zur Algenflora des Balaton-Sees. III. Algologische Untersuchungen im Aufwuchs der Makrovegetation des Sees im Jahre 1963. Annal. Biol. Tihány **31**, 255–272 (1964)

Vollenweider, R. A. ed.: A Manual on Methods for Measuring Primary Production in Aquatic Environments. (IBP Handb. No. 12.) Oxford: Blackwells Sci. Publ. (1969)

Wetzel, R. G.: A comparative study of the primary productivity of higher aquatic plants, periphyton and the phytoplankton in a large shallow lake. Int. Rev. Ges. Hydrobiol. Hydrogr. **49**, 1–64 (1964)

Fig. I. Winter aspect of a marsh in a terrestrialized bay of a South Bohemian fishpond, occupied by a large amount of detritus and standing dead material in communities of *Caricetum elatae* and *Phragmitetum communis*. (Photo: J. Ševčík)

Section 5

Decomposition Processes in the Fishpond Littoral

The generally high level of primary production in the fishpond littoral ecosystems is usually linked with an intense decomposition of the organic matter produced. The decomposition processes taking place in wetlands are still little understood but great differences, both in space and time, are known to exist between various fishpond biotopes, even within the littoral of one fishpond. The decomposing microorganisms are varied, numerous and difficult to inventarize. The Section shows some methodological aspects of evaluating the decomposition and of estimating the spatial and temporal variation in the microbial production and biomass. The microbial transformations of dead organic matter and the microbes' role in detritus formation and mineralization are characterized. Most of these studies were conducted in the South Moravian Nesyt fishpond but an analogous situation exists at the Opatovický and other fishponds in South Bohemia (see Fig. I).

5. Decomposition Processes in the Fishpond Littoral

B. ÚLEHLOVÁ

5.1 Biomass and Production of Microorganisms

One of the main goals of the IBP was to present quantitative data on the production of biotic components of ecosystems. But estimates of microbial populations in ecosystems are usually rather difficult and open to all sorts of criticism. Despite all inherent uncertainties, in the course of the IBP several authors tried to estimate the microbial biomass in various ecosystems (Kuznetsov and Romanenko, 1966; Clark and Paul, 1970; Aristovskaya, 1972; Parkinson, 1970). Different ways have been suggested of estimating the biomass of bacteria, fungi and actinomycetes from data on their population densities. Such methods are discussed by Parkinson et al. (1971) and Sorokin and Kadota (1972) for terrestrial and aquatic ecosystems respectively.

The procedure recommended by Alexander (1961) has been used for calculating, from the respective plate counts, the highest and lowest bacterial biomass in the different materials studied in the Nesyt fishpond littoral: bacterial biomass = density × volume, whereby density of an average bacterial cell = 1.5, and volume of an average bacterial cell = 1 μm^3.

The results of the calculations are given in Table 1. The table includes the bacterial biomass values per 1 g of the respective dry material and per 1 m^2 ground area.

The smallest bacterial biomass was found on standing dead plant materials (in an aerial environment), the largest was on the submerged plant materials. Although the data on the maximum biomass provide a basis for comparing instantaneous trophic suitability of various substrates, they give no information on the birth and death rates of bacteria.

Romanenko (1964), Sorokin (1965), and Kuznetsov and Romanenko (1966) emphasized the importance of heterotrophic CO_2 fixation for bacterial production in waters. They consider the fixation and transformation of dissolved organic matter into bacterial biomass to be one of the basic processes in the turnover of matter in aquatic environments. A number of organic substances present in water are frequently of allochthonous origin. Examples are humic substances, dead plants and animals, and other detritic materials.

According to Kuznetsov and Romanenko (1966), the production of bacterial biomass exceeded the net production of autotrophic phytoplankton in the Ry-

Table 1. Ranges of bacterial biomass in different materials present in fishpond littoral biotopes

		$mg\ g^{-1}$		$g\ m^{-2}$	
		min.	max.	min.	max.
Standing dead	*Phragmites communis*	0.06–	0.17	0.001 –	0.033
	Typha angustifolia	0.02–	0.60	0.0004–	0.118
Floating plant litter	*Typha angustifolia*	0.02–	10.40	—	
Litter—terrestrial environment	Upper layer	0.02–	1.20	0.02 –	1.00
	Lower layer 5 cm	0.03–	1.64	1.4 –	64.3
Litter—aquatic environment	Upper layer	0.08–	4.35	0.07 –	3.50
	Lower layer 5 cm	0.08–	1.50	0.8 –	59.60
Sapropel	5 cm layer	0.01–	0.47	0.4 –	16.3

binsk reservoir in 1964. Overbeck (1972a), working on the Plussee lakes (GDR), arrived at similar values. All authors calculated the bacterial and phytoplanktonic production from data on CO_2 fixation in light and in the dark, presuming that heterotrophic fixation represents about 6% of the total production of bacterial biomass. Overbeck (1972a) calculated bacterial biomass: (a) from bacterial counts on bacteriological filters, fixing the volume of bacterial cells at 0.5 μm^3, their dry weight at 15% of fresh weight and their carbon content at 50% of dry weight; (b) from both autotrophic (in light) and heterotrophic (in the dark) CO_2 fixation. The primary production by the phytoplankton in Lake Plussee appeared to exceed the bacterial production. In view of the complexity of these phenomena, the author marked out factors pertinent to sample sedimentation and to ecological relationships such as symbiosis, metabiosis, parasitisms between bacteria, algae and protozoa. During certain periods and in certain habitats (for example in shallow water in winter), when the primary production is small, close correlation can be established between heterotrophic CO_2 fixation and bacterial biomass (Overbeck, 1972a). Data by Punčochář (1975) indicate that the growth characteristics of bacteria may be strongly species-dependent. This is another aspect complicating the interpretation of data on bacterial biomass in such complex systems as wetlands.

No values are available on the biomass of other taxonomical groups of microorganisms than bacteria, such as actinomycetes, fungi and protozoa. The fungal populations are important on standing dead plant materials and the optimum conditions for their development occur in fall.

The seasonal changes in biomass of individual microbial groups are important for the estimation of microbial production. The trends in changes of bacterial biomass in water may be expressed by three different parameters estimated by different methods. Figure 1 shows the changes in bacterial counts and contents of bacterial P and N in water samples from the four vegetational zones of the Nesyt fishpond littoral taken during the 1971 vegetation period. The peaks of all the characteristics occur more or less simultaneously. But the relationships between individual characteristics are not straightforward and appear to depend on the developmental stages of the microbial populations.

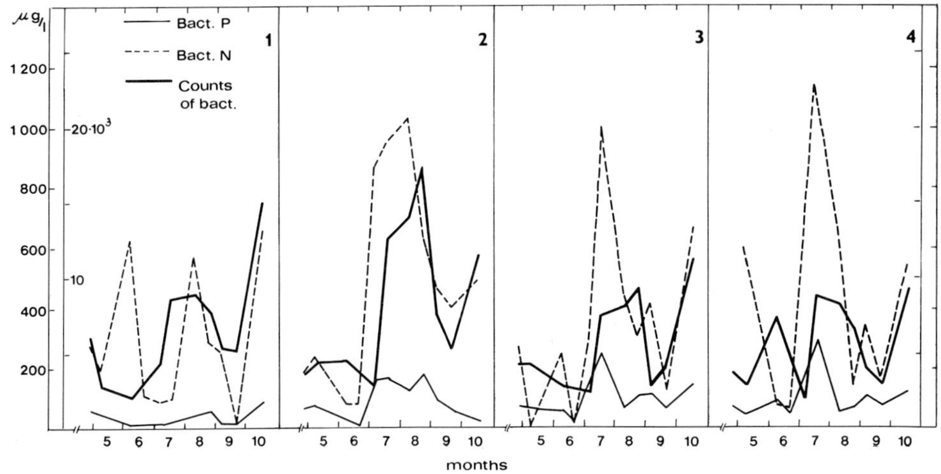

Fig. 1. Changes in counts of microorganism, and concentrations of bacterial nitrogen and phosphorus in water from 4 different zones of the Nesyt fishpond littoral: (1) Mixed community of *Phragmites communis*, *Carex riparia*, and *Typha angustifolia;* (2) Pure *Phragmites communis* stand; (3) "Lagoon" without emergent vegetation; (4) Pure *Typha angustifolia*

The conclusion of this paragraph is rather pessimistic. We still know very little about the biomass and production of microorganisms in the littoral, and to improve this situation will require much effort. Further integrated microbiological studies are greatly needed on the densities, population dynamics, and composition of microbial populations inhabiting the complicated littoral ecosystems.

5.2 Methods of Evaluating the Decomposition Processes

A number of methods are available for the study of microbial activities in ecosystems (Parkinson et al., 1971; Sorokin and Kadota, 1972). In the investigations of decomposition processes in the littoral vegetation of the Nesyt fishpond, data on the following features and processes were considered desirable:

1. Chemical composition of plant remnants at various stages of decomposition;

2. Changes in chemical composition of plant remnants decomposing on the soil surface as well as submerged in the course of different seasons of the year;

3. Decomposition rates of plant remnants of different origin;

4. Rates of cellulose decomposition in different habitats in the fishpond and at different periods of the year.

The respective methods applied included:

1. Chemical analyses of plant remnants at different stages of decomposition for nitrogen, phosphorus, potassium, sodium, calcium, and magnesium, according to Koppová et al. (1955); during 1971 and 1972 the standing dead matter, litter, semidecomposed litter and the soil sediment were sampled at monthly intervals from a terrestrial *Phragmitetum communis* stand and from aquatic stands of

Phragmitetum and *Typhetum angustifoliae*, all situated along a transect across the littoral.

2. Litter decomposition was studied using the litter mesh-bag method (Witkamp and van der Drift, 1961). Plant samples of 25 g were placed in silon mesh-bags (mesh size 1 mm^2) in three to five replicates on the soil surface or in shallow water at the sites of interest for three, six or twelve months. Subsequently the samples were brought to the laboratory, rinsed with water, weighed, and analyzed by standard chemical methods (Koppová et al., 1955).

3. Decomposition rates were calculated from the loss in weight during exposure, assuming a linear relationship between the loss in weight and time.

4. Seasonal and spatial variation in cellulose decomposition rates was also studied using mesh-bags, but weighed 1 dm^2 squares (about 0.8 g) of filter paper Whatman 2 were the substrate. Five replicates were exposed at four sites along the littoral transect described (see Sects. 1.6, 2.3) for 10–14 days in summer and for 50–100 days in winter each time. Back in the laboratory the mesh-bags were rinsed with water, the remaining filter paper was carefully collected, dried and weighed. The cellulose decomposition rates were calculated as those of litter, i.e., assuming linearity between the loss in weight and time.

5.3 Microbial Transformations of Dead Organic Matter and Their Role in Detritus Formation and Mineralization

Two basic processes are usually distinguished when considering organic matter decomposition:

1. Mineralization, accompanying the assimilation of organic matter by different organisms, varying according to the level of organization, and using some of the substances for building their own bodies (secondary production), releasing the rest as mineral or simple organic compounds, i.e., CO_2, CH_4, H_2O, NH_4, NO_3, SO_4, $CO(NH_2)_2$, H_2S, PO_4, etc.

2. Humification, i.e., a relatively long series of various fermentative biochemical processes, wherein different organic compounds (end products or intermetabolites) become gradually transformed into complex organic heteropolycondensates with bonds of different strength, called humus. The formation as well as the decomposition and stability of humic substances is affected to a considerable extent by bonding with certain mineral constituents of the environment such as ions and clay minerals.

In wetlands, the detritus food chain dominates because relatively few animals graze directly on the plants. A number of animals feed on materials pretreated by microorganisms, on decomposition products, and on microorganisms responsible for the decomposition. Quantitative data pertaining to the food webs and to the processes related to the detritus food chain are rather scarce, as contrasting with all the high diversity of forms and functions of heterotrophs in wetlands.

Figure 2 shows the long-term changes in mineral composition of certain plant remnants undergoing decomposition in the Nesyt fishpond. The appropriate

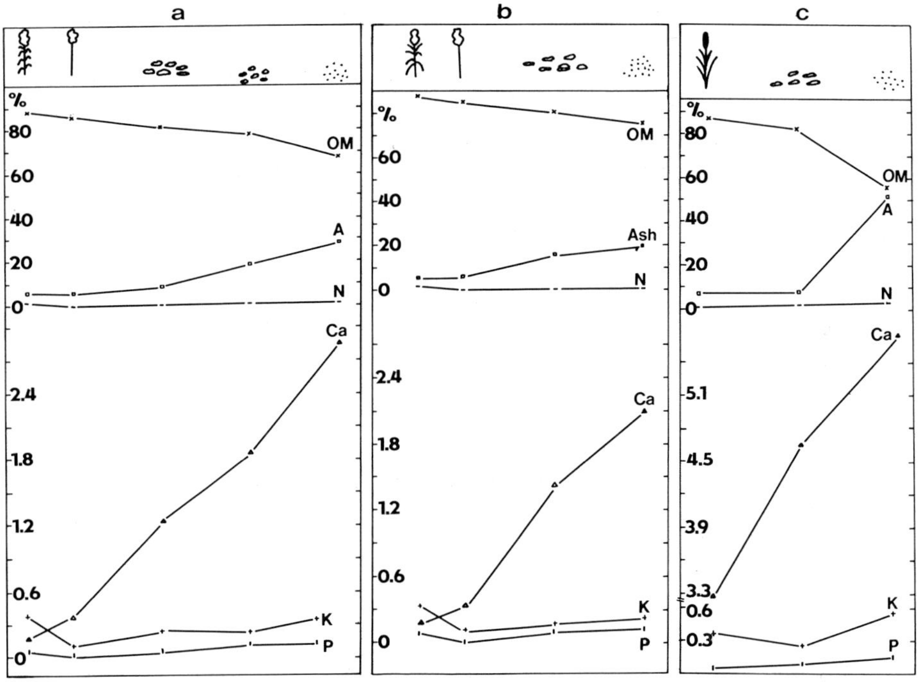

Fig. 2a–c. Changes in the chemical composition of remnants of: (a) Terrestrial *Phragmites communis* during decomposition; (b) Littoral *Phragmites communis* during decomposition; (c) *Typha angustifolia* during decomposition

stages of decomposition of plant remnants sampled for analyses are illustrated in the upper part of each graph.

The organic matter content (OM) generally decreases while the contents of ash (A) and individual mineral constituents increase with advancing stages of decomposition; the increase in calcium (from 0.3 to 3%) and in nitrogen (from 0.3 to 1.5%) is the most conspicuous. Interesting are the differences in the chemical composition between dying plants and standing dead plant material. Despite the relatively short time interval separating these stages, a marked decrease was recorded in the nitrogen, phosphorus and potassium contents. The following factors possibly bring about such changes: (1) loss of leaves, which are considerably richer in nitrogen and somewhat richer in phosphorus than the culms (Table 2, see also Sect. 3.4); (2) translocation of certain nutrients from the aboveground organs to the roots and rhizomes, accompanied by possible reverse transport of surplus calcium; and (3) leaching of the plant structures.

Noteworthy is a comparison between the changes in chemical composition of the *Phragmites* remnants in the terrestrial and littoral ecophases as shown in Figure 2a and b. The organic matter content decreases less in plant remnants in the littoral habitat where anaerobic conditions prevail and the concentrations of mineral substances in the plant remnants are smaller. In contrast the nitrogen content increased in the plant remnants by 1.7% concurrently with a 6% decrease in the organic matter content in the littoral habitat, as compared with a 1.2%

Table 2. Chemical composition of dead leaves, shoots and roots of *Phragmites communis*

Material	Org. matter %	N %	PO_4	K	Na	Ca	Mg
			mg/100 g dry weight				
Leaves	80.4	1.13	274	195.5	105.5	1,980	93.4
	70.4–92.2	0.8 –1.4	225–320	157 –250	65 –137	1,250–2,700	75–122
Shoots	90.5	0.47	231	376.5	122.0	700	91.2
	84.5–95.5	0.3 –0.5	150–495	117 –835	72 –200	600–2,000	63–137
Roots	73.6	1.73	655	735.0	242.5	6,400	109.5
	70 –77	1.6 –1.8	585–725	700 –770	220 –265	5,200–7,600	100–119

increase in the remnants in terrestrial ecophase, where the decrease in organic matter content represented 18%, all related to the original contents. An active accumulation of nitrogen apparently accompanies the decomposition in the littoral ecophase. The results agree with similar observations quoted by Odum and de la Cruz (1967) and de la Cruz (1973).

The chemical changes in the community of *Typhetum angustifoliae* (Fig. 2c) somewhat differ from those in both zones of the *Phragmitetum*. First of all, standing dead *Typha* does not persist as long as dead *Phragmites* culms do; this is partly due to different mechanical properties of the two materials, partly to greater mechanical damage by waves to the *Typha* stand, adjoining open water. The broken-off plant parts are mostly drifted away, mixed with other materials and finally deposited and decomposed elsewhere in the pond. The difference between the initial and final organic matter contents (32.4%) is therefore much greater here than the comparable decrease in either *Phragmitetum*, while the nitrogen content increases by only 1.1%.

At the beginning of the winter of 1972, *Phragmites* stems one or two years old, protruding above the water surface, were collected for chemical analyses. The resulting data are presented in Table 3. In addition to the decrease in weight, the standing dead material lost, in situ, about 7% of its organic matter in one year. The nitrogen content of the plant material increased by nearly two thirds during the same period. A plant material with a wide C/N ratio appears to represent a suitable energy source for organisms capable of fixing atmospheric nitrogen; their activities then narrow down the C/N ratio. Kaushik and Hynes (1968, 1971) and Mathews and Kowalczewski (1969) arrived at similar conclusions.

On the other hand, the content of monovalent cations decreased to even less than one third, apparently by leaching, although direct exchange reactions are not excluded. Planter (1970) observed in laboratory experiments that potassium was washed out most rapidly from *Phragmites* litter. The same apparently holds true under field conditions. According to Table 3, the aging submerged standing dead material of *Phragmites* further shows a considerable relative increase in calcium content, and a less marked increase in magnesium content. This may result from ionic reactions between the bivalent cations of the fishpond water and the plant remnants, and/or their sessile microorganisms.

Short-term chemical changes taking place during the decomposition of dead stems of *Phragmites* enclosed in litter mesh-bags and submerged in the fishpond

Table 3. Chemical composition of above-water standing dead shoots of littoral *Phragmites communis* and of detritus on bottom

Material	Org. matter %	N %	PO_4	K	Na	Ca	Mg
			mg/100 g dry weight				
Shoots of the current year	90.46 84.5 –96.6	0.47 0.35–0.53	231 150–495	376.5 117 –835	122 72 –200	700 600– 800	74.0 63 – 94
Shoots of the previous year	83.77 88 –88	0.77 0.59–0.88	228 185–260	140.8 110 –180	80.8 77.5– 85	1,683 1,400–2,000	84.7 75 – 96
Difference	−6.69	+0.30%	− 3	−235.7	− 21.2	+ 983	+10
Detritus	67.36 61.5 –74.8	1.76 1.5 –1.9	441 140–605	283.0 170 –415	226.0 70 –330	6,090 5,200–8.400	95.2 48 –119

Table 4. Chemical composition of *Phragmites communis* shoots decomposing in the fishpond littoral during summer and winter

Material	Org. matter %	N %	PO_4	K	Na	Ca	Mg
			mg/100 g				
Control shoots	85.6	0.13	102	68.7	28.7	255	23
Shoots exposed for 190 days in summer	I 80.9 II 81.8	0.34 0.46	70 160	47.5 42.5	37.5 45.0	400 825	20 34
Shoots exposed for 180 days in winter	I 57.8 II 62.2	0.49 0.32	37.0 42.5	45.0 37.5	40.0 27.5	400 400	25 19

I Exposed on bottom below 10 to 15 cm water layer in stand of *Phramites communis.*
II Exposed on bottom below 25 cm water layer in lagoon.

water either within the *Phragmites* stand or in the lagoon were studied both in summer and winter. The results of the chemical analyses are presented in Table 4. The loss of organic matter was hardly 10% in summer and about 30% in winter. Analogous data for phosphorus reveal practically insignificant changes in summer, but a strong deprivation (by 60%) in winter. The submerged *Phragmites* remnants, in analogy to the standing dead material in situ, also accumulated nitrogen and calcium to a considerable degree.

The differences between the decomposition rates of plant remnants in winter and summer may result from different oxygen solubilities in water as well as from differences in the prevailing microflora.

Differences in the composition of microbial populations inhabiting different substrates indicate that metabiosis as well as succession of different microbial populations, micro- and mesofauna accompany the decomposition of plant material. While the advanced mechanical disintegration and the concurrent increase in specific surface area of most plant materials result, above all, from the feeding processes of both micro- and mesofauna, the direct enzymatic attack on chemical components of the exposed surfaces is mostly mediated by the microflora of the animal digestive tracts and by the sessile microorganisms. The passage of plant remnants through the digestive tracts of animals transfers them to the start of the coprophagous food chain. The sessile microflora may also serve as food for various animals; the mechanical removal of sessile microflora by animals may be followed, in turn, by a fast regrowth of the microflora.

Disintegrated plant materials may by translocated within a fishpond by the movements of water masses. At the same time, they may be subjected to mixing, or separated according to particle size, weight, etc., to local sedimentation and accumulation as well as to successive sedimentation in layers differing in specific properties of their components. Along windward fishpond shores, considerable amounts of plant remnants are drifted ashore and decompose in heaps which may be several decimeters high. These heaps usually contain a mixture of remnants of emergent, submerged and floating macrophytes and macroalgae. They sometimes form a barrier between the inner and outer littoral. In limosal or terrestrial ecophase, they host specific types of vegetation (see also Sects. 1.2, 4.2).

5.4 Spatial and Temporal Variation in Litter and Cellulose Decomposition in the Nesyt Fishpond Littoral

The information on the decomposition rates of plant materials is required for the overall balance of energy flow, and for a quantification of mineral cycles. This section demonstrates the spatial and seasonal dynamics of decomposition processes taking place on selected sites in the Nesyt fishpond. The decomposition rates of three different kinds of plant material as estimated during 1973 and 1974 are given in Table 5. The decomposition rates are about three times higher in summer than in winter, and the variation in ranges is also greater in summer. The *Typha angustifolia* leaves decompose most rapidly, *Phragmites* culms do extremely slowly. A comparison between the decomposition rates on different sites has revealed that in winter, the highest decomposition rates can be expected in *Typhetum angustifoliae*, where the relatively deep water provides favorable temperature conditions.

Table 5. Decomposition rates (mg g^{-1} day^{-1}) of plant materials in Nesyt fishpond; average values and ranges of variation during 1973 to 1974

Plant material	*Carex riparia*	*Phragmites communis*		*Typha angustifolia*
		Stems	Leaves[a]	
Summer average 1973 to 1974	3.19	2.03	5.79	3.03
Range	2.26–4.49	1.80–3.87	3.87–7.59	2.67–3.37
Winter average	0.99	0.80	—	1.22
Range	0.83–1.21	0.69–0.89	—	1.10–1.30
Yearly average	2.50	1.81	—	2.81

[a] Values measured in Opatovický fishpond.

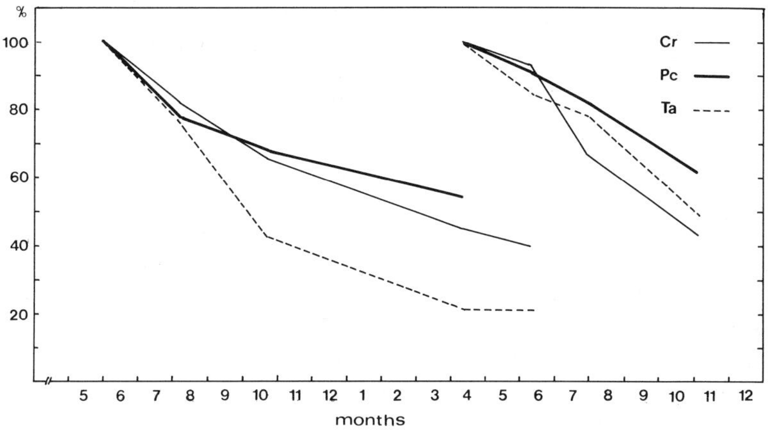

Fig. 3. Decomposition of different plant materials in Nesyt fishpond. *Cr*, *Carex riparia; Pc*, *Phragmites communis; Ta*, *Typha angustifolia*

Figure 3 and Table 6 describe the time course of decomposition of the three kinds of plant material exposed in the fishpond for up to 12 months. The absolute amounts of the plant material decomposed are rather small after the first three months of exposure. These represent the first phase of decomposition during which the most easily destructible components or structures are metabolized. The differences increase later, when the more resistant constituents come under attack. During this stage, the remnants of *Typha angustifolia* and *Carex riparia* exposed in mesh-bags lose weight faster than the *Phragmites* culms. The results for the years 1973 and 1974 differ somewhat because of differences in pond management. While the decomposition of all samples remained submerged in 1973, the *Phragmitetum communis* with *Carex riparia* undergrowth remained above the water level in 1974, and the remnants of *Carex riparia* were decomposed more rapidly than those of *Typha angustifolia*.

Table 6 presents the percentage of plant remnants decomposed during different periods on different sites along the transect across the littoral (see Sect. 5.3). Replicate samples exposed in different periods demonstrated clearly that the remnants of *Carex riparia* and *Phragmites communis* were losing weight at a

Table 6. Percentage of decomposed plant materials in litter mesh-bags exposed in Nesyt fishpond during different periods

Period of exposure	Site	(Cr + Pc)	Pc	L	T
	Material	*Carex riparia*	*Phragmites communis*		*Typha angustifolia*
31 May–7 Aug. 1973		18.00	26.31	20.8	22.93
7 Aug.–26 Oct. 1973		24.70	14.40	22.32	22.33
31 May–26 Oct. 1973		42.70	40.71	43.12	45.26
		34.76	31.16	32.76	46.73
Difference:		− 7.94	− 9.55	−10.37	+ 1.47

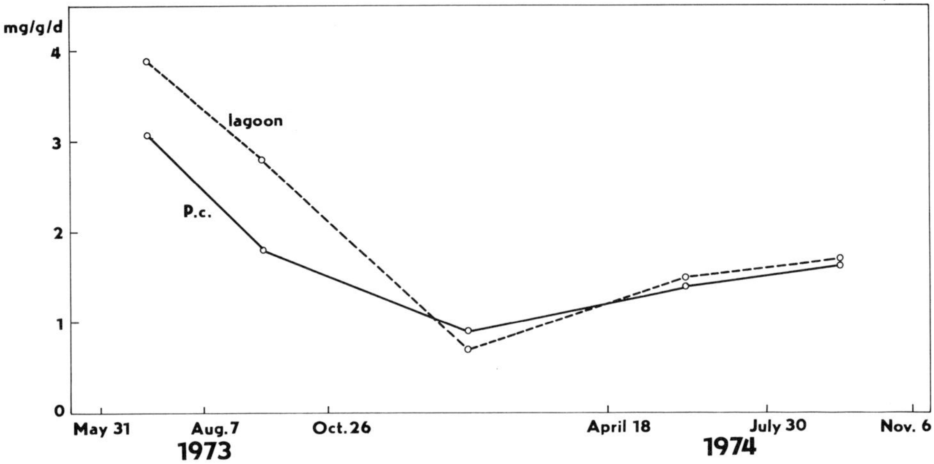

Fig. 4. Decomposition rates(mg g^{-1} d^{-1})of *Phragmites communis* stems during 1973 and 1974 in two different habitats: the community of *Phragmitetum communis (P.c.)* and in a lagoon

decreasing rate during later stages of decomposition. A linear relationship between loss in weight and time in calculating decomposition rates may therefore be used only as an approximation. On the other hand, the leaves of *Typha angustifolia* were losing weight at a nearly constant rate.

Figure 4 contains data on the spatial variability in the decomposition rate of one kind of plant material, namely of *Phragmites* culms, exposed in the *Phragmitetum* and in the lagoon during 1973 and 1974. Different environmental conditions control the decomposition rates in the two biotopes differing in water depth, temperature and aeration.

Ketner (1973) also used the litter mesh-bag method for estimating decomposition in a *Junco-Caricetum extensae.* The decomposition rates of 1–14 mg g^{-1} d^{-1} were encountered in the course of a year.

Mathews and Kovalczewski (1969) measuring leaf litter disappearance by the same method in the river Thames have observed an increase in the nitrogen content in litter between the first and the third months of the experiment. This increase was most probably associated with the establishment of an active microbial community in or on the litter. Increased nitrogen contents in litter have also been observed by other authors (Gilbert and Bocock, 1960; Saito, 1957) and by Dykyjová (1973) in two- and four-year-old decomposing stubbles and rhizomes of *Typha angustifolia* in the Opatovický fishpond. Pieczynska (1972) measured the decomposition of plant material by the same method in the eulittoral of Mikołajskie Lake. The rates of decomposition were generally higher there than those recorded either by Mathews and Kowalczewski (1969) or by Úlehlová in the Nesyt fishpond.

Pure cellulose, as a well-defined material representing the main organic constituent of most plants, was also used in mesh-bags instead of litter. Such "cellulose tests" provide comparable data on cellulose decomposing efficiency of the biota present in various habitats. The data on cellulose decomposition rates along

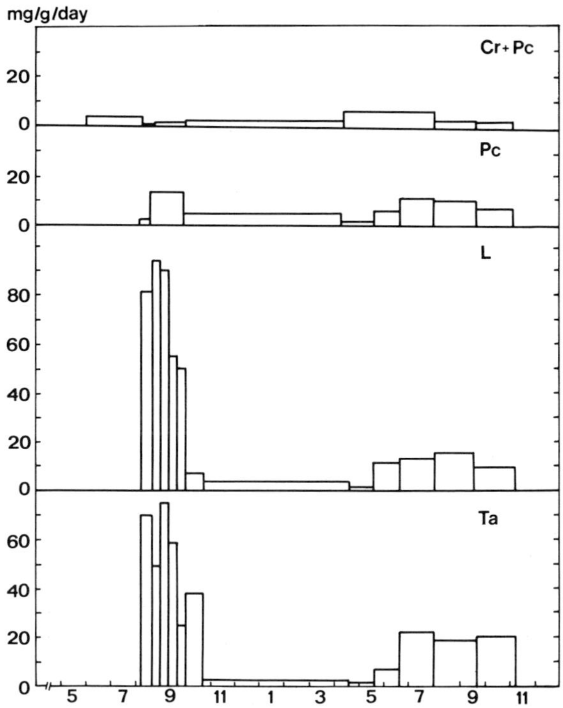

Fig. 5. Spatial and temporal variations in cellulose decomposition rates in transect across littoral of Nesyt fishpond. Species abbreviations as in Figure 3. *L*, lagoon

the transect across the littoral in 1973 and 1974 are shown in Figure 5. Higher decomposition rates were found in the lagoon and *Typhetum angustifoliae*, and differences were observed between the two years. The cellulose decomposition proceeded at the highest rate in the lagoon in the summer of 1974, otherwise the decomposition rate decreased from the *Typhetum angustifoliae* to the *Phragmitetum* with *Carex riparia* undergrowth. In the winter of 1973/74 the decomposition of cellulose proceeded at the highest rate in the *Phragmitetum* and from here it decreased in either direction along the transect.

The rates of cellulose decomposition were also studied by Hundt and Unger (1968) in several wetland plant communities in Germany, and were found to be highest in the wettest communities.

Tesařová and Úlehlová (1968) published data on the cellulose decomposition in wetland grassland communities of *Phalaridetum arundinaceae* and *Glycerietum maximae* in the course of the vegetation period. They used filter paper as a standard source of cellulose in the mesh-bags. Filter paper offers further advantages in connection with the mesh-bag method by providing a nearly constant surface: bulk ratio and more uniform contact with the surrounding medium.

The results of the study on the decomposition processes in the pond littoral may be summarized as follows:

1. Definite temporal and spatial patterns characterize the decomposition rate of organic materials

2. The decomposition rate of plant remnants depends on the quality of plant material and on the time of exposure. The decomposition rate of plant materials decreases with the advancing stage of their decomposition

3. The temperature, moisture content and aeration are recognized as driving variables of the decomposition processes

4. The half-time of decomposition of the plant remnants varies from 4 to 12 months according to the kind of plant material

5. During the decomposition of plant remnants, their carbon content decreases concurrently with the loss in weight

6. In winter, the loss of carbon and phosphorus from the decomposing plant material proceeds at a higher rate than in summer, reflecting changes in predominance of either mineralization or humification

7. Nitrogen content usually increases in the decomposing plant material. The increase in nitrogen content is greater than that expected from carbon loss, pointing to an active accumulation of nitrogen.

References

Alexander, M.: Introduction to Soil Microbiology. — New York: Wiley, 1961

Aristovskaya, T.V. (ed.): Problems of Abundance, Biomass and Productivity of Microorganisms in Soil. Leningrad: Nauka, 1972

Clark, F.E., Paul, E.A.: The microflora of grassland. Advan. Agron. **22**, 375–432 (1970)

Cruz, A.A. de la: The role of tidal marshes in the productivity of coastal waters. Assoc. Southeastern Biologists Bull. **20**, 147–156 (1973)

Dykyjová, D.: Content of mineral macronutrients in emergent macrophytes during their seasonal growth and decomposition. In: Ecosystem Study on Wetland Biome in Czechoslovakia Hejný, S. (ed.), Czechosl. IBP/PT-PP Rep. No **3**, Třeboň, 1973, pp. 163–172

Gilbert, O., Bocock, K.L.: Changes in leaf litter when placed on the surface of soil with contrasting humus types. II. Changes in the nitrogen content of oak and ash litter. J. Soil. Sci. **11**, 10–19 (1960)

Hundt, R., Unger, H.: Untersuchungen über die zellulolytische Aktivität unter Grünlandgesellschaften. In: Mineralisation der Zellulose. Tagungsberichte DAL **98**, 263–275 (1968)

Kaushik, N.K., Hynes, H.B.N.: Experimental study on the role of autumn-shed leaves in aquatic environments. J. Ecol. **56**, 229–243 (1968)

Kaushik, N.K., Hynes, H.B.N.: The fate of the dead leaves that fall into streams. Arch. Hydrobiol. **68**, 465–515 (1971)

Ketner, P.: Primary production of salt-marsh communities on the island of Terschelling in the Netherlands. Verh. Rijksinst. Natuurbeheer. **5**, 1–181 (1973)

Koppová, A., Pirkl, J., Kalina, J.: The determination of mineral elements in plant material by quick accurate methods. — Věd. Pr. Výzk Úst. Rostl. Výroby v Praze-Ruzyni 1955, 119–127 (1955)

Kuznetsov, S.J., Romanenko, V.J.: Produktion der Biomasse heterotropher Bakterien und die Geschwindigkeit ihrer Vermehrung im Rybinsk-Stausee. Verh. Int. Ver. Theor. Angew. Limnol. **16**, 1493–1500 (1966)

Mathews, C.P., Kowalczewski, A.: The disappearance of leaf litter and its contribution to production in the river Thames. J. Ecol. **57**, 543–552 (1969)

Odum, E.P., Cruz, A.S. de la: Particulate detritus in a Georgia salt marsh—estuarine ecosystem. Estuaries AAAS: 383–388 (1967)

Overbeck, J.: Experimentelle Untersuchungen zur Bestimmung der bakteriellen Produktion im See. Verh. Int. Ver. Theor. Angew. Limnol. **18**, 176–187 (1972a)

Overbeck, J.: Zur Struktur und Funktion des aquatischen Ökosystems. Ber. Deutsch. Bot. Ges. **85**, 553–577 (1972b)
Parkinson, D. (ed.): Progress in soil microbiological studies in IBP/PT projects. Report of UNECSO-IBP meeting, UNESCO, Paris, 1970
Parkinson, D., Gray, T. R. G., Williams, S. T.: Methods for studying the ecology of soil microorganisms. (IBP Handbook No. 19) Oxford: Blackwells Sci. Publ., 1971
Pieczyńska, E.: Ecology of the eulittoral zone of lakes. Ekol. Pol. **20**, 637–732 (1972)
Planter, M.: Elution of mineral components out of dead reed *Phragmites communis* Trin. Pol. Arch. Hydrobiol. **17**, 357–362 (1970)
Punčochář, P.: Measurement of various characteristics of bacterial isolates from surface waters. Folia Microbiol. **20**, 75 (1975)
Romanenko, W. I.: Heterotrophic assimilation of CO_2 by the aquatic microflora (Transl. from Russian). Microbiologia **33**, 679–683 (1964)
Saito, T.: The soil fungi of a salt marsh and its neigbourhood. Ecol. Rev. (Japan) **14**, 111–119 (1957)
Sorokin, Y. I.: On the trophic role of chemosynthesis and bacterial biosynthesis in water bodies. Mem. Ist. Ital. Idrobiol. Suppl **18**, 187–205 (1965)
Sorokin, Y. I., Kadota, H.: Technique for the assessment of microbial production and decomposition in fresh waters. (IBP Handbook No. 23) Oxford: Blackwells Sci. Publ., 1972
Tesařová, M., Úlehlová, B.: Abbau der Zellulose unter einigen Wiesengesellschaften. In: Mineralisation der Zellulose. Tagungsberichte DAL **98**, 277–287 (1968)
Witkamp, M. J., van der Drift, J.: Breakdown of forest litter in relation to environmental factors. Plant and Soil **15**, 295–311 (1961)

Fig. I. Nesting family of great crested grebe *(Podiceps cristatus)* (Photo: J. Formánek)

Section 6

Structure and Role of Animal Populations in Fishpond Littorals

The animal constituents of pond littoral ecosystems are far more diverse than the plant constituents. This variety of species composition has stimulated several studies on the structure, density, and production in selected populations or groupings of animals closely associated, as secondary producers, with the pond littoral biotopes. As it was clearly not possible to cover all aspects of animal life in the pond littoral, this section presents the principal results of zoological studies on the species variety of mammals in reedswamps and on the energy flow through some important populations of small mammals; on the role, species composition and densities of bird populations; on the soil-surface arthropods in the outer zone of the fishpond reed-belt; and on phytophagous insects attacking the common reed. A brief survey is also given and biomass estimated of the invertebrate macrofauna inhabiting fishpond reed-belts.

6.1 Mammals in the Reedswamp Ecosystem

J. PELIKÁN

Investigations of the secondary productivity and the function of mammals in wetlands of Czechoslovakia were carried out to discover (1) what is the species diversity of mammals in wetland communities; (2) what is the ecological dominance of species in successive trophic levels of the grazing predatory food chain; and (3) what is the amount of secondary production, energy flow and role of mammalian populations in ecosystems of this type?

The research was carried out in the South Moravian lowlands, especially in the reed-belt of the Nesyt fishpond, in reedswamps of the former lake "Kobylské jezero" and, to a small extent, in reed-belts of varions fishponds near Lednice. Particular attention was paid to the water vole *(Arvicola terrestris)* with respect to its negative effects on gardens and orchards.

6.1.1 Species Diversity

The results from the Nesyt fishpond can serve as a basis for evaluation (Pelikán, 1973, 1975). In the reed-belt around the pond, a temporary or permanent occurrence of 31 species of mammals has been ascertained (Tables 1–3). This number also includes stray dogs and cats, which exert a rather high predation pressure on the population of reed-belt animals. Even the North American mink *(Lutreola vison)* was found in this locality, as single animals escape occasionally from nearby fur animal farms and can temporarily live in the reed-belts of ponds or along rivers.

The values of the mean annual population densities from the period 1971 to 1973 (see first columns in Tables 1–3) were used for evaluation of the species diversity. For this purpose, the density was only approximated in some species, as indicated in the corresponding consumption columns. The species diversity is expressed by the Shannon-Wiener index with the formula $H_S = -\sum p_i \cdot \log_2 p_i$, where S = the number of species ascertained, p_i = the probability proportion of the ith species in the number of individuals of all species.

In this case of 31 species, the index of species diversity $H_{31} = 3.2148$. Species diversity under conditions of maximum equitability (or evenness of allotment among individuals) would be $H_{max.} = \log_2 31 = 4.9542$. Equitability $E = H_S/H_{max.} = 0.649$, i.e., the equitability in this case represents 64.9% of ideal evenness.

Besides the species ascertained in the reed-belt of Nesyt pond, the wetlands of Czechoslovakia are also inhabited by the northern vole *(Microtus oeconomus)* and the striped mouse *(Apodemus agrarius)*. The first of these occurs only in a few localities in southwest Slovakia in lowlands of the Danube where its distribution was recently reviewed by Binder and Štollmann (1975). *Microtus oeconomus* lives mainly in sedge communities *(Cariceta)* and is relatively rare.

Apodemus agrarius is common along the whole northern frontier of Czechoslovakia and in the East Slovakian lowlands (Kratochvíl and Rosický, 1954). The importance of the striped mouse for reedswamp ecosystems in the area of its occurrence is evidently much greater than that of *Apodemus sylvaticus*, because it is the most common in reedswamps of all *Apodemus* species. Quantitative data on both species are lacking.

Other species of mammals may occur rarely in reedswamps. The water shrew *(Neomys anomalus)* is very scarce (Kratochvíl, 1954; Zbytkovský, 1975). Bauer (1960) and Böck (1974) ascertained in reedswamps of Neusiedler See Lake the incidental occurrence of other species of mammals, e.g., *Cervus elaphus* and *Sus scrofa* of the herbivores; *Martes foina*, *Meles meles*, and *Lutra lutra* of the carnivores.

6.1.2 Distribution of Species

The populations of mammals are spatially distributed across the reed-belt in dependence on the water level and on the depth of the ground water. The local distribution of each species is conditioned by the degree of its adaptation to the aquatic or semiaquatic life habits. According to the trophic specializations of the species, the corresponding ecological niches are filled (see p. 75).

The littoral ecophase, permanently flooded, is occupied by *Ondatra zibethicus*. At the border between the littoral and the limosal ecophase, there is the greatest concentration of the herbivorous water vole *(Arvicola terrestris)*. The maximum population density of *Neomys fodiens* is in the limosal ecophase where the water alternates with patches of dry land. Nearer to the outer margin of the reed-belt, between the limosal zone and the permanently dry terrestrial ecophases, there is maximum occurrence of *Sorex araneus* and *S. minutus*. *Micromys minutus*, a typical stem-climber, lives across the whole width of the reed-belt.

All the other mammalian species occur permanently or temporarily in the terrestrial outer margin of the reed-belt. Here, the most common herbivores (Table 94) include *Microtus arvalis* and *Clethrionomys glareolus;* less common are the mice of the genus *Apodemus*, as well as *Mus musculus* and *Rattus norvegicus*. Insectivorous mammals (Table 95) are represented by *Crocidura suaveolens*, by the eastern hedgehog and by the mole. The mammalian fauna of the reed-belt also comprises insectivorous bats, five species of which were ascertained in South Moravia. The carnivores also occur predominantly in the terrestrial ecophase; eight species of them were found in Nesyt (Table 96). In the hydrosere of mammalian species living in the reed-belt from water to land, the muskrat and the mole are at opposite extremes of the row.

6.1.3 Herbivores

At the start of the grazing-predatory food chain, the muskrat has the greatest ecological importance. The mean density of muskrat *(Ondatra zibethicus)* lodges in the reed-belt of the Nesyt pond was 8.7 lodges per hectare (maximum 18 lodges; Pelikán, 1975). In spring, only 2 lodges ha^{-1} are occupied as a rule, so that about two pairs of animals per hectare is the approximate mean spring density of muskrats.

The mean annual density, biomass, and production are shown in Table 1. Holišová (1975) analyzed the stomach contents of animals from the reed-belt of Nesyt pond; there was only vegetable food present in the diet of the muskrats. Shells *(Anodonta)* are nearly absent in Nesyt; muskrats therefore consume the plant biomass evidently even in winter. About 60% of the food comprises the aboveground parts of *Phragmites* and *Typha*, about 40% white inner parts of their rhizomes (Akkermann, 1975). The consumption figures are given in Table 1.

In the littoral ecophase, roughly equal shares of *Phragmites* and *Typha* are assumed in the vegetation. The average net production of the aboveground plant biomass amounts to about 2.8 kg dry weight according to data by Husák and Květ (1973), and the production of the rhizomes is about half, i.e., 1.4 kg m^{-2} $year^{-1}$ (Fiala, 1973). Consumption of muskrats represents about 0.9% of the aboveground production and 1.3% of the underground production, totalling 2.2%. Destruction of reed stands by muskrats is about two or three times greater than the consumption itself, therefore the total share of the biomass destroyed is 6.6 to 8.8% of primary production (Fig. 1).

Very high population densities were ascertained in reedswamps of the former lake Kobylské jezero (Pelikán et al., 1970), where 9 to 31 muskrat lodges (mean 16 lodges) were found per hectare. The spring population density was four pairs as a maximum, the fall density was about 55 individuals per hectare. The estimated mean annual biomass was approximately 27 kg and production 31 kg ha^{-1} $year^{-1}$ (fresh weight). These values represent probably the maximum which a muskrat population may attain and they correspond with the peak densities ascertained by Errington (1963).

The size of muskrat lodges ranges from 160 by 30 cm (diameter at base by height, size group I) up to 285 by 90 cm. Net dry weight of one lodge ranges from 9 up to 180 kg. With the mean density of 16 muskrat lodges per hectare, the amount of plant material deposited in the lodges attained nearly 700 kg dry weight, which is a quite exceptional case (Pelikán et al., 1970).

The stand of *Typha latifolia* around a muskrat family lodge was damaged within a circle with a radius of 14 m. The destruction of the aboveground biomass in that area amounted to 206 kg dry weight, representing 20% of the net production on the same area. The destruction of the *Typha* stands per hectare was estimated at 5–10% of its net aboveground production per year.

The ecological importance of other herbivorous mammals for the energy flow in the reedswamp ecosystem is evident from the values of consumption in Table 1. The data for *Arvicola terrestris* originate from the littoral and limosal zones of the swampy lake Kobylské jezero (Pelikán, 1974) and represent the situation at medium population densities (mean values for the period 1966–1967). The trophic

Table 1. Herbivorous mammals in the reed-belt of Nesyt fishpond. Survey of data; mean values obtained from three years investigation (1971–1973)

Species	Mean annual density	Mean annual biomass		Production per yr		Calorific input kcal ha^{-1}	Consumption per yr Food intake in g dry wt m^{-2}				Egested energy in kcal ha^{-1} yr^{-1}	
	ind. ha^{-1}	fresh wt. kg ha^{-1}	dry wt. g m^{-2}	fresh wt. kg ha^{-1}	dry wt. g m^{-2}		above-ground plant biomass	under-ground plant biomass	seeds	animals	feces	urine
Ondatra zibethicus	18.40	10.300	0.2971	24.32	0.7020	1,824,000	25.88	17.50	—	—	684,000	124,000
Arvicola terrestris[a]	7.11	0.778	0.0224	2.28	0.0658	128,050	2.39	0.61	0.026	—	48,020	8,700
Microtus arvalis	21.10	0.581	0.0168	3.27	0.0942	132,730	2.32	0.54	0.039	—	24,830	5,070
Clethrionomys glareolus	16.30	0.338	0.0098	0.91	0.0263	66,030	1.28	—	0.207	0.024	8,730	2,600
Micromys minutus	10.00	0.076	0.0022	0.25	0.0072	35,970	0.08	—	0.427	0.192	4,250	750
Apodemus sylvaticus	2.13	0.042	0.0012	0.14	0.0039	10,170	0.01	0.01	0.163	0.019	1,200	210
Apodemus flavicollis	0.21	0.004	0.0001	0.014	0.0004	1,020	—	—	0.005	0.013	120	20
Apodemus microps	0.16	0.003	0.0001	0.001	0.0003	740	—	—	0.015	—	90	10
Mus musculus	0.32	0.006	0.0002	0.020	0.0006	1,530	—	0.01	0.014	0.007	180	30
Total	75.73	12.128	0.3499	31.205	0.9007	2,200,240	31.96	18.67	1.096	0.255	771,420	141,390

[a] All data from reedswamps of the "Kobylské jezero" (Pelikán, 1974).

Fig. 1a and b. Muskrats building their lodges can destroy large areas of reedswamp stands: (a) Material of *Schoenoplectus lacustris;* (b) *Typha angustifolia*, both in Opatovický fishpond at autumnal drawdown. (Photos: D. Dykyjová)

requirements of water voles in reed stands were analyzed by Holišová (1970b, 1975); the food consists of about 79% aboveground plant biomass, 20% underground biomass and 1% seeds. The energetic budget and assimilation of food were also studied in animals originating from the South Moravian lowlands (Droźdź et al., 1971). The population density is about two to ten times lower in the terrestrial zone of the reed-belt and in places where banks do not permit underground burrowing of the animals.

In *Microtus arvalis*, the energy flow may be higher than in the preceding species, especially in peak years, as suggested by the data from the terrestrial zone of the Nesyt pond (Table 1).

The importance of the remaining herbivores decreases successively, as they are listed in Table 1. In *Clethrionomys glareolus*, whose occurrence in reed stands is connected with the presence of willows, the consumption equals only half that of *Microtus arvalis*. The proportions between four food components were evaluated according to the analyses of stomach contents of captured animals (Holišová, 1972, 1975; Obrtel, 1975). In *Micromys*, *Apodemus*, and *Mus*, the proportion of the green components decreases successively but the proportion of seeds and animals increases in the food.

In the following three species, which are not included in Table 1, approximations of population densities are used for evaluation of the species diversity index. In *Rattus norvegicus*, only a third of the density of *Mus musculus* is assumed (i.e., 0.1 individual ha^{-1}), in *Lepus europaeus* about 1 animal per 30 ha of the terrestrial ecophase (0.03 individual ha^{-1}), and in *Capreolus capreolus* about one third of that value (0.01 individual ha^{-1}). The remaining values are not given for these three species.

The animal component in stomach contents of captured rodents was analyzed by Obrtel (1975). It consists of larvae, pupae, and adults of insects, especially *Coleoptera* (Carabidae, Staphylinidae, Dytiscidae), *Lepidoptera* (Noctuidae), *Diptera* (Limoniidae, Itoniidae) as well as of spiders (Araneidea, especially Lycosidae). The earthworms, snails as well as tissues from dead mammals and birds are a less common food of rodents.

Summarizing the results from Table 1 the conclusion can be drawn that the mean density per hectare of reed stands in the South Moravian lowlands amounts to 76 individuals of mammals, the mean annual biomass is 0.35 g and the mean annual net production is 0.90 g (dry weight) per m^{-2}. The corresponding values for rodents, excluding the muskrat, are: density 57.3 individual ha^{-1}, biomass 0.05 g and production 0.20 g m^{-2} $year^{-1}$ (dry weight). It is necessary to underline the fact that these values are higher than those recorded in the reed-belt of the Nesyt pond (Pelikán, 1975), because the high density of *Arvicola terrestris* from Lake Kobylské jezero is included in the calculation.

The total consumption by herbivorous mammals corresponds with an energy input of $2.2 \cdot 10^6$ kcal ha^{-1} $year^{-1}$. It represents, in terms of dry weight per 1 m^2, the consumption of nearly 32 g of green aboveground plant biomass, 18.7 g of underground plant biomass, 1.1 g of seeds and 0.255 g of animal biomass per year. Consumption by rodents (except the muskrat population) amounts to some 6.1 of aboveground and 1.2 g underground plant biomass.

In the terrestrial and limosal ecophases, the mean annual aboveground plant production amounts to some 1.3 kg dry weight per m^2 and underground production is about half of that value (Husák and Květ, 1973). Thus, the rodents (excluding the muskrat) consume 0.55% of the aboveground and about 0.18% of the underground plant production. The destruction of vegetation by small rodents can be estimated at 2.2–2.9% of total primary production.

6.1.4 Insectivores

This group includes the species of the order *Insectivora* and *Chiroptera*. Also, the animal component in the food of several rodents is taken into account in this group (see Table 2). The species of the order *Insectivora* are placed in Table 2 according to their ecological importance (their population density and amount of consumption). In *Talpa europaea*, only about one animal is assumed per 10 ha of the terrestrial ecophase when standing dry; at the most about 0.05 individual ha^{-1}. It is of no practical importance for the reed-belt ecosystem, therefore no other values are estimated (all data from Pelikán, 1975).

The consumption of insectivores, including the mole, represents an energy input of about 100,100 kcal ha^{-1} $year^{-1}$. Approximately half of the food includes larvae, pupae, and adults of insects with a higher energy content (5.8 kcal per g dry weight). The other half is composed of worms, molluscs, and other invertebrates with a lower energy content (4.6 kcal per g dry weight; data from Golley, 1961; Hawkins, and Jewell, 1962). The insectivores consume, per m^2 per year, 0.86 g insects and 1.09 g dry weight of other invertebrates. The consumption of insects by rodents, amounting to 0.255 g, should be added so that the total intake of insects and other invertebrates represents 2.20 g m^{-2} $year^{-1}$ (dry weight).

At present, it is impossible to evaluate the proportion which is consumed from the actual standing crop or production of invertebrate animals. Obrtel (1972, 1973) and Miller and Obrtel (1975) investigated the soil-surface fauna of inverte-

Table 2. Insectivores in the reed-belt of Nesyt fishpond; survey of data

Species	Mean annual density ind. ha^{-1}	Mean annual biomass		Production per yr		Consumption per yr			Egested energy feces and urine kcal ha^{-1} yr^{-1}
		fresh wt. kg ha^{-1}	dry wt. g m^{-2}	fresh wt. kg ha^{-1}	dry wt. g m^{-2}	input in kcal ha^{-1}	dry wt. g m^{-2}		
							arthropods	other in-vertebrates	
Sorex araneus	10.07	0.092	0.0026	0.400	0.0115	39,800	0.343	0.433	2,770
Neomys fodiens	8.00	0.064	0.0018	0.280	0.0080	27,860	0.240	0.303	1,950
Sorex minutus	7.40	0.027	0.0008	0.177	0.0051	17,670	0.152	0.192	1,240
Erinaceus concolor	0.17	0.069	0.0020	0.113	0.0033	14,180	0.122	0.154	1,420
Crocidura suaveolens	0.15	0.001	0.0001	0.006	0.0002	560	0.005	0.006	40
Total	25.79	0.253	0.0073	0.976	0.0281	100,070	0.862	1.088	7,420

brates in the terrestrial ecophase. They estimated the mean annual biomass of spiders to be about 0.167 g m^{-2} dry weight. Similarly, Dvořák (1970) studied the soil macrofauna in the littoral ecophase of a pond in southwest Bohemia and estimated the mean biomass per m^2 as 53 g of the formaline weight. The amount of soil macrofauna is probably smaller in the dry terrestrial zone, nevertheless, the consumption of 2.2 g dry weight by small mammals evidently represents only a small proportion of the actual standing crop.

An approximation of the population density of bats is about 2.5 individual/ha with respect to the data given by Gaisler (1975). The densities of the five species ascertained can be approximated with respect to their abundance as follows: *Nyctalus noctula* 0.90, *Pipistrellus pipistrellus* 0.65, *Myotis daubentoni* 0.65, *M. mystacinus* 0.25, *Pipistrellus nathusii* 0.05 individual/ha. These values enabled us only to calculate the index of species diversity. No other values can be estimated for bats at present.

6.1.5 Carnivores

In the reed-belt of Nesyt pond, a total of eight species of carnivores was ascertained. In Table 3, the North American mink *(Lutreola vison)*, escaped from fur animal farms, is not indicated. Its density is arbitrarily estimated at about 0.0001 individual ha^{-1} for calculating the diversity index. Of both species of polecats, *Putorius eversmanni* is much more common at Nesyt. As evidenced by Table 3, most of the energy flow passes through the population of foxes and stray cats.

The consumption of carnivores represents an energy input of 16,760 kcal ha^{-1} $year^{-1}$. Most of their food consists of small mammals and birds and even of frogs, in some carnivores, but to a lesser extent. In fall and winter, the foxes gather in the reed-belt, where they prey upon rodents, pheasants, ducks wounded by shooting, and on other birds. Approximately two-thirds of the total annual consumption is assumed to be covered by the production of small rodents. The remaining one-third consists of birds, fish, and to a lesser extent even of invertebrates.

6.1.6 Final Energy Budget

Data on the energy flow through the mammalian populations allow approximations to be given of the input and output of the successive links of the grazing-predatory food chain. All values are indicated in kcal m^{-2} $year^{-1}$.

The energetic input amounts to 220 in herbivores, 10 in insectivores, and in carnivores to only 1.7 kcal. The greatest part of the energy intake is dissipated in the form of respiratory losses. The amount of energy fixed in the secondary production of animal biomass is only very small. In herbivores, the production amounts to 4.68, in insectivores to 0.15, and in carnivores to 0.014 kcal, representing 2.1, 1.5, and 0.8% of the energy received by consumption. In terms of the dry weight of animal biomass, the values of the annual secondary production amount to 0.900, 0.028, and 0.0027 g respectively.

Table 3. Carnivores in the reed-belt of Nesyt fishpond

Species	Mean annual density ind. ha^{-1}	Mean annual biomass		Production per yr		Con-sumption per yr dry wt. $g\ m^{-2}$	Egested energy feces and urine kcal $ha^{-1}\ yr^{-1}$
		fresh wt. kg ha^{-1}	dry wt. $g\ m^{-2}$	fresh wt. kg ha^{-1}	dry wt. $g\ m^{-2}$		
Mustela nivalis	0.120	0.007	0.00021	0.008	0.00023	0.045	235
Mustela erminea	0.020	0.004	0.00011	0.004	0.00011	0.016	85
Putorius eversmanni	0.025	0.020	0.00058	0.018	0.00052	0.063	480
Putorius putorius	0.001	0.020	0.00058	0.018	0.00052	0.063	480
Vulpes vulpes	0.009	0.043	0.00124	0.027	0.00078	0.092	1,170
Felis catus	0.060	0.175	0.00505	0.032	0.00092	0.093	1,195
Canis familiaris	0.008	0.083	0.00239	0.006	0.00017	0.017	225
Total	0.243	0.332	0.00958	0.095	0.00273	0.326	3,390

The energy output in the form of feces and urine is several times greater than the energy fixed in the secondary production. By feces, the losses of energy amount in herbivores to 77.1, in insectivores to 0.6, and in carnivores to 0.3 kcal; the output of energy by urine is 14.14, 0.11, and 0.05 kcal. These energy transfers into the decompository food chain represent 41.5, 7.4, and 20.2% of the energy received by consumption, and they are not negligible, especially with herbivores.

References see pp. 393—395.

6.2 Birds in the Reedswamp Ecosystem

K. HUDEC and K. ŠŤASTNÝ

Investigations on the secondary production and role of birds in the reedswamp ecosystems in Czechoslovakia were mainly aimed at (1) obtaining basic data on the species composition and numbers of birds breeding in various types of reedswamps (i.e., on their species composition and diversity, density and biomass per unit area); (2) ascertaining the basic parameters of the population dynamics of the species breeding in the reeds (clutch size, number of young produced); and (3) ascertaining the impact of the dominant bird species on the reedswamp vegetation.

The investigations were carried out in two regions, in South Moravia and South Bohemia. The following were the South Moravian localities investigated:

(a) Kobylské jezero (Lake of Kobylí) near Brumovice (district of Břeclav), a continuous reedswamp developed in a depression on flooded arable land on the bottom of a former shallow lake. The whole extensive (some 1.5 km^2) swamp was practically without any open water surface. Most investigations were carried out here in 1966 and 1967

(b) Mlýnský, Prostřední a Hlohovecký fishponds in the Lednické rybníky State Nature Reserve (also district of Břeclav). Here, the investigations involved narrow belts of littoral reedswamp edging with open water surface of the ponds. These investigations continued previous investigations carried out here since 1959, and ended in 1968

(c) The Nesyt fishpond in the same State nature reserve. Here, the investigations involved an extensive continuous reedswamp edging with open water surface, with developed littoral to terrestrial ecophases; some of the work was also done in narrow littoral reed-belts. The investigations were concentrated mainly in the period of 1969–1971.

In the South Bohemian Třeboň basin, the investigations were carried out in the littoral reed-belts of the Opatovický fishpond in 1972–1973.

6.2.1 Species Composition of Breeding Birds

Table 1 contains a list of bird species whose breeding was recorded in the above study areas. In all, 29 breeding species were recorded. The table does not contain two additional species *(Ardea purpurea, Aythya nyroca)* which breed rarely at Nesyt fishpond but which have not been observed breeding directly in the plots under study. Moreover, the table does not comprise species which forage in the reeds but do not breed in them, such as swallow-like birds which hunt for insects over the reeds *(Hirundo rustica, Delichon urbica, Riparia riparia, Apus apus)*, or herons *(Ardea cinerea, Nycticorax nycticorax)* which forage in open spaces within extensive reed stands.

The dominant species (i.e., those with dominace over 5% in at least one experimental plot) include *Podiceps cristatus*, (Fig. I in the Introduction to this Sect.) *P. nigricollis, P. ruficollis, Rallus aquaticus, Fulica atra, Larus ridibundus, Acrocephalus arundinaceus, A. scirpaceus, A. schoenobaenus, Locustella luscinioides, Panurus biarmicus* and *Emberiza schoeniclus*. Except *R. aquaticus* and *P. biarmicus*, all these species showed a high dominance in all plots.

Certain plots harbor breeding colonies of certain species whose density attains such values that all the remaining species are suppressed. This is particularly true of *Larus ridibundus*. The dominance of this species amounted to 92% in Kobylské jezero (Hudec, 1975b) and to 90% in the littoral ecophase of Nesyt fishpond (Hudec, 1975a). In these areas, *L. ridibundus* participates even in the avian biomass in the same decisive way: in Kobylské jezero, the share of the gulls in the total biomass of breeding birds amounted likewise to 92% (Hudec,1975b).

6.2.2 Birds Occurring Outside the Breeding Season

The qualitative and quantitative structure of the avian community occurring in the reeds outside the breeding season has not been investigated with such precision. During various periods, the dominant breeding species populations are supplemented by additional non-breeding populations as follows:

Migration periods (February-May; July-November): Occurrence of individuals of species breeding in more northern regions, especially geese and ducks, the numbers of which amount to tens or hundreds in any one locality. In addition small passerines, above all *Acrocephalus* sp. div. and *Emberiza schoeniclus*, are present in tens of individuals, as well as passerines that do not breed in the ecosystem *(Erithacus rubecula, Luscinia suecica, Saxicola torquata, S. rubetra, Remiz pendulinus)*, whose numbers range from individual birds to tens of individuals. During the autumnal drawdown, the bare mud flats are visited by various waders (Limicolae).

Moulting period (June–Aug.): Occurrence of moulting and flightless individuals of breeding duck-like birds (Anatidae), the numbers of which vary between individuals and hundreds of individuals according to locality and species.

Summer and fall roosting places (June–Oct.): The reeds provide roosting places namely for *Sturnus vulgaris* (hundreds to hundreds of thousands), swallow-like birds (*Hirundo rustica, Delichon urbica, Riparia riparia* — tens to hundreds), wagtails (*Motacilla alba*, tens of individuals).

Winter period (Oct.–Feb.): *Parus caeruleus* (individuals to tens) and *Phasianus colchicus* (individuals to hundreds) are the dominant and almost the sole species frequenting the reeds in this period.

Table 1. List of species of birds ascertained in experimental plots in reedswamps in South Moravia (Lake of Kobylí and Lednice fishpond including Nesyt) and South Bohemia (Opatovický fishpond), their topic and trophic characteristics and density. (Compiled by K. Hudec and K. Šťastný)

Species	Ecophase					Food			Lake of Kobylí	Lednice fishpond	Nesyt/T	Nesyt/L	Opatovický/T	Opatovický/L
	W	L	M	T	S	H	C_1	C_2	adp	adn	adp	adp	adp	adp
Podiceps cristatus	x	X	x				X	·		1.78				
Podiceps nigricollis	x	X	x				X			2.44		16.24		
Podiceps ruficollis		X	x				X		0.05	1.45		0.23		
Ixobrychus minutus		x	x	x			X	·	0.08	0.76	0.47	0.70		0.38
Cygnus olor	x	X			x	X	·			0.04				
Anser anser		X	x	x	x	X			0.08	0.09		0.93		
Anas platyrhynchos		x	x	x	x	X	x		0.14	0.19				
Anas strepera				x	x	X	·			0.09			(0.83)	
Anas querquedula				x	x	X	X		0.02					
Anas clypeata				x	x	X	X		0.05					
Netta rufina		x	x		x	X	·			0.09				
Aythya ferina		x	x		x	X	X			1.42				
Aythya fuligula		x	x		x	X	X			0.05				
Circus aeruginosus		x	x	x				X		0.05				
Phasianus colchicus				x	x	X	X	·	0.05	0.05				
Rallus aquaticus			x	x		x	X	·	0.06	0.09	0.94		1.25	1.54
Porzana parva		x	x			·	X			0.06				
Gallinula chloropus		x	x	x		x	X		0.13	1.07	0.47	0.23		
Fulica atra	x	X	x			X	x		0.73	10.67		0.70		(2.31)
Larus ridibundus	x	X	x			x	X	·	29.64			232.02		
Cuculus canorus		x	x	x	x		X			1.57				
Acrocephalus arundinaceus		x	x				X		0.36	5.38		2.09		3.08
Acrocephalus scirpaceus		x	x	x			X		0.18	6.92	3.77	1.86	6.66	8.08
Acrocephalus palustris				x	x		X		0.03					
Acrocephalus schoenobaenus			x	X			X		0.16		6.13		1.25	
Locustella luscinioides			x	X			X		0.01		2.83		1.25	
Panurus biarmicus		x	X	X		X	X		0.13		2.36	1.62		
Emberiza schoeniclus		x	x	X	x	X	x		0.16	0.05	5.65	0.70	1.25	2.31
Motacilla flava				x	x		X		0.05					
Total									32.11	34.31	22.62	257.32	11.66	15.39

W, open water; L, littoral ecophase; M, limosal ecophase; T, terrestrial ecophase; S, shore; X, main environment. Trophic characteristics: H, herbivorous species; C_1, consumer of invertebrates; C_2, consumer of vertebrates (carnivorous species); X, primary food; x, secondary food; period (·) indicates insignificant quantity. Quantitative characteristics: adp, average density of pairs; adn, average density of nests; densities stated per one ha of area.

6.2.3 Spatial Distribution

Table 1 indicates the spatial distribution of the species breeding in the reeds. The presentation is simplified and related mainly to the nest site. Some of the reed breeders take advantage of the vegetation acting as shelter but they depend, partly or entirely, on the surrounding area for food. These species include, for example, *Anser anser* or *Larus ridibundus*. The importance of such species for the energy flow through the ecosystem is thereby strongly decreased; on the other hand, such species import materials into the reedswamp ecosystem from its surroundings in the form of feces or of material for nest construction or as food for their young.

A considerable number of species, however, are specifically adapted to life in the reedswamps, such as *Locustella luscinioides* or *Ixobrychus minutus*. Birds foraging in open water are typical inhabitants of the littoral ecophase in the reedswamp, but they sometimes even build their nests in open water outside the reeds, e.g., *Podiceps* spp. In contrast, typical of the terrestrial ecophase are such species that invade from the surrounding shores and frequently breed outside the reeds in fields, ridges, and meadows (*Motacilla flava*, certain *Anas* spp.).

The various species inhabiting each ecophase differ by their ecological niches, that is, mainly by the nest site as well as by the kind of food and way of foraging. In the terrestrial ecophase, for example, the birds forage in the stratum of the reed shoots (*Parus caeruleus*), in dense vegetation covering the ground (*Locustella luscinioides*) or on the ground (*Emberiza schoeniclus*). *Anser anser*, on the other hand, builds its nests mainly near an open water surface, at the edge of the reedbelt or near pools and "lagoons" inside the stands (Hudec, 1971b).

6.2.4 Interaction Between Birds and Plant Communities

The relations between birds and the helophyte communities are of two kinds: (i) trophic, i.e., direct consumption of the plant matter or of secondary products (mainly small invertebrates); (ii) destructive, i.e., mechanical damage to the stands in obtaining material for nest construction, or inadvertent damage resulting from the movements of the birds in the stands.

In Nesyt fishpond, the relations of *Anser anser*, one of the common and most important avian species, to the reed communities were investigated in detail (Hudec, 1971a, b; Hudec and Kux, 1971, 1972; Kux and Hudec, 1971). Three kinds of interaction exist between *A. anser* and the reed communities during a year (Hudec, 1973a): (a) destruction and accumulation of plant material for nest building (March to May); (b) mechanical damage to the stands by the moving geese (mainly March to June); (c) grazing (February to November, and mainly April to June). Grazing exerts the greatest pressure on the reed stands (Květ and Hudec, 1971). The preference of the grazing geese for *Phragmites communis* over *Typha angustifolia* even results in changes in the composition of the plant community. Permanent grazing of individual reed shoots results in various growth malformations, prolonged growth and, hence, exhaustion of reserves stored in rhizomes. (Fig. 1).

Fig. 1. A *Phragmites communis* stand in Nesyt fishpond destroyed by grazing greylag geese. (Photo: K. Hudec)

The breeding season of all reed breeders lasts from early March till late August, culminating in late May or early June in coincidence with the full development of the aerial shoots stand (Hudec, 1975a; Fig. 2).

In the Nesyt fishpond, investigations were also made on the composition of nests of six species of birds (*Anser anser, Fulica atra, Ixobrychus minutus, Acrocephalus arundinaceus, A. scirpaceus, A. schoenobaenus*) and on the impact of removal of the nest material on the primary production of the ecosystem (Toušková, 1973; Kožená-Toušková, 1973). Four of the above species were found to use green plant material for nest construction; of these, three are numerous (*A. anser, F. atra, A. arundinaceus*). The total removal of plant material for nest building by these three species averaged 0.914 kg ha^{-1} of dry matter, which is a negligible part of the total primary production of the ecosystem.

Trophic relations were investigated in detail only in two reed warbler species, *Acrocephalus arundinaceus* and *A. scirpaceus* (Mayer, 1971). The food of *A. scirpaceus* was found to consist mainly of *Diptera*, above all adult *Culicidae*, while that of *A. arundinaceus* consisted of several groups of invertebrates, above all insects, spiders, and molluscs. From the nests situated inside continuous reed stands, the birds flew to forage mostly in the surrounding areas, which suggests a poor food supply inside the reed stands.

The trophic relations of the remaining species were not evaluated. In Table 1 the species are divided into three blocks according to the nature of their food:

Fig. 2. (a) Flock of greylag geese in Lednice fishponds (Photo: K. Hudec). (b) Greylag goose in its nest. (Photo: Z. Kux)

H: herbivorous species. Most of these feed on seeds and small particles of soft aquatic plants, mainly found submerged in the open water. *Phragmites* is grazed almost exclusively by *Anser anser* (Hudec, 1973b; Květ and Hudec, 1971).

C_1: species subsisting largely or exclusively on insects, zooplankton, or other small invertebrates. This group comprises most reed-inhabiting avian species.

C_2: carnivorous species. Only *Circus aeruginosus* belongs fully to this category. Several other species consume small vertebrates to a negligible extent. Of the species not listed in Table 1, this group also comprises herons which, however, feed mostly in open water or in sparser parts of the reedswamp.

6.2.5 Density and Production

Table 1 summarizes data on the breeding density of birds per hectare of reed stand, as recorded in the various study areas (Hudec, 1975a, b; Šťastný, 1973, unpublished). Data on breeding of the various species (clutch size, number of young, losses during breeding) are given by Hudec (1975a). The date on one of the dominant reed breeders in the Opatovický fishpond *Acrocephalus scirpaceus*, are reported by Šťastný (1973); its annual production amounted to 303.3 g ha^{-1} of fresh weight.

For a synthesis of investigations on Nesyt fishpond, Hudec and Pelikán (in press, 1977) estimated the production of the avian component of the reedswamp ecosystem. This estimate is based on empirical data (density, breeding parameters) supplemented by data adopted from the literature.

The average number of birds is estimated at 20 pairs per hectare of the reedswamp (about 13 passerines, 3 grebes, 1.7 ralliforms, 1 *Anser anser*, 0.5 ducks, 0.04 *Circus aeruginosus*). In addition, the local colony of *Larus ridibundus* represents 83 breeding pairs per hectare on the average. The ascertained maximum values were about double. The mean biomass of breeding birds was 44.4 kg ha^{-1} for *L. ridibundus* and 11.2 kg ha^{-1} for the remaining species, both data fresh weight. The mean egg production amounted to 246 eggs ha^{-1} for the gulls and 171 eggs ha^{-1} for the remaining species, i.e., 8.15 and 2.26 kg ha^{-1} fresh weight, or about 8320 and 2300 kcal ha^{-1} respectively. The mean annual net secondary production of fledged young was 20.14 kg ha^{-1} for *L. ridibundus* and 13.61 kg ha^{-1} for the remaining species. In addition, the annual losses among the young amounted to 12.31 kg ha^{-1} for *L. ridibundus* and 4.08 kg ha^{-1} for the remaining species. The resulting annual mean young production is about 32.5 kg ha^{-1} for *L. ridibundus* and 17.7 kg ha^{-1} for the remaining species, all data fresh weight, i.e., 10.93 and 5.96 kg ha^{-1} respectively. The total annual mean production of biomass (of eggs and young) was about 40.6 kg ha^{-1} fresh weight (or 6088 kcal ha^{-1}) for *L. ridibundus* and 20.0 kg ha^{-1} fresh weight (or 3096 kcal ha^{-1}) for the remaining species. The maximum values are about double.

References see pp. 393—395.

6.3 Soil Surface Arthropods

R. OBRTEL

Material of surface-active arthropods was obtained by operating five formalin pitfall traps in the terrestrial ecophase of the reedswamp covering the southern shore of the western bay of the Nesyt fishpond continuously between April 17 and November 18, 1969.

The material consisted of isopods *(Isopoda)*, spiders *(Araneidea)*, mites *(Acarina)*, harvestmen *(Opilionidea)*, and insects *(Insecta)*, both at their adult stage and developmental stages. Of the insects, the material comprised springtails *(Collembola)*, beetles *(Coleoptera)* various forms of dipterous insects *(Diptera)*, aphids *(Homoptera Aphidioidea)*, leafhoppers *(Homoptera Auchenorhyncha)*, various forms of hymenopterous insects (*Hymenoptera*, especially ants, *Formicoidea*) and various groups of parasitic *Hymenoptera Apocrita* (Obrtel, unpublished).

So far, only the material of *Coleoptera* and *Araneidea* has been evaluated.

6.3.1 Coleoptera

The results of the investigation have been presented in two papers (Obrtel, 1972, 1973). The material of 3451 adult *Coleoptera* comprises members of 17 families, 80 genera and 116 species. The largest number of species belongs to Staphylinidae (36 spp.), Carabidae (32 spp.) and Hydrophilidae (12 spp.); these species make up almost 70% of the whole number of species present, the rest falling to the species of the remaining 14 coleopterous families.

In terms of number of specimens, the most numerous were Carabidae (48.2%), Hydrophilidae (21%), Staphylinidae (15%), Silphidae (7%), and Ptiliidae (6.8%). The remaining families were represented by far smaller numbers of specimens. *Agonum moestum, Anacaena limbata, Ptenidium pusillum* and *Olophrum assimile* were dominant. *Bembidion inoptatum, Oodes helopoides, Silpha tristis, Pterostichus nigrita, Catops fuscus, Bembidion assimile, Pterostichus vernalis, Cercyon tristis, Pterostichus niger* and *Paederus riparius* were subdominant. Of the recedent species, one should mention *Carabus granulatus* (1.97%), *Agonum fuliginosum* (1.94%), *Pterostichus strenuus* (1.74%), *P. anthracinus* (1.56%), and *P. melanarius* (1.27%).

Thus the soil surface in the terrestrial reedswamp under study is inhabited by a beetle community in which strongly hygrophilous and more or less psychrophilous species predominate.

The largest number of species (61, or 53%) and specimens (43%) belong to carnivorous predators, namely Carabidae and Staphylinidae, besides several species of Dytiscidae. The group of scavengers comprises 12 species of Carabidae (nearly 20% of material). The third trophic group consists of six necrophagous species of Silphidae (7% of material). The fourth group comprises saprophagous species of Cryptophagidae, Lathridiidae, Anthicidae, Ptiliidae and Byrrhidae, besides *Oxytelus* spp. of the Staphylinidae; in all, there are 16 species (14%) whose larvae and adults feed on microorganisms decomposing organic matter. The saprophagous species make up the second greatest trophic group of the material (28%). The fifth trophic group consists of *Harpalus rufipes* (Carabidae), a ground beetle feeding preferably on vegetable food. Besides the above eucenic species of *Coleoptera* there is a group of 17 tychocenic species which belong to the entomocenoses of the herb stratum of the reedswamp or of the surrounding ecosystems.

Predators (first or second order carnivores), saprophagous and necrophagous beetles evidently prevail over the remaining trophic groups in the soil-surface beetle community of the reedswamp.

In the growing season of 1969, the catch curve of species and specimens of soil surface *Coleoptera* in the reedswamp under study showed three peaks (in May; in mid-June to late August; in November) and two distinct depressions (in late June and early July; in late September and early October).

At present, the collected relative and semi-quantitative data cannot be converted to actual population densities or biomass per unit of area. This problem requires additional investigation.

6.3.2 Araneidea

Miller and Obrtel (1975) evaluated material of spiders collected in the same way and place as the material of soil surface *Coleoptera*. The community of soil surface spiders consists of members of 10 families. In terms of number of species, the material was dominated by Linyphiinae (26.3%), Micryphantinae (23.7%), and Lycosidae (21.1%). In terms of number of specimens, most of the material belonged to Lycosidae (67.2%), *Antistea elegans* (*Hahniidae*, 16.6%), and *Linyphiinae* (7.2%).

As regards moisture requirements, 75% of the present species are hygrobiotic or hygrophilous. As regards their light requirements, 60% of the species are photophilous and hemiphotophilous.

The species diversity of the spider community was highest in the vernal and toward the end of the aestival aspects. The amount of activity (i.e., number of individuals having crossed the effective edge of the pitfall traps within one day) was markedly high toward the end of the vernal aspect. Another peak was observed toward the end of the aestival aspect. The activity of the spiders decreased gradually by the end of the trapping season, except for a slight increase at the beginning of the autumnal aspect.

In the vernal aspect, the spider community under study was dominated by *Pardosa*, *Trochosa*, and *Pirata* spp. as well as *Antistea elegans;* in the aestival aspect, by *Pirata*, *Trochosa*, and *Pardosa* spp. as well a *Zelotes lutetianus*, *Bathy-*

phantes approximatus, *B. gracilis*, and *Antistea elegans;* in the serotinal aspect, by *Pirata* spp., *Bathyphanthes gracilis*, *B. approximatus*, *Pachygnatha clercki*, and *Antistea elegans;* in the autumnal aspect, by *Antistea elegans*, *Centromerus sylvaticus*, *C. incilium*, *Bathyphantes approximatus* as well as *Pardosa*, *Alopecosa*, and *Pirata* spp.

As regards reproduction types, the community comprises eurychronous *(Trochosa ruricola*, *Pachygnatha clercki*, *Antistea elegans*, *Alopecosa pulverulenta*, *Pirata latitans*, *P. moravicus)*, stenochronous *(Pardosa paludicola*, *P. prativaga*, *Arctosa leopardus*, *Zelotes lutetianus*, *Pocadicnemis pumila*, *Oedothorax retusus*, *Centromerus sylvaticus)* and diplochronous *(Bathyphantes approximatus*, *B. gracilis*, *Meioneta rurestris)* species. No unambiguous conclusions can be drawn on the reproduction type of the remaining present species.

On the basis of data on activity of the spider species showing great, medium and slight spatial activity, and the corresponding estimates of their abundance, the mean annual biomass of soil surface spiders in the terrestrial reedswamp under study is estimated at 25.4 kg ha^{-1} fresh weight, and their biomass in the periods of their peak activity at 24.5 kg ha^{-1} fresh weight.

Estimating the dry matter content of spiders at 25% of their fresh weight, and the energy content of dry matter at 5.8 kcal g^{-1}, the mean annual biomass of the above spiders amounted to 1675 kg ha^{-1} dry weight, or 0.167 g m^{-2} dry weight (972 kcal m^{-2}). The standing crop biomass during peaks of the spiders' activity amounted to 6.35 kg ha^{-1} dry weight, or 0.635 g m^{-2} dry weight (3680 cal m^{-2}), which represents a considerable resource of food for shrews and other insect-eating mammals.

References see pp. 393—395.

6.4 Invertebrates: Destroyers of Common Reed

V. SKUHRAVÝ

Common reed *(Phragmites communis)* stands harbor a multitude of invertebrates. Studies conducted in Czechoslovakia in 1968–1975 made it possible for the first time to draw a picture of the distribution over a relatively extended territory of invertebrates which either are important as primary pests of the common reed, or merely make use of the reed stem as a hiding place, or rather form part of wetland ecosystems in which reed constitutes one of the most important plants.

Attention was focused on species causing damage to common reed and depleting its reproduction rate. According to the type of injury the following breakdown was made:

1. Species damaging the growing point of the reed and checking the growth of reed stems (so that the reed does not form any inflorescences) and initiating lateral branching of the shoots

(a) Species destroying the growing point without causing the formation of galls

(b) Species causing gall formation on the growth tip

2. Species developing inside the common reed stems and causing them to break

3. Species causing secondary damage by attacking lateral shoots

4. Species living on common reed leaves.

6.4.1 Species Injuring the Growing Point of the Common Reed Without Transforming it into a Gall

These species include the moth *Archanara geminipuncta* HW. (*Lepidoptera*, Noctuidae) and the fly *Platycephala planifrons* (Fabr.) (*Diptera*, Chloropidae).

Archanara geminipuncta (*Lepidoptera*, Noctuidae)

The caterpillars of this moth are the earliest destroyers of the growing point in common reed. As a consequence, infested stems have only four to seven normally developed internodes. The shoot apex is growing dry outside and becomes foul inside. Damage is slightly discernible as early as the end of May and becomes conspicuous at the beginning of June. The infested apical part of the shoot falls off

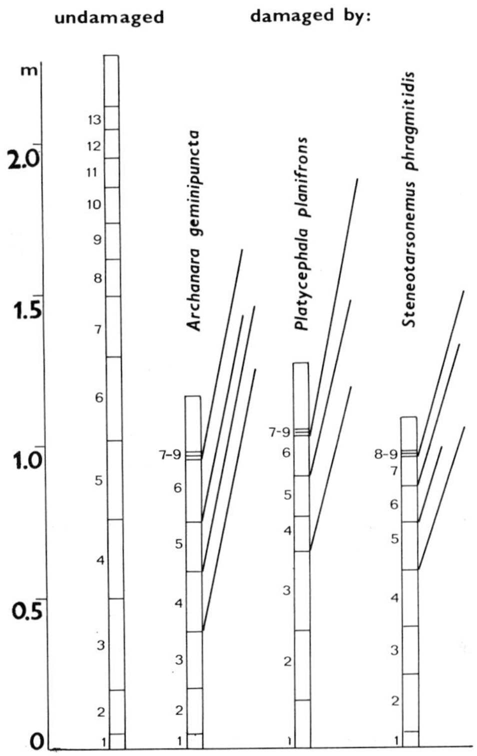

Fig. 1. Differences in number of internodes and stem length between undamaged and damaged *Phragmites* shoots

completely after mid-July. Inside the stem, the larva continues to grow from mid-May to mid-June, feeding first on the inner part of the growth tip and later eating its way through the inner portion of the 2nd to 4th upper internodes of the stem. Finally, having damaged and destroyed the stem apex as well as several internodes, it leaves and attacks another stem. Altogether, one larva may damage up to seven stems. A grown larva is 4–4.5 cm long. The infested internodes are filled with excreta of the caterpillars. As pupation approaches the larva retreats from the outside into a lower internode (2nd–4th). Here it drills an exit hole for the future adult and undergoes pupation. According to temperature conditions, the pupation period may take two to four weeks, so that the adult is ready to leave the stem between July 23 and August 15. In the field, adults were captured in light traps between August 12 and September 7 (Figs. 1, 2).

Following the injury and destruction of its growth tip, a reed stem proceeds to form lateral shoots which attain a length of 5–15 cm by mid-June, and 40–70 cm by the end of July. One stem produces one to four lateral shoots, mostly from the 3rd to 7th internodes. Stands along river banks, as a rule, suffer heaviest infestation, whereas littoral stands, in fishponds or oxbows, 3–4 m offshore, are invaded

Fig. 2. *Archanara geminipuncta.* Damaged inside of stem; pupa in the stem; opening through which the adult left

substantially less or not at all (Skuhravý in press, 1977). The dry weight of the infested stems only amounts to 43–55% of that of healthy stems.

Platycephala planifrons (*Diptera*, Chloropidae)

External damage to the common reed stem is similar to that caused by *A. geminipuncta:* desiccation of the apical leaves and destruction of the inner part of the stem by the larva. Infested stems form only four to six normal internodes. These are followed by one internode which is 1.4–3 cm long, and a further two which are only 1–3 cm long.

Inside the growing point, the larva of *P. planifrons* develops from mid-May to mid- or late June. The larva then becomes transformed into the puparium in which metamorphosis takes place in 14–20 days according to temperature conditions. The adults emerge between June 30 and July 21. Without drilling exit holes the adults tunnel up through the center of the destroyed part of the stem apex to its very end. When feeding on sugar solution they survive over one month. No second generation of this fly has been found on common reed. The fly presumably hibernates in adult stage, as proved by repeated findings of adults hidden within galls of the genus *Lipara*, where they were mistaken for inquilins using the gall of this fly for their own development. The infested stems develop lateral shoots similar to those initiated by *A. geminipuncta* (Skuhravá and Skuhravý, in press, 1977a; Fig. 1). In contrast to *A. geminipuncta*, one larva of *P. planifrons* destroys only one stem (Skuhravá and Skuhravý, in press, 1977a). The dry weight of infested stems amounts only to 40–60% of that of uninfested ones.

The damage caused by both species mentioned above is inconspicuous and therefore escaped attention for a long time. Presumably, the species are much more widespread in Central Europe than has been believed so far.

6.4.2 Species Transforming the Growing Points of the Common Reed into a Gall

This group of pests includes the acarid *Steneotarsonemus phragmitidis*, and species of the genus *Lipara*.

Steneotarsonemus phragmitidis

This acarid causes the common reed to form conspicuous galls on the shoot apex. All internodes higher than the 7th to 9th are markedly stunted, thickened and in most cases partially curled. The gall is 29–129 cm long and 0.5–3 cm wide (Fig. 3). The galls make their first appearance in mid-June and attain marked proportions by the end of June and, in particular, during July and August. No inflorescence will develop on infested stems. Only exceptionally, if the attack by the pest has been delayed, the gall does not develop until shortly before flowering time when at least a part of the inflorescence can be seen protruding out of the uppermost leaf sheath. (Skuhravý et al., 1975). Like the moth *A. geminipuncta*, and the fly *P. planifrons*, *S. phragmitidis* also causes the infested stems to form lateral shoots (Fig. 1).

Flies of the genus *Lipara* (*Diptera*, Chloropidae)

In Czechoslovakia, four species of the genus *Lipara* have been found to initiate gall formation on the common reed. Out of these, the newly identified species *Lipara pullitarsis* occurs in the largest numbers (Doskočil and Chvála, 1971). A description of the galls is given here for all four species; the most important data on the life history and damage to reed stems are summarized in Table 1. The galls initiated through the action of the *Lipara* species are formed by bundling of the leaves which grow from the stunted internodes: this results in the typical cigar-like shape of the galls.

Lipara lucens. It initiates the largest and stoutest terminal gall in the form of a cigar which is two to three times thicker than the stem. It develops from leaves growing from 10–13 stunted internodes. Their walls become ligneous, six to eight of the highest internodes become brownish with a continuous cover of fine hair.

Fig. 3. (a) Gall of *Steneotarsonemus phragmitidis;* (b) *Giraudiella inclusa*, galls in stem; (c) exit openings

The nodes between the stunted internodes bear hairy garlands. The larva lives in a tunnel eaten through the uppermost six to nine internodes.

Lipara pullitarsis. The gall is as large as that of the foregoing species, or only slightly smaller. It usually grows on massive stems, developing from leaves growing from five to six stunted internodes. The latter are of much the same width and colour as the non-stunted internodes, but are devoid of hair and do not form ligneous walls. The larvae live within curled leaves above the remnants of the growth tip.

Lipara rufitarsis. The gall, which is quite similar to that caused by the foregoing species, develops from leaves growing from five to six stunted internodes. The

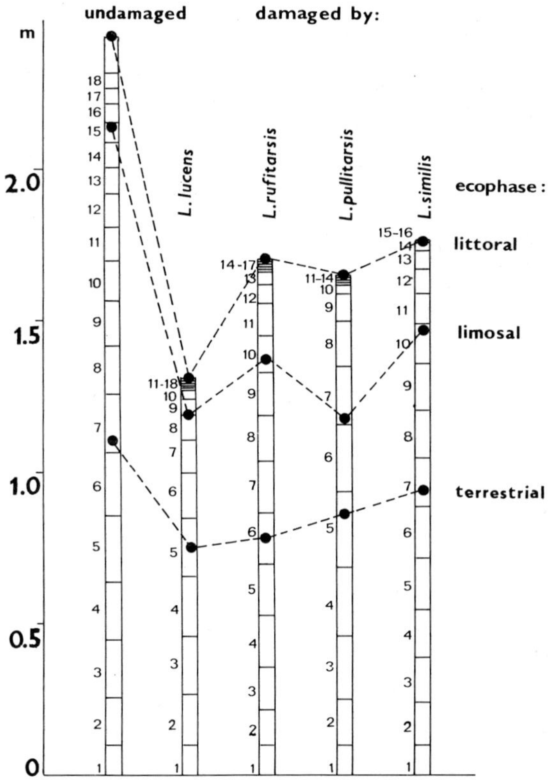

Fig. 4. Differences in number of internodes and stem length between *Phragmites* shoots undamaged and damaged by 4 *Lipara* species in three ecophases at Nesyt fishpond

Table 1. Summarized life history of 4 *Lipara* species infesting *Phragmites* in Czechoslovakia

Data	*L. lucens*	*L. rufitarsis*	*L. pullitarsis*	*L. similis*
Pupation	mid-April	late March	mid-March	early May
Emergence of adults	late May	mid-May	mid-May	early June
Development of larvae until following year	early June–mid-April	early June–late March	early June–mid-March	early June–late April
Average length of infested stems as % of healthy stems	43	59	59	66
Average proportion of internodes on infested stems compared with healthy ones	more by 2–3	less by 1–2	less by 1–2	less by 1–3
Number of stunted internodes forming a gall	10–13	5–6	5–6	3–4
Share of individual species (%) in total infestation by *Lipara*	11	30	43	16

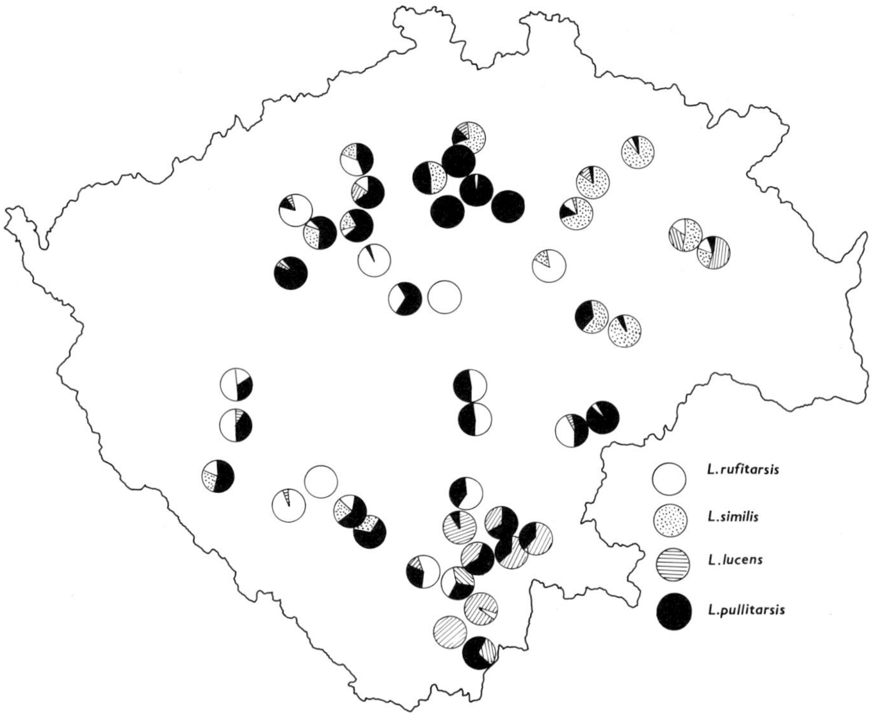

Fig. 5. Occurrence of four *Lipara* species in Bohemia

main difference consists in the smaller size of the galls as well as in that the larvae live in tunnels excavated within the uppermost two to four stunted internodes.

Lipara similis. Inconspicuous and slender, the gall is often indistinguishable from a non-flowering shoot. Only three to four internodes are stunted. They are neither thickened, nor are their walls ligneous. Compared with the normal internodes, the stunted ones have a more telescopic shape. The larva lives within a tube made of leaf sheaths which grow out of the stunted internodes from the residues of curled leaves above the growth tip.

Larvae of all species of the genus *Lipara* hibernate in galls situated at the highest point of the apex of a desiccated reed stem. *L. pullitarsis* enters the pupation period by mid-March (the earliest species). The adults emerge at the end of May. Shortly after emergence, they deposit their eggs on the leaf apexes. The larvae make their first appearance at the beginning of June and the galls assume conspicuous size by the beginning of July. The development of the larva takes place from the beginning of June until mid-March of the following year. The infested stems are about 40% shorter than the healthy stems and the number of internodes is reduced by one to three. The gall is formed of five to six stunted internodes (Pokorný, 1970b, 1971; Rychnovský, 1973a). The dry weight of the infested stems amounts to only 40–50% of that of the healthy ones. The distribution is shown in Figure 5.

Comparison of data obtained for the individual species in Czechoslovakia is given in Table 1 and Figure 4.

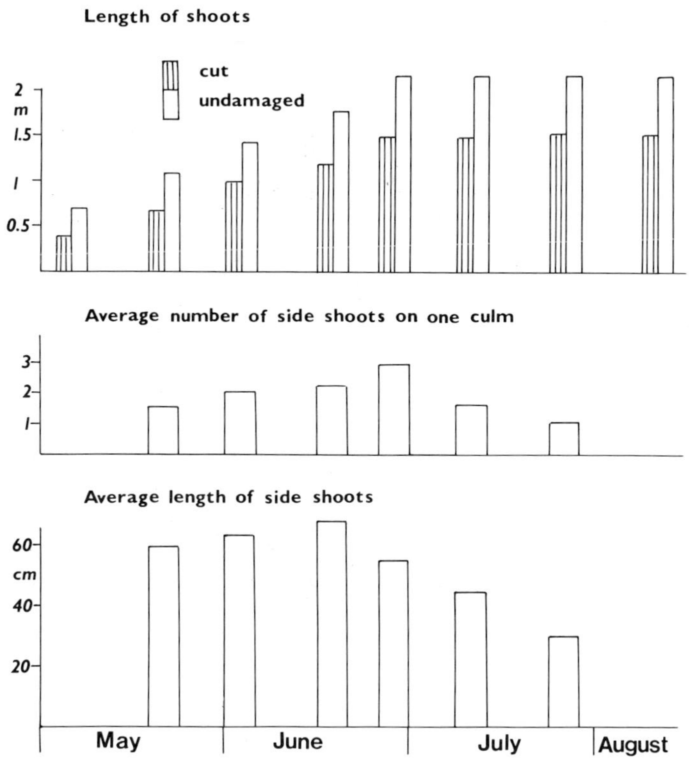

Fig. 6. Effect of cutting off shoot apex on stem development in common reed

6.4.3 Experimental Simulation of Common-Reed Infestation by Pests Destroying the Phragmites Growing Point

All the aforementioned insect species and the acarid *Steneotarsonemus phragmitidis* are capable of injuring or even totally destroying the growing point of the common reed. Total destruction occurs after attack by *A. geminipuncta*, *P. planifrons*, and sometimes also by *L. similis*. The stem apex is transformed into a gall following the attack by *S. phragmitidis*, and by *L. lucens*, *L. rufitarsis*, *L. pullitarsis*, and *L. similis*. Following the destruction of the growing point, the reed stems proceed to form lateral shoots. In Czechoslovakia, the development of lateral shoots is initiated almost exclusively by some insect attack on reed. Only a small proportion of lateral shoots are formed after a mechanical removal of the reed apex or after its destruction by birds (Květ and Hudec, 1971).

The length of the lateral shoots, their number and location on the stem vary according to the injuring agent. In order to elucidate these differences a series of experiments was conducted in which infestation by various insect species was simulated by cutting the reed apex at two-week intervals from May to August. Evaluation of the experiment included the assessment of the numbers of lateral shoots growing from each stem and their lengths, and the number of lateral shoots formed per main stem. The development of *Lasioptera arundinis*, a gall-midge causing secondary infestation of the lateral shoots, was assessed at the same time. The experiments were repeated several times at three localities in Bohemia; the results are given in Figure 6.

Removal of the apex at the beginning of May clearly killed the reed stems. When it was cut off in mid- to late June, the main stems formed the largest number of lateral shoots. No lateral shoots were produced, however, after cutting off the apex later than in early August. The largest lateral shoots were formed, on an average, after cutting off the apex during the second half of June. In this case, the lateral shoots developed on all the cut stems. The number of main stems developing lateral shoots decreased rapidly with the cutting performed successively during July. At the same locality, in fully sunlit stands two to three times more lateral shoots were formed per main stem than in shaded stands.

The experiments showed the dependence of the lateral shoot development (their lengths and numbers) on the time of destruction or injury to the growing point. Early injury by *A. geminipuncta*, *P. planifrons*, and *S. phragmitidis* entails the development of numerous and long lateral shoots. As for attacks by the *Lipara* species, lateral shoot growth is mainly promoted by the earlier developing species such as *L. pullitarsis* and *L. rufitarsis*. Initiation of lateral shoot formation by *L. lucens* and, in particular, *L. similis*, is only exceptional. The secondary pest of lateral side shoots, the gall-midge *Lasioptera arundinis*, thus mainly depends on the primary infestation by *A. geminipuncta*, *P. planifrons*, and *S. phragmitidis;* it indeed occurs only rarely on stems infested by *Lipara* species (Skuhravá and Skuhravý, in press, 1977b).

6.4.4 Insect Species Developing on the Stems of Phragmites

Giraudiella inclusa is a gall-midge whose first-generation larvae from grain-like galls inside the reed stems between the 2nd to 6th internodes, with maximum occurence in the 4th (Fig. 3b, c). The females of the gall-midge lay their eggs within the upper part of the internodes underneath the rim of the wrapping leaf, in a place where the stem is left exposed (Fig. 7). By the time of emergence of the adults, around mid-May, only one to four internodes at the most are available as appropriate receptacles for egg deposition. The development of the larvae inside the internodes is relatively rapid, the larvae attaining maturity by the end of June or in early July. Under higher temperatures, one more generation may develop, whose larvae attack the 8th to 10th internodes of the main stem, or internodes of the lateral shoots. Infested stems are clearly distinguished by the exit holes, which are rectangular, in contrast to the smaller and circular holes drilled by parasites. The infested stems become thickened at places where galls have developed, and they tend to break off during winter-time.

Infestation by *G. inclusa* varies in different regions of Bohemia. It was found to be smallest in South Bohemia, with 0–10% of stems infested, and heaviest in Central Bohemia, particularly in areas northeast of Prague (up to 30–50%) (Fig. 8). The second-generation galls found within the slender lateral shoots of reed differed to some extent from the first-generation galls on the main stems. If

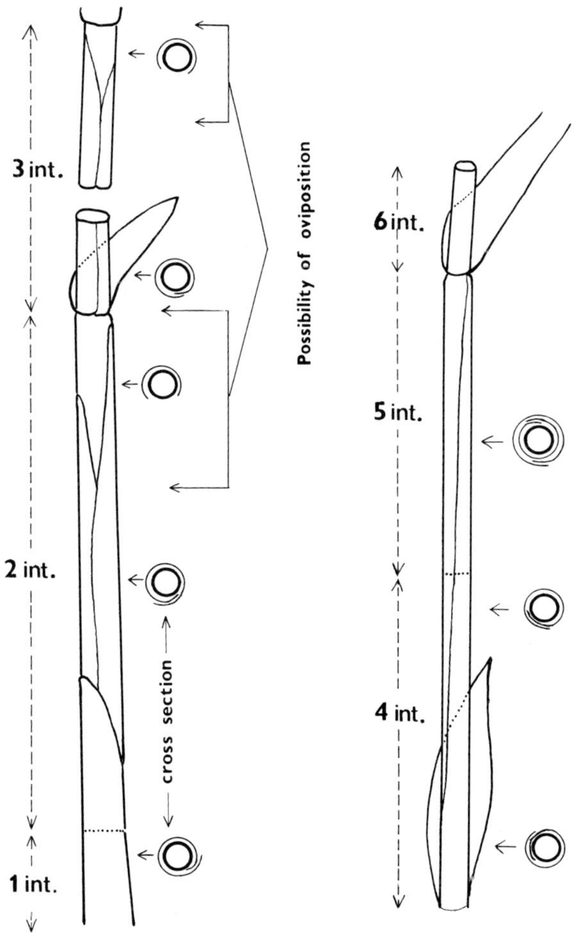

Fig. 7. Possibility of oviposition by *Giraudiella inclusa* on a reed stem at places not surrounded by leaves

the lateral shoot is attacked by *G. inclusa* in addition to the gall-midge *Lasioptera* the galls develop superposed.

Stems infested by the gall-midge *L. arundinis* react with heavy swelling of the tissue in which the development of the larva takes place. The attack by this gall-midge takes its course from the basis towards the apical end of a lateral shoot, whereas the second-generation larvae of *G. inclusa* move basipetally from the apical part of an internode. The second-generation galls of *G. inclusa* show great seasonal variation (Skuhravá and Skuhravý, in press, 1977 b).

Lasioptera flexuosa

A gall-midge whose females lay eggs into the 7th to 11th internode of the main stem in mid-July. Larvae hatched from the eggs finish their development after mid-September. They develop within an internode along with a fungal tissue sheathing the walls inside and harboring the larvae. A group of 20–30 larvae may

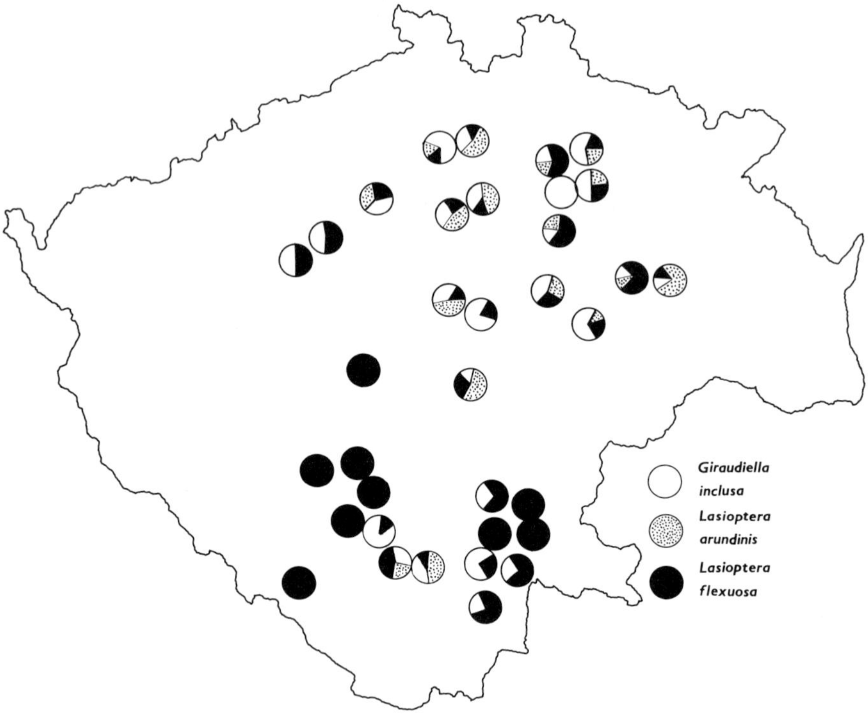

Fig. 8. Occurrence of three gall-midge species in Bohemia

be present within the internode without influencing the development of the reed. With the beginning of September the stems infested by *L. arundinis* fall prey to pecking birds, particularly to the *Parus* species. More than a third of a stem may be cut open at the spot where the larvae occur, and the stem becomes prone to breaking off. This tendency is enhanced by *L. flexuosa* infesting the two subsequent internodes in almost 45% of instances (Skuhravá and Skuhravý, in press, 1976, 1977 b).

6.4.5 Species Causing Secondary Attack in Lateral Shoots of Common Reed

The gall-midge *Lasioptera arundinis* causes gall formation on lateral shoots of the common reed. The female lays 40–90 eggs under the basal leaves of lateral shoots at a time when their length is only 3–11 cm. The first-instar larvae migrate towards the apical end of the lateral shoot and form a gall. Simultaneously, a fungus of the genus *Sporothrix* is developing within the infested internode. Development of the larvae to the third instar takes four to eight weeks. The infested lateral shoot internodes are 1–3 cm long on average, while the healthy internodes attain 3.8–4.8 cm im length. Beginning in September, grown third-instar larvae of this gall-midge are present in tiny enclosures within the galls. In the following

Fig. 9. *Lasioptera arundinis.* Lateral shoot at time of oviposition; damaged young internodes; cross-section of a damaged lateral shoot

spring, they enter the pupation stage by the end of May and the adults are ready to emerge after another two to three weeks (Fig. 9).

The dry weight of the infested lateral shoots including the galls is two to three times the weight of the healthy shoots although the latter are substantially longer (Skuhravý, 1975; Skuhravá and Skuhravý, in press, 1977 b)

In Czechoslovakia, this gall-midge occurs only in reed stands producing lateral shoots after an attack by the moth *A. geminipuncta*, the fly *P. planifrons*, and the acarid *S. phragmitidis*. The gall-midge lays eggs synchronously with the development of the lateral shoots.

6.4.6 Species Living on Leaves of Phragmites

The only species living on leaves of *Phragmites* is the aphid *Hyalopterus pruni*, which uses common reed as its secondary host plant during late spring and in summer (primary hosts of this species are shrubs and trees of the genus *Prunus*).

Winged females of this aphid migrate to the reed during the second half of June. The aphids stay on the reed for 60–70 days, i.e., from late June to early September. Individual generations partly superpose one another, which complicates the assessment of their actual numbers on reed. The maximum seems to be six generations. The larva of *H. pruni* develops on the reed in nine to eleven days. Immediately after reaching maturity, it is capable of giving birth to living larvae. The aphids are mainly found on the upper sides of reed leaves, simultaneously in the uppermost and lowermost canopy layers. Leaves growing about the middle are usually least infested, or are even free of attack. The influence of the aphids on reed growth is not substantial. Shoots carrying great numbers of aphids only tend to be lighter and to grow slowly. *H. pruni* occurs throughout Czechoslovakia. In some years, its occurrence increases substantially, but a harmful effect of its gradation becomes manifest only on *Prunus* (Pintera, 1971).

Summary. Among the invertebrate destroyers of *Phragmites* occurring in Czechoslovakia *Lipara pullitarsis* ranks first, *Steneotarsonemus phragmitidis* second, and *Lipara rufitarsis* third. Next in order follow: *Archanara geminipuncta*, *Giraudiella inclusa*, *Lipara similis*, *Lipara lucens*, *Platycephala planifrons*, *Lasioptera flexuosa*, *Hyalopterus pruni*, and *Lasioptera arundinis*. Several years' studies have revealed substantial differences in the occurrence of the individual species both between localities—even closely adjacent ones—and subsequent years. In Czechoslovakia, about one third of stems of *Phragmites* suffers injury from invertebrates which results in a loss of 10–20% of its annual net production.

References see pp. 393—395.

6.5 Macrofauna of Invertebrates in Helophyte Communities

J. Dvořák

The aquatic fauna of fishpond littorals was studied independently of the IBP project in fishponds of the South Bohemian basins of Vodňany and Blatná, some 25 km and 50 km, respectively, to the northwest of České Budějovice. But most results of this study briefly reviewed in this section are also applicable to other fishponds, and complement the data acquired within the IBP project itself. They also form a link with hydrobiological studies of fishponds undertaken, also within the IBP, by Straškraba et al. (1967) and Kořínková (1971).

6.5.1 Communities and Their Biotopes

The community of aquatic invertebrates represents a basic component of the whole set of secondary producers in a pond littoral. The methodology of estimating the biomass and production of this community is complicated by the taxonomic, dimensional, and life-form variety of the constituent animal species populations. In a simplified manner, the aquatic invertebrate fauna is classified into zooplankton, epiphytic fauna and macrofauna. This section presents the principal results of studies on the macrofauna inhabiting emergent macrophyte communities in South Bohemian fishponds. Most attention was paid to the following taxonomic groups: *Oligochaeta*, *Hirudinea*, *Mollusca*, *Coleoptera*, *Trichoptera*, and Chironomidae.

South Bohemian fishpond littorals, unless they have been altered by drastic management practices, are mostly shallow with a smooth transition from land to water. The environmental gradients along transects perpendicular to the shoreline are described in Section 2.3. The food supply to the invertebrates and the feeding activities of fish consuming them vary along these gradients as well (see Sect. 6.5.3).

6.5.2 Inventarization

Certain data on the macrofauna were obtained in original and undisturbed littoral plant communities of *Glycerietum maximae* with dominant *Glyceria maxima* or in *Sparganietum erecti* and in *Magnocariceta* (*Caricetum elatae*, *Caricetum gracilis, Caricetum vesicariae)*, all comprising a rich synusium of lemnids on the water surface, mainly with *Lemna minor*, *L. gibba*, and *Spirodela polyrhiza*, occasionally also with *Lemna trisulca*, *Ricciocarpus natans* and *Riccia fluitans*. The detailed results have been published by Dvořák (1970, 1971) and by Dvořák and Lišková (1970). As yet unpublished data presented in this section originate from *Phragmition* communities, namely from monospecific stands of either *Phragmitetum communis* or *Typhetum angustifoliae*, and from an extensive homogeneous *Caricetum elatae*. The macrofauna was followed only during the warm season, from April to September. The abundance and biomass of animals were expressed in the respective terms of number of individuals per 1 m^2 and grams of formalin-fixed weight per 1 m^2 bottom area. Formalin-fixed weight is that ascertained after more than six months of conservation in 4–5% formaldehyde.

Both the quality and quantity of the macrofauna are subject to marked seasonal changes in each habitat followed. This seasonal succession reflects the pattern of life cycles and migration of the populations constituting the respective animal communities. Gradients from shoreline to open water pertain not only to the environmental factors but also to the quantitative structure and dynamics of animal communites. Such gradients develop even within one helophyte community type, particularly in the *Glycerietum maximae* (Dvořák, 1970 and other new data). The animal communities seem to constitute themselves in relation to a whole complex of ecological factors, without specific bonds to any particular plant community type. More pronounced effects of the vegetation cover have only been observed in communities dominated by *Carex* spp., *Equisetum limosum*, and *Phragmites*, while *Glyceria maxima* and *Sparganium erectum* have little effect on the animal communities.

The animal communities of shallow outer parts of the reed-belt, largely isolated from the rest of a fishpond, are dominated by small molluscs (*Segmentina nitida* O.F. Müll., *Bathyomphalus contortus* L., *Gyraulus albus* O.F. Müll., *Spiralina vortex*, *Aplexa hypnorum* L.) followed by *Oligochaeta* (Enchytraeidae g. sp., *Eiseniella tetraedra* Savigny and some Tubificidae). Abundant are both larvae and adult stages of aquatic Coleoptera, especially of the small Dytiscidae and Hydrophilidae (*Noterus crassicornis* Müll.), certain species of the genera *Hydroporus* and *Phylidrus*, *Spercheus emarginatus* Schall., *Coelostoma orbiculare* F., and others, and larvae of the family Helodidae. An important component of the macrofauna living at the boundary between land and water (limosal ecophase) are larvae of *Diptera*, particularly of Culicidae (e.g., *Theobaldia annulata* Schrank, *Culex pipiens* L.), Chironomidae (especially *Orthocladinae*) semiterrestrial species of the genera *Metriocnemus*, *Limnophyes*, and *Smittia*, further certain Heleidae, Tabanidae, and Syrphidae. On the whole, this relatively isolated limosal community consists of rather small animals either depending on atmospheric oxygen or possessing adaptations to oxygen deficit in water. Some of these animals (molluscs) can survive periodically occuring dry conditions.

The invertebrate macrofauna inhabiting the littoral ecophase near to the open-water area of a fishpond is different. Here, the community is composed of *Oligochaeta* (*Stylaria lacustris* L., *Enchytraeidae* g. sp., *Limnodrilus hoffmeisteri* Claparede, *Lumbriculus variegatus* Müll., and others), several mollusc species (*Lymnaea palustris* O. F. Müll., *Radix auricularia*, *Bathyomphalus contortus* L.,

Physa fontinalis L., *Acroloxus lacustris* L., and others), and *Hirudinea*, especially *Helobdella stagnalis* L., *Glossiphonia heteroclita* L., *Herpobdella testacea* Savigny. Chironomid larvae also represent an important component living mainly near the edges of the helophyte communities. Their varied species composition comprises mainly phytophilous types such as the mining *Glyptotendipes gripekoveni* Kieff., representatives of the genera *Endochironomus*, *Phytochironomus*, *Limnochironomus*, and *Eucricotopus*, and detritus feeders such as *Microtendipes* of the group *Chloris*, *Chironomus* of the larvae group *Plumosus*, the predatory species *Pelopia Kraatzi* Kieff., *Procladius*, and *Psilotanypus*. Conspicuous are aquatic *Hemiptera* of the genus *Corixa* (*C. linnei* Fieb., C. falleni Fieb., *C. striata* L., and others) and the large predatory species *Ilyocoris cimicoides* L. and *Notonecta glauca* L. Frequent is the occurrence of the large predatory *Coleoptera* of the family Dytiscidae (e.g., *Rhantus notatus* F., *Graphoderes cinereus* L.). The community of macrofauna inhabiting littoral helophyte communities at their boundary with open water thus appears to consist predominantly of animals taking up oxygen dissolved in water (with the exception of *Pulmonata* and of certain insects). Bonds to the aquatic environment appear stronger here than in the limosal community.

6.5.3 Biomass Estimates

Besides the qualitative differences in the composition of animal communities inhabiting the two distinct zones of pond littorals, differences exist in their quantitative characteristics. The quantitative differences also pertain to the developmental dynamics of these communities. The Radov fishpond near Blatná, with communities of *Glyceria maxima*, *Carex elata*, and *Sparganium erectum*, may serve as an example (Dvořák and Lišková, 1970). The quantity of the animals was characterized by their abundance and biomass. Their production was not estimated as the growth curves and numbers of generations could not be defined in several important species populations. Moreover, the standing crop of the animals had been strongly affected by water-level fluctuations in the pond, and carp grazing depressed, to an unknown degree, only the animal populations inhabiting the inner littoral, while those living in the outer littoral remained unaffected. The data presented in Table 1 illustrate the situation in these two biotopes. The isolation of the outer littoral is evident from the physical and chemical characteristics of the water. A feature common to both biotopes was a peak in both abundance and biomass as well as species diversity of the macrofauna in June and July.

The quantity of the macrofauna was greater in the littoral biotope affected by the water from the fishpond pelagial than in the rather isolated limosal biotope. This difference persisted despite permanent fish-grazing only in the littoral biotope. Several factors appear to be responsible for the difference:

(a) Trophic conditions. The pelagial phytoplankton penetrates the inner littoral, sediments here and increases the food supply to animals feeding on fine detritus. This does not take place in the outer littoral.

(b) Hydrochemical conditions. They are correlated with the trophic conditions. The inner littoral may be denoted as eutrophic while the isolated outer littoral is somewhat dystrophic in character, its water being richer in dissolved

Table 1. Invertebrate macrofauna in a transect across helophyte communities in littoral of Radov fishpond, Blatná basin, South Bohemia, altitude about 460 m. Average data for April to September 1959

	Isolated outer littoral; community of *Glyceria maxima* and *Carex elata*; distance from open water 8 m		Inner littoral in contact with fishpond pelagial; community of *Sparganium erectum* and *Glyceria maxima*; distance from open water 1 m	
	maxima	averages per season	maxima	averages per season
Quantity of macrofauna				
biomass	65.00 g m^{-2}	42.18 g m^{-2}	72.63 g m^{-2}	66.04 g m^{-2}
abundance	31,885 ind. m^{-2}	18,393 ind. m^{-2}	30,060 ind. m^{-2}	20,346 ind. m^{-2}
Dominant groups				
biomass	Mollusca	Oligochaeta		Chironomidae
by abundance	Oligochaeta	Mollusca		Chironomidae

humic substances. Differences exist in the concentrations of oxygen dissolved in the water: 5–20 mg l^{-1} in the inner and 0–4 mg l^{-1} in the outer littoral, and in pH: 6.8–9.8 and 6.6–6.8 respectively. The oxygen gradient determines the distribution of animals with definite demands on oxygen.

(c) Living space and water-level fluctuations. These two factors modify the formation of the animal communities as well as their production. The share of relatively big and mobile animal forms is greater in the looser macrophyte communities in deeper water than in the shallow or limosal stands with a denser vegetation cover often consisting of sedges, and with a large amount of coarse detritus. Similar differences exist in the share of predators in the total biomass of the macrofauna. Data on the macrofauna of the *Glycerietum maximae* community illustrate these differences (Table 1). The greater share of predators in the deeper inner littoral reflects the abundant food supply to them provided by the zooplankton originating in the fishpond pelagial. Fluctuations of the water level are linked with marked changes in the physical and chemical environment in the outer littoral. These, in turn, affect strongly both the quality and quantity of the macrofauna (Dvořák and Lišková, 1970).

(d) Feeding activity of the fishes. This is selective; hence it not only reduces the total biomass of the macrofauna but also suppresses certain animal types more than others. The large fish such as carp usually have easy access to the inner littoral, while the outer littoral is practically inaccessible to them.

(e) Structure of the emergent macrophyte communities. The presence or absence of sedges in a community brings about the most marked differences. Communities of certain sedges (*Carex elata, C. gracilis, C. vesicaria*) host unproductive invertebrate communities of a specific species composition.

References see pp. 393—395.

References

Akkermann, R.: Untersuchungen zur Ökologie und Populationsdynamik des Bisams (*Ondatra zibethicus* L.). II. Nahrung und Nahrungsaufnahme. Z. angew. Zool. **62**, 173–218 (1975)

Bauer, K.: Die Säugetiere des Neusiedlersee-Gebietes. Bonn. Zool. Beitr. **11**, 141–344 (1960)

Binder, P., Štollmann, A.: Beitrag zur Verbreitung der nordischen Wühlmaus (*Microtus oeconomus* méhelyi Ehik, 1928) in der Slowakei. Lynx, Praha, N. S. **17**, 19–22 (1975)

Böck, F.: Säugetiere im Seegebiet. In: Der Neusiedlersee. Löffler, H. (ed.). Wien-München-Zürich: Molden, 1974

Doskočil, J., Chvála, M.: A revision of *Lipara* Meigen *(Diptera, Chloropidae)*, including the description of a new species from Europe. Acta Ent. Bohemoslov. **68**, 100–107 (1971)

Dvořák, J.: Horizontal zonation of macrovegetation, water properties and macrofauna in a littoral stand of *Glyceria aquatica* (L.) Wahlb. in a pond in South Bohemia. Hydrobiologia **35**, 17–30 (1970)

Dvořák, J.: The zonation of environmental factors and the macrofauna of littoral emergent vegetation in ponds in South Bohemia. Hidrobiologia **12**, 325–329 (1971)

Dvořák, J., Lišková, E.: A quantitative study on the macrofauna of stands of emergent vegetation in a carp pond of South-West Bohemia. Rozpr. Česk. Akad. Věd. Řada Mat. Přír. Věd. **80**, 63–114 (1970)

Errington, P. L.: Muskrat populations. Iowa: Ames 1963

Fiala, K.: Underground biomass and estimation of annual rhizome increments in two polycormones of *Phragmites communis* Trin. In: Littoral of the Nesyt Fishpond. Květ, J. (ed.). Studie ČSAV **15**, Praha, 1973, pp. 83–87

Gaisler, J.: A quantitative study of some populations of bats in Czechoslovakia. Acta Sci. Nat. (Brno) **9**, 1–44 (1975)

Golley, F. B.: Energy values of ecological material. Ecology **42**, 581–584 (1961)

Hawkins, A. E., Jewell, P. A.: Food consumption and energy requirements of captive British shrews and moles. Proc. Zool. Soc. Lond. **138**, 137–155 (1962)

Holišová, V.: Trophic requirements of water vole (*Arvicola terrestris* Linn.) on the edge of stagnant waters. Zool. Listy **19**, 221–233 (1970a)

Holišová, V.: Trophic requirements of water vole (*Arvicola terrestris* Linn.) in reed stands. In: Productivity of Terrestrial Ecosystems. Production Processes. Dykyjová, D. (ed.). Czechosl. IBP/PT-PP Rep. No. 1, Praha, 1970b, pp. 129–130

Holišová, V.: The food of *Clethrionomys glareolus* in a reedswamp. Zool. Listy **21**, 293–307 (1972)

Holišová, V.: The food eaten by rodents in reedswamps of Nesyt fishpond. Zool. Listy **24**, 223–237 (1975)

Holišová, V., Pelikán, J.: Analysis of the catch of *Arvicola terrestris* in live-traps. Symp. theriol. II, Brno **1971**, 367–372 (1974)

Hudec, K.: Breeding distribution and numbers of the greylag goose *(Anser anser)* in Czechoslovakia. (In Czech, with English summary). Československá ochrana Přírody Bratislava **12**, 105–151 (1971a)

Hudec, K.: The breeding environment of the greylag goose *(Anser anser)* in Czechoslovakia. Zool. Listy **20**, 177–194 (1971b)

Hudec, K.: Occurrence and food of the greylag goose *(Anser anser)* at the Nesyt Fishpond. In: Littoral of the Nesyt Fishpond. Květ, J. (ed.). Studie ČSAV **15**, Praha, 1973a, pp. 121–124

Hudec, K.: Die Nahrung der Graugans, *Anser anser*, in Südmähren. Zool. Listy **22**, 41–58 (1973b)

Hudec, K.: Density and breeding of birds in the reed swamps of Southern Moravian ponds. Acta Sci. Nat. (Brno) **9** (6), 1–40 (1975a)

Hudec, K.: The bird synusia of the "Kobylské jezero" reed swamp (Southern Moravia) during the breeding season. Zool. Listy **24**, 65–80 (1975b)

Hudec, K., Kux, Z.: The clutch size of greylag goose *(Anser anser)* in Czechoslovakia. Zool. Listy **20**, 365–376 (1971)

Hudec, K., Kux, Z.: Passage migration of the greylag goose *(Anser anser)* through Southern Moravia. Zool. Listy **21**, 245–262 (1972)

Hudec, K., Pelikán, J.: Reedswamp communities—secondary productivity. Birds. In: Ecology of Wetlands. Květ, J., Szczepański, A., Westlake, D.F. (eds.). Cambridge Univ. Press, In Press 1978

Husák, Š., Květ, J.: Zonation of higher-plant shoot biomass in the reed belt of the Nesyt fishpond. In: Littoral of the Nesyt Fishpond. Květ, J. (ed.). Studie ČSAV **15**, Praha, 1973, pp. 73–81

Kořínková, J.: Quantitative relations between submerged macrophytes and populations of invertebrates in carp ponds. Hidrobiologia **12**, 377–382 (1971)

Kožená-Toušková, I.: Composition of nests of birds breeding in the *Phragmition* plant community. Acta Sci. Nat. (Brno) **7** (7), 1–36 (1973)

Kratochvíl, J.: Beitrag zur Frage der Rassenangehörigkeit unserer Populationen von *Neomys anomalus*. Zool. Ent. Listy **3**, 167–168 (1954)

Kratochvíl, J., Rosický, B.: Zur Verbreitung und Vermehrung der Brandmaus *(Apodemus agrarius)* in der Tschechoslowakei. Zool. Ent. Listy **17**, 97–108 (1954)

Kux, Z., Hudec, K.: Der Legebeginn bei der Graugans (*Anser anser* L.) in der Tschechoslowakei. Acta Musei Moraviae (Brno) **55**, 233–246 (1971)

Květ, J., Hudec, K.: Effects of grazing by greylag goose on reedswamp plant communities. Hidrobiologia **12**, 351–359 (1971)

Mayer, J.: Ecological relations of the great reed warbler (*Acrocephalus arundinaceus* L.) and reed warbler (*Acrocephalus scirpaceus* L.) to the biocoenosis during the breeding season. (In Czech). Thesis J. E. Purkyně Univ., Brno, 1971

Miller, F., Obrtel, R.: Soil surface spiders *(Araneidea)* in a terrestrial reed swamp in southern Moravia (Czechoslovakia). Acta Ent. Bohemoslov. **72**, 272–285 (1975)

Obrtel, R.: Soil surface *Coleoptera* in a reed swamp. Acta Sci. Nat. (Brno) **6** (6), 1–35 (1972)

Obrtel, R.: Surface-active *Coleoptera* in the terrestrial reed-belt of the Nesyt fishpond. In: Littoral of the Nesyt Fishpond. Květ, J. (ed.). Studie ČSAV **15**, Praha, 1973, pp. 133–134

Obrtel, R.: Animal foods eaten by rodents in a terrestrial reed swamp. Zool. Listy **24**, 325–334 (1975)

Pelikán, J.: Mammals in reed stands of the Nesyt fishpond. In: Littoral of the Nesyt Fishpond. Květ, J. (ed.). Studie ČSAV **15**, Praha, 1973, pp. 117–120

Pelikán, J.: Dynamics and energetics of a reedswamp population of *Arvicola terrestris* (L.). Zool. Listy **23**, 321–334 (1974)

Pelikán, J.: Mammals of Nesyt fishpond, their ecology and production. Acta Sci. Nat. (Brno) **9**, 1–40 (1975)

Pelikán, J., Svoboda, J., Květ, J.: On some relations between the production of *Typha latifolia* and a muskrat population. Zool. Listy **19**, 303–320 (1970)

Pintera, A.: Some observations on male plum aphid, *Hyalopterus pruni* Geoffr., occurring on reeds. Hidrobiologia **12**, 293–295 (1971)

Pokorný, V.: Bionomics of *Lipara lucens* Meigen and *L. similis* Schin. on the common reed. (In Czech). Praha: Živa 1970a

Pokorný, V.: The influencing of reed (*Phragmites communis* Trin.) by gall-forming *Diptera* of the genus *Lipara* Meigen. In: Productivity of Terrestrial Ecosystems Production Processes. Dykyjová, D. (ed.). Czechosl. IBP/PT-PP Rep. No. **1**, Praha, 1970b, pp. 135–136

Pokorný, V.: Flies of the genus *Lipara* Meigen on common reed. Hidrobiologia **12**, 287–292 (1971)

Rychnovský, B.: The influence of larvae of the genus Lipara Meigen *(Diptera, Chloropidae)* on the productivity of reed. (In Czech. Ms.) Thesis, J. E. Purkyně Univ., Brno, 1973a

Rychnovský, B.: Occurrence of *Lipara* spp. at the Nesyt fishpond and their influence on the productivity of reed. In: Littoral of the Nesyt Fishpond. Květ, J. (ed.). Studie ČSAV 15. Praha, 1973b, pp. 129–132

Skuhravá, M., Skuhravý, V.: Das Schadbild und Entwicklung der Art *Platycephala planifrons* (F.) *(Diptera, Chloropidae)* an Schilf *Phragmites communis*. IV. Dipter. Sem., Bratislava, 11.–14. Nov. 1975. In press, 1977a

Skuhravá, M., Skuhravý, V.: Die Gallmücken *(Diptera, Cecidomyiidae)* an Schilf (*Phragmites communis* Trin.). Studie ČSAV. Praha: Academia. In press, 1977b

Skuhravý, V.: Die Entwicklung der Gallmücke *Lasioptera arundinis* (Schiner) *(Cecidomyidae, Diptera)* an Schilf (*Phragmites communis* Trin.). Marcellia **38**, 287–298 (1975)

Skuhravý, V.: Bionomie und Schädlichkeit der Art *Archanara geminipuncta* Hw. *(Lepidoptera, Noctuidae)* an Schilf (*Phragmites communis* Trin.). Acta Ent. Bohemoslov., In press, 1977

Skuhravý, V., Pokorný, V., Skuhravá, M.: Die Gliederfüßler (*Lipara* spp., *Steneotarsonemus* sp. und *Lepidoptera*-Larvae) als Ursache der Nichtbildung des Blütenstandes von Schilf *(Phragmites communis)*. Acta Ent. Bohemoslov. **72**, 87–91 (1975)

Straškraba, M., Kořínková, J., Poštolková, M.: Contribution to the productivity of the littoral region of ponds and pools. Rozpr. Česk. Akad. Věd. Řada Mat. Přír. Věd. **77** (7), 1–80 (1967)

Šťastný, K.: The quantity of birds frequenting the reed-beds of the Opatovický pond in 1972 (a preliminary report). In: Ecosystem Study on Wetland Biome in Czechoslovakia. Hejný, S. (ed.). Czechosl. IBP/PT-PP Rep. No. **3**, Treboň, 1973, pp. 225–229

Toušková, I.: Composition of birds' nests in a reedswamp community. In: Littoral of the Nesyt Fishpond. Květ, J. (ed.). Studie ČSAV 15, Praha, 1973, pp. 125–127

Zbytkovský, P.: Occurrence of *Neomys anomalus* Cabrera, 1907 in South Bohemia. Lynx (Praha) N.S. **17**, 39–41 (1975)

Fig. I. Bottom of a summer-drained fishpond with young plants of both annual and perennial macrophytes. *Eleocharis ovata*, *Typha latifolia*, *Sagittaria sagittifolia*, *Alisma plantago-aquatica*, *Batrachium* sp., and terrestrial form of *Lemna minor*. (Photo: K. Gregor)

Section 7

Effect of Fishpond Management on the Littoral Communities. Exploitation of Reed

Fishpond management practices have been evolving for several centuries and have strongly affected the biotic communities in fishponds and along their shores. Some communities may benefit but others become suppressed to such an extent that the presence of certain species populations in fishpond habitats is threatened. Plant communities both integrate and clearly indicate the effects of fishpond management; hence their response receives the most attention. Exploitation of some fishpond reedswamp species, especially of the common reed for the building industry and various crafts, has stimulated studies and practical experiments on the propagation, cultivation and fertilizer treatment of common reed; their results are summarized here.

7.1 Management Aspects of Fishpond Drainage

S. HEJNÝ

The ultimate goal of all fishpond management is the production of the largest possible amount of fish meat and, recently, also of duck or goose meat in some fishponds. Other purposes of fishpond management are secondary or complementary to its principal aim. Water retention, nature conservation, recreation and, locally, reed cultivation interfere to some extent with the present-day standard management of fishponds for maximum yields of fish. Certain limitations may be imposed on manuring and fertilizing of ponds used for recreational activities, or, exceptionally, a wide reed-belt may be maintained in fishponds serving as wildfowl refuges and declared as nature reserves.

All fishpond management practices have had a profound effect on the communities of biota inhabiting the fishponds, both in their littoral and central parts. The effects of fishpond management on the structure of the macrophyte communities have been studied most intensely and represent by far the best understood effects, especially with respect to the fishpond littoral ecosystems. In Section 7, attention is particularly paid to the effects of the following management practices on fishpond vegetation:

(a) Drainage, during the growing season (summer drainage) and winter drainage, with the associated water-level oscillations.

(b) Cutting of reed and reed-mace applied as a control measure in summer or as a harvesting technique in winter.

(c) Amelioration of fishpond littorals with scrapers applied as a control measure drastically reducing the size of the reed-belt.

(d) Duck-farming introduced additionally to fish-farming.

(e) Cultivation of common reed practiced in some selected ponds and regarded as a prospective land use in certain wetland areas.

(f) Fertilizer application and fish-feeding, both of which have now become routine practices applied to practically all fishponds.

A survey of fertilizer doses applied to the Opatovický fishpond is presented in Section 1.1. The effects of fertilizer application on water and bottom chemistry and on the mineral nutrient uptake by the vegetation are discussed in Sections 3.2.4 and 4.2 respectively. The response of various types of fishpond plant communities to eutrophication and fertilizer application is indicated in the survey of the communities in Section 1.2.

All management practices, both traditional or more recently introduced and intensified, modify the energy flow and budgets of water and mineral elements in various components of a fishpond ecosystem and particularly in its most sensitive component system, the littoral. These modifications still require assessment and evaluation.

7.1.1 Summer Drainage

Summer drainage was introduced into fishpond management quite early and became a regular practice. It had its periodicity, being applied each time after cropping of the fish stock, i.e., mostly every three years. A fishpond was then left empty for a whole growing season. This management practice thus had important consequences for all biota and their communities in a pond, and its role may be regarded as "sanitary". It controlled those processes which could not be controlled if the pond were continuously filled with water. The effects of summer drainage may be summarized as follows:

1. Summer drainage reduces the sedimentation of silt and accumulation of both autochthonous and allochthonous organogenous materials.

2. Rapid mineralization of the whole complex of organic matter (accumulation and sedimentation along margins of the reed-belts) induced by summer drainage slows down the penetration of helophytes into the open water and thus prevents filling of the pond and its terrestrialization.

3. Amelioration works are performed during summer drainage of a pond, especially in the reed-belt. These control measures nowadays involve the scraping off (with bulldozers) of the helophyte communities with their rhizome systems (see Sect. 7.3.1). All other technical work in the fishpond and on the dam can also be performed during summer drainage.

4. Summer drainage enables the control by fire of rapidly advancing littoral plant communities. This used to be practiced in late summer and applied especially to communities belonging to the foed. *Phragmition communis*, *Caricion elatae* and *Caricion gracilis* (see Sect. 1.2). This rather primitive, severe, and biologically unsuitable practice was pursued until the middle of this century, when it was replaced by amelioration through scraping off the whole reed-belt.

5. The period of summer drainage fulfills a sanitary role by controlling the fish parasites if a network of ditches drains even the most swampy and muddy parts of a fishpond bottom.

6. Summer drainage favors the aeration of the entire soil complex of a fishpond bottom, especially in its accumulation and sedimentary areas, as well as the aeration of the soils in the whole set of littoral communities of perennial plants.

7. Summer drainage is important for exploited or cultivated reed-beds as it brings about a retreat of the muskrat (*Ondatra zibethicus*) populations for several years to follow.

8. Summer drainage is also important for reed-beds in that it suppresses the initial reed populations, while developed invasion populations of *Phragmites* and of both *Typha* species profit from the self-thinning induced by the relatively dry conditions. Summer drainage thus acts as a positive factor in the selection of types suitable for cultivation.

From the fishpond managers' viewpoint, summer drainage may acquire two forms: (1) fallow management without any crop cultivation; or (2) farming man-

Fig. 1. Summer-drained fishpond with polycormones of *Schoenoplectus lacustris*, *Acorus calamus*, and *Glyceria maxima*. (Photo: K. Gregor)

Fig. 2. Regenerated community of *Sparganietum erecti* in a fishpond after summer drainage. (Photo: K. Gregor)

agement when crops are cultivated on the dry bottom. The latter form of management is applicable only to those ponds where the bottom can be fully drained as rapidly as possible and in time for the application of the necessary agrotechnical soil treatment. The fallow management is applied to ponds whose summer drainage starts rather late in spring and where complete drainage of the bottom is slow or difficult to achieve. In practice, both kinds of management are frequently combined in one pond. The fallow management is thus mostly applied to those sites in the sedimentary areas of fishponds where deep layers of mud are accumu-

Fig. 3. A *Glycerio fluitantis-Oenanthetum aquaticae* community at flowering time, intermingled with isolated loose polycormones of *Schoenoplectus lacustris*. Záblatský fishpond, Třeboň basin, 2nd year after summer drainage. (Photo: K. Gregor)

lated and to sites in the erosion (asedimentary) areas of summer-drained ponds or of emerged fishpond shores where poor and rapidly drying sandy soils prevail. The farming management is applied to areas with suitable soil conditions in the sedimentary area and to the relatively fertile parts of the erosion zones.

The ecological effects of summer drainage are also illustrated in Figs. 2 and 3 and briefly mentioned in Section 1.2. At present, summer drainage is only rarely applied to the "main" productive fishponds and may be regarded as a nearly abandoned management practice here; it has been replaced by an occasional partial drainage of a pond. Complete summer drainage is, however, still applied regularly to the small fry and fingerling ponds; in other fishpond types, it is only applied when large-scale repairs of the dam or other reconstruction works are taking place.

7.1.2 Winter Drainage

This is also an ancient fishpond management practice. It takes place during the late fall to early spring period (Fig. 4) and has the following aims:

1. Control of fish parasites; hence it is mostly combined with liming.
2. Improved aeration of the bottom soils.
3. Control of certain aquatic and littoral macrophytes by frost damage to their rhizomes situated near the bottom surface. (*Nymphaea candida*, *Glyceria maxima*, and *Acorus calamus* are among the most sensitive plants).

Fig. 4. Winter-drained fishpond after autumnal drawdown. (Photo: K. Gregor)

4. Suppression of muskrat populations.
5. Creation of favorable and safe conditions for reed harvesting on large areas. (In filled ponds, this is possible only during periods of long frost when firm ice is formed; such periods may be only short or may not occur at all during a mild winter).

Winter drainage usually lasts only several weeks to a few months; for this reason, as well as because it is practiced during the cold season, its ecological effects are less profound than those of summer drainage. Its advantage is that it does not upset the continuity of the use of a pond for fish production.

Nevertheless, winter drainage alone, if it makes the bottom sediments freeze to a sufficient depth and for a sufficiently long period, affects and damages the rhizome systems of perennial macrophytes to such an extent that the structure of their communities becomes altered. Particularly a winter drainage repeated at short intervals (e.g., every winter) induces substantial structural changes in communities of the ass. *Glycerietum maximae*, *Potamogetono natantis-Nymphaeetum candidae*, and others.

At present, winter drainage is applied regularly to small and medium-sized fishponds. Large fishponds are winter-drained quite rarely and mostly only when the risk of water shortage due to a delayed filling of the pond is small. A substitute practice is a partial filling of a pond for the winter following its autumnal drawdown for fish cropping, and its complete filling in the early spring.

References see pp. 424–425.

7.2 Control of Reed and Reed Mace Stands by Cutting

Š. HUSÁK

Both winter and summer cutting represent management practices frequently applied to helophyte communities in pond littorals.

Winter cutting of *Phragmites* culms, which are used for various purposes (see Sect. 7.4), has a definite favorable effect on the development of a reed stand in the subsequent growing season. The removal of the standing dead reeds improves and levels out the spring microclimate in reed stands (Šmíd, 1973; see also Sect. 2.1.2), although the risk of late spring frosts is increased as well. In any case, this treatment results in an even development of most of the shoots constituting a reed stand. Insect larvae hibernating in the standing dead reed culms and fungi which sometimes colonize their surfaces are also removed if a reed stand is cut in winter; the subsequent infestation of the reeds by insect and fungus pests is therefore strongly reduced (see also Sect. 6.4). The most important favorable long-term effect of winter cutting on a reed stand is a reduction of the amount of litter accumulating on the bottom: terrestrialization of the habitat and the associated undesirable succession of plant communities is thus retarded or prevented. The effects of winter cutting may thus be compared with those of reed burning but are less drastic.

Winter-cut *Phragmites* stands mostly acquire a more uniform structure, both vertical and horizontal, and are more productive than uncut stands (see Sects. 3.1.7). Krotkevič (1959) has proved the benefical effects of winter cutting on the phenological, structural and production characteristics of *Phragmites* stands in the Dnieper delta. This finding is supported by similar observations made by Haslam (1969), van der Toorn and Mook (1975), and others.

Reed mace *(Typha angustifolia)* is sometimes cut in fishpond reed-belts in early winter: its dry dead leaves are used for making wall-mats, baskets, etc. There is, however, practically no difference between winter-cut and uncut reed mace stands as their fragile standing dead leaves are mostly broken off by waves and wind during the winter season anyway.

Summer cutting is a traditional management practice aimed at controlling the size and area of helophyte communities in a fishpond and at maintaining a relatively large area of open water required for fish production. Despite the relatively small efficiency of its application summer cutting is ecologically sounder than other contemporary control practices employed for achieving the same aims, such as herbicide application or mechanical removal of the entire littoral

Table 1. *Phragmitetum communis* and *Typhetum angustifoliae* at Nesyt, treated on June 23 and 29 July 1969. Seasonal maximum shoot biomass (g m^{-2}) in treated and untreated variants in successive years

Type of stand	*Phragmites communis*			*Typha angustifolia*			
Treatment	control	180 cm	120 cm	control	infloresc. removed	180 cm	80 cm
28 Aug. 1969							
biomass[a]	1,388	1,026	943	2,039	1,942	1,177	1,814
removed by treatment	—	268	572	—	76	772	1,195
total harvest	1,388	1,294	1,515	2,039	2,018	1,949	3,009
n shoots m^{-2}	130	136	158	62	69	58	62
15 Aug. 1970							
biomass	1,763	2,660	527	2,371	2,281	1,473	1,466
n shoots m^{-2}	145	200	76	70	66	54	44
of which flowering	67	106	5	16	18	10	22
9 Aug. 1971							
biomass	2,227	2,100	955	2,004	2,173	2,106	2,805
n shoots m^{-2}	131	130	137	49	53	52	61
of which flowering	43	59	17	8	6	21	19

[a] Comprising both remaining old and regenerated new plant parts.

helophyte communities with scrapers. When cutting littoral helophytes the level of cut has to be below the water surface: the truncated shoot bases become flooded with water from the top and mostly die and rot off, especially when the cut is applied twice during one growing season.

The regeneration of summer-cut *Phragmites* in littoral ecophase was followed at the Opatovický fishpond in 1965 (Dykyjová and Husák, 1973). At Nesyt, reed-belt stands repeatedly clipped by goose-grazing, whose effects are similar to those of summer cutting, were observed by Květ and Hudec (1971). Special investigations aimed at assessing the response of *Typha angustifolia* and *Phragmites communis* to the prevention of flowering and to varying degrees of control by cutting were undertaken at the Nesyt fishpond in 1969 to 1972 (Husák and Květ, 1970; Husák, 1971, 1973).

Two levels of cut, at 180 cm and 80 cm above bottom level, were applied to an invasion stand of *Typhetum angustifoliae* on 23 June 1969; in addition, only the inflorescences (spadices) were removed in one treatment. Two levels of cut, at 180 cm and 120 cm above bottom level, were also applied to an invasion stand of *Phragmitetum communis*, on 29 July 1969. The treatments were timed so as to coincide with the onset of flowering. The treated plots of 30–100 m^2 in size were compared with untreated parts of the stands. During the 1969 growing season, the data on shoot biomass and leaf area index (LAI) were obtained by an indirect method in which non-destructive measurements on live shoots in permanent plots (4 quadrates of 0.5 m × 1 m and transect belts of 0.25 m × 4 m for each treatment in *Phragmites* and *Typha* respectively) were calibrated by destructive assessments of biomass in shoots harvested from parallel plots; see Section 3.1.2, and Ondok (1971). In order to follow the after-effects of the control treatments, the seasonal maximum shoot biomass and LAI were assessed in both the treated and untreated plots in both stands in the subsequent two years (1970 and 1971); the direct harvest method (4 square plots of 1 m^2 in each species and treatment) was employed. In order to evaluate the effect of the treatments on the underground stand parts, underground biomass was first assessed in both stands prior to the application of the treatments, i.e., in the fall of 1968, and again in both untreated stands and in all treatments in the fall of 1969. On each occasion, four replicate monoliths were excavated below a ground surface area of 0.3 m × 0.3 m down to the depth of 0.35 m in *Typhetum* and 0.45 m in *Phragmitetum* (for details see Husák, 1971).

The results of this experiment are evident from Table 1 and Figure 1. *Phragmites* cut at 180 cm formed only few small new leaves on a small number of axillary branches by the end of the 1969 growing season. In this first year, the cut apparently did not initiate shoot regeneration and underground reserve materials were saved, while the life of the remaining leaves was prolonged and their photosynthetic activity enhanced in comparison with uncut *Phragmites*. This probably explains why in the subsequent year (1970) this stand produced the highest aboveground biomass of all treatments: 2.3 kg m^{-2}. In 1971, no significant differences were detected in aboveground biomass between this treatment and untreated parts of the *Phragmites* stand. In *Phragmites* cut at 120 cm, the stand density increased by 46 shoots per 1 m^2 by the end of the 1969 growing season and numerous axillary branches were formed from the uppermost remaining nodes. In 1970, the aboveground biomass was very small here, partly because the underground reserve pool had been considerably depleted, partly because of damage inflicted by goose grazing in the spring. In 1971, i.e., two years after cutting, a compact stand was formed, with the highest shoot density of all treatments. But the shoots were short and their biomass and LAI were about half those of the untreated control stand (see Table 1). The increasing seasonal maximum biomass in the control *Phragmites* stand observed from 1969 to 1971 most probably reflected its gradual recovery from the unfavorable effect of the limosal and terrestrial ecophases during two successive years (1967 and 1968) preceding the experiment. Increased fertilizer doses to both the fishpond and adjacent land may have contributed to this increase as well (see Sect. 7.3).

In *Typha*, the removal of inflorescences altered neither the aboveground biomass nor LAI significantly in 1969, but the underground biomass increased substantially in comparison with the control stand. In subsequent years, however, *Typha* treated in this way practically did not differ from the untreated control stand. Cutting at 180 cm did not induce any regeneration in the *Typha* shoots in 1969 as their growth tips had not been affected. The unfavorable effect on rhizome biomass was more pronounced: it probably resulted from the severe reduction of the assimilatory surface area by the treatment. The stand needed two years to attain the same aboveground biomass as the untreated stand (see Table 1). Most strongly affected was the *Typha* stand cut at 80 cm. This treatment destroyed its assimilatory apparatus as well as shoot growth tips almost completely. This resulted in the translocation of a large proportion of the rhizome reserves into new offshoots which replaced the original stand. The depleted rhizomes could hardly subsidize the early growth of the shoots in the subsequent spring of 1970; the seasonal maximum aboveground biomass was accordingly small (see Table 1).

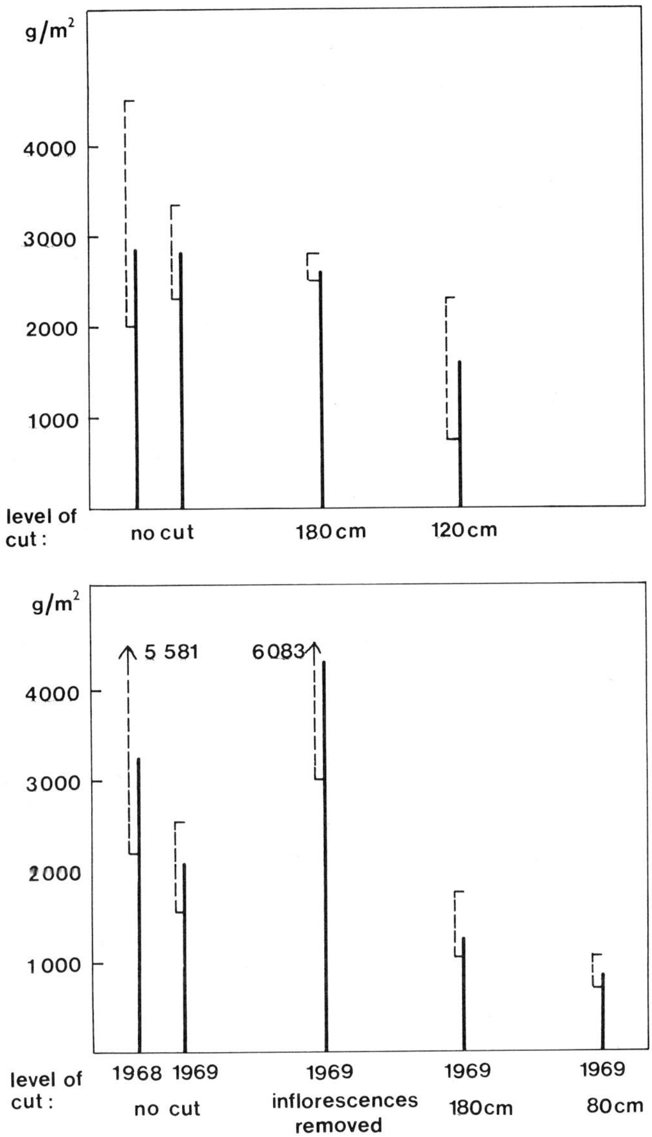

Fig. 1. Rhizome biomass in control and cut stands of *Phragmites communis* and *Typha angustifolia*. Control sampled in fall 1968 and 1969, i.e., prior to and after the application of cut to those parts of the stands, in which rhizome biomass was assessed only in fall 1969 (--- ranges)

But this stand was remarkably productive in the second season following the treatment; its aboveground biomass surpassed that of both the control stand and of the other treatments. A rejuvenation of the *Typha* polycormones by the severe cut might be proposed as a possible explanation of this sequence of events.

Typha angustifolia and *Phragmites communis* reacted to the destructive control treatments rather differently. *Typha* was evidently quicker both to react and to

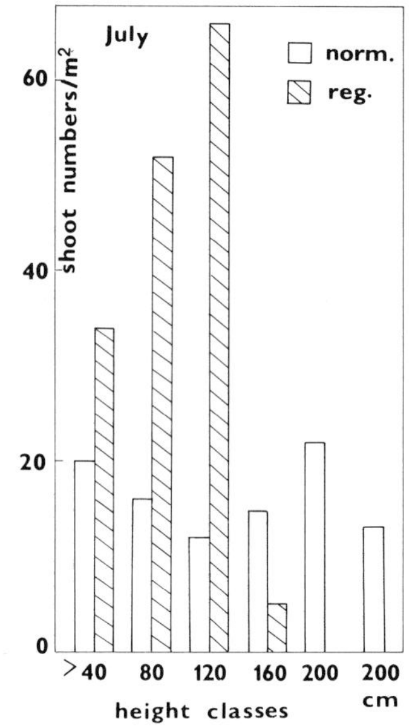

Fig. 2. Frequency diagram of stem height classes of an intact (norm.) and regenerated *Phragmites* stand in July, five weeks after cutting. (From Dykyjová and Husák, 1973)

regenerate after cutting. In *Phragmites*, the cut induced an increase in stand density and homogeneity. The cut was applied to the *Phragmites* stand somewhat late to induce any drastic changes in its performance. A more pronounced response would have been induced if the treatment had been applied earlier: best in early June (Fig. 2), at a time of lowest reserve content in the rhizomes (see Sect. 3.1.3) but immediately before the onset of panicle emergence at the latest. Any control treatment applied later will prove less effective; the growth curves for the first experimental year support this conclusion. Similarly important is the timing of the cut for the control of *Typha angustifolia*. It must be made before the end of June; a repeated cut applied a month or six weeks later reinforces the previous treatment.

Květ and Rejmánková (unpublished data) induced changes in the production of experimental sand cultures of *Typha latifolia* at Třeboň by removing, respectively, only the biggest, and the biggest and medium-sized shoots at the peak of the growing season. In the latter treatment, only a few small shoots were left in each experimental vessel of 1 m × 1 m × 0.7 m the rhizome and root biomass was drastically reduced here at the end of the first year as compared with both the untreated control vessels and the less severe removal treatment. This difference persisted during the subsequent year while the much smaller difference between the control and the less severe treatment became nearly obliterated. These findings confirm that *Typha latifolia* reacts to a partial removal of its aboveground parts in a similar way to *Typha angustifolia*.

References see pp. 424—425.

7.3 Ecological Effects of Fishpond Amelioration

S. Hejný and S. Husák

In modern pond management, amelioration is mainly performed in two ways:

a. the mechanical way, using heavy machinery

b. the biotechnical way, employing duck- and goose-farming

Both types were introduced more or less simultaneously in the early sixties. The following pertains almost exclusively to South Bohemian fishponds, where the effects of pond amelioration on plant life were followed in the years 1961–1971.

7.3.1 Amelioration Employing Heavy Machinery

7.3.1.1 Principles

The main purpose is a complete removal of rooted perennial plant communities—with their rhizome systems—from shallow fishpond littorals in order to increase the open-water area and to check the advance of the communities with the associated land formation.

The technique employs scrapers (bulldozers) for removal of the littoral vegetation during winter or summer drainage of a fishpond. The scraped-off mixture of bottom soil and plant matter is piled up either as low (1–1.5 m high) islets within the pond, or as low walls along the spring shoreline or higher up in the epilittoral.

This practice has the following consequences:

1. Radical change in the character and physiognomy of a fishpond (Fig. 1). The treatment frequently gives a pond a new shape and obliterates the gentle slope from water up through wet to dry land. The buffer zone of the reed-belt and tall sedge communities thus disappears, and so do its distinct ecological functions (Fig. 1); this has both positive and negative consequences for fish-farming. A freshly ameliorated fishpond littoral comprises two new and structurally entirely different complexes: (a) open water covering the newly shaped littoral; (b) set of deponia, i.e., the scraped-off and piled-up material. Under the former extensive fishpond management these two complexes existed in rudimentary form and on negligibly small areas.

2. Changes in the bottom soils and in the trophic, mineral nutrient and hydrological régimes of a pond. The amelioration practically removes the organogenous horizons of the soil profiles in the littoral. Impermeable and biologically inactive clay or deeper sediments such as sand or gravelly sand form the new

Fig. 1. Fishpond littoral with piled-up deponia of scraped-off bottom material. (Photo: K. Gregor)

bottom. If the clay layer is preserved, the water-retention capacity of the fishpond will not change, but if this layer is punched or scraped off the pond will lose water by percolation through the permeable sandy sediments. In the latter case, the amelioration does not achieve its aim because the cleared area does not remain flooded and turns into wet or, possibly, even dry land. The removal of the biologically highly active and nutrient-rich organogenous bottom sediments brings about temporary changes in the nutrient régime of a fishpond, mostly from a man-induced eutrophy to meso- or oligotrophy.

From the economical point of view, the mechanical amelioration is hardly effective because of the high costs of both the technical operation itself and of the subsequent fertilization required.

From the hydrobotanical point of view, this technique frequently creates favorable conditions for the regeneration of macrophyte communities characteristic of meso- to oligotrophic habitats. This represents a reverse of the previous oligo- to mesotrophic succession which had been enhanced *ad absurdum* by the forced eutrophication.

7.3.1.2 Regeneration of Macrophyte Communities and Their Further Succession

1. Communities in newly reclaimed areas in a fishpond.

In accordance with the analysis given above, three categories of newly reclaimed area may be distinguished in an ameliorated pond:

(a) The reclaimed areas are flooded with some 0.4–0.6 m of water carried by impermeable clay. The pond littoral is therefore more or less permanently flooded (= littoral ecoperiods prevail). The regeneration of plant communities begins with an initial *Eleocharitetum acicularis* which becomes gradually invaded by hydatophytes adapted to relatively nutrient-poor substrates: *Potamogeton acutifolius*, *P. trichoides*, *P. gramineus*, *Batrachium trichophyllum*. The aquatic vegetation is then easily controlled by machine cutting and further succession of macrophyte communities is checked. The amelioration has thus fulfilled its aim. The regeneration of macrophyte cenoses is much the same here as in areas where perennial emergent vegetation has never been present. The reduced initial succession sere proceeds as follows: *Litorellion* (initial) → *Potamogetonion pusilli* → *Oenanthion* →.

(b) The reclaimed area is exposed to a fluctuating water-level régime with periods of flooding (with up to 0.2–0.3 m of water) alternating with a fall of the water level below bottom level, the latter giving rise to littoral-limosal ecoperiods. The *Potamogeton* species populations do not participate in the regeneration of macrophyte cenoses, but the participation of initial communities of the foed. *Batrachion aquatilis* combined with *Litorellion* is enforced. Both *Eleocharis acicularis* and *Litorella uniflora* grow vigorously under such conditions, while the initial communities of the foed. *Oenanthion aquaticae* are deprived of certain species. They occur in loamy clays, and those of the foed. *Veronico-Juncion bulbosi* in sandy soils, particularly if these also contain some peat. All the above initial communities are especially characteristic of littoral ecoperiods. Limosal ecoperiods favor an ample development of initial communities colonizing emerged oligotrophic bottoms.

The following succession schemes apply:

During littoral ecoperiods:

Litorellion ——————→ *Veronico-Juncion bulbosi*
↓
Batrachion aquatilis
↓
Oenanthion aquaticae

During limosal ecoperiods:

Litorellion ——————→ *Veronico-Juncion bulbosi*
↓ ↓
Elatini-Eleocharition *Radiolion linoidis*

The vegetation colonizing the reclaimed areas attains maximum diversity within two to three years after the amelioration treatment. The various community types are present only in rudimentary form as cenotic nuclei, which is due to the great variation of habitat conditions in both time, due to water-level fluctuations, and space, due to the variable soil substrate (clay, sand, gravelly sand, loam).

Control of the aquatic vegetation by motor cutting is, however, impracticable in the shallow water, so that the unwanted rooted perennials eventually re-invade the reclaimed area. This may take as little time as four to seven years. The character of the prevailing ecophases and ecoperiods then determines the prevailing vegetation types in the following manner:

(a) Littoral ecophases and ecoperiods promote the establishment of initial stages of reedswamp communities of dominant *Glyceria maxima* (in all kinds of substrate including sand) or *Typha latifolia* (in loamy to clayey soils).

(b) Littoral-limosal ecoperiods promote the establishment of initital stages of tall sedge communities dominated by *Carex vesicaria* and *C. gracilis*.

(c) Predominantly limosal ecoperiods favor the establishment of initial stages of wet grassland communities of the order *Molinietalia*, with *Deschampsia caespitosa* or *Molinia coerulea*, or of the foed. *Agropyro-Rumicion crispi*, with dominant *Juncus effusus* (see also Blažková, 1973). Communities of the last type tend to be invaded by juvenile willow populations, especially of *Salix cinerea* (Jeník and Větvička, 1973).

In any case, within some ten years, the reclaimed area again becomes vegetated as densely as it was before amelioration, but the previous zonation of biotopes and communities is replaced by a mosaic-like pattern of cenotic nuclei and polycormones developing into initial stages of perennial helophyte communities. The pattern of the vegetation corresponds with that of soil conditions.

A fluctuating water-level régime thus makes the amelioration ineffective, the succession of rooted perennials having been only slowed down and altered but not stopped.

3. The newly reclaimed area falls dry or is only waterlogged for most of the time, becoming flooded, at the most, only for short periods mainly in early spring. Limoso-terrestrial ecoperiods clearly prevail.

The early formed initial communities of *Eleocharitetum acicularis* or *Ranunculo-Juncetum bulbosi* gradually disappear. Initial communities of emerged bottoms survive for only two to three years. Characteristic of both this and the previous type (see *b*) is the complete absence of communities belonging to the foed. *Bidention tripartiti*, which colonize organogenous substrates.

Later, the prevalence of the limosal ecophase with a short-term littoral ecophase in early spring promotes the establishment of *Phalaris arundinacea* populations, while the prevalence of the terrestrial ecophase favors the establishment of *Calamagrostis epigeios*. The terrestrial habitats, especially when colonized by the latter species, become invaded by anemochorous biennials or perennials belonging to weeds or ruderal plants (e.g., *Cirsium palustre*, *Gnaphalium silvaticum*, *Carlina vulgaris*, *Centaurium umbellatum*). The eventual vegetation acquires an entirely terrestrial character. Littoral plant communities disappear completely but the amelioration has failed and the pond has been reduced in size.

4. Communities colonizing deponia (scraped-off material piled up).

The presence of deponia in a fishpond has the following ecological effects:

(a) The relatively bulky and conspicuous deponia have introduced types of biotopes into a fishpond which were before present only on a minute scale in composts consisting of manually removed tussocks of tall sedges.

(b) By contrast to purely organogenous substrates, the deponia contain both organic and inorganic materials.

(c) A microzonation develops on the surface of a deponium, comprising three zones: (i) waterlogged at the base; (ii) inclined terrestrial on the slopes; (iii) flat terrestrial on the top plateau. The substrate is usually thoroughly mixed, hence the diversity of the vegetation depends primarily on this microzonation and on the location of a deponium, either within a pond or along its shores.

(d) The maximum diversity of plant species populations as well as of their various ecological types occurs in the deponia, as in the pond itself, during the first few years following the amelioration treatment.

(e) The colonization of the deponia by vegetation is characterized by a more or less equal participation of two groups of plant species populations: (i) autochthonous, originating from both generative and vegetative propagules present in the deponium. These are seeds of various plants and piled up rhizomes of the perennial helophytes; (ii) allochthonous, whose generative propagules have been brought in by wind or by animals (hares, pheasants).

(f) The vegetation succession is usually irreversible on the deponia, unless they are dug up like composts.

(g) The vegetation succession is rapid and usually passes through the following stages in South Bohemian fishponds, with the dominants named in this very general scheme:

Polygono–Bidentetum	*Galeopsido–Chenopodietum*	(1st–2nd years)
↓	↓	
Calamagrostis canescens	*Cirsium* sp. div. stage	(2nd–3rd years)
	↓	
Glyceria aquatica–Phalaris arundinacea	*Urtica dioica–Calamagrostis epigeios*	(4th–6th years)
↓	↓	
Frangulo–Salicetum cinereae	*Sambuco–Salicetum capreae*	(7th–10th years)
↓	↓	
Populus tremula–Betula verrucosa	*Betula verrucosa–Pinus silvestris*	

For South Moravia, the succession of plant communities on deponia of scraped-off bottom material is illustrated by examples from the Nesyt and Hlohovecký fishponds.

The sequence of dominants colonizing newly created deponia at Nesyt (made in the winter of 1971–1972) depended on their position in the pond. The deponia, 1.5–2 m high and forming islets in the pond, were first colonized by nearly pure *Persicaria lapathifolia* with some *Schoenoplectus tabernaemontani*, and were then gradually washed away; within three to four years they disappeared, replaced by colonies of *Typha angustifolia* growing in shallow water. The deponia situated along the shores, also 1.5–2 m high, became successively colonized by the following dominants:

Chenopodium ficifolium + *Atriplex hastata* + *Melilotus dentatus*
↓ (1st year after amelioration)

Atriplex hastata + *Lactuca seriola* + *Phragmites communis*
↓ (2nd year after amelioration)

Lactuca seriola + *Cirsium arvense* + *C. lanceolatum* + *Phragmites communis*
(3rd and 4th years after amelioration)

Within the four years, the height of the deponia fell to 1.2–1.5 m. Further development of the vegetation on these deponia will probably be similar to that recorded at Hlohovecký in 1965–1975, on deponia made in 1960. Two types of habitats were distinguished there: the top plateaux and the slopes of the deponia. The plateaux became colonized by willows: *Salix alba*, *S. cinerea*, *S. fragilis*, and *S. triandra*, while the slopes provided habitats for communities belonging to the following foederationes: *Bidention tripartiti* (in accumulation biotopes) or *Bolboschoenion maritimi* (in erosion biotopes) in the 6th to 10th years after amelioration. The same two community types were joined by *Cyperion flavescentis*, *Chenopodion glauci*, *Loto-Trifolion*, and *Agropyro-Rumicion crispi*, according to the actual position of the water level in the 10th to 15th years.

Although the same principles apply to the succession on deponia in both South Bohemia and South Moravia, the differences in the species outfit of the plant communities involved are quite marked and correspond with the general ecological characteristics of the two regions (see Sect. 1.1).

(h) When dominated by anemochorous weeds (especially *Cirsium* spp.), the deponia may become a dangerous source of weed infestation for the surrounding fields.

(i) Deponia which are situated in the outer littoral and form an impermeable wall around a pond may give rise to waterlogged habitats along their outer edges. These habitats become colonized by communities of inundated or wet grasslands (order *Molinietalia*, foed. *Caricion canescenti–fuscae)*.

(j) A positive aesthetic feature of the deponia results from their colonization by shrubs and trees, starting from the 6th to 7th years following amelioration treatment. In this way, the deponia contribute to the variety of the landscape surrounding ponds.

7.3.2 Amelioration by Waterfowl Farming

Its main effect is economic: a combined cultivation of carp and ducks in a fishpond approximately doubles its yield in meat. The utilization of the fodder supplied to the carp and ducks is complementary, and the duck excrements fertilize the water and thus promote the development of the plankton serving as food for carp. In addition, the ducks ameliorate fishponds by:

(a) Feeding on plants occurring both in the open water and the littoral zone. Young sprouts and whole juvenile populations are grazed on by the ducks systematically. The ducks avoid only a few species containing special repelling substances (e.g., *Nymphoides peltata*, *Nymphaea candida*).

(b) Locally suppressing the development of certain plant communities by their accumulated excrements. The amelioration of fishponds by duck-farming appears ecologically sounder than the amelioration employing heavy machinery: "The thousands of ducks' bills represent an efficient and inexpensive amelioration machine which can remove the littoral vegetation more effectively than heavy machinery can, without destroying the fertile sapropel layer covering the bottom" (Hejný, 1967, p. 88). The associated manuring of the fishpond water multiplies this advantageous effect of duck-farming. However, this seemingly advantageous method of biotechnical amelioration also has several drawbacks:

(i) Long-term duck-farming in one pond profoundly changes its entire set of macrophyte cenoses, from open water to the outer littoral. Larger areas are eventually affected in this way than by mechanical amelioration with heavy machinery. Unless the duck-farm is confined to a short stretch of the shore and to a small area in a pond (like in Nesyt), the fishpond water and soils will become stressed with excessive nitrogen and phosphate. In the long run, deterioration will take place instead of amelioration.

(ii) The littoral biotic communities gradually become destroyed during long-term duck-farming. Simultaneously, they are replaced first by nitrophilous communities of emerged bottoms (e.g., of the foed. *Bidention*), and later by ruderal communities (of the foed. *Sisymbrion officinalis*, *Euarction*, and *Agropyro-Rumicion crispi*) which are tolerant to the effects of the duck guano. Ruderal communities of hypertrophic soils thus eventually succeed the destroyed mesotrophic communities in both the inner and outer littoral.

(iii) The danger of infestation of surrounding fields with both local and alien weeds is increased around the depos and feeding sites where duck fodder (poor-quality grain, etc.) is handled. The danger of ducks acting as vectors of parasitic diseases is not to be underestimated.

References see pp. 424—425.

7.4 Propagation, Cultivation and Exploitation of Common Reed in Czechoslovakia

K. VÉBER

Among the helophytes colonizing fishpond shores, there is only one species that is exploited and/or cultivated on a large scale: the common reed, *Phragmites communis* Trin. In the Czechoslovak IBP Wetlands project, attention was paid to applied aspects of the ecology of this species, which may also be looked upon as a technical crop with a long tradition as well as promising prospects.

The only Czechoslovak enterprise specializing in large-scale processing of common reed is the "Rákosárna" (Reedcraft) at Městec Králové in East Bohemia, which forms part of the Czechoslovak State Fisheries Organization. Products made of reed are mainly used in the building industry (insulating mats, etc.), followed by horticulture (shades and other screens), and by various domestic crafts. Reed sometimes represents a substitute for other materials but, in other instances, cannot be replaced by any substitute. In Czechoslovakia, the demand for reed products exceeds their production because the supply of high-quality reed culms is insufficient. Efforts are being made to increase reed production by effectively exploiting the existing reed resources as well as by establishing new reed-beds (Table 1).

A survey is given here of methods of reed propagation, cultivation, and introduction required to achieve this aim. The present state of commercial reed production in Czechoslovakia is also reviewed.

7.4.1 Cultivation and Propagation

Reed stands conquer new areas relatively slowly (see Sect. 3.1.3). It takes three to five years or longer to achieve a 90% participation of *Phragmites* in a mixed community containing up to 70% of *Phragmites* (Rudescu et al., 1965). A shorter period required would, of course, be economically desirable. Hence the importance of artificial propagation of reed. The techniques proved under conditions of the temperate zones are listed here.

7.4.1.1 Propagation by Dividing a Reed Polycormone

This technique has good results but it is laborious (Bittmann, 1953; Arens, 1958). Planting out separate parts of a reed polycormone is effective as long as these parts remain embedded in the soil substrate. Individual polycormone parts

Table 1. Average annual reed crops (in thousands of commercial bundles per year) from the principal reed-producing districts in Czechoslovakia

District	Crop
Northeast Bohemia	150–160
South Bohemia	40– 50
Moravia (mainly South)	55– 60
Southwest Slovakia	260–300
East Slovakian lowlands	25– 30

are planted out immediately on the site selected at 5 to 10 m distances in rows which are 5 m apart. Clusters originating from each new plant will cover a weed-free plot continuously within three to four years; a slightly weed-infested plot will be covered within four to six years. The density of the resulting reed plantation will be some 35–40 shoots per 1 m^2. This technique is relatively expensive in view of the laborious recovery and transport of the polycormones from the original stands.

7.4.1.2 Propagation by Layering

Layering not only yields vigorous new plants but reed plantations obtained by this technique rapidly form a closed canopy (Véber, 1973). In natural reed stands, the frequency distribution of basal culm diameter measured in the basal internodes shows two to three peaks (Fig. 1). Each of these peaks approximately indicates culms having developed from terminal buds, lateral buds of the 1st order and lateral buds of subsequent orders respectively.

The root-taking capacity is excellent, up to 100% in the terminal and 1st order lateral shoots, but it is very poor or lacking in the other shoots. In Bohemia and perhaps over most of Central Europe, the period between 10 June and mid-July is suitable for reed propagation by layering.

The reed shoots chosen for propagation are bent off and placed into 8–10 cm deep ridges, fixed to the ground and heeled in with soil. The apical leaf-bearing parts, 25–30 cm long, remain uncovered. The layered shoots become rooted within three to four weeks and new tillers begin to sprout after another two to three weeks. New separate reed plants are then obtained by cutting off the layered shoots. By the end of the growing season, they produce only roots and no new rhizomes below ground and, on an average, four aerial shoots above ground, up to 1.1 m in length and 3.7 mm in basal culm diameter.

7.4.1.3 Propagation by Stem Cuttings

Young reed stems are capable of root-taking if cut off at a time when their tissues are rich in easily soluble assimilates. In Czechoslovakia, they usually reach this stage in May (Véber and Dykyjová, 1971).

This technique of reed propagation is simple: the apical part, 20–30 cm long, of a reed shoot is cut off; the level of the cut is located at 0.5–1 cm below the lowest node of the cutting. The nodal meristems thus remain intact. The leaf blades are shortened. The cuttings are then placed in a vessel containing either fishpond water or a diluted nutrient solution (1/1); the vessel is placed either in a

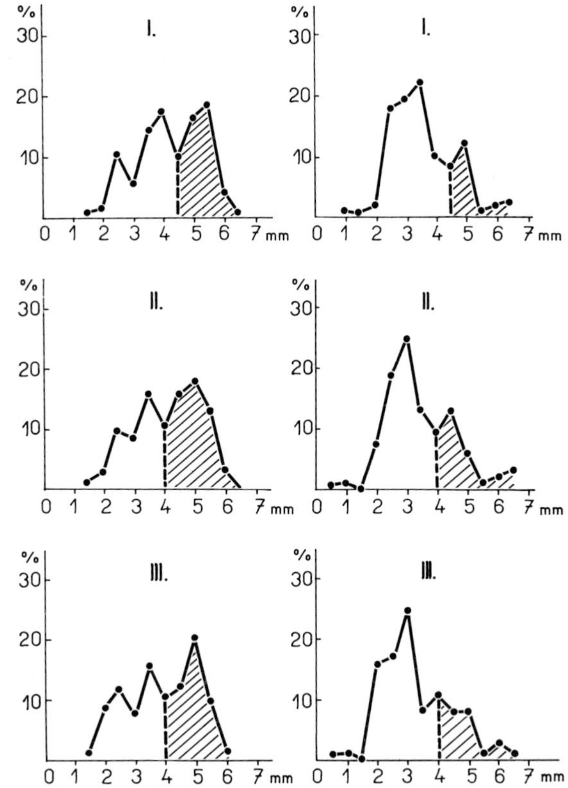

Fig. 1. Root-taking capacity of layered reed culms originating from two-year-old stands of different shoot density cultivated in sand hydroponics. Abscissa: culm diameter in nodes, mm. Ordinate: root-taking capacity of nodes (%). *I–III:* 3 lowest basal nodes. *Hatched area:* Nodes of root-taking culms. *Left:* Stand of medium shoot density (140 sh m^{-2}). *Right:* Dense stand (432 sh m^{-2})

Table 2a. Characteristics of growth and development of common reed plants propagated by shoot cuttings for the subsequent 3 years (mean value per plant)

Age of cutting (years)	0	1	2	3
Number of tillers	22	38	56	95
Maximum culm length, cm	90	110	145	232
Average culm length, cm	42	51	81	136
Leaf area, cm^2	289	1,240	5,000	17,860
Aboveground fresh weight, g	22.20	48.65	125	350
Aboveground dry weight, g	6.62	18.63	44.60	160
Maximum tiller distance, cm	—	47	109	250

greenhouse or in a bright room. At a temperature of 16–18° C and relative air humidity of 80–90% the cuttings take roots within 10–20 days, without application of any growth stimulators. New reed plants originating from the cuttings are planted out in the field either directly or after precultivation in pots. Table 2a, b provides a more detailed evaluation of this technique of reed propagation.

Table 3 shows the results of an experiment with cuttings of reed sampled at weekly intervals from the beginning of May to 11 June 1972. In each batch of 300

Table 2b. Root-taking capacity of shoot cuttings of littoral common reed "V" form from Opatovický fishpond; % of all cuttings planted out and cultivated (1 and 2) or immediately planted out in a fishpond (3)

	Phragnites communis (littoral "V"). Root-taking capacity (%)					
Week in 1971	1.–7. May	8.–14. May	15.–21. May	22.–28. May	29. May –4. June	5.–11. June
1. Tray with mud	0	0	3	9	6	1
2. Sand + peat 1:1	0	0	0	1	3	1
3. Fishpond littoral	0	0	1	2	2	1

cuttings, the root-taking capacity was tested under three different sets of conditions: (1) cultivation in trays 20 cm deep filled with fishpond mud and water; (2) cultivation in a watered mixture of sand and peat; (3) cultivation at the boundary of water and dry land (limosal ecophase) in a fishpond littoral.

In the first two treatments 1 and 2, the cuttings were given highly favorable conditions. The greatest percentage of rooted cuttings was found among those samples between 22 May and 4 June and planted out in the trays (treatment 1).

7.4.1.4 Propagation by Rhizome Cuttings

One-year-old rhizomes bearing either terminal and/or lateral buds are used for the propagation. The success of this technique depends on the developmental stage of the buds and on the degree of damage to them during sampling and hibernation. The rhizomes and buds, adapted to low winter temperatures, require a cold period for normal growth and development of aerial shoots. They are therefore sampled either from natural reed stands or from rhizomes stored in a cold greenhouse.

A rhizome cutting bearing two buds or a single terminal bud is cut about 1 cm below a node and is placed about 1.5 cm below soil surface in a hothouse or about 4 cm deep in the field. The cutting must not be damaged in any way; planting out cuttings bearing more than two buds did not prove successful. The propagation is done in late winter or early spring and potted young plants capable of transplantation are obtained within eight to ten weeks. A disadvantage of this technique is the high mortality of the young plants: 20–25%, which is due to even slight and inconspicuous damage to the rhizomes.

7.4.1.5 Propagation by Seed

Whole *Phragmites* panicles are sampled in late fall before the caryopses fall out. The panicles are stored at -2–$0°$ C; they are then sown out in a greenhouse in February to March or outdoors in May. At 18–20° C, germination will take place within 10 to 30 days; the last caryopses will germinate as late as after three to four months.

After having formed three to five leaves, i.e., after five to six weeks, the seedlings are transplanted into pots or garden beds. Transplantation favorably affects their further growth and tillering. The young plants are further transplanted into the field when their height has reached some 0.6–0.8 m and their rhizome systems have become fully differentiated.

Direct sowing of *Phragmites* in the field can meet with success only under rather warm conditions in limosal ecophase in water shallower than 4 cm (van der Toorn, 1972). This technique is therefore applied only rarely. An exception is reed sowing on large areas in newly reclaimed polders in the Netherlands. Here, however, a poor germination is taken into account with some two surviving seedlings per 1 m^2 (van der Toorn and Mook, 1975).

7.4.2 Recultivation of Reed Plantations

New reed plantations are set up in recultivated "degenerated" old reedswamps or on other abandoned and weed-infested waterlogged sites. The young *Phragmites* plants are obtained by one of the propagation techniques listed in Section 7.4.1. The recultivation comprises the removal of weeds, adjustment of the water régime to the requirements of the reed plants, fertilizer application and, in most instances, the introduction of new productive types of reed. The developing reed polycormones will form a continuous cover within three to six years. The rate of spreading depends on the vigor of the young plants (Haslam, 1969; Rudescu et al., 1965; Véber and Dykyjová, 1971; Fiala, 1973a, and others), on the plantation density, on the degree of infestation with weeds (Véber and Bartoš, 1973), and on numerous other environmental factors. (For their effects on the production of *Phragmites*, see Sect. 3). The time for setting up a reed plantation has to be adjusted to the vigor of the seedlings or to the root-taking capacity of the vegetative plant parts used for propagation.

7.4.3 Introduction of Reed Ecotypes and Forms

The introduction of highly productive ecotypes and forms of *Phragmites* (see also Björk, 1967; Rudescu et al., 1965; Bernatowicz, 1969; van der Toorn, 1972, and others) meets with varying success in dependence on habitat. The adaptation of the reed ecotypes or forms to the new environmental conditions has to be tested experimentally. In the following, the uncommitted term "form" is used though in many instances the described differences will probably be fixed genetically.

In experiments, vegetatively propagated imported reed forms at Třeboň were compared with the local "S" form from the limosal accumulation biotope in the Opatovický fishpond (see Sects. 2.4, 3.1.1, 3.4.2). The experiments were conducted on an abandoned and recultivated waterlogged site near Třeboň in South Bohemia. Apart from the local S form, the five cultivated imported reed forms were of the following origin; (a) Romania, Danube delta; (b) South Bulgaria, Rupite; (c) GDR, Moritzburg; (d) East Austria, Neusiedler See; (e) GDR, form "pseudodonax" from Saxony. All these reed forms had been propagated vegetatively from stem cuttings (see Fig. 2) in May 1971. They were planted out in the field in September 1971.

The root-taking capacity was good in all forms tested except that from Bulgaria. During subsequent growth of the young plants, the differences between the forms manifested themselves in the structure both of the aboveground and underground organs. The more productive forms produced more robust shoots with taller culms, more suitable for industrial processing. After four years of cultiva-

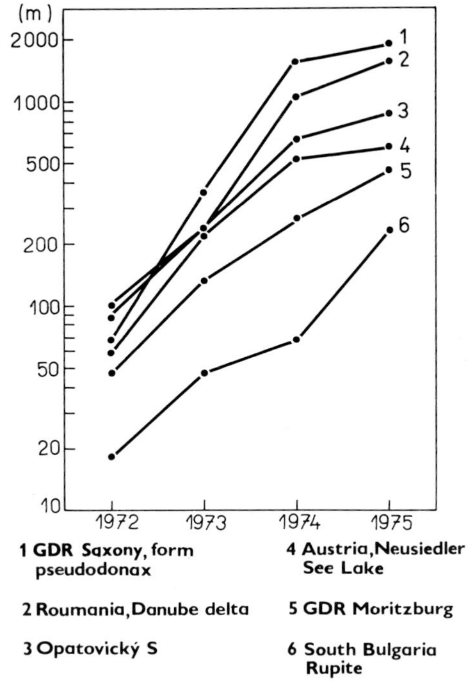

Fig. 2. Total length of reed culms (average culm length × shoot density) per five plants cultivated in the field for four years (see also Table 3)

tion, in 1974, all forms but one (that from Bulgaria) produced dense canopies analogous to those of natural reed stand (Table 3). The processing qualities of the reeds are evident from Figure 3. The product of stem length × shoot density clearly may serve as an adequate index of production in reed plantations. Moreover, the order of all six reed forms compared, arranged according to the values of this index, was the same during the last three years of the experiments.

7.4.4 Survey of Exploited Reed Beds

The reed crop is expressed in terms of commercial bundles per hectare. A commercial bundle (CB) consists of tightly bound dead reed culms harvested in winter; its circumference is 105 cm at 40 cm from the cut ends of the reeds (see also Véber and Kupka, 1973).

The yields of reed widely fluctuate from year to year (between 350 and 1000 CB per hectare). The total reed crop shows a decreasing tendency as more and more land is being drained and reclaimed for agricultural production. Table 1 gives a survey of average reed yields in individual reed-production districts of Czechoslovakia. The actual reed production is somewhat higher because not all reed harvested is handed over to the "Reedcraft" enterprise; some is processed locally. Small areas colinized by reed usually remain unexploited, but there are about 190 more or less regularly harvested reed stands in the whole of Czechoslovakia. Hence it is an uneasy task to estimate the country's total reed crop. Figure 3 illustrates the processing characteristics of the reeds originating from the most important exploited sites in the country.

Table 3. Characteristics of growth and development in six forms of common reed propagated by shoot cuttings, ascertained at end of 4th season of cultivation in the field

	Phragmites communis Trin.					
Form	Roumania, Danube delta	Opato-vický "S"	Bulgaria, Rupite	GDR Moritz-burg	Austria Neusied-ler See	GDR pseudo-donax
Number of tillers	750	590	200	295	305	*1,140*
Maximum culm length, cm	*318*	245	216	200	288	248
Average culm length, cm	*204*	147	114	152	195	165
Leaf area, m^2	14.82	10.85	1.92	7.18	12.50	*20.37*
Aboveground fresh weight, g	*9,800*	2,300	800	1.800	5,200	5,850
Aboveground dry weight, g	*3,750*	1,200	340	1,100	1,680	2,550
Maximum tiller distance, cm	*480*	375	160	320	395	400
Average tiller distance, cm	177	179	71	112	*204*	175

Maximum characteristics italics; average values for stands, each originating from five plants.

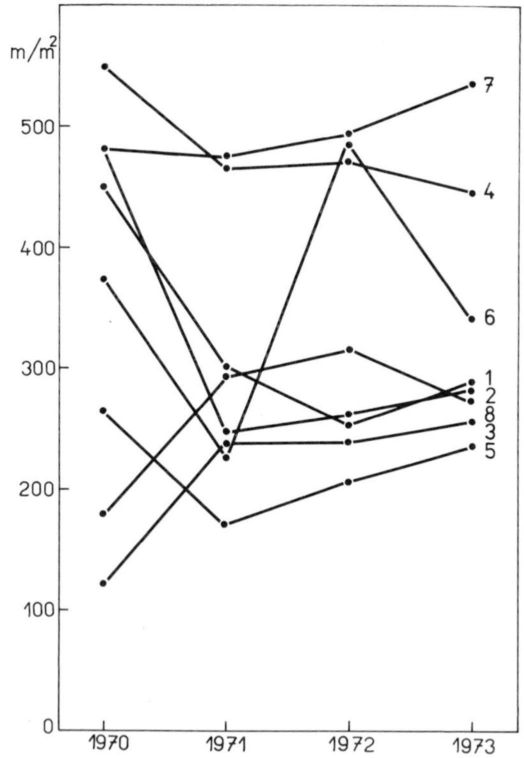

Fig. 3. Total length of reed culms (4 years' averages) from exploited reed beds in Czechoslovakia. Northeast Bohemia: *1*, Bohdaneč; *2*, Třesice; *3*, Žehuň; Southwest Slovakia: *4*, Dolní Saliby; *5*, Gabčíkovo; *6*, Gbelce; *7*, Landor; *8*, Velké Kosihy

The following text briefly characterizes the reed habitats and their reed stands in Czechoslovakia's most important reed-production districts (see also Fig. 4).

Northeast Bohemia: This district, with mostly small fishponds concentrated in the surroundings of Chlumec and Cidlinou, Bohdaneč, Opočno, and Jičín, is distinguished by a warmer and drier climate than South Bohemia and by mostly

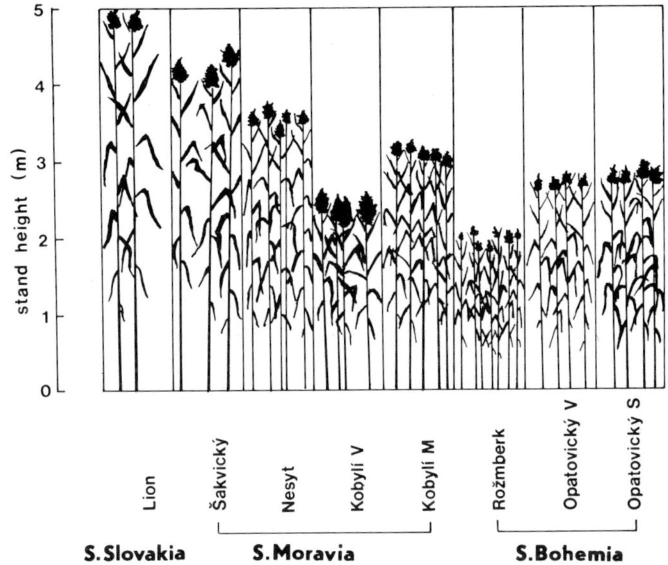

Fig. 4. Schematic representation of aboveground organs in 8 stands of *Phragmites communis* from 3 regions. (After Fiala, 1973)

base-saturated heavy soils. *Phragmites* occurs here above all in fishpond reed-belts. Its shoots attain a maximum height of up to 4 m and the golden-coloured "ripe" culms are relatively thick: it is more productive here than in South Bohemia.

South Bohemia: *Phragmites* occurs here predominantly in reed-belts on fishpond shores. Its shoots do not exceed 3.4 m in height, the slender culms are brownish when "ripe", the leaves are mostly narrow and the panicles are loose. The littoral *Phragmites* form "V" from the Opatovický fishpond is a typical representative of South Bohemian reed (see Sect. 3.1.1). The production characteristics of South Bohemian *Phragmites* stands are described in Section 3.1.7.

Southwest Slovakia and South Moravia: In the South Moravian alluvia, reed occurs in fishpond reed-belts and in oxbow lakes and backwaters of larger rivers (Dyje and Morava). In Southwest Slovakia, the principal habitats occupied by reed are oxbow lakes and backwaters of the Danube and reedswamps or water-logged fields supplied with ground water. Climatically, the district is warmer than the preceding two, especially during the growing season (mean annual and July temperatures mostly between 9 and 10° C and 19 and 21° C, respectively). The temperatures increase from the Northwest (South Moravia) to the Southeast (Komárno and Štúrovo in South Slovakia). The alluvial soils are greatly variable, from fertile clay loams to relatively poor sands; some habitats receive additional mineral nutrients either through fertilizer application to ponds and adjacent fields, or through silt deposition during floods. The reed production and yield are correspondingly variable. The local reed forms are mostly vigorous, up to 4.8 m high, with thick culms (see Fig. 4), which are pale yellowish when "ripe". The leaves are relatively broad and thick, dense panicles prevail. Most of the thick rhizomes are concentrated in the upper soil horizons.

References see pp. 424—425.

References

Arens, H.: Schilfrohr im niederländischen Wasserbau. Die Wasserwirtschaft **48** (1958)

Bernatowicz, S.: Macrophytes in the lake Warniak and their chemical composition. Ekologia Polska A **17**, 447–467 (1969)

Bittmann, E.: Das Schilf (*Phragmites communis* Trin.) und seine Verwendung im Wasserbau. Pflanzensoziologie, Stolzenau/Weser 7, 5–47 (1953)

Björk, S.: Ecological investigations of *Phragmites communis*. Studies in theoretical and applied limnology. Folia Limnol. Scand. **14** (1967)

Blažková, D.: Pflanzensoziologische Studie über die Wiesen der südböhmischen Becken. Studie ČSAV, Praha, 1973, **10**, pp. 1–170

Dykyjová, D., Husák, Š.: The influence of summer cutting on the regeneration of reed. In: Ecosystem Study on Wetland Biome in Czechoslovakia. Hejný, S. (ed.). Czechosl. IBP/PT-PP Rep. No. **3**, Třeboň, 1973, pp. 245–250

Haslam, S. M.: A study of *Phragmites* in relation to its cultivation and harvesting in East Anglia for the thatching industry. Norfolk Reed Growers Association **1**, 1–42 (1969)

Hejný, S.: Outline of macrophyte communities in water reservoirs with special reference to fishponds in the South Bohemian basins. Inst. of Botany, Czechosl. Acad. Sci., Průhonice, 1967

Husák, Š.: Productivity and structure of intact and cut invasion stands of *Phragmites communis* Trin. and *Typha angustifolia* L. at the Nesyt fishpond. (Czech. Ms.). Thesis, Bot. Dept., Fac. Sci., J. E. Purkyně Univ., Brno, 1971

Husák, Š.: Destructive control of stands of *Phragmites communis* and *Typha angustifolia* and its effects on shoot production followed for three seasons. In: Littoral of the Nesyt Fishpond. Květ, J. (ed.): Studie ČSAV 15, Praha, 1973, pp. 89–91

Husák, Š., Květ, J.: Productive structure of *Phragmites communis* and *Typha angustifolia* stands after cutting at two different levels. In: Productivity of Terrestrial Ecosystems. Production Processes. Dykyjová, D. (ed.). Czechosl. IBP/PT-PP Rep. No. 1, Praha, 1970, pp. 117–119

Jeník, J., Větvička, V.: Ecology and structure in stands of *Salix* spp. in the Třeboň basin. In: Ecosystem study on Wetland Biome in Czechoslovakia. Hejný, S. (ed.). Czechosl. IBP/PT-PP Rep. No. **3**, Třeboň, 1973, pp. 39–46

Krotkevič, P. G.: On the effects of the timing and height of reed cutting on its regeneration and growth. (In Russian). Sb. Tr. UKS. Nauchno-Issled. Inst. Tsellyul. Bumaz. Promsti. **3**, 8–17 (1959)

Květ, J., Hudec, K.: Effect of grazing by grey-lag geese on reedswamp plant communities. Hidrobiologia **12**, 351–358 (1971)

Ondok, J. P.: Indirect estimation of primary values used in growth analysis. In: Plant Photosynthetic Production. Manual of Methods, Šesták, Z., Čatský, J., Jarvis, P. (eds.). The Hague, Dr. W. Junk N.Y., Publishers, 1971, pp. 392–411

Rudescu, L., Niculescu, C., Chivu, I. P.: Monografia Stufului din Delta Dunarii, Bucureşti, 1965

Šmíd, P.: Microclimatological characteristics of reedswamps at the Nesyt fishpond. In: Littoral of the Nesyt Fishpond. Květ, J. (ed.). Studie ČSAV, 15, Praha, 1973, pp. 29–38

Toorn, J. van der: Variability of *Phragmites australis* (Cav.) Trin. ex Steudel in relation to the environment. Van See tot Land No. **48**, 1–122, Staatsuitgeverij, ś-Gravenhage, 1972

Toorn, J. van der, Mook, J. H.: Experiment on the development of reed vegetation in the Zuid-Flevoland polder. Verh. Kon. Ned. Akad. Wetensch. Afd. Natuurk., 2e Reeks, **66**, 11–15 (1975)

Véber, K.: Propagation of reed (*Phragmites communis* Trin.) by layering; In: Ecosystem Study on Wetland Biome in Czechoslovakia. Hejný, S. (ed.). Czechosl. IBP/PT-PP Rep. No. **3**, Třeboň, 1973, pp. 251–254

Véber, K., Bartoš, J.: Growing of common reed, a new technical plant. I. Selection of culms for layering. (In Czech with Engl. Summary). Sb. Vys. Šk. Zeměd. vPraze Provozně Ekon. Fak. vCesk. Budějovicích Řada Biol. **594**, 17–26 (1973)

Véber, K., Dykyjová, D.: Growing and propagation of common reed, a new technical plant. (In Czech with English summary). Rostl. Výroba **17** (XLIV) 1, 97–109 (1971)

Véber, K., Kupka, Z.: Stands and crops of reed (*Phragmites communis* Trin.) in Czechoslovakia. In: Ecosystem Study on Wetland Biome in Czechoslovakia. Hejný, S. (ed.). Czechosl. IBP/PT-PP Rep. No. **3**, Třeboň, 1973, pp. 255–256

Fig. I. Nesting female of the pochard. *(Netta rufina)* in the reed belt of a South Bohemian fishpond. (Photo: J. Formánek)

Section 8

Conservation of Plant Communities and Waterfowl in Wetlands of Czechoslovakia

Some of the South Bohemian fishponds, and the South Moravian Lednice fishponds represent internationally important wetland nature reserves included in the IUCN list. The Třeboň basin and the South Moravian floodplains have also been proposed as research areas in the UNESCO/MAB Project No. 2, and the Třeboň basin has been a biosphere reserve in the MAB Project No. 8. This chapter briefly discusses some aspects of the conservation both of plant communities and wildfowl in fishpond biotopes, which are inevitably greatly influenced by man. A brief rewiew is also presented of the most prominent wildfowl populations and endangered species and communities occurring both in the South Bohemian and South Moravian fishponds.

8.1 Conservation of Plant Communities in Fishpond Littorals

S. HEJNÝ

The protection and successive regeneration of the gene pool contained in the temperate aquatic and wetland vegetation nowadays represents a much more urgent task than its control and suppression. This is only justified in waterways, in irrigation and drainage systems, in certain reservoirs, etc. The rather drastic amelioration techniques, intense fertilizing, and herbicide application have brought numerous sensitive plant species and their communities near to extinction (Tables 1, 2).

In fishponds, the vegetation of aquatic vascular plants has been changing rapidly during the last 20 years because of the combined effects of intense fertilizing and manuring, duck-farming and mechanical amelioration (see Sect. 7). The basically correct concept of rational control of macrophytes should therefore be replaced by a regulation of macrophyte development, combined with conservation. The conservation aspect is particularly important because the ponds still contain the largest pool of macrophyte species populations and communities in Central Europe, though somewhat altered. Most alluvial wetland habitats, by contrast, have been changed irreversibly or destroyed completely.

In Czechoslovakia, a specimen management scheme has been worked out for certain South Bohemian fishponds whose total area amounts to some 800 ha. This scheme comprises the division of the fishponds, i.e., an outline of their allocation to certain main purposes. The fishponds have been divided into four groups (with the percentages of total fishpond area indicated in parentheses):

1. Intensely managed fishponds used, first of all, for fish- and duck-farming, with precisely defined measures of macrophyte regulation or control, including the amelioration with heavy machinery (about 80%). The heavy fertilizing and liming and the indirect fertilizing through fish- and duck-feeding affect mainly the open-water vegetation. Communities belonging to the foederatio *Nymphaeion albae* are largely replaced by those of the foed. *Potamogetonion lucentis* and *Potamogetonion pusilli*. In the large "main" production ponds, the macrophyte vegetation frequently becomes suppressed almost completely in favor of algal vegetation with water-blooms of blue-green algae during the summer season. One of the most persistent aquatic macrophytes, under these conditions, is *Persicaria amphibia*. Excessive eutrophication also favors the development of duckweed communities, especially of the ass. *Lemnetum gibbae* and *Lemno-Spirodeletum*: the

Table 1. List of retreating and/or endangered species of higher plants in Czechoslovak fresh waters and wetlands

Category	Species	Habitat
0:	*Pilularia globulifera* L.	l
1:	*Marsilea quadrifolia* L.	b
	Nuphar pumilum (Hoffm.) DC.	a
	Ranunculus polyphyllus W.K.	a, l
	Utricularia bremii Heer.	a
	Utricularia ochroleuca Hartm.	a
	Aldrovanda vesiculosa L.	a
	Pinguicula bohemica Kraj.	l
	Ludwigia palustris L.	a, l
	Blackstonia perfoliata (L.) Huds.	b
	Blackstonia serotina Beck.	b
	Acorellus pannonicus (Jacq.) Palla	b, l
	Mariscus serratus Gilib.	l
	Beckmania eruciformis (L.) Host	l
	Potamogeton coloratus Hornem.	a
2:	*Isoetes lacustris* L.	a
	Isoetes echinospora Durieu	a
	Batrachium baudotii (Godron) van der Bosch	a
	Batrachium rionii (Lagger) Nyman	a
	Chenopodium chenopodioides (L.) Aellen	b, l
	Cardamine parviflora (L.)	b
	Hippuris vulgaris (L.)	a, l
	Sonchus paluster (L.)	l
	Urtica radicans Bolla non All.	l
	Juncus sphaerocarpus Nees	b
	Juncus capitatus Weig	b, l
	Schoenoplectus triqueter (L.) Palla	l
	Schoenoplectus mucronatus (L.) Palla	l, b
	Chlorocyperus glomeratus (L.) Palla	l
	Sparganium affine Schnitzl.	b
	Typha minima Hoppe	b
3:	*Lycopodiella inundata* L.	l
	Thelypteris palustris Schott	l
	Salvinia natans (L.) All.	a
	Nymphaea candida Presl	a
	Nymphaea alba L.	a
	Ceratophyllum submersum L.	a
	Ranunculus lingua L.	l
	Ranunculus lateriflorus DC.	l
	Elatine alsinastrum L.	b, a
	Hottonia palustris L.	a
	Utricularia intermedia Hayne	a
	Utricularia minor L.	a
	Menyanthes trifoliata L.	l
	Radiola linoides Roth	b
	Illecebrum verticillatum L.	b, l
	Tillaea aquatica L.	b, l
	Litorella uniflora L. (Asch.)	b
	Centunculus minimus L.	b
	Senecio fluviatilis Wallr.	l
	Chrysanthemum serotinum L.	l
	Trapa natans L.	a
	Stratiotes aloides L.	a
	Juncus tenageia Ehrh.	b
	Juncus atratus Krock.	l
	Leucojum aestivum L.	l
	Carex melanostachya Willd.	l
	Calla palustris L.	l
	Wolffia arhiza (L.) Wimm.	a
	Sparganium minimum Fr.	b

Takhtajan's (1975) classification is used: 0, extinct; 1, only a few localities exist; 2, rare and retreating; 3, potentially endangered. Habitats: a, aquatic; l, littoral; b, emerged bottoms and shores.

Table 2. Numbers of plant species colonizing 3 types of wetland habitats and belonging to categories 0–3 (see Table 104) as related to corresponding total numbers of species in the Czechoslovak flora

Habitat	Origin		Total	0–3	
	Indigenous	Alien		No.	%
Aquatic	73	3	76	22	29
Littoral	118	12	130	26	20
Emerged bottoms	71	3	74	15	21

latter community type indicates pollution of the water with domestic or farm sewage. Fertilizer application, liming and amelioration of pond littorals have caused the greatest damage to the following community types: *Potamogetoneto natantis—Nymphaeetum candidae*, *Nymphoidetum peltatae*, and *Trapetum natantis*, *Nymphaea candida* being one of the most sensitive species. Scraping of the bottom

Fig. 1. Grey heron *(Ardea cinerea)* reposing on a tussock of *Carex elata* in a South Bohemian fishpond. (Photo: J. Ševčík)

sediments has nearly exterminated the communities of *Potamogetonetum graminei* and *Nupharetum pumili;* both require strict protection. The emergent reedswamp communities are more resistant to eutrophication and pollution; most of their dominants grow luxuriantly in fertilized ponds (see Sect. 3). But the amelioration of fishpond littorals with scrapers may remove them completely: the waterfowl thus becomes deprived of suitable breeding sites (Fig. 1, see also Sect. 8.2).

Annual species are also threatened, particularly those whose seeds germinate during summer drainage, after a long anabiosis under anerobic conditions in the flooded bottom sediments. Some of these species are quite rare, occurring hardly elsewhere than in summer-drained ponds (see Table 1). The recent techniques of fishpond management have suppressed summer drainage as a mangement tool: hence the threat to these species.

The most threatened communities of the pond outer littorals are the following: *Rhynchosporetum albae*, *Caricetum diandrae*, *Caricetum lasiocarpae*, *Pinguiculo vulgaris—Cratoneuretum commutati*, and *Caricetum davallianae.* The communities belonging to the foederatio *Sphagnion fusci* and to the ass. *Sparganietum minimi* also require protection.

Waterlogged sites and swamps occurring beyong the deponia around ameliorated ponds represent refuge biotopes for certain wetland plant species. The contemporary systematic drainage of these swamps and their subsequent conversion into arable land endanger these species and their communities as well.

2. Fishponds used mainly for recreation, with certain conservation measures adopted (about 5%). This category is at present acquiring considerable importance for social reasons. Limited control of the reed-belt with scrapers is permissible here, but only rather low fertilizer doses may be applied (no organic manure and no duck-farming, of course). Certain beautiful and rare plant species (but indigenous in the area) may be reintroduced into these ponds, e.g., *Nymphaea candida*, *Nymphoides peltata*, *Butomus umbellatus* (but not *Trapa natans*). The present numbers of the recreation ponds are insufficient: in addition to the hygienic risk, excessive concentration of people on a small area in and around a pond destroys the vegetation in much the same way as a duck farm does. As a rule of thumb, some 50 swimmers per day represent the maximum permissible number of visitors to a 5 ha pond during the summer season.

3. Fishponds suitable for reed and reed mace cultivation (see also Sect. 7.4), some 5–10%. In view of the presently increasing demand for *Phragmites* and *Typha* as resources for the building industry and domestic crafts, certain ponds will become equally important for the cultivation of these plants as for fish-farming. This, however, will not be excluded from these ponds in which wide littoral reed-belts are to be maintained, with the inner edges carefully controlled by cutting. The littoral *Phragmiteta* and *Typheta* suitable for exploitation have to consist of nearly pure stands comprising healthy and evenly developed shoots, little infested by insects and other pests. Relatively small ponds are best suited for this kind of management which, at the same time, perpetuates and preserves communities of the föed. *Phragmition communis*. The perpetuation of communities of *Acorus calamus* (a medical plant) in certain littoral habitats should not be overlooked.

4. Fishpond nature reserves, either managed or undisturbed, with strict conservation measures applied (2–5%), hosting typically developed plant communities requiring protection. Any fishpond nature reserve must be managed in order to preserve its conservation value. The following conservation measures have to be considered in each particular case:

(a) Each part of the littoral *Phragmition* communities has to be cut every other summer in order to control their excessive development. It is especially the communities of the foed. *Rhynchosporion albae* colonizing transitional bogs that cannot survive the invasion of *Phragmites*. The dead reed crop ought to be harvested every other winter.

(b) Land formation due to a succession of wetland plant communities, from reedswamp, through sedge swamp and willow carr to alder swamp forest or another alluvial forest, has to be retarded to a minimum.

This is best achieved by regular cutting of the reedswamp and sedge stands and by a skilful manipulation of the water level. Where fishpond biotopes have already become excessively terrestrialized the accumulated organic material is to be removed and shallow aquatic biotopes regenerated, perhaps even by dredging.

These measures cannot be applied to fishpond reserves with floating islets hosting a typical vegetation of the foed. *Cicution virosae*. Nevertheless, the land-forming succession has to be watched even here.

(c) In fishpond reserves where both plants and waterfowl deserve protection, a balance has to be set up between these two components. If the vegetation is more important the numbers of ducks and gulls may be somewhat reduced. Otherwise, too much wildfowl in a pond may act as too many domestic ducks. Plant communities of slightly dystrophic habitats, sensitive to eutrophication, are then in the greatest danger (foed. *Utricularion vulgaris* (most associations) and *Caricion canescenti-fuscae*).

(d) If a fishpond reserve comprises communities of rare plants colonizing emerged bottoms (see Table 1), the conservation measures have to be adjusted accordingly. For instance, the rare communities of the ass. *Radiolo-Juncetum tenageiae* are confined to periodically disturbed peaty sands. The disturbances, therefore, must continue in a reserve comprising these communities. Relic communities or rare species appearing on bare fishpond bottoms can only be saved by applying summer drainage every ten years or more frequently. For maintenance of communities of the foed. *Litorellion uniflorae*, and especially of their rare species components such as *Litorella uniflora* or *Pilularia globulifera*, summer drainage must be applied at least every three years.

In the present situation, critical for the survival of many plant species populations and community types, it is an urgent task to review and revaluate the existing system and statutes of nature reserves in Czechoslovak wetlands, to increase their number as well as to improve their representativness. Use can be made of various international recommendations for the conservation of wetlands.

International projects for wetland conservation (AQUA, MAR, TELMA, etc.), as well as those for the protection of waterfowl (IUCN/WWF) have stimulated the elaboration of lists and statutes of conservation for wetlands of international importance. Most of the earlier conservation measures were worked out primarily with respect to waterfowl. In 1974, at the Heiligenhafen conference, new and broadly defined criteria were elaborated for selecting internationally important wetlands. These criteria accentuate the conservation of whole waterfowl biotopes of which the plant communities form an essential constituent. Both regions investigated within the IBP project, the Třeboň basin and the Lednice region, have been proposed for inclusion in the list of wetlands of international importance (Heiligenhafen, 1974). Within the UNESCO program on "Man and Biosphere" (MAB), the Třeboň basin has been proposed as a biosphere reserve in project No. 8, and both the Třeboň basin and Lednice region are being investigated in project No. 2 (temperate forest biome). The "Třeboň basin" project (Czechoslovak contribution No. 7 to MAB project No. 2) is entitled "The role of wetlands within the temperate forest biome", and has also outputs to the MAB projects No. 3 (grasslands) and 5 (fresh water, etc.). Such studies increase the value of the protected wetland areas.

This text has been compiled on the basis of data contained in the papers by Hejný (1967, 1969), Fiala and Květ (1971), and Luther and Rzóska (1971).

References see pp. 438.

8.2 Conservation of Wildfowl in Fishpond Regions

K. Hudec and K. Šťastný

Animal conservation in wetlands, and particularly that of wildfowl, depends entirely on the conservation of the whole wetland ecosystems with the surrounding waterlogged grasslands and woodlands. Both the South Bohemian and South Moravian fishpond regions include several nature reserves that are of international importance for wildfowl conservation, being included in the lists of Projects Aqua (Luther and Rzóska, 1971) and Mar (Hudec et al., 1967; Hudec and Randík, 1972) and in IUCN's Check List of Wetlands of International Importance.

The amelioration of fishpond littorals (see Sect. 7.3) strongly affects the wildfowl, and especially those bird populations which breed in reedswamp biotopes. The gradual reclamation of wetlands by drainage, mechanical control of the fishpond reed-belts by scraping, and large-scale duck- and goose-farming in ponds have had a pronounced negative effect, so that the populations both of waterfowl and waders have been deprived of an appreciable proportion of suitable breeding sites. The result has been a marked decline in the numbers of certain rare species populations of birds.

According to Hudec (1975), conserving large areas covered with reed is of importance for the occurrence of certain species of birds, namely the large and typically reed-inhabiting ones. For a numerous occurrence of the major species, rather narrow reed-belts of sufficient length are satisfactory. A very narrow belt of pure reed, less than about 10 m in width, is satisfactory for just a small number of species. Diversity and structure of stands control the number of species and individuals of birds, which increases with increasing variety of stand structure. The presence of low dense stands is of importance. Sedge stands and, in deeper water, areas with a dense bottom layer of old plant material are most satisfactory. Of advantage is the presence of small islets inside the stands, providing firm support for nests of certain species, above all ducks. Pools of open water, even though small in area, are essential inside the stands, as well as the presence of at least narrow water passages.

The areas adjacent to the reed stands on the shore are very important for some of the reed-inhabiting species of birds. It is particularly advantageous if the reed stands pass gradually into permanent low waterlogged vegetation, at least on a minor part of the whole area. Presence of water in stands is essential. Those stands which are dry during the breeding season do not enable most species of birds to breed in them.

The birds breeding directly in the stands can be divided into five groups:

1. Typical reed-inhabiting species, confined to the reed stands as regards the situation of their nests and their food. These species hardly ever leave the reed stands. Various species prefer various types and/or various structures of the stands: denser bottom layer, old stands, polymorphous stands with pools of open water, etc. According to specific environmental requirements, there is a gradual transition from the species of the terrestrial ecophase which sometimes fly to forage outside the reed stands, such as reed bunting *(Emberiza schoeniclus)*, up to the species mainly occurring in the littoral ecophase, as great reed warbler *(Acrocephalus arundinaceus)*. Besides the above two species, this group includes reed warbler *(Acrocephalus scirpaceus)*, sedge warbler *(A. schoenobaenus)*, Savi's warbler *(Locustella luscinioides)*, bearded tit *(Panurus biarmicus)*, moorhen *(Gallinula chloropus)*, little crake *(Porzana parva)* and water rail *(Rallus aquaticus)*

2. Birds foraging largely in water and in reed stands sheltering their nests. These species sometimes build their nests quite unsheltered in open water. This group includes, above all, coot *(Fulica atra)*, all grebes (*Podiceps* spp.) and mute swan *(Cygnus olor)*. Inside the reed stands, they build nests in places which they can reach by swimming

3. Birds foraging mainly in water and frequenting the reed stands mainly for nesting, like group (2). The reed stands, however, are not essential for the situation of their nests, as they can equally well or preferably nest on shores and/or islets. In the reed stands, they build their nests on sites which provide a firm support (as a rule lodging old stands, muskrat, houses, etc.) This group includes all duck species, *Aythya* spp. being the most typical representatives

4. Mallard (*Anas platyrhynchos*) represents a transition between group (3) and a further group of birds which forage not only in water but frequently (or mainly) outside the water and the reed stands. This group includes the black-headed gull *(Larus ridibundus)*, the greylag goose *(Anser anser)*, and the marsh harrier *(Circus aeruginosus)*. These species mostly build their nests anew and are hence dependent on the selection of a safe site. *Anser anser* and *Larus ridibundus* build their nests quite apart from the stands, on sites surrounded by water islets

5. The terrestrial ecophase of the reed stands is utilized by certain terrestrial bird species building their nests in dense low vegetation, such as *Phasianus colchicus*.

In both regions of IBP investigations the birds unter protection were distributed as follows:

The Třeboň Basin. Here, the most outstanding locality of waterfowl is the State nature reserve of the Velký and Malý Tisý ponds, acknowledged as being of international importance. The area of the reserve with its buffer zone is 706 ha. The nature reserve represents an important station for migrating birds. It is situated around the two fishponds named above, also comprising a few adjacent smaller fishponds. Out of the bird species breeding in the reserve, the following are most notable: night heron (*Nycticorax nycticorax*, about 100 pairs), purple heron (*Ardea purpurea*, 5–10 pairs), greylag goose (*Anser anser*, 15 pairs). In recent years, the total number of greylag geese breeding in the whole Třeboň basin has increased to more than 30 pairs; several hundred migrating birds visit the area every year. The numbers of mute swans *(Cygnus olor)* have also increased recently, these swans hibernate regularly in the Třeboň Basin; whooper swans *(Cygnus cygnus)* hibernate here occassionally. Grey herons *(Ardea cinerea)* occur commonly in this area (Fig. 1, page 431). Some 100 pairs used to nest on a large breeding site along wet woodlands south of Třeboň, but they have been expelled by large-scale timber felling and peat-digging. The numbers of purple herons *(Ardea purpurea)* are now declining as a result of a large-scale destruction of fishpond reed-belts, and the species is becoming scarce in the Třeboň basin. The large local population of night heron *(Nycticorax nycticorax)* has also been reduced by the management of the ponds and their surroundings. The numbers of bitterns *(Botaurus stellaris)* have also declined in the whole Třeboň basin. On the other hand, the populations of certain birds adapted or attracted to human settlements (by food, etc.) have increased in numbers, such as those of gulls breeding in large colonies, of mute swans and, to some extent, of greylag geese which graze, among others, on cereal fields adjacent to fishponds. This observation also applies to South Moravia (Hudec, 1973).

The Třeboň basin is the only breeding area of the golden-eye *(Bucephala clangula)* in Czechoslovakia; these birds nest in hollow oaks growing on fishpond dams. The red-crested pochard *(Netta rufina)* has recently been spreading in the northern part of the Treboň basin; quite recently, nesting red-necked grebe *(Podiceps griseigena)* has also been recorded there. Of the terns, the larger ponds host both the common and the black tern (*Sterna hirundo* and *Chlidonias niger*).

Of the waders, the lapwing *(Vanellus vanellus)* and common snipe *(Gallinago gallinago)* are commonly encountered in the Třeboň basin, while the redshank *(Tringa totanus)*, and black-tailed godwit *(Limosa limosa)* are rarer, and the curlew *(Numenius arquata)* occurs extremely rarely here. Wet grasslands are also inhabitated by the corncrake *(Crex crex)*, and swamps are by another three species of this group, as well as by the meadow pipit *(Anthus pratensis)* and grasshopper warbler *(Locustella naevia)*. The numbers of black grouse *(Lyrurus tetrix)* have fallen drastically. Out of the birds of prey, the fishpond reedswamps host the marsh harrier *(Circus aeruginosus)*; the outer margins of the reedswamps bordering on fields or grassland are inhabited by Montagu's harrier *(C. pygargus)*, and the black kite *(Milvus migrans)* is encountered in the Velký and Malý Tisý nature reserve. Pygmy owls *(Glaucidium passerinum)* and black storks *(Ciconia nigra)* inhabit swampy woodlands.

Fishpond dams represent biotopes that are richest in the perching and other small birds. Forty-five species have been recorded during two seasons on a 1 km long section of the dam along the Nová Řeka (New River) canal, a nature reserve (see Table 2 in Sect. 1.5).

Apart from the already mentioned golden-eye *(Bucephala clangula)*, the most typical birds of the dams are the middle and lesser spotted woodpeckers *(Dendrocopos medius, D. minor)*, collared flycatcher *(Ficedula albicollis)*, and river warbler *(Locustella fluviatilis)*. The very rare Savi's warbler *(L. luscinioides)* breeds in some extensive reedswamps. Penduline tits *(Remiz pendulinus)* sometimes breed along streams. The bearded tit *(Panurus biarmicus)* breeds in the reedswamps of the Třeboň basin extremely rarely.

The whole Třeboň basin represents an important resting area during migrations and for hibernation of bird populations. The Velký and Malý Tisý nature reserve has already been mentioned. Other large ponds are visited by bean geese (*Anser fabalis*). Thousands of migrating ducks pass through the Třeboň basin. In fall, cormorants, various divers, grebes, mergansers, gulls, skuas, may be seen there. At the peak of the summer season, most waders stop here during their migration. Osprey (*Pandion haliaetus*) remains here regularly until the beginning of its breeding season.

Among the hibernating aquatic birds, *Anseriformes* are most prominent. Flocks of ducks leave the frozen ponds and gather along rivers, streams, and canals. The grey heron, bittern, mute swan and, sometimes, whooper swan hibernate here as well. A regular winter visitor is the white-tailed eagle (*Haliaeetus albicilla*), which may be encountered till the end of May.

The Lednice Fishponds. A warmer region, they host mainly more thermophilous species penetrating from Southeast Europe across the lowlands of Hungary. The South Moravian ornithologically important areas comprise some 1500 ha, of which the Lednice Fishponds nature reserve itself accounts for 635 ha. The following species of hercynian distribution never breed in South Moravia: *Bucephala clangula*, *Numenius arquata*, *Lyrurus tetrix*, *Carduelis flammea*, *Glaucidium passerinum*, and *Casmerodius albus*. Wide-spread, on the other hand, is the bearded tit (*Panurus biarmicus*). Other occasional bird species breed here as well: the avocet (*Recurvirostra avosetta*, some six pairs in the Lednice Fishponds), black-winged stilt (*Himantopus himantopus*, Nesyt fishpond), mediterranean gull (*Larus melanocephalus*, Mlýnský fishpond). Nesting has been recorded of the cormorant (*Phalacrocorax carbo*) and spoonbill (*Platalea leucorodia*). More abundant are the black-necked grebe (*Podiceps nigricollis*, some 400 pairs) and greylag goose (*Anser anser*, over 100 pairs, of which some 25 pairs are in the Lednice Fishponds reserve). The mallard (*Anas platyrhynchos*) occurs in hundreds of pairs, while some 50 pairs of the red-crested pochard (*Netta rufina*) live in the Lednice fishponds. The black-headed gull (*Larus ridibundus*) is extremely abundant (tens of thousands of individuals). Relatively abundant are the grey, purple and night herons (*Ardea cinerea*, *A. purpurea*, *Nycticorax nycticorax*) as well as the white and black storks (*Ciconia ciconia*, *C. nigra*); the black-tailed godwit (*Limosa limosa*) and pintail (*Anas acuta*) inhabit the meadows and ponds. The Lednice region is also of great importance for migrating birds; in the fall some 60000 birds of various ducks and numerous waders gather here.

This brief survey has only illustrated the wealth of bird life both in the South Bohemian and South Moravian fishponds. If not for anything else, they deserve full protection as waterfowl biotopes under the Ramsar convention and other international agreements and programs of wetland conservation. The system of protection of their mammals, reptiles, amphibians, and arthropods still remains to be worked out.

References see p. 438.

References

Fiala, K., Květ, J.: Dynamic balance between plant species in South Moravian reedswamps. In: The Scientific Management of Animal and Plant Communities for Conservation. Duffey, E., Watt, A. S. (eds.). Oxford: Blackwells Sci. Publ., 1971, pp. 241–269

Hejný, S.: Problems of protection and the regional system of water basins from the hydrobotanical aspects. (In Czech with English summary). Ochr. Přírody **22**(6), 83–90 (1967)

Hejný, S.: Conservation, management and control of aquatic higher plants and of their communities in the temperate zones. UNESCO-IHD Meeting on Control and Ecology of Aquatic Plants, Paris, 16–18 Dec., 1969

Hudec, K.: Die Nahrung der Graugans, *Anser anser* in Südmähren. Zool. Listy **22**, 41–58 (1973)

Hudec, K.: Density and breeding of birds in the reed swamps of Southern Moravian ponds. Acta Sci. Nat. (Brno) **6**, 1–41 (1975)

Hudec, K., Leiský, K., Randík, A.: Status and trends of Czechoslovakian wetlands classified in the MAR list. Proc. 2nd Europ. Meet. Wildf. Conservation, Noordwijk aan Zee, 41–42 (1967)

Hudec, K., Randík, A.: Projekt MAR. (In Slovakian). Ochrana Fauny, Bratislava **6**, 3–4 (1972)

Luther, H., Rzóska, J.: Project Aqua. (IBP Handbook No. 2) Oxford: Blackwells Sci. Publ. 1971

Smart, M. (ed.): Proceedings Internat. Conf. on the Conservation of Wetlands and Waterfowl. Heiligenhafen, 2—6 Dec. 1974. Int. Waterfowl Res. Bureau, Slimbridge, 1976

Takhtajan, A. L. (ed.): Red Book. Native plant species to be protected in the U.S.S.R. (In Russian). Leningrad: Nauka 1975

Taxonomic Index

Higher Plants

Algae and Other Microorganisms

Higher — Plant Communities

Evertebrata

Vertebrata

Subject Index

Ecological Studies
Analysis and Synthesis

Editors: W.D. Billings, F. Golley, O.L. Lange, J.S. Olson

Volume 4:

Physical Aspects of Soil Water and Salt in Ecosystems

Editors: A. Hadas, D. Swartzendruber, P.E. Rijtema, M. Fuchs, B. Yaron
221 figures, 61 tables. XVI, 460 pages. 1973
ISBN 3-540-06109-6
Contents: Water Status and Flow in Soils: Water Movement in Soils. Energy of Soil Water and Soil-Water Interactions. – Evapotranspiration and Crop-Water Requirements: Evaporation from Soils and Plants. Crop-Water Requirements. – Salinity Control.

Volume 9:
B. Slavík

Methods of Studying Plant Water Relations

With contributions by B. Slavík, J. Catský, J. Solárová, H.R. Oppenheimer, J. Hrbáček, J. Slavíková, V. Kozinka, J. Úlehla, P.G. Jarvis, M.S. Jarvis
181 figures. XVIII, 449 pages. 1974
ISBN 3-540-06686-1
Distributions rights for the Socialist countries: Academia Publishing House of the Czechoslovak Academy of Sciences, Prague
Contents: Water in Cells and Tissues. – Water Content. – Water Exchange between Plant Roots and Soil. – Liquid Water Movement in Plants. – Water Exchange between Plant and Atmosphere. – Table Appendix.

Volume 10:

Coupling of Land and Water Systems

Editor: A.D. Hasler
95 figures. XVII, 309 pages. 1975
ISBN 3-540-06707-8
Contents: Nutrient-Hydrologic Interaction (Eastern USA). – Balances in Man-Made Lakes (Bohemia). – Bog Mires and their Influence on Landscapes. – Effects of Marshes on Water Quality. – Mangrove Forests and Aquatic Productivity. – Food Relations and Behavior of Salmonid Fishes. – Natural and Modified Plant Communities as Related to Funoff and Sediment Yields. – Landscapes of River Basins. – Trophic Conditions of Italian Lakes as a Consequence of Human Pressures. – The Supply of Minerals to Tropical Rivers and Lakes (Uganda). – Ecological Interactions between Land and the Littoral Zones of Lakes (Masurian Lakeland, Poland). – Silica and Nitrate Depletion as Related to Rate of Eutrophication in Lakes Michigan, Huron, and Superior.

Volume 14:

Primary Productivity of the Biosphere

Editors: H. Lieth, R.H. Whittaker
With contributions by E. Box, J.S. Bunt, C.A.S. Hall, H. Lieth, G.E. Likens, P.L. Marks, R. Moll, P.G. Murphy, D.D. Sharp, D. Sharpe, D. Whigham, R.H. Whittaker
67 figures, 46 tables. VIII, 339 pages. 1975
ISBN 3-540-07083-4
Contents: Introduction. – Methods of Productivity Measurements. – Global Productivity Patterns. – Utilizing the Knowledge of Primary Productivity. – Index.

Volume 19:

Water and Plant Life

Problems and Modern Approaches
Editors: O.L. Lange, L. Kappen. E.-D. Schulze
178 figures, 66 tables. XX, 536 pages. 1976
ISBN 3-540-07838-X
Contents: Fundamentals of Plant Water Relations. – Water Uptake and Soil Water Relations. – Transpiration and ist Regulation. – Direct and Indirect Water Stress. – Water Relations and CO_2 Fixation Types. – Water Relations and Productivity. – Water and Vegetation Patterns.

Volume 20:
F.B. Christiansen, T.M. Fenchel

Theories of Populations in Biological Communities

68 figures, 4 tables. X, 144 pages. 1977
ISBN 3-540-08010-4
Contents: Population Growth and Fitness. – Interactions Between Species. – The Niche and Limiting Similarity. – Geographically-Structured Populations and Migration-Extinction Equilibria. – The Concept of Biological Communities.

Springer-Verlag
Berlin Heidelberg NewYork

The Early Life History of Fish

The Proceedings of an International Symposium Held at the Dunstaffnage Marine Research Laboratory of the Scottish Marine Biological Association at Oban, Scotland, from May 17–23, 1973

Editor: J.H.S. Blaxter, Dunstaffnage Marine Research Laboratory, Oban, Scotland

299 figures. X, 765 pages. 1974
ISBN 3-540-06719-1
Distributions rights for India:
Allied Publishers Ltd., New Delhi

Contents: Population Studies. – Distribution. – Feeding and Metabolism. – Physiological Ecology. – Developmental Events. – Behaviour. – Taxonomy. – Rearing.

From the reviews: "That animals change from egg to adult is a truism, but one we too often neglect when dealing with physiological, ecological, or evolutionary problems. Eggs, embryos, and developing young are all functioning individuals that interact with their environment, are subject to natural selection, and evolve. Too often we focus all our attention on the reproductive adult when the entire life cycle ought to be scrutinized.... This timely volume will provide invaluable supplemental reading and browsing for all students of ichthyology, occeanography, and limnology and for many nascent experimental zoologists as well. Professionals can hardly afford to be without it."

Bioscience

"....be useful in representing the current status of research into marine fish eggs and larvae. The coverage of the various branches of the research is wide and the presence together at the Symposium of so many of the active research workers provided an excellent opportunity to crystallize opinion and plan for the future." *Scientia*

Springer-Verlag
Berlin
Heidelberg
New York